© 1949 by Earle Buckingham.
reserved under Pan American and Inter-
yright Conventions.

in Canada by General Publishing Com-
30 Lesmill Road, Don Mills, Toronto,

in the United Kingdom by Constable and
Ltd., 10 Orange Street, London WC2H 7EG.

tion, first published in 1988, is an un-
nd slightly corrected republication of the
published by the McGraw-Hill Publishing
1949 and first reprinted by Dover Publica-
in 1963.

tured in the United States of America
Publications, Inc., 31 East 2nd Street,
.Y. 11501

nal Standard Book Number: 0-486-65712-4

MECH

Prof

BUCKINGHAM ASS

i

DOVER PUB

PREFACE

This text is the gratification, to some degree at least, of a suppressed desire of some twenty-five years' standing: this has been the development of a reasonably complete mathematical analysis of the mechanics of gearing.

The subtitle of this treatise could well be "Final Report of the ASME Special Research Committees on Worm Gears and the Strength of Gears." The information and the inspiration from the work of these committees and from personal contact with their individual members has been the major motive force behind this work.

To quote from a note to one of my colleagues: "If one understands the significance of the calculus, then the subject of conjugate gear-tooth action should be mastered with ease. On the other hand, if one has a visual imagination and can grasp the significance of this conjugate gear-tooth action, then this presentation should be of material aid to help him to master the significance of the calculus. If however, as with myself, he has neither, then continued application to the problem may give him some slight appreciation of both."

This is not a text on gear design. It should form, however, a sound foundation upon which logical design practices and design data can be erected. These design structures, however, must be erected by the specialized engineering groups that have need of them. In the final analysis, there are so many unknown and uncertain factors that in critical cases our only answer is to "try and see." An analysis such as this should be of material aid in the interpretation and in the application of the information that is made available by definite tests and by actual service experience.

The trends of today indicate a growing demand for the seemingly opposed requirements of higher speeds and greater loads with more reliability and quietness of operation. These demands are met in part by improved materials, better balancing, more nearly perfect machined surfaces, and more intensive attention to many details of design. This last should also include a rigorous mathematical analysis of both the kinematic and dynamic conditions of operation.

The major objection to the thorough analysis of any mechanism is the amount of time required to make it, and also the dearth of individuals

v

who have both the ability and the inherent urge to do such work. If each new project must be started from the beginning, the time required to carry it through might be excessive for any single application. If, however, we can build up a foundation of general solutions of different basic types of mechanisms, then the specific task for any given project is to arrange the general solution for the particular case and solve. Even so, considerable time must be spent in computing the results, and a calculating machine is one of the essentials.

Among the machine elements that have received much less analytical attention than they deserve are linkages, cams, and gears. The book attempts to give in usable form such a general analysis of gears.

As noted before, this is not a text on gear design. The major effort has been to make apparent the nature of the action and contact between the contacting teeth of the different types of gears. A clear understanding of these features is essential if the designs are to exploit the full possibilities of these mechanical elements.

The first chapters give an analysis of conjugate gear-tooth action, nature of the contact, and resulting gear-tooth profiles of the several types of gears. These include spur gears, internal gears, helical gears, spiral gears, worm gears, bevel gears, and hypoid or skew bevel gears. Spur, internal, and helical gears are used to drive parallel axes. Bevel gears are used to drive intersecting axes. Spiral, worm, and hypoid gears are used to drive nonparallel, nonintersecting axes.

The last chapters give an analysis of gear teeth in action. These include frictional heat of operation and its dissipation, friction losses and efficiencies, dynamic loads in operation, beam strength or resistance of the teeth to breakage and fatigue, surface-endurance limits of materials, and the limiting wear loads or the potential resistance to surface disintegration and excessive wear.

No claim is made that this analysis is complete. In the first place, space limitations prevent the full development of many interesting and pertinent factors of this general problem. While in the second and more important place, the writer has much to learn about the subject. It is probably much larger than the capacity of any individual to master completely. Even this incomplete analysis is the result of the original work of many individuals.

The effort has been made to give due credit in the text to the many sources of information that have made possible this analysis. In addition to this, much help both direct and indirect has been obtained from close personal contact with Wilfred Lewis, Carl G. Barth, Charles H. Logue, Ernest Wildhaber, and all the other members of the ASME Special Research Committees on Worm Gears and the Strength of Gear Teeth, as well as from many other individuals who are struggling with various types of gear problems. And last, but not least, a debt of gratitude is owed to Guy L. Talbourdet, who has studied this manuscript critically and who has checked carefully all the derivations of the equations.

EARLE BUCKINGHAM

CAMBRIDGE, MASS.
March, 1949

CONTENTS

CHAPTER 1

CONJUGATE ACTION ON SPUR GEARS

The essential purpose of gear-tooth profiles is to transmit rotary motion from one shaft to another. In the majority of cases, the addiional requirement of uniform rotary motion also exists.

There is an almost infinite number of forms that can be used as gear-tooth profiles. Although the involute profile is the one most commonly used today for gear-tooth forms that are used to transmit power, occasions arise when some other form of profile can be used to advantage. In addition, there are also other problems than the transmission of rotary motion, where a thorough knowledge of the theory of gearing will assist to the most direct solution. One of such problems is the hobbing of spline shafts.

Again, in order to appreciate fully the great simplicity of the involute form both in its theory and in its production, it is necessary to have a clear understanding of the principles of conjugate gear-tooth action. We will therefore consider now the characteristics of gear-tooth profiles that will transmit through each other uniform rotary motion. The action between such teeth is called *conjugate gear-tooth action*.

In essence, a pair of mating gear-tooth profiles are cams, the one acting against the other to produce the relative motion desired. With certain restrictions, one profile can be chosen at random and a correct mating profile can be developed. On the other hand, if both profiles are selected arbitrarily, the nature of the resulting action can be determined, but it will seldom if ever be uniform motion.

In all cases when two curved surfaces act against each other, the line of action between them will be along the common normal to the two curves at the point of tangency between them. If these two curved profiles are mounted on pivots, as shown in Fig. 1-1, this line of action a-a will intersect the line of centers c-c at A. Then the angular-velocity ratio between the two arms will be inversely proportional to the radii of the respective arms to the point A. These radii are the momentary pitch radii of the two forms.

In order to transmit uniform rotary motion, the values of the momentary pitch radii must remain constant for all operating positions of the contacting profiles. This gives the basic law of conjugate gear-tooth action, which may be expressed as follows:

1

Law of Conjugate Gear-tooth Action. To transmit uniform rotary motion from one shaft to another by means of gear teeth, the normals to the profiles of these teeth at all points of contact must pass through a fixed point in the common center line of the two shafts.

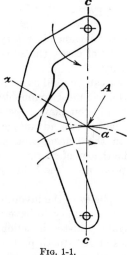

Fig. 1-1.

Pitch Point. This fixed point in the common center line is called the *pitch point.* With every gear-tooth form except the involute, there is a definite pitch line or pitch circle from which the conjugate gear-tooth profiles must be developed. The pitch circles of mating spur gears must be tangent to each other. The point of tangency of these two pitch circles is the pitch point.

These pitch circles are of such size that if they were to drive each other by friction without slipping, they would transmit the relative motion required. The sizes of these pitch circles are inversely proportional to the angular velocities of the two gears. For equal velocities these sizes are equal. For double speed, the pitch circle of the slower gear is twice the size of that of the faster gear. The tooth proportions may be symmetrical or unsymmetrical in respect to the pitch line: the tooth profile may be all above or all below the pitch line or it may be partly above and partly below.

Basic-rack Form. As stated before, the tooth profile for one gear may be chosen arbitrarily, and the conjugate profile for the mating gear can be developed. For every pair of conjugate gear-tooth profiles, there is also a definite basic rack form. This basic-rack form is the profile of the conjugate gear of infinite diameter. Its pitch line is a straight line.

Path of Contact. When conjugate gear-tooth profiles act together, the point of contact between them will travel along a definite path, which is called the *path of contact.* In other words, the path of contact is the locus of all points of contact between the conjugate gear-tooth profiles.

Line of Action. From any point of contact, a straight line, normal to both of the mating profiles at the point of contact, can be drawn from it to the pitch point. This straight line or common normal to the profiles is called the *line of action.*

Pressure Angle. The angle between this line of action and the common tangent to the two pitch circles at the pitch point is called the *pressure angle* of the specific points of the teeth that are in contact. With all gear-tooth forms except the involute form, this pressure angle changes from point to point.

Once a pitch line has been established for any given tooth profile, a definite path of contact exists along which contact with all other conjugate gear-tooth profiles is made regardless of the number of teeth in these gears. The path of contact for any given conjugate gear-tooth system is the same for any two gears as it is for any one gear and the basic rack of the system.

The simplest way to define any definite conjugate gear-tooth system is to specify the form and size of its basic rack. There is a definite relation between a gear-tooth profile and its path of contact so that if either one is given, the other is fixed.

BASIC-RACK FORM GIVEN

We shall start our analysis with a specified form for the basic rack. We shall use the pitch point as the origin of the coordinate system for the basic rack and for the path of contact. We shall use the center of the conjugate gear as the origin of the polar coordinate system for the gear-tooth profile, and the vectorial angle will be zero at the pitch point. All angles will be plus when they are measured in a counterclockwise direction. In all these calculations, great care must be exercised to use the correct sign (plus or minus) of these angular values.

We shall use the following symbols, and subscripts will be used when needed to identify the different gears of a pair or train:

x = abscissa of basic-rack profile

y = ordinate of basic-rack profile and of path of contact

x_p = abscissa of path of contact

ϕ = pressure angle

r = length of radius vector of conjugate gear-tooth profile

R = pitch radius of gear

θ = vectorial angle of radius vector

ψ = angle between tangent to tooth profile and radius vector

ϵ = angle of rotation of gear from zero position

X = abscissa of conjugate gear-tooth profile (origin at center of gear)

Y = ordinate of conjugate gear-tooth profile (origin at center of gear)

Path of Contact. *Given the values of x, y, and ϕ; to determine the path of contact.*

$$\tan \phi = dx/dy$$

Referring to Fig. 1-2, we have from the geometrical conditions shown there

$$x_p = -y/\tan \phi \qquad (1\text{-}1)$$

The value of x_p will be negative when the value of x is positive, because the two points will always be on opposite sides of the origin. In order to establish the path of contact for any given basic-rack profile, a series of points on the given profile, together with the values of the tangents at those points, are selected or determined, for which the equivalent points on the path of contact can be determined by the use of Eq. (1-1). These values can then be plotted to any desired scale.

The first step is to determine the values of x, y, and tan ϕ for a selected series of points on the basic-rack profile. Generally the equation of this

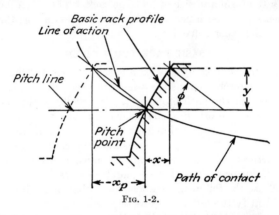

FIG. 1-2.

rack profile can be readily established. In other cases, values measured from an enlarged layout may be used.

Conjugate Gear-tooth Profile. When we have the coordinates of the basic-rack profile and those of the path of contact, we can then calculate the coordinates of the conjugate gear-tooth profile for any specific pitch radius. Referring to Fig. 1-3, we have the following from the geometrical conditions shown there:

$$r = \sqrt{(R - y)^2 + x_p^2} \tag{1-2}$$
$$\epsilon = (x_p - x)/R$$
$$\theta = \{\tan^{-1}[x_p/(R - y)]\} - \epsilon$$
$$\theta = [(x - x_p)/R] + \tan^{-1}[x_p/(R - y)] \tag{1-3}$$

As noted before, the values of x and x_p will always be of opposite sign, so that the numerical value of their difference will always be the sum of their values with the sign of x.

Limitations to Conjugate Action. Whenever there is a cusp in the form of the conjugate gear-tooth profile, the extent of the useful profile is limited to the bottom of the cusp. If the tooth profiles engage beyond this point, there will be interference in the action, or undercut. These conditions will be analyzed in detail later in connection with the study of the form of the fillet or trochoid below the conjugate profile of the gear-tooth.

Mating Gear-tooth Profile. For the coordinates of the mating gear-tooth profile, which must be conjugate to the first gear-tooth profile, we must use the coordinates of the inside surface of the original basic rack, or we must determine them from the values of the first gear-tooth profile. When the basic-rack form is given, it is best always to work directly from it.

The values for the coordinates of the inside surface of the basic rack and its path of contact will be the same as before but of opposite sign. With these values, we proceed as before to determine the coordinates of the mating gear-tooth profile.

Cartesian Coordinates for Conjugate Gear-tooth Profiles. When it is desired to use rectangular coordinates to plot the form of the gear tooth or tooth space, we can shift the origin as may be desired. Generally it is desirable to make the y axis at the center line of the tooth or space. Then when

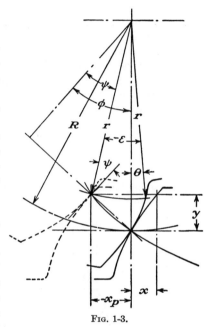

Fig. 1-3.

θ' = original vectorial angle at center of tooth or space
θ = calculated vectorial angle
θ'' = vectorial angle with Y axis at center line of tooth or space

$$\theta'' = \theta' - \theta \qquad (1\text{-}4)$$
$$X = r \sin \theta'' \qquad (1\text{-}5)$$
$$Y = r \cos \theta'' \qquad (1\text{-}6)$$

Arc of Action. The arc of action is the arc through which one tooth travels from the time it first makes contact with the mating tooth along the path of contact until contact ceases. For smooth continuous action, the arc of action must be something greater than the arc between successive teeth of the gear.

The arc of approach is the arc through which the tooth travels from the time it first makes contact with the mating tooth along the path of contact until the contact has reached the pitch point.

The arc of recess is the arc through which the tooth travels from the time contact is at the pitch point until contact ceases along the path of contact.

Thus when

ϵ_a = arc of rotation of pitch point of driving gear to position of first contact near root of tooth[1]

ϵ_r = arc of rotation of pitch point of driving gear to position of last contact at tip of tooth[1]

β_a = arc of approach of driving gear

β_r = arc of recess of driving gear

β = arc of contact of driving gear

$$\beta_a = \epsilon_a \qquad (1\text{-}7)$$
$$\beta_r = \epsilon_r \qquad (1\text{-}8)$$
$$\beta = \beta_a + \beta_r \qquad (1\text{-}9)$$

The foregoing values are independent of their signs. Here we are interested only in the numerical sums.

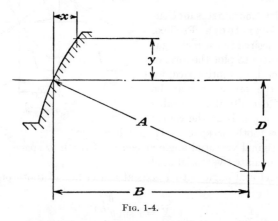

Fig. 1-4.

Example with Given Form of Basic Rack. As a definite example we shall use a basic-rack profile that is formed by the arc of a circle as shown in Fig. (1-4). For this example, we have when

A = radius of rack profile

B = distance along x axis to center of radius, A

D = distance along y axis to center of radius, A

---

[1] These values are the values calculated for the respective contact points.

$$x = B - \sqrt{A^2 - (D + y)^2} \tag{1-10}$$

$$\frac{dx}{dy} = \tan \phi = \frac{D + y}{\sqrt{A^2 - (D + y)^2}} \tag{1-11}$$

$$\frac{d^2x}{dy^2} = \frac{A^2}{[A^2 - (D + y)^2]^{3/2}} = \frac{1}{A \cos^3 \phi} \tag{1-12}$$

We will let

Then

$$A = 5.00 \qquad B = 4.5315 \qquad D = 2.1131$$

$$x = 4.5315 - \sqrt{25 - (2.1131 + y)^2} \tag{1-10}$$

$$\tan \phi = \frac{(2.1131 + y)}{\sqrt{25 - (2.1131 + y)^2}} \tag{1-11}$$

Example of Path of Contact. We shall use increments of 0.10 for values of y from 0 to ± 1.00 with which we obtain the values of x, y, and $\tan \phi$ tabulated in Table 1-1. By the use of these values in Eq. (1-1), we obtain the values of x_p that are tabulated in Table 1-1 and plotted in Fig. 1-5.

TABLE 1-1. COORDINATES OF GIVEN BASIC-RACK FORM AND OF CONJUGATE GEAR
TOOTH
(Plotted in Fig. 1-5)

y, in.	x, in.	$\tan \phi$	x_p, in.	r, in.	θ, rad	X, in.	Y, in.
1.00	0.6189	0.79566	−1.2568	19.0415	0.0277	0.2203	19.0402
0.90	0.5414	0.75514	−1.1918	19.1371	0.0244	0.2846	19.1350
0.80	0.4678	0.71686	−1.1160	19.2324	0.0211	0.3494	19.2282
0.70	0.3979	0.68054	−1.0286	19.3274	0.0181	0.4092	19.3218
0.60	0.3316	0.64599	−0.9288	19.4222	0.0152	0.4675	19.4166
0.50	0.2687	0.61300	−0.8157	19.5171	0.0124	0.5242	19.5101
0.40	0.2090	0.58140	−0.6880	19.6120	0.0098	0.5788	19.6036
0.30	0.1524	0.55105	−0.5444	19.7075	0.0072	0.6316	19.6974
0.20	0.0987	0.52181	−0.3833	19.8037	0.0047	0.6848	19.7918
0.10	0.0480	0.49361	−0.2026	19.9010	0.0024	0.7333	19.8875
0.00	0.0000	0.46631	0.0000	20.0000	0.0000	0.7852	19.9846
−0.10	−0.0453	0.43985	0.2274	20.1013	−0.0023	0.8358	20.0840
−0.20	−0.0880	0.41414	0.4829	20.2058	−0.0046	0.8858	20.1864
−0.30	−0.1282	0.38910	0.7710	20.3146	−0.0070	0.9395	20.2929
−0.40	−0.1659	0.36469	1.0968	20.4295	−0.0094	0.9941	20.4052
−0.50	−0.2011	0.34085	1.4669	20.5524	−0.0120	1.0535	20.5253
−0.60	−0.2341	0.31750	1.8898	20.6865	−0.0147	1.1158	20.6565
−0.70	−0.2647	0.29463	2.3759	20.8359	−0.0178	1.1885	20.8018
−0.80	−0.2930	0.27217	2.9393	21.0067	−0.0212	1.2696	20.9683
−0.00	−0.3191	0.25009	3.5987	21.2076	−0.0254	1.3706	21.1633
−1.00	−0.3430	0.22835	4.3792	21.4517	−0.0305	1.4956	21.3994

Example of Conjugate Gear-tooth Profile. We shall use a value of $R = 20.00$ for the first gear. With this value and the tabulated ones used in Eqs. (1-2) and (1-3), we obtain the values tabulated in Table 1-1 for r and θ.   These values are also plotted in Fig. 1-5.

For the mating gear to run with the first gear, we shall also use a value of $R = 20.00$. For this gear, the values of x, y, and x_p are the same as before but of opposite sign. These values are tabulated in Table 1-2 together with the computed values for r and θ for the second gear.   This second profile has a cusp at its bottom, which limits its active profile below the radius of about 19.4237.   This corresponds to a maximum radius of about 21.0067 on the mating gear.

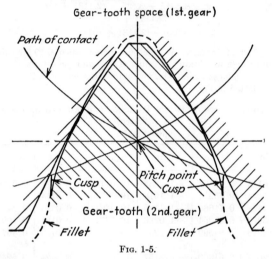

Fig. 1-5.

Arc of Action.   We shall assume that the outside radius of the second gear is 21.0376 and that this point makes contact with the first gear at a radius of 19.0415. We shall also use an outside radius for the first gear of 21.0067.

At the start of contact

$$x = 0.6189 \qquad x_p = -1.2568 \qquad y = 1.00$$
$$\epsilon_a = \frac{(x - x_p)}{R} = \frac{1.8757}{20} = 0.09373 \text{ radian}$$

Here, as noted before, we are interested only in the numerical values and not the signs; hence $\beta_a = 0.09373$ radian.

At the end of contact

$$x = -0.2930 \qquad x_p = 2.9393 \qquad y = -0.80$$
$$\epsilon_r = -\frac{3.2323}{20} = -0.16161 \text{ radian}$$

Hence

$$\beta_r = 0.16161 \text{ radian}$$
$$\beta = 0.09373 + 0.16161 = 0.25534 \text{ radian}$$

These gears must have $(6.2832/0.25534) = 25$ or more teeth for continuous action.

Rectangular Coordinates.   We shall assume that both of the gears have 40 teeth, and that we want their rectangular coordinates in reference to the center line of the tooth space of the first gear, and the center line of the gear tooth of the second gear. We shall also assume that the arc tooth thickness of each gear at its pitch line is equal to one-half of the circular pitch, or 1.5708. For the first gear we have

TABLE 1-2. COORDINATES OF INSIDE SURFACE OF BASIC-RACK FORM AND OF CONJU-
GATE GEAR TOOTH
(Plotted in Fig. 1-5)

y, in.	x, in.	$\tan \phi$	x_p, in.	r, in.	θ, rad	X, in.	Y, in.
1.00	0.3430	0.22835	−4.3792	19.4981	0.0096	0.9525	19.4749
0.90	0.3191	0.25009	−3.5987	19.4361	0.0097	0.9514	19.4128
0.80	0.2930	0.27217	−2.9393	19.4237	0.0097	0.9508	19.4004
0.70	0.2647	0.29463	−2.3759	19.4457	0.0095	0.9478	19.4226
0.60	0.2341	0.31750	−1.8898	19.4918	0.0091	0.9422	19.4690
0.50	0.2011	0.34085	−1.4669	19.5551	0.0083	0.9300	19.5330
0.40	0.1659	0.36469	−1.0968	19.6307	0.0072	0.9120	19.6095
0.30	0.1282	0.38910	−0.7710	19.7151	0.0058	0.8882	19.6950
0.20	0.0880	0.41414	−0.4829	19.8059	0.0042	0.8608	19.7873
0.10	0.0453	0.43985	−0.2274	19.9013	0.0022	0.8249	19.8842
0.00	0.0000	0.46631	0.0000	20.0000	0.0000	0.7854	19.9846
−0.10	−0.0480	0.49361	0.2026	20.1010	−0.0025	0.7389	20.0873
−0.20	−0.0987	0.52181	0.3833	20.2036	−0.0051	0.6904	20.1919
−0.30	−0.1524	0.55105	0.5444	20.3073	−0.0080	0.6350	20.2973
−0.40	−0.2090	0.58140	0.6880	20.4116	−0.0111	0.5748	20.4034
−0.50	−0.2687	0.61300	0.8157	20.5162	−0.0145	0.5080	20.5098
−0.60	−0.3316	0.64599	0.9288	20.6209	−0.0180	0.4386	20.6162
−0.70	−0.3979	0.68054	1.0286	20.7255	−0.0217	0.3641	20.7222
−0.80	−0.4678	0.71683	1.1160	20.8299	−0.0256	0.2845	20.8280
−0.90	−0.5414	0.75514	1.1918	20.9340	−0.0297	0.2001	20.9330
−1.00	−0.6189	0.79566	1.2568	21.0376	−0.0340	0.1104	21.0373

$$\theta' = \frac{1.5708}{40} = 0.03927 \text{ radian} = 2.250°$$

Using this value in Eq. (1-4), we obtain values of θ'', from which, using Eqs. (1-5) and (1-6), we obtain the values of X and Y, which are tabulated in Table 1-1 and plotted in Fig. 1-5.

For the second gear we have

$$\theta' = -\frac{1.5708}{40} = -0.03927 \text{ radian} = -2.250°$$

Proceeding as before, we obtain the values of X and Y for the second gear, which are tabulated in Table 1-2 and plotted in Fig. 1-5.

ONE GEAR-TOOTH FORM GIVEN

Now we shall start with a specified form for one of the gear-tooth profiles. We shall use the center of the gear as the origin of the polar coordinate system and the pitch point as the place where the vectorial angle is equal to zero. All symbols and other conventions will be the same as before. Angles are plus when measured in a counterclockwise direction. We will use the subscript $_1$ on all symbols for the original gear and the subscript $_2$ on all symbols for the mating gear. It is assumed that we have or can obtain the values for R_1, r_1, θ_1, ϵ_1, and ψ_1, where

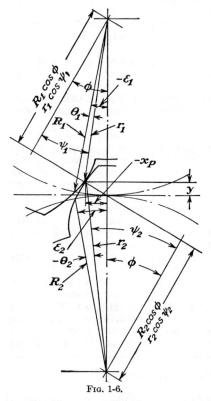

$$\tan \psi_1 = -(r_1 \, d\theta_1/dr_1) \quad (1\text{-}13)[1]$$

If we need the sine or cosine of this angle, we have

$$\sin \psi_1 = \frac{\tan \psi_1}{\sqrt{1 + \tan^2 \psi_1}}$$

$$\cos \psi_1 = \frac{1}{\sqrt{1 + \tan^2 \psi_1}}$$

Path of Contact. Referring again to Fig. 1-3, we have from the geometrical conditions shown there the following:

$$\cos \phi = (r_1 \cos \psi_1)/R_1 \quad (1\text{-}14)$$
$$\epsilon_1 = \psi_1 - \phi - \theta_1 \quad (1\text{-}15)$$
$$x_p = r_1 \sin (\psi_1 - \phi) \quad (1\text{-}16)$$
$$y = R_1$$
$$\qquad - r_1 \cos (\psi_1 - \phi) \quad (1\text{-}17)$$

Basic-rack Form. If we need the form of the basic rack, we obtain the value of x as follows:

FIG. 1-6.

$$x = x_p - R_1\epsilon_1 \quad (1\text{-}18)$$

When we have the coordinates of the basic-rack form and of its path of contact, we can proceed as before to determine the coordinates of the mating gear-tooth profile. Or we can determine these coordinates directly from those of the original gear and the path of contact.

Mating Gear-tooth Profile. Referring to Fig. 1-6, we have from the geometrical conditions shown there the following:

$$r_2 = \sqrt{(R_2 + y)^2 + x_p{}^2} \quad (1\text{-}19)$$
$$\cos \psi_2 = R_2 \cos \phi/r_2 \quad (1\text{-}20)$$

[1] The sign here is minus because of the special coordinate system used. When the value of r is increasing, the value of θ is decreasing.

$$\epsilon_2 = -(R_1/R_2)\epsilon_1 \qquad (1\text{-}21)[1]$$
$$\theta_2 = \psi_2 - \phi - \epsilon_2 \qquad (1\text{-}22)$$

Examples of Mating Tooth Profile from Given Tooth Profile. The first step toward the solution of any of these problems is to set up an equation for the profile of the given tooth form. Then the values for the mating profile are obtained by the use of the values established for the first profile and the foregoing equations. As a definite example, we shall start with a pin-tooth gear where the teeth are cylinders or pins mounted between plates. Such members are often called *lantern pinions*.

First Example: Pin-tooth Gear. Referring to Fig. 1-7, we will use

A = radius of pins
B = radius to center of pins[2]
R_1 = pitch radius of lantern pinion.

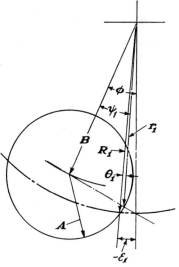

$$\theta_1 = \cos^{-1} \frac{B^2 + r_1{}^2 - A^2}{2Br_1}$$
$$\qquad - \cos^{-1} \frac{B^2 + R_1{}^2 - A^2}{2BR_1} \qquad (1\text{-}23)$$
$$\sin \psi_1 = \frac{A^2 + r_1{}^2 - B^2}{2Ar_1} \qquad (1\text{-}24)$$

First we select a series of values for r_1, then we determine the corresponding values of θ_1, and ψ_1. Next we substitute these values into Eqs. (1-14), (1-15), (1-16), and (1-17), to obtain values for the coordinates of the given tooth profile and its path of contact.

Fig. 1-7.

Last we substitute these values into Eqs. (1-19), (1-20), (1-21), and (1-22) to obtain the values of the coordinates of the mating tooth profile.

If we wish to plot the results and use rectangular coordinates, then we determine the corrected value for the vectorial angle to make the form symmetrical to the center line of the tooth or space, and use Eqs. (1-4), (1-5), and (1-6) to obtain these values.

As a definite example, we shall use the following values:

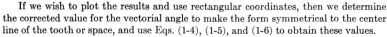

$$A = 0.750 \qquad B = 5.90$$
$$R_1 = 6.00 \qquad \text{with 12 pins or teeth}$$
$$R_2 = 18.00 \qquad \text{with 36 teeth}$$

We shall choose an increment of 0.05 for r_1 and use values ranging from 6.65 to 5.70. For the original pin-tooth gear and its path of contact, we obtain the values tabulated in Table 1-3. For the mating gear-tooth profile, we obtain the values tabulated in Table 1-4 and plotted in Fig. 1-8.

[1] Mating spur gears always run in opposite directions.
[2] The center of the pins must always be off the pitch line, on spur gears, to avoid a cusp in the form of the mating gear tooth.

TABLE 1-3. COORDINATES OF PIN-TOOTH GEAR AND ITS PATH OF CONTACT
(Plotted in Fig. 1-8)

r_1, in.	θ_1, deg	ψ_1, deg	ϕ, deg	ϵ_1, deg	x_p, in.	y, in.
5.70	−0.016	−11.862	21.609	−33.455	−3.1436	1.2453
5.75	0.070	− 7.897	18.334	−26.301	−2.5414	0.8421
5.80	0.122	− 4.004	15.354	−19.480	−1.9229	0.5279
5.85	0.140	− 0.163	12.839	−13.142	−1.3162	0.3000
5.90	0.125	3.644	11.084	− 7.565	−0.7640	0.1497
5.95	0.078	7.436	10.477	− 3.119	−0.3157	0.0584
6.00	0.000	11.229	11.229	0.000	0.0000	0.0000
6.05	−0.112	15.041	13.147	2.006	0.2000	−0.0467
6.10	−0.257	18.891	15.865	3.283	0.3220	−0.0915
6.15	−0.433	22.801	19.108	4.126	0.3961	−0.1372
6.20	−0.648	26.794	22.723	4.719	0.4401	−0.1844
6.25	−0.901	30.904	26.648	5.157	0.4638	−0.2328
6.30	−1.198	35.164	30.864	5.498	0.4724	−0.2823
6.35	−1.544	39.628	35.400	5.772	0.4682	−0.3327
6.40	−1.949	44.364	40.309	6.004	0.4525	−0.3840
6.45	−2.426	49.483	45.702	6.207	0.4253	−0.4359
6.50	−2.998	55.162	51.767	6.393	0.3849	−0.4886
6.55	−3.713	61.753	58.892	6.574	0.3269	−0.5418
6.60	−4.690	70.160	68.079	6.771	0.2397	−0.5956
6.65	−7.163	90.000	90.000	7.163	0.0000	−0.6500

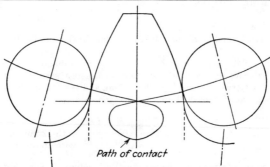

Path of contact

FIG. 1-8.

In practice, when pin-tooth gears are used, no attempt is made to obtain any of the dedendum contact on the mating gear. This part of the form is cut away as indicated by the dotted lines in Fig. 1-8.

Second Example. As a second example, we shall use a gear-tooth profile that consists of a straight line as shown in Fig. 1-9. For this we have the following:

A = radius of circle to which straight line profile is tangent

TABLE 1-4. COORDINATES OF GEAR MATING WITH PIN-TOOTH GEAR
(Plotted in Fig. 1-8)

ϕ, deg	ψ_2, deg	θ_2, deg	ϵ_2, deg	r_2, in.	X, in.	Y, in.
21.609	30.886	−1.875	11.152	19.5003	0.2508	19.4986
18.334	26.012	−1.089	8.767	19.0122	0.5053	19.0055
15.354	21.278	−0.469	6.293	18.6274	0.6965	18.6143
12.839	16.953	−0.267	4.381	18.3473	0.7508	18.3319
11.084	13.494	−0.112	2.522	18.1658	0.7924	18.1485
10.477	11.479	−0.038	1.040	18.0612	0.8111	18.0429
11.229	11.229	0.000	0.000	18.0000	0.8203	17.9813
13.147	12.510	0.032	−0.669	17.9533	0.8282	17.9352
15.865	14.835	0.064	−1.094	17.9085	0.8361	17.8918
19.108	17.838	0.105	−1.375	17.8628	9.8469	17.8472
22.723	21.307	0.157	−1.573	17.8156	0.8609	17.7993
26.648	25.153	0.224	−1.719	17.7673	0.8792	17.7516
30.864	29.337	0.306	−1.833	17.7177	0.9023	17.7009
35.400	33.882	0.406	−1.924	17.6673	0.9305	17.6489
40.309	38.836	0.528	−2.001	17.6160	0.9653	17.5954
45.702	44.314	0.681	−2.069	17.5641	1.0092	17.5402
51.767	50.507	0.871	−2.131	17.5114	1.0641	17.4833
58.892	57.751	1.050	−2.191	17.4582	1.1153	17.4256
68.079	67.290	1.468	−2.257	17.4044	1.2384	17.3618
90.000	90.000	2.388	−2.388	17.3500	1.5122	17.2839

From the geometrical conditions shown in this figure, we have

$$\theta_1 = \sin^{-1}(A/r_1) - \sin^{-1}(A/R_1) \qquad (1\text{-}25)$$
$$\sin \psi_1 = A/r_1 \qquad (1\text{-}26)$$

For the definite example, we shall use

$R_1 = 10.00$
$A = 5.00$ with 20 teeth
$R_2 = 20.00$ with 40 teeth

We shall select a series of values for r_1 varying from 9.00 to 11.00 in increments of 0.10, and determine the coordinates of the original profile and those of the path of contact by the use of Eqs. (1-14), (1-15), (1-16), and (1-17). These values are tabulated in Table 1-5. Then we shall determine the values of the coordinates of the mating gear-tooth profile by

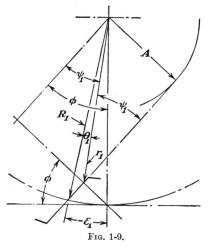

FIG. 1-9.

TABLE 1-5.   COORDINATES OF STRAIGHT-LINE GEAR-TOOTH PROFILE AND PATH OF
CONTACT
(Plotted in Fig. 1-10)

r_1, in.	θ_1, deg	ψ_1, deg	ϕ, deg	ϵ_1, deg	x_p, in.	y, in.
9.00	3.745	33.745	41.551	−11.551	−1.2224	1.0834
9.10	3.329	33.329	40.507	−10.507	−1.1370	0.9713
9.20	2.921	32.921	39.442	− 9.442	−1.0448	0.8595
9.30	2.523	32.523	38.357	− 8.357	−0.9453	0.7482
9.40	2.135	32.135	37.251	− 7.251	−0.8382	0.6374
9.50	1.757	31.757	36.121	− 6.121	−0.7229	0.5276
9.60	1.388	31.388	34.964	− 4.964	−0.5988	0.4187
9.70	1.028	31.028	33.777	− 3.777	−0.4652	0.3112
9.80	0.678	30.678	32.558	− 2.558	−0.3215	0.2053
9.90	0.335	30.335	31.300	− 1.300	−0.1667	0.1014
10.00	0.000	30.000	30.000	0.000	0.0000	0.0000
10.10	−0.327	29.673	28.652	1.348	0.1800	−0.0984
10.20	−0.647	29.353	27.246	2.754	0.3751	−0.1931
10.30	−0.959	29.041	25.776	4.224	0.5866	−0.2833
10.40	−1.266	28.734	24.226	5.774	0.8174	−0.3679
10.50	−1.563	28.437	22.586	7.414	1.0704	−0.4452
10.60	−1.855	28.145	20.826	9.174	1.3503	−0.5136
10.70	−2.141	27.859	18.917	11.083	1.6631	−0.5700
10.80	−2.422	27.578	16.806	13.194	2.0185	−0.6097
10.90	−2.695	27.305	14.410	15.590	2.4306	−0.6251
11.00	−2.964	27.036	11.484	18.516	2.9412	−0.5973

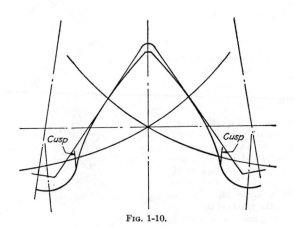

FIG. 1-10.

the use of Eqs. (1-19), (1-20), (1-21), and (1-22). The rectangular coordinates of this profile are found by the use of Eqs. (1-4), (1-5), and (1-6). These values are tabulated in Table 1-6 and plotted in Fig. 1-10.

There is a cusp at the root of the mating gear so that its active profile cannot extend below a radius of about 19.50, and the maximum radius of the first gear must be reduced to about 10.700.

TABLE 1-6. COORDINATES OF MATING GEAR-TOOTH PROFILE OF STRAIGHT-LINE
TOOTH PROFILE

(Plotted in Fig. 1-10)

ϕ, deg	ψ_2, deg	θ_2, deg	ϵ_2, deg	r_2, in.	X, in.	Y, in.
41.551	45.711	−2.459	5.775	21.1188	0.0439	21.1187
40.507	43.610	−2.150	5.253	21.0021	0.1569	21.0015
39.442	42.310	−1.853	4.721	20.8857	0.2642	20.8832
38.357	40.966	−1.569	4.178	20.7697	0.3658	20.7664
37.251	39.568	−1.308	3.625	20.6517	0.4576	20.6466
36.121	38.138	−1.043	3.060	20.5403	0.5503	20.5329
34.964	36.644	−0.802	2.482	20.4275	0.6294	20.4179
33.777	35.086	−0.580	1.889	20.3165	0.7082	20.3041
32.558	33.469	−0.368	1.279	20.2078	0.7792	20.1929
31.300	31.775	−0.175	0.650	20.1021	0.8429	20.0844
30.000	30.000	0.000	0.000	20.0000	0.8994	19.9798
28.652	28.133	0.155	−0.674	19.9024	0.9489	19.8797
27.246	26.161	0.292	−1.377	19.8105	0.9919	19.7857
25.776	24.072	0.408	−2.112	19.7254	1.0275	19.6986
24.226	21.841	0.502	−2.887	19.6492	1.0557	19.6209
22.586	19.451	0.572	−3.707	19.5841	1.0761	19.5545
20.826	16.849	0.610	−4.587	19.5331	1.0862	19.5028
18.917	14.025	0.649	−5.541	19.5010	1.0977	19.4700
16.806	10.862	0.653	−6.597	19.4951	1.0987	19.4641
14.410	7.246	0.655	−7.795	19.5268	1.1013	19.4957
11.484	2.889	0.663	−9.258	19.6244	1.1094	19.5930

PATH OF CONTACT GIVEN

When the coordinates of the path of contact are given, we can then determine the coordinates of the basic-rack profile. Then with this information, we can proceed as before to determine the coordinates of the other gear teeth.

We shall start with Eq. (1-1)

$$x_p = -y/\tan \phi = -y \, dy/dx$$

Then

$$\tan \phi = -y/x_p \tag{1-27}$$

At the origin

$$\tan \phi = -dy/dx_p \tag{1-28}$$

whence

$$x = -\int y \, dy/x_p \tag{1-29}$$

When we have the value of x_p in terms of y, we substitute this expression for x_p in the foregoing equation, simplify, and integrate. The constant of integration must be selected to bring the resulting curve through the origin or pitch point. This constant of integration will be the value of the indefinite integral when y is equal to zero.

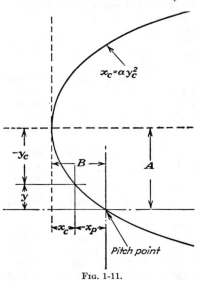

Example of Given Path of Contact. As a definite example, we shall use a portion of a parabola as our given path of contact, as shown in Fig. 1-11. The symbols are indicated there. From this figure, we have for the original equation of the parabola

$$x_c = ay_c^2 \tag{1-30}$$

When the origin is transferred to the pitch point, we have

FIG. 1-11.

$$B = aA^2 \qquad x_c = B + x_p = aA^2 + x_p \qquad y_c = y - A$$

Substituting these values into the original equation for the parabola, we have

$$aA^2 + x_p = a(y - A)^2$$
$$x_p = a(y - A)^2 - aA^2 = ay(y - 2A) \tag{1-31}$$

Then

$$x = -\int \frac{y \, dy}{ay(y - 2A)} = \frac{1}{a} \int \frac{dy}{(2A - y)} = -\frac{1}{a} [\log (2A - y) + C]$$

As x must equal zero when y equals zero, we have

$$C = -\log 2A$$

then

$$x = \frac{1}{a} [\log 2A - \log (2A - y)] \tag{1-32}$$

For the definite example, we will use[1]

$$a = 0.375 \qquad A = 2.3094$$

[1] These values have been selected to give a pressure angle of 30 deg at the pitch point.

TABLE 1-7.   COORDINATES OF PARABOLIC PATH OF CONTACT AND ITS BASIC-RACK
FORM
(Plotted in Fig. 1-12)

y, in.	x_p, in.	x, in.	tan ϕ	ϕ, deg
1.00	−1.3571	0.6507	0.73689	36.387
0.90	−1.2551	0.5780	0.71707	35.643
0.80	−1.1456	0.5072	0.69830	34.927
0.70	−1.0287	0.4383	0.68048	34.235
0.60	−0.9042	0.3711	0.66355	33.566
0.50	−0.7723	0.3055	0.64743	32.920
0.40	−0.6328	0.2416	0.63209	32.297
0.30	−0.4859	0.1791	0.61747	31.694
0.20	−0.3314	0.1180	0.60348	31.110
0.10	−0.1695	0.0584	0.59011	30.545
0.00	0.0000	0.0000	0.57735	30.000
−0.10	0.1770	−0.0571	0.56510	29.471
−0.20	0.3614	−0.1131	0.55339	28.960
−0.30	0.5534	−0.1678	0.54213	28.464
−0.40	0.7528	−0.2215	0.53133	27.983
−0.50	0.9598	−0.2741	0.52095	27.517
−0.60	1.1742	−0.3257	0.51097	27.066
−0.70	1.3962	−0.3763	0.50136	26.628
−0.80	1.6256	−0.4260	0.49211	26.202
−0.90	1.8626	−0.4747	0.48319	25.790
−1.00	2.1071	−0.5220	0.47459	25.389

Then

$$x_p = 0.375y(y - 4.6188)$$
$$x = 2.6667[\log^* 4.6188$$
$$- \log^* (4.6188 - y)]$$

At the origin

$$\tan \phi = -dy/dx_p = 1/2aA = 0.57735$$

Values for this path of contact and
the resulting basic-rack form are tabulated in Table 1-7 and plotted in Fig.
1-12.

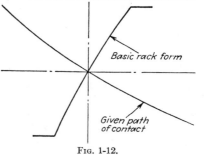

FIG. 1-12.

RADIUS OF CURVATURE OF TOOTH PROFILE

The radius of curvature of the gear-tooth profile is an essential value
that is needed for the determination of the intensity of the local stresses

* These are natural logarithms to base e.   All logarithms used in this text are to
this base.

set up between the loaded gear teeth. When R_c = radius of curvature, for Cartesian or rectangular coordinates

$$R_c = \frac{[1 + (dy/dx)^2]^{3/2}}{(d^2y/dx^2)} \qquad (1\text{-}33)$$

for polar coordinates

$$R_c = \frac{[r^2 + (dr/d\theta)^2]^{3/2}}{r^2 - r(d^2r/d\theta^2) + 2(dr/d\theta)^2} \qquad (1\text{-}34)$$

When the equation of the profile of the basic rack or of one of the gears is known, the value of the radius of curvature at any point of that profile can be determined readily by means of one of the two preceding equations.

The determination of the radius of curvature of one of the calculated profiles is a much more complex procedure because here we have one independent variable and several dependent ones. The radius of curvature at the pitch point is the simpler one to find because here several of these variables become equal to each other. We shall first consider the radius of curvature of the calculated profiles at any point.

Radius of Curvature of Gear-tooth Profile. We shall start from a given basic-rack form. For the conjugate gear-tooth profile, the problem is to obtain values for $(dr/d\theta)$ and $(d^2r/d\theta^2)$ in terms of the known values so that they may be substituted into the general polar equation for the radius of curvature. We have, to start,

$$dr/d\theta = -r/\tan\psi = -r\cot\psi$$

Differentiating in respect to θ, we obtain

$$d^2r/d\theta^2 = -(dr/d\theta)\cot\psi + r\csc^2\psi\,(d\psi/d\theta)$$

whence

$$d^2r/d\theta^2 = (r/\sin^2\psi)[\cos^2\psi + (d\psi/d\theta)]$$

The derivation of an expression for $d\psi/d\theta$ may be very complex and difficult. A reasonable approximation will be to use the ratio of the differences from the tabulated values. In other words, we shall substitute $\Delta\psi/\Delta\theta$ for $d\psi/d\theta$. This ratio will be the same whether we use the values of these angles in degrees or in radians.

Substituting the foregoing into Eq. (1-34), combining, and simplifying, we obtain

$$R_c = \frac{r}{\sin\psi\,[1 - (d\psi/d\theta)]} \qquad (1\text{-}35)$$

This relationship applies to both gears of the pair.

Example of Radius of Curvature of Gear-tooth Profile. As a definite example, we shall use the pin-tooth gear whose values are tabulated in Table 1-3, and determine its radius of curvature at a radius of 6.50. We shall use a radius difference of plus and minus one increment of the tables from the selected value of 6.50 (*i.e.*, 6.55 − 6.45). From the values in Table 1-3, we obtain

$$\Delta\psi = 12.270° \qquad \Delta\theta = -1.287° \qquad (\Delta\psi/\Delta\theta) = -9.534$$

Substituting this value into Eq. (1-35), we obtain $R_c = 0.7518$. This value is plus, and so the form is convex. The actual radius of the pin is 0.750, so that the error in this approximation is 0.0018 or 0.24 per cent.

Radius of Curvature of Rack-tooth Profile. When we start from a given form for one of the gears, then it may be difficult to find the values of (dy/dx) and (d^2y/dx^2). In this case we have, to start,

$$dy/dx = 1/\tan\,\phi = \cot\,\phi$$

Differentiating in respect to x, we obtain

$$d^2y/dx^2 = -\,\text{cosec}^2\,\phi(d\phi/dx) = -(d\phi/dx)/\sin^2\,\phi$$

If we use differences here instead of the differentials, then the value of $\Delta\phi$ must be in radians to obtain the correct value of the ratio $\Delta\phi/\Delta x$. Substituting these values into Eq. (1-33), combining, and simplifying, we obtain

$$R_c = -1/\sin\,\phi(d\phi/dx) \tag{1-36}$$

Example of Radius of Curvature of Rack-tooth Profile. As a definite example, we shall use the rack whose values are tabulated in Table 1-1. We shall take the value of y as 0.80, where $\tan\,\phi = 0.71686$ and $\phi = 35.635°$. Proceeding as before, we obtain $\Delta\phi = 2.821° = 0.04924$ radian, and $\Delta x = 0.1435$, whence $(\Delta\phi/\Delta x) = 0.343$, and $R_c = -5.004$. The sign here is minus, and the form is convex. This rack has a radius of curvature of 5.00.

Radius of Curvature at Pitch Point. We shall use the following symbols for the radius of curvature at the pitch point:

R_{cr} = radius of curvature of rack profile at pitch point
R_{c1} = radius of curvature of first gear at pitch point
R_{c2} = radius of curvature of second gear at pitch point

We have to start

$$\frac{dr_1}{d\theta_1} = -\,\frac{R_1}{\tan\,\psi_1} = -\,\frac{R_1}{\tan\,\phi}$$

$$\frac{d^2r_1}{d\theta_1{}^2} = \frac{-R_1[1 - R_1\tan\,\phi(d^2x/dy^2)]}{\tan^4\,\phi}$$

Because of the specific coordinate system used where the value of r decreases as the value of θ increases, these terms are minus.

When we substitute these values into Eq. (1-34) and simplify, we get

$$R_{c1} = \frac{R_1\sin\,\phi}{1 - R_1\sin\,\phi\cos^3\,\phi(d^2x/dy^2)} \tag{1-37}$$

For the basic-rack form at the pitch point, we obtain

$$R_{cr} = - \frac{1}{\cos^3 \phi(d^2x/dy^2)} \tag{1-38}$$

whence

$$\frac{d^2x}{dy^2} = - \frac{1}{R_{cr} \cos^3 \phi}$$

Substituting this value into Eq. (1-37), we obtain

$$R_{c1} = \frac{R_1 R_{cr} \sin \phi}{R_{cr} + R_1 \sin \phi} \tag{1-39}$$

For the mating gear, we obtain

$$R_{c2} = \frac{R_2 R_{cr} \sin \phi}{R_{cr} - R_2 \sin \phi} \tag{1-40}$$

With the specific coordinate system used here, on the rack form when the value of R_{cr} is plus, the form is concave. For the gears, on the other hand, when R_{c1} or R_{c2} is plus, the form is convex.

We now have the radius of curvature of either gear in terms of the radius of curvature of the basic rack. We shall now combine these equations and obtain expressions for the radius of curvature of either gear in terms of the radius of curvature of the other. For this we get

$$R_{c1} = \frac{R_1 R_2 \sin \phi}{(R_1 + R_2) - (R_1 R_2 \sin \phi/R_{c2})} \tag{1-41}$$

$$R_{c2} = \frac{R_1 R_2 \sin \phi}{(R_1 + R_2) - (R_1 R_2 \sin \phi/R_{c1})} \tag{1-42}$$

Examples of Radii of Curvature at Pitch Point. The foregoing equations give values of the radius of curvature at the pitch point of any one member of the system in terms of the radius of curvature of any other member. The first step is to determine the radius of curvature of the given member. Then this value is used to find the others.

First Example. For the first example, we shall use the basic-rack form shown in Fig. 1-12. The equation of this profile is given by Eq. (1-32) and is as follows:

$$x = \frac{1}{a} [\log 2A - \log (2A - y)] \tag{1-32}$$

where

$$a = 0.375 \quad \text{and} \quad A = 2.3094$$

$$\frac{dx}{dy} = \tan \phi = \frac{1}{a}\left[\frac{1}{(2A - y)} \right]$$

$$\frac{d^2x}{dy^2} = \frac{1}{a}\left[\frac{1}{(2A - y)^2} \right] = \frac{1}{8}$$

$$R_{cr} = \frac{-1}{\cos^3 \phi(d^2x/dy^2)} = \frac{-1}{(\cos^3 30°)(\frac{1}{8})} = \frac{-1}{0.08119} = -12.3167$$

This value is minus for the rack, and hence the form is convex.

We can also determine this value directly from the general equation (1-33), as follows:

$$\frac{dy}{dx} = \frac{1}{\tan \phi} = a(2A - y)$$

Differentiating in respect to x, we obtain

$$\frac{d^2y}{dx^2} = -a\frac{dy}{dx} = -a^2(2A - y) = -0.64952$$

$$R_{cr} = \frac{1}{\sin^3 \phi(d^2y/dx^2)} = \frac{-1}{0.64952 \sin^3 30°} = \frac{-1}{0.08119} = -12.3167$$

When the value of the radius of curvature of the basic rack at its pitch point is known, the radius of curvature of any other gears of the system at their pitch points can be readily determined.

Second Example. As a second example we shall use the basic-rack form shown in Fig. 1-4, which is an arc of a circle with a radius of 5.00. This form is convex so that

$$R_{cr} = -5.00 \qquad R_1 = 20 \qquad R_2 = 20$$
$$\tan \phi = 0.46631 \qquad \phi = 25.000° \qquad \sin 25° = 0.42262$$
$$R_{c1} = \frac{-20 \times 5 \times 0.42262}{-5 + (20 \times 0.42262)} = -12.2413$$

This value is minus so that the form of this first gear is concave at the pitch point.

$$R_{c2} = \frac{20 \times 5 \times 0.42262}{(20 \times 0.42262) + 5} = 3.1416$$

This value is plus so that the form of this second gear tooth is convex at the pitch point.

Enveloping Gear-tooth Form. When all other factors are identical except for the radii of curvature, then the intensity of the maximum specific compressive stress set up between curved contacting surfaces under load is given by the following equation:

When s = compressive stress, psi

A = value depending upon load, modulus of elasticity of materials, length of contacting surfaces, etc.

$$s^2 = A\,[(1/R_{c1}) + (1/R_{c2})] \tag{1-43}$$

In this equation, when the form is convex, the sign for the radius of curvature is plus. When the contacting surface is concave, then the sign is minus. This agrees with the conditions set up in this analysis for the gear-tooth profiles, but it is opposite to the conditions on the basic-rack profile as developed here.

As an example, if we have R_{c1} = 2.00, R_{c2} = 4.00, and if both forms are convex, then

$$s^2 = A(\tfrac{1}{2} + \tfrac{1}{4}) = \tfrac{3}{4}\,A$$

If the larger member is concave, while the smaller one is convex, then we have

$$s^2 = A(\tfrac{1}{2} - \tfrac{1}{4}) = \tfrac{1}{4}\,A$$

Thus when the values of all other factors are equal except for the direction of the relative curvatures, the stresses are less when the directions of the curvatures follow each other than they are when these directions depart from each other.

This condition has led to the belief that if the gear-tooth forms were made enveloping, *i.e.*, if one profile is convex and the other is concave, then the compressive stresses at the region of contact would be less than they are when the forms of the teeth of both members are convex.

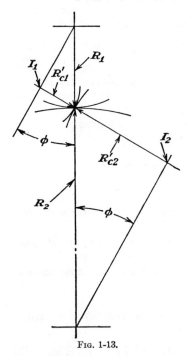

Equation (1-42) gives the relationship between the radii of curvature of conjugate gear-tooth profiles at their pitch point. We will now derive an equation for the sum of their reciprocals, which is the value needed for the determination of the contact stresses. Thus we have

$$\frac{1}{R_{c1}} + \frac{1}{R_{c2}} = \frac{1}{R_{c1}}$$
$$+ \frac{(R_1 + R_2) - (R_1 R_2 \sin \phi / R_{c1})}{R_1 R_2 \sin \phi}$$

Combining and simplifying, we obtain

$$\frac{1}{R_{c1}} + \frac{1}{R_{c2}} = \frac{1}{\sin \phi}\left(\frac{1}{R_1} + \frac{1}{R_2}\right) \tag{1-44}$$

Fig. 1-13.

From this last equation we can see that the sum of the reciprocals of the radii of curvature of conjugate gear-tooth profiles at their pitch points is independent of the radius of curvature of the tooth profile of either member and depends solely upon the sizes of the gears and the pressure angle at the pitch point. In other words, a certain amount of rocking action or change in relative curvature is needed at this point, and the resulting compressive stresses are identical regardless of the specific curvature of either member. Hence the belief in the lower compressive-stress conditions for enveloping gear-tooth forms is a fallacy.

This condition can be shown graphically as in Fig. 1-13, where I_1 and I_2 represent the instantaneous centers of the effective radii of curvature, R'_{c1} and R'_{c2}, respectively. Whatever the actual conditions may be, the relative rocking action must be the same as that given by the two effective radii of curvature.

Radius of Curvature of Internal Gears at Pitch Point. For internal gears, the sign for the curvature of the pitch line of the internal gear will be minus, and so the expression for the sum of the reciprocals of the radii of curvature of the tooth profiles at the pitch point is as follows:

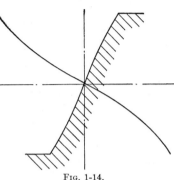

$$\frac{1}{R'_{c1}} + \frac{1}{R'_{c2}} = \frac{1}{\sin \phi}\left(\frac{1}{R_1} - \frac{1}{R_2}\right)$$

(1-45)

Fig. 1-14.

INTERCHANGEABLE GEAR-TOOTH FORMS

No consideration has been given to the factor of interchangeability in the preceding analyses. For example, the gear-tooth forms shown in Fig. 1-5 will mate and run together properly, but two gears of the form of either will not run together correctly. In order to obtain such interchangeability between gears so that all gears of all sizes conjugate to the same basic rack will mesh together correctly, the path of contact of the system must be symmetrical in relation to the pitch point. When this condition is met, then the profile or form of the basic rack of the system will also be symmetrical in relation to the pitch point. Then all gears of all numbers of teeth that are conjugate to such a basic rack will also be conjugate to each other.

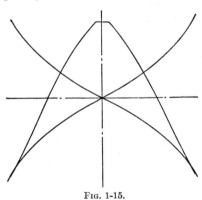

Fig. 1-15.

Example of Interchangeable Gear-tooth System. We shall use the rack form shown in Fig. 1-4 for the addendum of the basic rack and will reverse it for the dedendum, which gives us the basic-rack form and path of contact with the values tabulated in Table 1-8 and plotted in Fig. 1-14.

With 20 teeth in the gear and a pitch radius of 20, the form of the dedendum of the gear will be the same as that tabulated in Table 1-1, while the form of its addendum will be the same as that tabulated in Table 1-2. These values, together with the Cartesian coordinates adjusted to the center line of the tooth are tabulated in Table 1-8 and are plotted in Fig. 1-15.

CYCLOIDAL TOOTH FORMS

One of the first general forms to be used for gear-tooth profiles was the cycloidal form. Theoretically, in its kinematics, it has many points of advantage. The practical difficulties of producing it accurately, however, are largely responsible for its retirement from the field of commercial gears.

Wilfred Lewis has said:

The practical consideration of cost demands the formation of gear teeth upon some interchangeable system. The cycloidal system cannot compete with the involute, because its cutters are formed with greater difficulty and with less accuracy, and a further expense is entailed by the necessity for more accurate center distances. Cycloidal teeth must not only be accurately spaced and shaped, but their wheel centers must also be fixed with equal care to obtain satisfactory results.

TABLE 1-8. COORDINATES OF INTERCHANGEABLE BASIC-RACK FORM, ITS PATH OF
CONTACT, AND A CONJUGATE GEAR-TOOTH PROFILE
(Plotted in Figs. 1-14 and 1-15)

y, in.	x, in.	$\tan \phi$	x_p, in.	r, in.	θ, rad	X, in.	Y, in.
1.00	0.6189	0.79566	−1.2568	19.0415	0.0277	1.2743	18.9888
0.90	0.5414	0.75514	−1.1918	19.1371	0.0244	1.2177	19.0983
0.80	0.4678	0.71686	−1.1160	19.2324	0.0211	1.1603	19.1974
0.70	0.3879	0.68054	−1.0286	19.3274	0.0181	1.1082	19.2955
0.60	0.3316	0.64599	−0.9288	19.4222	0.0152	1.0573	19.3935
0.50	0.2687	0.61300	−0.8157	19.5171	0.0124	1.0079	19.4911
0.40	0.2090	0.58140	−0.6880	19.6120	0.0098	0.9618	19.5885
0.30	0.1524	0.55105	−0.5444	19.7075	0.0072	0.9156	19.6862
0.20	0.0987	0.52181	−0.3833	19.8037	0.0047	0.8704	19.7845
0.10	0.0480	0.49361	−0.2026	19.9010	0.0024	0.8291	19.8837
0.00	0.0000	0.46631	−0.0000	20.0000	0.0000	0.7854	19.9846
−0.10	−0.0480	0.49361	0.2026	20.1010	−0.0025	0.7389	20.0873
−0.20	−0.0987	0.52181	0.3833	20.2036	−0.0051	0.6904	20.1919
−0.30	−0.1524	0.55105	0.5444	20.3073	−0.0080	0.6350	20.2973
−0.40	−0.2090	0.58140	0.6880	20.4116	−0.0111	0.5748	20.4034
−0.50	−0.2687	0.61300	0.8157	20.5162	−0.0145	0.5080	20.5098
−0.60	−0.3316	0.64599	0.9288	20.6209	−0.0180	0.4386	20.6162
−0.70	−0.3979	0.68054	1.0286	20.7255	−0.0217	0.3641	20.7222
−0.80	−0.4678	0.71686	1.1160	20.8299	−0.0256	0.2845	20.8280
−0.90	−0.5414	0.75514	1.1918	29.9340	−0.0297	0.2001	20.9330
−1.00	−0.6189	0.79566	1.2568	21.0376	−0.0340	0.1104	21.0373

George B. Grant wrote in his excellent "Treatise on Gearing":[1]

There is no more need of two different kinds of tooth curves for gears of the same pitch than there is need for two different threads for standard screws, or two different coins of the same value, and the cycloidal tooth would never be missed if it were dropped altogether. But it was first in the field, is simple in theory, and has the recommendation of many well-meaning teachers, and holds its position by means of "human inertia," or the natural reluctance of the average human mind to adopt a change, particularly a change for the better.

Although cycloidal forms are seldom used today for gear-tooth profiles that are employed primarily for the transmission of power, they are widely used for impellers or rotors of pressure blowers and for other special

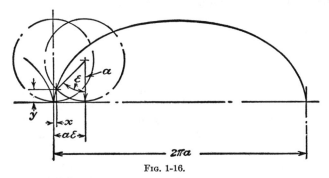

FIG. 1-16.

applications. Their interaction in these uses is conjugate gear-tooth action, although the actual rotation of their shafts may be controlled by other gears. An analysis of these cycloidal forms is, therefore, in order.

The Cycloid. The path described by a point on the circumference of a circle that rolls upon a straight line is called a *cycloid*. When the point where the curve meets the straight line is the origin of the coordinate system, and also its pitch point, the derivation of the equation of this curve is as follows: We shall let

a = radius of rolling circle

ϵ = angle of rotation of rolling circle

Referring to Fig. 1-16, the distance from the origin to the point of tangency of the rolling circle with the straight line is equal to the length of the arc ϵ at the radius a, which is equal to $a\epsilon$. The tracing point on the rolling circle is at a distance of $a \sin \epsilon$ from the vertical center line of the rolling circle, whence

$$x = a(\epsilon - \sin \epsilon) \tag{1-46}$$

[1] This is one of the classics on the subject. It was first published by Mr. Grant in 1890. It is now published by the Philadelphia Gear Works.

The tracing or generating point on the rolling circle is at a distance of $a \cos \epsilon$ below the center of this rolling circle, whence

$$y = a(1 - \cos \epsilon) \tag{1-47}$$

These two equations are the simplest form in which the equation of the cycloid can be given. They may be combined into a single equation with the third variable ϵ eliminated, which gives the following:

$$x = a\{\cos^{-1}[(a - y)/a] - \sqrt{2ay - y^2}/a\} \tag{1-48}$$
$$dx/dy = \tan \phi = y/\sqrt{2ay - y^2} \tag{1-49}$$

For the radius of curvature of the cycloid, we get

$$R_{cr} = -2\sqrt{2ay} \tag{1-50}$$

For the path of contact, we get

$$x_p = -y/\tan \phi = -\sqrt{2ay - y^2} \tag{1-51}$$

Equation (1-51) gives the value of the abscissa of the path of contact; the value of y is the same as that for the cycloid itself. We shall next determine the equation of a circle of radius a that is tangent to the x axis at the pitch point. We have for the equation of such a circle

$$x_p{}^2 + (a - y)^2 = a^2$$

whence

$$x_p = \pm\sqrt{a^2 - (a - y)^2} = \pm\sqrt{2ay - y^2}$$

This equation is the same as that for the abscissa of the path of contact; therefore the path of contact of a cycloid is the rolling circle itself in its starting position.

The Epicycloid. When a rolling circle, tangent externally to a fixed circle, rolls upon the fixed circle, the path described by a point on the rolling circle is called an *epicycloid*.[1] We shall take the origin of the polar coordinate system at the center of the fixed circle, and let

R = radius of fixed circle, which is also the pitch circle
a = radius of rolling circle
ϵ = angle of rotation of rolling circle
ϵ_1 = angle of rotation of gear

We have the following from the geometrical conditions shown in Fig. 1-17:

$$\cos \epsilon = \frac{(R + a)^2 + a^2 - r^2}{2a(R + a)}$$

---

[1] The epicycloid is the form of the addendum surface of a gear tooth that meshes with a cycloidal rack.

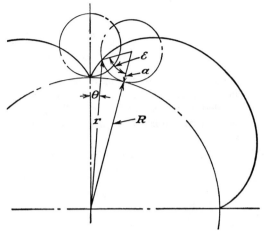

Fig. 1-17.

whence

$$\cos \epsilon = 1 - \frac{(r^2 - R^2)}{2a(R + a)} \tag{1-52}$$

$$\theta = \frac{a\epsilon}{R} - \sin^{-1}\left(\frac{a}{r} \sin \epsilon\right) \tag{1-53}$$

From the foregoing equations, we obtain

$$\tan \psi = \frac{r \, d\theta}{dr} = \frac{r^2(1 - \cos \epsilon) + aR \sin^2 \epsilon}{R \sin \epsilon[R + a(1 - \cos \epsilon)]} \tag{1-54}$$

$$\epsilon_1 = \frac{a\epsilon}{R} \tag{1-55}$$

$$\phi = \frac{\epsilon}{2} \tag{1-56}$$

Radius of Curvature. The radius of curvature of the epicycloid is given by the following equation:

$$R_c = \frac{4a(R + a) \sin (\epsilon/2)}{R + 2a} \tag{1-57}$$

Examples of Radius of Curvature of Epicycloid. As a definite example of the radius of curvature, we shall let

$$R = 10.00 \qquad a = 2.00 \qquad r = 10.00 = \text{radius of pitch circle}$$

whence

$$\epsilon = 0° \qquad \sin (\epsilon/2) = 0.00$$
$$R_c = 0.00$$

The radius of curvature of all cycloidal forms at the pitch point is equal to zero. This is one feature of this form of curve that makes it unsatisfactory for a gear-tooth form to transmit power.

As a second example, we shall determine the radius of curvature of this same epicycloid at its maximum radius of 14 in. Here

$$\epsilon = 180° \qquad \epsilon/2 = 90° \qquad \sin{(\epsilon/2)} = 1.00$$
$$R_c = \frac{4 \times 2 \times 12 \times 1}{14} = \frac{96}{14} = 6.8571$$

The Hypocycloid. When the rolling circle, tangent internally to the fixed circle, rolls on the fixed circle, the path described by a point on the rolling circle is called a *hypocycloid*.[1] We shall again take the origin of the coordinate system at the center of the fixed circle, and shall use the same symbols as before.

We have the following from the geometrical conditions shown in Fig. 1-18:

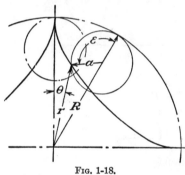

$$\cos \epsilon = \frac{r^2 - (R - a)^2 - a^2}{2a(R - a)}$$

$$= 1 - \frac{R^2 - r^2}{2a(R - a)} \qquad (1\text{-}58)$$

$$\theta = \frac{a\epsilon}{R} - \sin^{-1}\left(\frac{a}{r} \sin \epsilon\right) \qquad (1\text{-}59)$$

$$\tan \psi = \frac{r \, d\theta}{dr}$$

$$= \frac{r^2(1 - \cos \epsilon) + aR \sin^2 \epsilon}{R \sin \epsilon[R - a(1 - \cos \epsilon)]} \qquad (1\text{-}60)$$

Fig. 1-18.

Radius of Curvature. The radius of curvature of the hypocycloid is given by the following equation:

$$R_c = \frac{-4a(R - a) \sin{(\epsilon/2)}}{R - 2a} \qquad (1\text{-}61)$$

Example of Radius of Curvature of Hypocycloid. As a definite example, we shall determine the radius of curvature of a hypocycloid at its minimum radius when

$$R = 10.00 \qquad a = 2.00 \qquad r = 6.00 = \text{minimum radius}$$
$$\epsilon = 180° \qquad \epsilon/2 = 90° \qquad \sin{(\epsilon/2)} = 1.000$$
$$R_c = \frac{-8 \times 8 \times 1.0}{10 - 4} = \frac{-64}{6} = -10.6667$$

The sign for this radius of curvature is minus and the form of this curve is concave.

Application of Cycloid to Gear-tooth Forms. When cycloidal curves are used for gear-tooth profiles, the addendum of the tooth is an epicycloid and the dedendum is a hypocycloid. For interchangeable tooth forms,

[1] The hypocycloid is the form of the dedendum surface of a gear tooth that meshes with a cycloidal rack.

the size of the rolling circle must be identical for both parts of the tooth form.

When the size of the rolling circle for a hypocycloid is one-half the size of the pitch circle, or fixed circle, the form becomes a radial line to the fixed circle. When the gear-tooth system is based on a pinion of 12 teeth for the smallest of the system, this pinion is made with radial flanks (or form of dedendum). Then the size of the rolling circle for all gears of the system will be one-half the pitch diameter of the 12-tooth pinion.

Cycloidal Rotors. The cycloidal form is well adapted for use as the form of rotors in blowers and rotary pumps, because of the closed path of contact for the full-cycloidal form, which eliminates the possibility of trapping between the lobes of the rotors. It is widely used in that type of blower known as the *Root* type. In many other places it, or its equivalent, would be a more effective form than the involute, which is now used extensively in oil and water pumps with rotors of gear-tooth form, but it would require a more elaborate construction than the present one in many cases. Most of such pumps consist of two meshing and self-driving rotors of gear-tooth form. The action between two full-cycloidal-form rotors is conjugate gear-tooth action, but the pressure angle between them rises from 0 to 90 deg and then becomes negative in effect, so that one rotor or spur of straight tooth form will not drive the other rotor through the whole cycle of operation. This requires the provision of other gears than the rotors themselves to drive them and to keep them in the correct angular relationship to each other.

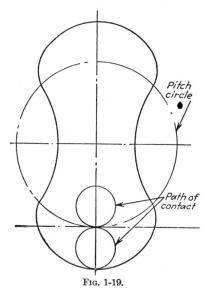

Fig. 1-19.

Example of Cycloidal Rotors. As an example, we shall determine the form of a pair of 2-lobed rotors operating at a center distance of 4 in. In this example the diameter of the rolling circle must be one-quarter the size of the pitch circle. This gives the following values for these rotors: $R = 2.000$ in. and $a = 0.500$ in. With these values and the use of the foregoing equations, we obtain the values of the coordinates for these rotors, which are tabulated in Table 1-9 and are plotted in Fig. 1-19.

SEGMENTAL FORM FOR ROTORS

For most purposes a segmental form can be substituted for the cycloidal form. If the form is to be generated, this is a more practical and better form to use than the cycloidal. It is a reasonably close approximation to the cycloidal, has a closed path of contact, and its basic-rack form is composed of arcs of circles whose centers are off the pitch line. The equations for such a basic-rack form and its path of contact have already been established. These are as follows:

TABLE 1-9. COORDINATES OF CYCLOIDAL ROTOR
(Plotted in Fig. 1-19)

r, in.	θ, rad	ϕ, deg	ϵ_1, deg	X, in.	Y, in.
3.00	−0.78540	90.000	45.000	0.0000	3.0000
2.90	−0.49743	69.909	34.945	0.8234	2.7806
2.80	−0.38289	61.214	30.607	1.0968	2.5762
2.70	−0.29647	54.210	27.105	1.2674	2.3840
2.60	−0.22629	47.985	23.992	1.3791	2.2041
2.50	−0.16733	42.130	21.065	1.4487	2.0375
2.40	−0.11724	36.391	18.195	1.4869	1.8839
2.30	−0.07498	30.526	15.263	1.4999	1.7436
2.20	−0.04040	24.197	12.098	1.4915	1.6172
2.10	−0.01419	16.640	8.320	1.4637	1.5058
2.00	0.00000	0.000	0.000	1.4142	1.4142
1.90	0.00649	21.134	−10.567	1.3522	1.3347
1.80	0.01963	30.220	−15.110	1.2975	1.2476
1.70	0.03523	37.035	−18.517	1.2437	1.1590
1.60	0.06514	43.854	−21.927	1.2026	1.0553
1.50	0.09889	49.707	−24.853	1.1602	0.9508
1.40	0.14510	55.550	−27.775	1.1227	0.8364
1.30	0.20567	61.342	−30.621	1.0876	0.7121
1.20	0.28964	67.482	−33.741	1.0555	0.5708
1.10	0.41748	74.658	−37.329	1.0264	0.3956
1.00	0.78540	90.000	−45.000	1.0000	0.0000

$$x = B - \sqrt{A^2 - (D + y)^2} \qquad (1\text{-}10)$$

$$\frac{dx}{dy} = \tan \phi = \frac{D + y}{\sqrt{A^2 - (D + y)^2}} \qquad (1\text{-}11)$$

For the dedendum of this rack, we obtain

$$x = \sqrt{A^2 - (D - y)^2} - B \qquad (1\text{-}62)$$

$$\frac{dx}{dy} = \tan \phi = \frac{D - y}{\sqrt{A^2 - (D - y)^2}} \qquad (1\text{-}63)$$

TABLE 1-10.   COORDINATES OF THE SEGMENTAL BASIC RACK, ITS PATH OF CONTACT,
AND SEGMENTAL ROTOR

(Plotted in Figs. 1-20 and 1-21)

y, in.	x, in.	tan ϕ	x_p, in.	r, in.	θ, rad	X, in.	Y, in.
1.00	1.5708	∞	0.0000	1.0000	0.7854	1.0000	0.0000
0.90	0.9905	2.81531	−0.3197	1.1455	0.3723	1.0491	0.4599
0.80	0.7624	1.89725	−0.4217	1.2719	0.2514	1.0966	0.6444
0.70	0.5960	1.46327	−0.4784	1.3852	0.1846	1.1426	0.7831
0.60	0.4631	1.20404	−0.4983	1.4860	0.1388	1.1860	0.8954
0.50	0.3527	1.01283	−0.4937	1.5792	0.1052	1.2222	0.9886
0.40	0.2520	0.86433	−0.4628	1.6656	0.0758	1.2635	1.0852
0.30	0.1790	0.74269	−0.4039	1.7473	0.0582	1.3053	1.1616
0.20	0.1100	0.63917	−0.3129	1.8270	0.0393	1.3417	1.2401
0.10	0.0507	0.54845	−0.1823	1.9087	0.0209	1.3775	1.3212
0.00	0.0000	0.46708	0.0000	2.0000	0.0000	1.4142	1.4142
−0.10	−0.0507	0.54845	0.1823	2.1079	−0.0299	1.4453	1.5344
−0.20	−0.1100	0.63917	0.3129	2.2221	−0.0702	1.4572	1.6776
−0.30	−0.1790	0.74269	0.4039	2.3352	−0.1176	1.4461	1.8336
−0.40	−0.2520	0.86433	0.4628	2.4442	−0.1669	1.4173	1.9914
−0.50	−0.3527	1.01283	0.4937	2.5483	−0.2282	1.3473	2.1629
−0.60	−0.4631	1.20404	0.4983	2.6473	−0.2914	1.2553	2.3308
−0.70	−0.5960	1.46327	0.4784	2.7421	−0.3618	1.1270	2.4997
−0.80	−0.7624	1.89725	0.4217	2.8316	−0.4426	0.9519	2.6668
−0.90	−0.9905	2.81531	0.3197	2.9176	−0.5453	0.6938	2.8339
−1.00	−1.5708	∞	0.0000	3.0000	−0.7854	0.0000	3.0000

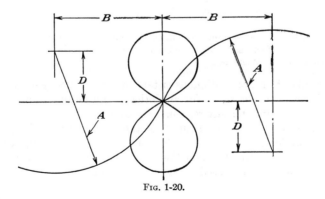

FIG. 1-20.

With these coordinates, we are in a position to proceed as before to determine the coordinates of the path of contact and those of the conjugate gear-tooth profiles.

Example of Segmental Rotor.   For a basic rack of segmental form to be used as a substitute for the cycloidal rack of the rotor shown in Fig. 1-19, the height of the addendum must be equal to 1.00 and the circular pitch must be equal to 2π.   Hence when $x = 1.5708$ and $y = 1.00$, the value of B must be equal to 1.5708 so as to have a

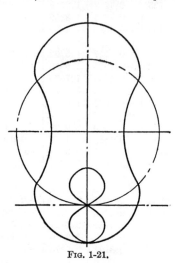

Fig. 1-21.

closed path of contact.   Then

$$A = 1 + D = \sqrt{B^2 + D^2}$$

whence

$$D = 0.73370 \quad \text{and} \quad A = 1.73370$$

With the use of these constants, we obtain the values for the coordinates of the basic rack, its path of contact, and the conjugate rotor form tabulated in Table 1-10 and plotted in Figs. 1-20 and 1-21.

Many other forms of continuous or tangent curves can be used as the basic-rack form for rotors, or for any other types of gear-tooth forms, when they are desired for any reason.   For example, a sine curve or two tangent ellipses could be substituted for the form of the basic rack in place of the cycloid.   The equations for such forms can be set up readily, and then the several equations for the path of contact and the conjugate gear-tooth forms would be used as before.   The choice here is practically unlimited.

HOB FORM FOR SPLINE SHAFTS

The problems of determining the forms of hobs that are to be used to generate spline shafts, ratchets, sprocket wheels, etc., are essentially problems of conjugate gear-tooth forms. The hob form must represent

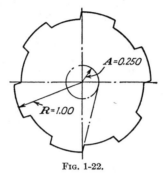

Fig. 1-22.

in its cutting edges and cutting action the form of the basic rack that is conjugate to the particular form required.

Example. We shall determine the form of the basic rack that will generate the form of the spline shaft shown in Fig. 1-22.

The first step is to select the pitch circle. In this case we shall use the outside diameter of the shaft for the pitch diameter. The next step is to set up the equation of the gear-tooth form. In this example it is a straight line tangent to a circle. Such equations have already been set up together with the equations for the path of contact.

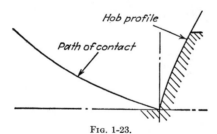

Fig. 1-23.

We shall also rearrange the equations for the path of contact and conjugate rack profile to suit this particular problem. These equations are as follows:

$$\theta = \sin^{-1}(A/r) - \sin^{-1}(A/R) \tag{1-25}$$
$$\cos \phi = \sqrt{r^2 - A^2}/R \tag{1-64}$$
$$x_p = (\sqrt{r^2 - A^2}/R)(A - \sqrt{R^2 - r^2 + A^2}) \tag{1-65}$$
$$y = (\sqrt{R^2 - r^2 + A^2}/R)(\sqrt{R^2 - r^2 + A^2} - A) \tag{1-66}$$
$$x = R\left(\theta - \tan^{-1}\frac{x_p}{R - y}\right) + x_p \tag{1-67}$$

For the example shown, we have the following values for a 6-spline shaft 2 in. in diameter:

$$R = 1.00 \qquad A = 0.250 \qquad \text{Depth} = 0.100$$

The values obtained are tabulated in Table 1-11 and plotted in Fig. 1-23.

An examination of these tabulated values shows that the height of the fillet at the bottom of the spline will be about 0.020-in. because when the height of the rack form is nearly 0.100 in., it makes contact with the spline at a radius of 0.920 in., or 0.020 in. above the bottom of the spline.

TABLE 1-11.   COORDINATES OF SPLINE SHAFT, ITS PATH OF CONTACT, AND BASIC-RACK
FORM
(Plotted in Fig. 1-23)

r, in.	θ, rad	y, in.	x_p, in.	x, in.
1.000	0.00000	0.00000	0.00000	0.00000
0.995	0.00130	0.00517	−0.01850	0.00139
0.990	0.00261	0.01064	−0.03549	0.00298
0.985	0.00393	0.01633	−0.05123	0.00473
0.980	0.00527	0.02222	−0.06588	0.00666
0.975	0.00663	0.02825	−0.07961	0.00876
0.970	0.00799	0.03442	−0.09251	0.01100
0.965	0.00938	0.04070	−0.10469	0.01339
0.960	0.01077	0.04706	−0.11620	0.01591
0.955	0.01219	0.05350	−0.12711	0.01857
0.950	0.01361	0.06000	−0.13748	0.02136
0.945	0.01506	0.06655	−0.14733	0.02427
0.940	0.01652	0.07315	−0.15673	0.02730
0.935	0.01799	0.07980	−0.16570	0.03045
0.930	0.01948	0.08647	−0.17425	0.03371
0.925	0.02099	0.09317	−0.18243	0.03708
0.920	0.02252	0.09988	−0.19023	0.04056
0.915	0.02406	0.10662	−0.19771	0.04414
0.910	0.02562	0.11336	−0.20488	0.04783
0.905	0.02720	0.12011	−0.21173	0.05161
0.900	0.02880	0.12695	−0.21838	0.05551

CHAPTER 2

CONJUGATE ACTION ON INTERNAL GEARS

In principle, the conjugate gear-tooth action on internal spur gears is the same as that for external spur gears. Any of the basic-rack forms used for spur gears may be used also for internal gears. However the basic-rack form is not, in effect, revolved 180 deg for calculating the form of either member of the pair.

We shall use the same symbols as before. The subscript $_1$ will be used on symbols for the spur pinion, and the subscript $_2$ will be used on symbols for the internal gear. These symbols are as follows:

x = abscissa of basic-rack profile

y = ordinate of basic-rack profile and of path of contact

x_p = abscissa of path of contact

ϕ = pressure angle

r = length of radius vector of conjugate gear-tooth profiles

R = pitch radius of gears

θ = vectorial angle of radius vector

ψ = angle between tangent to tooth profile and radius vector

ϵ = angle of rotation of gear from zero position

X = abscissa of gear-tooth profile (origin at center of gear)

Y = ordinate of gear-tooth profile (origin at center of gear)

We shall use the pitch point as the origin of the coordinate system for the basic rack and its path of contact. We shall use the center of the gear as the origin of the polar coordinate system for the gear-tooth profiles, and the vectorial angle will be zero at the pitch point. The angles θ and ϵ will be plus when they are measured in a counterclockwise direction from the pitch point. In all these calculations, great care must be exercised to use the correct signs (plus or minus) of these angular values.

BASIC-RACK FORM GIVEN

We shall start this analysis with a known or given form for the basic rack. For internal-gear drives, the same coordinates of the basic-rack profile and the same equations for the gear-tooth form apply to both the spur pinion and the internal gear. On internal-gear drives, both gears of the pair rotate in the same direction. The equations for this problem have already been derived in Chap. 1 and are as follows:

$$r = \sqrt{(R - y)^2 + x_p{}^2} \tag{1-2}$$
$$\theta = [(x - x_p)/R] + \tan^{-1}[x_p/(R - y)] \tag{1-3}$$

LIMITATIONS TO CONJUGATE ACTION

The same limitations to conjugate action that exist on external spur-gear drives also exist on internal-gear drives. Whenever a tangent circle, with its center at the axis of either gear, can be drawn to the path of contact, a cusp will exist in the form of the gear-tooth profile if the actual action extends beyond that point. Conjugate action will cease at the point of tangency of this circle and the path of contact. Whenever the generating tool cuts below this point, unless it is rounded or relieved in form, the tooth form will be undercut as the generating tool rocks through the tooth space. The shape of the path of the corner of the generating tool as it travels through the tooth space is a trochoid. When no under-cut exists, this trochoid will be tangent to the tooth profile and will be the form of the fillet that joins the tooth profile to the land at the bottom of the tooth space. These trochoids will be analyzed in the next chapter.

There are more possible limitations to an internal-gear drive than there are for a spur-gear drive, particularly when the difference between the number of teeth in the internal gear and the number of teeth in the spur pinion is small. Hence the design of the tooth forms for internal-gear drives is more critical and more exacting than that for external- or spur-gear drives. Among the possible limitations are the following:

1. Undercut gear-tooth profiles
2. Interference between the corner of the spur-pinion tooth and the trochoidal fillet of the internal gear
3. Interference between the corner of the internal-gear tooth and the trochoidal fillet at the bottom of the tooth space of the spur pinion
4. Interference between the tips of the spur- and the internal-gear teeth as they come into and go out of mesh. (In special cases, such as with full continuous-form rotor gear teeth, a secondary action may exist here that can be used effectively for pumping action.)

INTERNAL-GEAR-TOOTH FORM GIVEN

When the form of the internal-gear-tooth profile is known, the forms of the path of contact, the basic-rack profile, and the conjugate pinion-tooth profile may be determined by analysis. Referring to Fig. 2-1, we can obtain these values from the geometrical conditions shown there.

Path of Contact

$$\tan \psi_2 = -r_2 \, d\theta_2/dr_2 \qquad (2\text{-}1)[1]$$
$$\cos \phi = (r_2 \cos \psi_2)/R_2 \qquad (2\text{-}2)$$
$$\epsilon_2 = \psi_2 - \phi - \theta_2 \qquad (2\text{-}3)$$
$$x_p = r_2 \sin (\psi_2 - \phi) \qquad (2\text{-}4)$$
$$y = R_2 - r_2 \cos (\psi_2 - \phi) \qquad (2\text{-}5)$$

[1] Because of the specific coordinate system used here, where the value of r_2 decreases as the value of θ_2 increases, this value is minus.

Basic-rack Profile. If the form of the basic rack is required, its ordinate y is the same as that for the path of contact. Its abscissa x is given by the following equation:

$$x = x_p - R_2\epsilon_2 \tag{2-6}$$

Conjugate Pinion-tooth Form. For the coordinates of the conjugate pinion-tooth form, we have the following:

$$r_1 = \sqrt{(R_1 - y)^2 + x_p{}^2} \quad (1\text{-}2)$$
$$\epsilon_1 = (R_2/R_1)\epsilon_2 \quad (2\text{-}7)$$
$$\cos \psi_1 = R_1 (\cos \phi)/r_1 \quad (2\text{-}8)$$
$$\theta_1 = \psi_1 - \phi - \epsilon_1 \quad (2\text{-}9)$$

Radius of Curvature. When R_c is the radius of curvature of profile, we have
For the basic rack

$$R_c = \frac{-1}{\sin \phi \, (d\phi/dx)} \quad (1\text{-}36)$$

For the spur pinion

$$R_c = \frac{r_1}{\sin \psi_1 \, [1 - (d\psi_1/d\theta_1)]} \quad (1\text{-}35)$$

For the internal gear

$$R_c = \frac{-r_2}{\sin \psi_2 \, [1 - (d\psi_2/d\theta_2)]} \quad (2\text{-}10)^1$$

In these equations, for the basic rack, when the sign is plus, the tooth surface is concave. For the two gears, when the sign is plus, the tooth surfaces are convex.

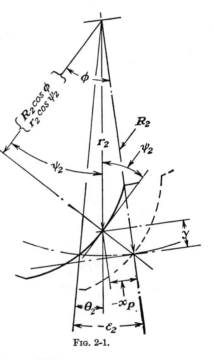

Fig. 2-1.

Radius of Curvature at Pitch Point. When
R_{cr} = radius of curvature of basic rack at pitch point
R_{c1} = radius of curvature of spur pinion at pitch point
R_{c2} = radius of curvature of internal gear at pitch point

$$R_{cr} = -1/\cos^3 \phi \, (d^2x/dy^2) \tag{1-38}$$
$$R_{c1} = \frac{R_1 R_{cr} \sin \phi}{R_{cr} + R_1 \sin \phi} \tag{1-39}$$
$$R_{c2} = \frac{-R_2 R_{cr} \sin \phi}{R_{cr} + R_2 \sin \phi} \tag{2-11}$$

[1] The sign for the curvature of the pitch circle of the internal gear is minus.

When the form of the internal gear is known, we have

$$R_{c1} = \frac{R_1 R_2 \sin \phi}{(R_2 - R_1) - (R_1 R_2 \sin \phi / R_{c2})} \qquad (2\text{-}12)$$

For the sum of the reciprocals of the radius of curvature of the tooth profiles at the pitch point, we have the following:

$$\frac{1}{R'_{c1}} + \frac{1}{R'_{c2}} = \frac{1}{\sin \phi} \left(\frac{1}{R_1} - \frac{1}{R_2} \right) \qquad (1\text{-}45)$$

Examples of Internal-gear Drives with Form of Internal Gear Given. For the first example we shall use a pin-tooth internal gear with 36 teeth or pins, 18-in. pitch radius, and a 12-tooth spur pinion, 6-in. pitch radius.

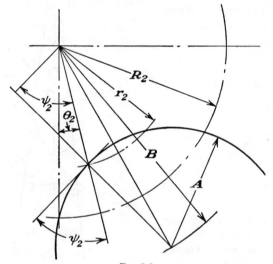

Fig. 2-2.

When A = radius of pins = 0.750
B = radius to center of pins = 18.10

referring to Fig. 2-2, we have

$$\theta_2 = \cos^{-1} \frac{B^2 + R_2{}^2 - A^2}{2BR_2} - \cos^{-1} \frac{B^2 + r_2{}^2 - A^2}{2Br_2} \qquad (2\text{-}13)$$

$$\sin \psi_2 = \frac{B^2 - A^2 - r_2{}^2}{2Ar_2} \qquad (2\text{-}14)$$

The values for the internal pin-tooth gear and its path of contact are tabulated in Table 2-1 and plotted in Fig. (2-3). The values for the conjugate spur pinion are tabulated in Table 2-2 and are also plotted in Fig. 2-3.

TABLE 2-1. COORDINATES OF PIN-TOOTH INTERNAL GEAR AND ITS PATH OF CONTACT
(Plotted in Fig. 2-3)

r_2, in.	θ_2, deg	ψ_2, deg	ϕ, deg	ϵ_2, deg	x_p, in.	y, in.
17.35	2.360	90.000	90.000	−2.360	0.0000	0.6500
17.40	1.490	68.521	69.269	−2.238	−0.2272	0.6015
17.45	1.153	59.676	60.695	−2.172	−0.3103	0.5528
17.50	0.910	52.400	53.615	−2.125	−0.3710	0.5038
17.55	0.722	46.341	47.693	−2.074	−0.4079	0.4547
17.60	0.565	40.907	42.356	−2.014	−0.4451	0.4056
17.65	0.435	35.902	37.414	−1.947	−0.4658	0.3562
17.70	0.330	31.210	32.753	−1.873	−0.4767	0.3064
17.75	0.241	26.753	28.291	−1.779	−0.4764	0.2564
17.80	0.165	22.353	23.853	−1.665	−0.4660	0.2060
17.85	0.105	18.340	19.729	−1.494	−0.4327	0.1554
17.90	0.058	14.312	15.512	−1.258	−0.3748	0.1039
17.95	0.024	10.367	11.202	−0.859	−0.2615	0.0520
18.00	0.000	6.481	6.481	0.000	0.0000	0.0000
18.0367	−0.010	3.656	0.000	3.666	1.1500	0.0000
18.00	0.000	6.481	− 6.481	12.962	4.0374	0.4586
17.95	0.024	10.367	−11.202	21.545	6.5988	1.3070
17.90	0.058	14.312	−15.512	29.766	8.9024	2.4607
17.85	0.105	18.340	−19.729	37.964	11.0065	3.9472

Williams Internal-gear Drive. As a second example, we shall use an
internal gear whose profiles consist of straight lines, as shown in Fig. 2-4.

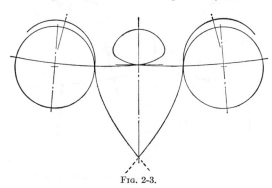

FIG. 2-3.

This form of internal gear is known as the *Williams internal gear.* It is a
simple form to produce and can be made on a vertical slotter or vertical
shaper. For large internal-gear drives where generating equipment is

TABLE 2-2. COORDINATES OF SPUR PINION CONJUGATE TO PIN-TOOTH INTERNAL GEAR

(Plotted in Fig. 2-3)

ϕ, deg	r_1, in.	ϕ_1, deg	ψ_1, deg	ϵ_1, deg	X, in.	Y, in.
90.000	5.3500	7.080	90.000	−7.080	1.3847	5.1677
69.269	5.4033	4.301	66.850	−6.714	1.1438	5.2808
60.695	5.4560	3.256	57.435	−6.516	1.0575	5.3526
53.615	5.5087	2.511	49.751	−6.375	0.9972	5.4176
47.693	5.5602	1.949	43.420	−6.222	0.9530	5.4779
42.356	5.6120	1.495	37.809	−6.042	0.9180	5.5364
37.414	5.6630	1.124	32.697	−5.841	0.8902	5.5926
32.753	5.7135	0.837	27.971	−5.619	0.8698	5.6469
28.291	5.7633	0.597	23.551	−5.337	0.8535	5.6998
23.853	5.8127	0.397	19.255	−4.995	0.8408	5.7515
19.729	5.8606	0.241	15.488	−4.482	0.8319	5.8013
15.512	5.9080	0.134	11.872	−3.774	0.8277	5.8497
11.202	5.9538	0.048	8.673	−2.577	0.8253	5.8963
6.481	6.0000	0.000	6.481	0.000	0.8267	5.9428
0.000	6.1092	− 0.148	10.850	10.998	0.8261	6.0531
− 6.481	6.8560	− 2.812	29.593	38.886	0.6104	6.8288
−11.202	8.0974	−10.057	43.376	64.635	−0.3020	8.0917

not available, it proves to be a very effective type. Only one curved-form cutter for the pinion is required, and the coordinates for this form

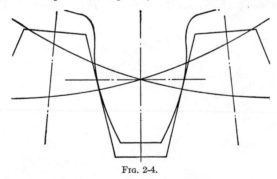

FIG. 2-4.

are readily determined as follows: When A is the radius of the circle to which the straight-line gear-tooth form is tangent,

$$\theta_2 = \sin^{-1} (A/r_2) - \sin^{-1} (A/R_2) \qquad (2\text{-}15)$$
$$\sin \psi_2 = A/r_2 \qquad (2\text{-}16)$$

TABLE 2-3. COORDINATES OF WILLIAMS INTERNAL GEAR AND ITS PATH OF CONTACT
(Plotted in Fig. 2-4)

r_2, in.	θ_2, deg	ψ_2, deg	ϕ, deg	ϵ_2, deg	x_p, in.	y, in.
17.00	0.976	17.104	25.432	−9.304	−2.4623	1.1792
17.10	0.873	17.001	24.703	−8.575	−2.2916	1.0542
17.20	0.772	16.900	23.895	−7.767	−2.0946	0.9280
17.30	0.673	16.801	23.062	−6.934	−1.8867	0.8031
17.40	0.572	16.700	22.197	−6.069	−1.6622	0.6795
17.50	0.474	16.602	21.300	−5.172	−1.4332	0.5588
17.60	0.376	16.504	20.365	−4.237	−1.1852	0.4400
17.70	0.281	16.409	19.390	−3.262	−0.9204	0.3239
17.80	0.186	16.314	18.364	−2.236	−0.6367	0.2114
17.90	0.092	16.220	17.280	−1.152	−0.3311	0.1030
18.00	0.000	16.128	16.128	0.000	0.0000	0.0000
18.10	−0.092	16.036	14.888	1.240	0.3627	−0.0964
18.20	−0.182	15.946	13.541	2.587	0.7637	−0.1840
18.30	−0.272	15.856	12.046	4.082	1.2160	−0.2596
18.40	−0.360	15.768	10.341	5.787	1.7403	−0.3176
18.50	−0.448	15.680	8.296	7.832	2.3776	−0.3466
18.60	−0.534	15.594	5.560	10.568	3.2407	−0.3154
18.60	−0.534	15.594	−5.560	21.688	6.7122	0.6615
18.50	−0.448	15.680	−8.296	24.424	7.5175	1.0962

TABLE 2-4. COORDINATES OF SPUR PINION CONJUGATE TO WILLIAMS INTERNAL GEAR
(Plotted in Fig. 2-4)

ϕ, deg	ψ_1, deg	θ_1, deg	ϵ_1, deg	r_1, in.	X, in.	Y, in.
23.895	1.470	0.876	−23.301	5.4875	0.7994	5.4290
23.062	3.139	0.869	−20.802	5.5288	0.8047	5.4699
22.197	4.701	0.811	−18.207	5.5741	0.8057	5.5156
21.300	6.544	0.760	−15.516	5.6269	0.8084	5.5684
20.365	8.347	0.693	−12.711	5.6852	0.8102	5.6271
19.390	10.181	0.577	− 9.786	5.7502	0.8079	5.6932
18.364	12.085	0.429	− 6.708	5.8235	0.8033	5.7678
17.280	14.065	0.241	− 3.456	5.9062	0.7955	5.8524
16.128	16.128	0.000	0.000	6.0000	0.7832	5.9486
14.888	18.292	−0.316	3.720	6.1072	0.7638	6.0592
13.541	20.582	−0.720	7.761	6.2310	0.7356	6.1874
12.046	23.041	−1.251	12.246	6.3766	0.6937	6.3387
10.341	25.743	−1.959	17.361	6.5529	0.6327	6.5223
8.296	28.831	−2.961	23.496	6.7774	0.5364	6.7561
5.560	32.713	−4.551	31.704	7.0983	0.3652	7.0889

We shall use a 36-tooth internal gear with an 18-in. pitch radius, and a 12-tooth spur pinion with a 6-in. pitch radius. The value of A will be 5 in.

The values for the Williams internal gear and its path of contact are tabulated in Table 2-3 and plotted in Fig. 2-4. The values for the conjugate spur pinion are tabulated in Table 2-4 and are also plotted in Fig. 2-4.

SECONDARY ACTION ON INTERNAL GEARS

It is possible to have secondary action between the teeth of an internal gear drive. Its most general practical application is for pump rotors where the tooth profile of one or both of the two members is formed by continuous curves and where the internal gear has one more tooth than the mating pinion. This secondary action will exist mostly between the addenda of the mating gear teeth while the primary action will exist between the addendum of one gear tooth and the dedendum of the mating gear tooth.

There is some secondary action on the pin-tooth internal-gear drive shown in Fig. 2-3. In that example it is a continuation of the primary action. Strictly speaking, this secondary action begins when the same point on the internal-gear-tooth profile makes a second contact with the mating pinion tooth. This will be when the pressure angle becomes zero and changes to a minus value.

In order to avoid cusps in the form of the pinion tooth and to have continuous action, the secondary path of contact must be a closed curve, and no part of it, except its extremes measured radially from either member of the pair, must be tangent to a circle concentric with either gear. Thus if an arc of a circle is used as the form of the internal-gear tooth, the center of that circle must be outside the pitch circle of the internal gear. If this center is inside the pitch circle of the internal gear, the pressure angle will not pass through a zero value. Secondary action may exist here, but it will not be a continuous action from primary to secondary and its path of contact will not be a closed curve. If the center of the circle of the internal-gear-tooth form is outside the pitch circle but is too close to it, there will be a cusp in the tooth form of the conjugate spur pinion at the beginning of the secondary action. In case of question, very small increments of radius for the internal gear should be used in this region to calculate the form of the conjugate spur pinion. An examination of the calculated coordinates, or an enlarged layout of these values, will soon show whether or not such a cusp exists.

The calculation of the tooth form of the conjugate spur pinion for secondary action is exactly the same as that for the primary action. The values for the internal gear are the same in both cases except for the pressure angle, which is of the same value at a given radius but is minus on the secondary action. The angle of rotation of the internal gear to the position of secondary contact must be calculated.

Example of Secondary Action on Internal Pump Rotors. As a definite example we shall use an internal gear with 4 teeth and a spur pinion with 3 teeth. The form of the internal-gear tooth will be an arc of a circle. In effect, this form is the same as that of the pin-tooth internal gear. Such internal-gear drives are often called *Gerotors*. With this construction we will have primary action between the addendum of the internal gear and the dedendum of the pinion. There will also be a small amount of primary action between part of the addendum of the pinion near its pitch line and the dedendum of the internal gear. There will also be secondary action between the greater part of the addendum of the pinion and the tooth profile of the internal gear. When

A = radius of internal-gear-tooth form,

B = radius on internal gear to center of A,

and all other symbols are the same as before, we have the following equations for the determination of the path of contact and of the conjugate pinion-tooth profile:

$$\theta_2 = \cos^{-1} \frac{B^2 + R_2{}^2 - A^2}{2BR_2} - \cos^{-1} \frac{B^2 + r_2{}^2 - A^2}{2Br_2} \tag{2-13}$$

$$\sin \psi_2 = \frac{B^2 - A^2 - r_2{}^2}{2Ar_2} \tag{2-14}$$

$$\cos \phi = \frac{r_2 \cos \psi_2}{R_2} \tag{2-2}$$

$$\epsilon_2 = \psi_2 - \phi - \theta_2 \tag{2-3}$$

$$x_p = r_2 \sin (\psi_2 - \phi) \tag{2-4}$$

$$y = R_2 - r_2 \cos (\psi_2 - \phi) \tag{2-5}$$

$$r_1 = \sqrt{(R_1 - y)^2 + x_p{}^2} \tag{1-2}$$

$$\epsilon_1 = \frac{R_2}{R_1} \epsilon_2 \tag{2-7}$$

$$\cos \psi_1 = \frac{R_1 \cos \phi}{r_1} \tag{2-8}$$

$$\theta_1 = \psi_1 - \phi - \epsilon_1 \tag{2-9}$$

To obtain the Cartesian coordinates for the pinion-tooth profile we have, when

θ' = original vectorial angle at center of tooth or space

θ = calculated vectorial angle

θ'' = vectorial angle with Y axis at center of tooth or space

$$\theta'' = \theta' - \theta \tag{1-4}$$

$$X = r \sin \theta'' \tag{1-5}$$

$$Y = r \cos \theta'' \tag{1-6}$$

For the definite example, we shall use the following values:

$$R_1 = 1.500 \qquad R_2 = 2.000 \qquad A = 1.500 \qquad B = 2.750$$

In this example, the pressure angle will be zero when the value of r_2 is equal to 2.03717. We shall use increments of 0.05 for the radius of the internal gear from 1.25 to 2.00, and increments of 0.01 for the radius of the gear from 2.00 to 2.03717. The values for the internal gear and its path of contact are tabulated in Table 2-5 and

TABLE 2-5. COORDINATES OF INTERNAL-GEAR PUMP ROTOR AND ITS PATH OF CONTACT

(Plotted in Fig. 2-5)

r_2, in.	θ_2, deg	ψ_2, deg	ϕ, deg	ϵ_2, deg	x_p, in.	y, in.
			Primary Action			
1.25	32.156	90.000	90.000	−32.156	0.0000	0.7500
1.30	20.498	68.256	76.066	−28.308	−0.1767	0.7121
1.35	16.090	59.511	69.972	−26.551	−0.2451	0.6724
1.40	12.974	52.960	65.061	−25.075	−0.2935	0.6311
1.45	10.557	47.557	60.708	−23.708	−0.3299	0.5880
1.50	8.599	42.887	56.666	−22.378	−0.3573	0.5432
1.55	6.974	38.741	52.808	−21.041	−0.3767	0.4965
1.60	5.624	34.990	49.050	−19.684	−0.3887	0.4479
1.65	4.457	31.549	45.327	−18.235	−0.3930	0.3975
1.70	3.471	28.360	41.584	−16.695	−0.3889	0.3451
1.75	2.630	25.377	37.761	−15.014	−0.3757	0.2907
1.80	1.912	22.569	33.790	−13.133	−0.3503	0.2344
1.85	1.302	19.910	29.575	−10.967	−0.3106	0.1763
1.90	0.786	17.379	24.956	− 8.363	−0.2502	0.1166
1.95	0.355	14.958	19.619	− 5.014	−0.1584	0.0565
2.00	0.000	12.635	12.635	0.000	0.0000	0.0000
2.01	−0.064	12.182	10.776	1.470	0.0493	−0.0094
2.02	−0.124	11.731	8.544	3.311	0.1123	−0.0169
2.03	−0.182	11.284	5.510	5.956	0.2042	−0.0197
2.03717	−0.222	10.988	0.000	11.210	0.3883	0.0000
			Secondary Action			
2.03717	−0.222	10.988	0.000	11.210	0.3883	0.0000
2.03	−0.182	11.284	− 5.510	16.976	0.5865	0.0566
2.02	−0.124	11.731	− 8.544	20.399	0.7000	0.1052
2.01	−0.064	12.182	−10.776	23.022	0.7840	0.1492
2.00	0.000	12.635	−12.635	25.270	0.8538	0.1914
1.95	0.355	14.958	−19.619	34.222	1.1066	0.3945
1.90	0.786	17.379	−24.956	41.549	1.2796	0.5955
1.85	1.302	19.910	−29.575	48.182	1.4064	0.7982
1.80	1.912	22.569	−33.790	54.447	1.4986	1.0028
1.75	2.630	25.377	−37.761	60.508	1.5612	1.2094
1.70	3.471	28.360	−41.584	66.473	1.5969	1.4170
1.65	4.457	31.549	−45.327	72.419	1.6069	1.6254
1.60	5.624	34.990	−49.050	78.416	1.5913	1.8339
1.55	6.974	38.741	−52.808	84.575	1.5494	2.0419
1.50	8.599	42.887	−56.666	90.954	1.4792	2.2489
1.45	10.577	47.557	−60.708	97.708	1.3770	2.4544
1.40	12.974	52.960	−65.061	105.047	1.2359	2.6577
1.35	16.090	59.511	−69.972	113.393	1.0419	2.8584
1.30	20.498	68.256	−76.066	123.827	0.7582	3.0560
1.25	32.156	90.000	−90.000	147.844	0.0000	3.2500

TABLE 2-6. COORDINATES OF 3-TOOTH ROTOR CONJUGATE TO INTERNAL GEAR
(Plotted in Fig. 2-5)

ϕ, deg	r_1, in.	θ_1, deg	ψ_1, deg	ϵ_1, deg	X, in.	Y, in.
			Primary Action			
90.000	0.7500	42.875	90.000	−42.875	0.6495	0.3750
76.066	0.8075	25.107	63.429	−37.744	0.5427	0.5979
69.972	0.8631	18.903	53.474	−35.401	0.5077	0.6980
65.061	0.9171	14.770	46.398	−33.433	0.4846	0.7787
60.708	0.9698	11.724	40.821	−31.611	0.4679	0.8495
56.666	1.0214	9.362	36.191	−29.837	0.4555	0.9141
52.808	1.0719	7.478	32.231	−28.055	0.4463	0.9746
49.050	1.1216	5.967	28.772	−26.245	0.4399	1.0317
45.327	1.1705	4.696	25.710	−24.313	0.4351	1.0866
41.584	1.2186	3.649	22.973	−22.260	0.4322	1.1394
37.761	1.2663	2.788	20.530	−20.019	0.4313	1.1906
33.790	1.3132	2.038	18.317	−17.511	0.4310	1.2404
29.575	1.3597	1.418	16.370	−14.623	0.4324	1.2891
24.956	1.4059	0.883	14.688	−11.151	0.4346	1.3370
19.619	1.4522	0.422	13.354	− 6.685	0.4378	1.3846
12.635	1.5000	0.000	12.635	0.000	0.4417	1.4335
10.776	1.5102	− 0.088	12.648	1.960	0.4425	1.4439
8.544	1.5210	− 0.181	12.778	4.415	0.4433	1.4550
5.510	1.5334	− 0.286	13.165	7.941	0.4442	1.4676
0.000	1.5494	− 0.434	14.513	14.847	0.4450	1.4842
			Secondary Action			
0.000	1.5494	− 0.434	14.513	14.847	0.4450	1.4842
− 5.510	1.5580	− 0.522	16.603	22.635	0.4451	1.4958
− 8.544	1.5606	− 0.549	18.106	27.199	0.4452	1.4958
−10.776	1.5618	− 0.564	19.356	30.696	0.4452	1.4970
−12.635	1.5625	− 0.571	20.487	33.693	0.4451	1.4977
−19.619	1.5643	− 0.601	25.409	45.629	0.4449	1.4997
−24.956	1.5670	− 0.655	29.788	55.399	0.4443	1.5027
−29.575	1.5718	− 0.764	33.904	64.243	0.4428	1.5082
−33.790	1.5789	− 0.951	37.855	72.596	0.4398	1.5164
−37.761	1.5880	− 1.225	41.691	80.677	0.4351	1.5273
−41.584	1.5991	− 1.605	45.441	88.630	0.4279	1.5408
−45.327	1.6118	− 2.097	49.134	96.558	0.4179	1.5567
−49.050	1.6260	− 2.706	52.798	104.554	0.4049	1.5748
−52.808	1.6415	− 3.489	56.469	112.766	0.3870	1.5952
−56.666	1.6580	− 4.418	60.188	121.272	0.3649	1.6213
−60.708	1.6754	− 5.552	64.020	130.277	0.3361	1.6413
−65.061	1.6934	− 6.932	68.069	140.062	0.2997	1.6667
−69.972	1.7120	− 8.680	72.538	151.190	0.2514	1.6934
−76.066	1.7309	−11.081	77.955	165.102	0.1822	1.7213
−90.000	1.7500	−17.125	90.000	197.125	0.0000	1.7500

plotted in Fig. 2-5. The values for the conjugate spur pinion are tabulated in Table 2-6 and are also plotted in Fig. 2-5.

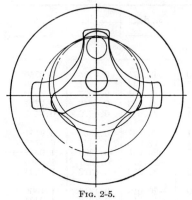

Fig. 2-5.

SECONDARY ACTION ON CYCLOIDAL INTERNAL GEARS

We shall examine next the conditions that exist between the epicycloidal addendum of a spur pinion and the hypocycloidal dedendum of an internal gear when there is a difference of 1 tooth in their respective tooth numbers, and when the diameter of the rolling circles of these cycloidal forms is equal to the center distance.

Referring to Fig. 2-6, at any pressure angle ϕ, the line of action passes through the points of tangency of the rolling circles of both gears

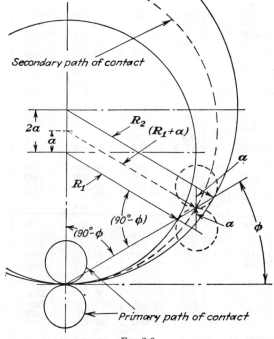

Fig. 2-6.

with their respective pitch circles, and also through the point of tangency (which is the point of contact) of the two tooth profiles, which last point is also the point of contact of the two rolling circles.

The radial lines from the centers of the two gears to the centers of their respective rolling circles pass through the points of tangency of their rolling circles and their pitch circles, and are parallel to each other, because the triangles formed by the line of centers, these radial lines, and the line of action are similar isosceles triangles, as shown in Fig. 2-6.

The length of the line between the centers of the two rolling circles is equal to $2a$, and the center distance between the axes of the two gears is also equal to $2a$. The length of each radial line to the center of its rolling circle is the same for both gears. Therefore the geometrical figure whose sides are the line of centers of the two gears, the line of centers of the two rolling circles, and the two radial lines is a parallelogram.

As noted before, the line of centers of the two rolling circles passes through the point of tangency of the two rolling circles, which point is also the point of contact between the two gear-tooth profiles. This point is at the middle of the line of centers of the two rolling circles.

Hence the form of the secondary path of contact of these two cycloidal forms is the path described by the middle of the link whose length is equal to $2a$, a link that is pivoted on two parallel arms whose lengths are each equal to $R_1 + a$, as this linkage is revolved. Such a form is a circle of radius equal to $R_1 + a$, whose center is on the line of centers of the two gears, at a distance of $R_1 + a$ from the pitch point. This secondary path of contact is shown as a dotted line in Fig. 2-6.

CHAPTER 3

TROCHOIDS, TOOTH FILLETS, AND UNDERCUT

As noted in Chaps. 1 and 2, whenever a tangent circle can be drawn from the center of the gear to the path of contact, as shown in Fig. 3-1, conjugate action cannot take place below the radius R_u. In addition, if the mating profile extends so that it would reach beyond the point of tangency of the path of contact and the tangent circle, a cusp will exist

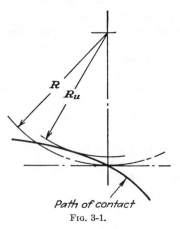

in the theoretical form of the tooth profile because two points of contact should exist for the same radial distance on the gear. Under such conditions, the corner of the mating gear will interfere or make improper contact with the incomplete profile. If the interfering member is a generating tool, the corner of its tooth, which travels in a trochoidal path in relation to the gear being generated, will sweep out its path, remove some of the conjugate profile, and produce an undercut tooth form.

We shall therefore analyze these trochoids so as to determine both the forms of the undercut when it exists and also to determine the form of the fillet of the tooth whether undercut exists or not. We sometimes have the condition where the generating tool is smaller than the mating gear. Here the generating tool cuts deeper than the mating tooth extends, but we must be sure that the fillet made by the generating tool does not interfere with the tip of the tooth of the mating gear. On internal gears another factor is present. Here we must be sure that the trochoidal path of the corner of one tooth does not interfere with the tip of its mating tooth as it comes into and goes out of mesh.

We have five relative conditions or types of trochoids to consider. These are as follows:

1. Corner of rack tooth in relation to root of gear tooth
2. Corner of gear tooth in relation to root of rack tooth
3. Corner of one gear tooth in relation to root of second gear

48

4. Corner of pinion tooth in relation to internal gear

5. Corner of internal-gear tooth in relation to pinion

TROCHOID OF CORNER OF RACK TOOTH AT ROOT OF GEAR

When the rack tooth represents the form of the generating tool, then this trochoid gives the form of the fillet of the gear tooth. When no undercut is present, this trochoid will be tangent to the generated gear-tooth profile. The equations for this trochoidal path are derived as follows:

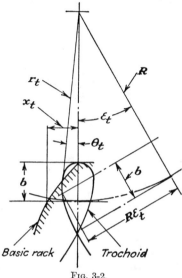

FIG. 3-2.

Let R = pitch radius of gear

b = distance from pitch line of rack to sharp corner of rack tooth

r_t = any radius of trochoid

θ_t = vectorial angle of trochoid

ψ_t = angle between tangent to trochoid and radius vector

ϵ_t = angle of rotation of gear

x_t = abscissa of corner of rack tooth measured from the pitch point

δ = angle between origins[1] of the trochoid and the gear-tooth profile

We have the following from the geometrical conditions shown in Fig. 3-2:

$$\theta_t = \tan^{-1}\left[\frac{\sqrt{r_t^2 - (R-b)^2}}{(R-b)}\right] - \frac{\sqrt{r_t^2 - (R-b)^2}}{R} \tag{3-1}$$

$$\tan \psi_t = \frac{r_t \, d\theta_t}{dr_t} = \frac{R(R-b) - r_t^2}{R \sqrt{r_t^2 - (R-b)^2}} \tag{3-2}$$

To plot this trochoid in its proper relation to a definite gear-tooth profile, we must first determine the angle between the origins of the two curves. For this we have

$$\text{arc } \delta = r_b/R \tag{3-3}$$

To obtain the Cartesian coordinates of the trochoid in relation to

---

[1] The origin of the coordinate system for the trochoid is the center of the gear. The vectorial angle is zero when the corner of the rack tooth is at its deepest position in the gear.

those of the gear-tooth profile, we have when

θ' = original vectorial angle at center of tooth or space

θ_t = calculated vectorial angle of trochoid

θ''_t = vectorial angle of trochoid with Y axis at center of tooth or space

$$\theta''_t = \theta' - \delta \pm \theta_t \qquad (3\text{-}4)[1]$$

Then we have as before

$$X_t = r_t \sin \theta''_t \qquad (3\text{-}5)$$
$$Y_t = r_t \cos \theta''_t \qquad (3\text{-}6)$$

TABLE 3-1.   COORDINATES OF TROCHIOD OF CORNER OF RACK TOOTH
(Plotted in Fig. 3-3)

r_t, in.	θ_t, rad	θ_t, deg	θ''_t, deg	X_t, in.	Y_t, in.
20.0000	0.00655	0.375	−2.949	−1.0290	19.9736
19.9013	0.00734	0.421	−2.903	−1.0078	19.8758
19.8059	0.00802	0.460	−2.864	−0.9895	19.7809
19.7151	0.00853	0.489	−2.835	−0.9751	19.6910
19.6307	0.00892	0.511	−2.813	−0.9637	19.6071
19.5551	0.00917	0.525	−2.799	−0.9549	19.5318
19.4918	0.00930	0.533	−2.791	−0.9491	19.4686
19.4457	0.00934	0.535	−2.789	−0.9462	19.4226
19.4237	0.00935	0.536	−2.788	−0.9448	19.4008
19.3000	0.00920	0.527	−2.747	−0.9418	19.2770
19.2000	0.00877	0.502	−2.822	−0.9452	19.1768
19.1000	0.00796	0.456	−2.868	−0.9560	19.0761
19.0000	0.00660	0.378	−2.946	−0.9764	18.9749
18.8500	0.00000	0.000	−3.324	−1.0931	18.8183
19.0000	−0.00660	−0.378	−3.702	−1.2268	18.9603
19.1000	−0.00796	−0.456	−3.780	−1.2593	19.0584
19.2000	−0.00877	−0.502	−3.826	−1.2810	19.1572
19.3000	−0.00920	−0.527	−3.851	−1.2962	19.2564
19.4237	−0.00935	−0.536	−3.860	−1.3076	19.3796
19.4457	−0.00934	−0.535	−3.859	−1.3087	19.4016
19.4918	−0.00930	−0.533	−3.857	−1.3112	19.4477
19.5551	−0.00917	−0.525	−3.849	−1.3127	19.5109
19.6307	−0.00892	−0.511	−3.835	−1.3129	19.5863
19.7151	−0.00853	−0.489	−3.813	−1.3111	19.6715
19.8059	−0.00802	−0.460	−3.784	−1.3070	19.7627
19.9013	−0.00734	−0.421	−3.745	−1.2998	19.8587
20.0000	−0.00655	−0.375	−3.699	−1.2902	19.9584

[1] Because the trochoid is symmetrical about its origin, the last term in Eq. (3-4) is plus and minus.

Example of Trochoid of Corner of Rack Tooth. As a definite example we shall use the gear-tooth form whose coordinates are tabulated in Table 1-2. We shall assume a clearance of 0.150, which will give the following values: $R = 20.00$, $b = 1.150$. The value of x_t must be calculated from Eq. (1-10), where $y = -1.150$, $A = 5.00$, $B = 4.5315$, and $D = 2.1131$. This gives the value $x_t = -0.3748$. This profile is conjugate to the inside surface of the basic-rack form so that the signs of these coordinates are reversed. Hence $x_t = 0.3748$ and

$$\delta = \frac{0.3748}{20} = 0.01874 \text{ radian} = 1.074°$$

The values for this trochoid are tabulated in Table 3-1 and are plotted in Fig. 3-3. Here the trochoid produces undercut.

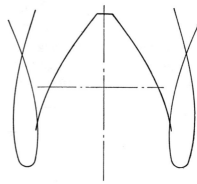

Fig. 3-3.

This same trochoid is at the root of the gear-tooth form whose coordinates are tabulated in Table 1-1. In this case, as there is no cusp, the trochoid will be tangent to the gear-tooth profile.

FILLET FORM OF ROUNDED CORNER OF RACK TOOTH

When the rack is represented by the cutting tooth of a rack-shaped cut-

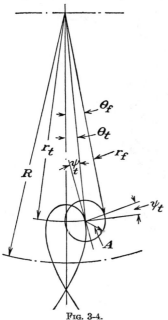

Fig. 3-4.

ter or of a hob, the corners of these cutting teeth are generally rounded. In such cases, the center of the rounding will follow the trochoidal path as given by Eqs. (3-1) and (3-2), and located by Eq. (3-3), but the actual form of the fillet will be the envelope of the path of a series of circles equal in size to the rounding of the corner, and with their centers on the trochoidal path. We can establish this fillet form also by analysis. Thus when all other symbols are the same as before except

x_t = abscissa of center of rounding of corner of rack tooth, measured from pitch point

b = ordinate of center of rounding, measured from pitch line

A = radius of rounding

r_f = any radius of fillet form

θ_f = vectorial angle of fillet form

the values for the trochoid of the center

of the rounding will be calculated as before. Then, referring to Fig. 3-4, we have for the fillet form

$$r_f = \sqrt{r_t^2 + A^2 - 2Ar_t \sin \psi_t} \qquad (3\text{-}7)$$
$$\theta_f = \theta_t + \cos^{-1}\left[(r_t - A \sin \psi_t)/r_f\right] \qquad (3\text{-}8)$$

TROCHOID OF CORNER OF GEAR TOOTH AT ROOT OF RACK

Some racks are generated from a pinion-shaped cutter. These cutters have substantially sharp corners. The form of the fillet at the roots of

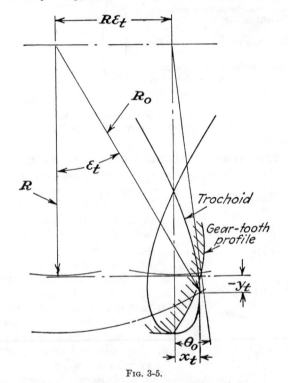

FIG. 3-5.

such rack teeth is the form of the trochoid developed by the path of the corner of the pinion tooth as it sweeps out of mesh. The equations of such a trochoid are as follows:

Let x_t = abscissa of trochoid

$\quad y_t$ = ordinate of trochoid

$\quad x_0$ = distance from pitch point to origin of trochoid

$\quad X_t$ = abscissa of trochoid with pitch point as origin

R_0 = outside radius of pinion

R = pitch radius of pinion

θ_0 = vectorial angle to corner of pinion-tooth

ϵ_t = angular rotation of pinion

Referring to Fig. 3-5, we have from the geometrical conditions shown there

$$y_t = R - R_0 \cos \epsilon_t$$
$$x_t = R_0 \sin \epsilon_t - R\epsilon_t$$
$$\cos \epsilon_t = (R - y_t)/R_0$$
$$\sin \epsilon_t = \sqrt{R_0{}^2 - (R - y_t)^2}/R_0$$

Whence

$$x_t = \sqrt{R_0{}^2 - (R - y_t)^2} - R \sin^{-1} [\sqrt{R_0{}^2 - (R - y_t)^2}/R_0] \quad (3\text{-}9)$$
$$x_0 = R\theta_0 \quad (3\text{-}10)$$
$$X_t = x_t - x_0 \quad (3\text{-}11)$$

TROCHOID OF CORNER OF ONE GEAR TOOTH AT ROOT OF SECOND GEAR TOOTH

When a gear is generated from a pinion-shaped cutter, and the corner of the cutter tooth is sharp, the form of the fillet produced is that of the trochoid of the corner of the tooth of the pinion-shaped cutter.

Let R_1 = pitch radius of first gear, or cutter

R_2 = pitch radius of second gear

C = center distance

R_0 = outside radius of first gear

r_t = any radius to trochoid on second gear

θ_t = vectorial angle of trochoid

ϵ_1 = angle of rotation of first gear

ϵ_2 = angle of rotation of second gear

θ_0 = vectorial angle to corner of tooth of first gear

δ = angle between origin of second gear-tooth profile and trochoid

Referring to Fig. 3-6, we have from the geometrical conditions shown there

$$\epsilon_2 = -(R_1/R_2)\epsilon_1 \quad (3\text{-}12$$
$$r_t = \sqrt{C^2 + R_0{}^2 - 2CR_0 \cos \epsilon_1} \quad (3\text{-}13)$$
$$\sin (\epsilon_2 + \theta_t) = \frac{R_0 \sin \epsilon_1}{r_t}$$

whence

$$\theta_t = \sin^{-1} \frac{R_0 \sin \epsilon_1}{r_t} - \epsilon_2 \quad (3\text{-}14)$$
$$\delta = (R_1/R_2)\theta_0 \quad (3\text{-}15)$$

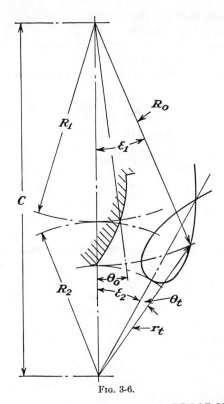

Fig. 3-6.

FILLET OF ROUNDED TOOTH OF FIRST GEAR AT ROOT OF SECOND GEAR

Sometimes the tip of the tooth of the pinion-shaped cutter is rounded to give a larger radius of fillet or to produce a full-rounded-root form to reduce the stress concentration at the root section of the tooth on critical and heavily loaded gear drives. We already have Eqs. (3-7) and (3-8) for the resulting fillet form, and Eqs. (3-13) and (3-14) for the trochoid of the center of the rounding. Before Eqs. (3-7) and (3-8) can be used, however, we must derive equations for the value of the tangent to the trochoid of the center of the rounding.

It is possible to combine Eqs. (3-13) and (3-14) and obtain a single equation for the value of this tangent, but it will give a complex equation. A simpler form for calculation will be obtained by using the angle of rotation as the independent variable, and deriving expressions for the first derivative of r_t and θ_t in respect to it. Thus when ψ_t = angle between

tangent to the trochoid and the radius vector, and all other symbols[1] are the same as before,

$$\tan \psi_t = r_t \frac{d\theta_t}{dr_t} = r_t \frac{d\theta_t/d\epsilon_1}{dr_t/d\epsilon_1} \tag{3-16}$$

$$\frac{d\theta_t}{d\epsilon_1} = \frac{R_0(r_t^2 \cos \epsilon_1 - CR_0 \sin^2 \epsilon_1)}{r_t^2(C - R_0 \cos \epsilon_1)} - \frac{R_1}{R_2} \tag{3-17}$$

$$\frac{dr_t}{d\epsilon_1} = \frac{CR_0 \sin \epsilon_1}{r_t} \tag{3-18}$$

With these values, we then use Eqs. (3-7) and (3-8) to determine the values of the coordinates of the fillet, and Eq. (3-15) to determine the angle between the origins of the fillet and the tooth form. To find the Cartesian coordinates, we use Eqs. (3-4), (3-5), and (3-6) as before.

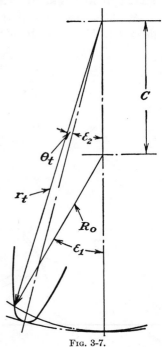

TROCHOID OF CORNER OF PINION TOOTH AT ROOT OF INTERNAL GEAR

When the internal gear is generated, it is usually done with a pinion-shaped cutter. This trochoid would then be the form of the fillet at the root of the internal-gear tooth.

When R_1 = pitch radius of pinion

$\quad R_2$ = pitch radius of internal gear

$\quad C$ = center distance

$\quad R_0$ = outside radius of pinion

$\quad \epsilon_1$ = angle of rotation of pinion

$\quad \epsilon_2$ = angle of rotation of internal gear

$\quad r_t$ = any radius to trochoid on internal gear

$\quad \theta_t$ = vectorial angle of trochoid

$\quad \delta$ = angle between origins of trochoid and internal-gear tooth

Fig. 3-7.

$\quad \theta_0$ = vectorial angle to corner of pinion tooth

then

$$\epsilon_2 = (R_1/R_2)\epsilon_1 \tag{3-12}$$

[1] Position of the center of the rounding is substituted for the position of the corner of the tooth of the first gear.

Referring to Fig. 3-7, we have from the geometrical conditions shown there

$$r_t = \sqrt{C^2 + R_0{}^2 + 2CR_0 \cos \epsilon_1} \tag{3-19}$$

$$\sin (\epsilon_2 + \theta_t) = \frac{R_0 \sin \epsilon_1}{r_t}$$

$$\theta_t = \sin^{-1}\left(\frac{R_0 \sin \epsilon_1}{r_t}\right) - \epsilon_2 \tag{3-14}$$

$$\delta = (R_1/R_2)\theta_0 \tag{3-15}$$

FILLET OF ROUNDED TOOTH OF PINION AT ROOT OF INTERNAL GEAR

We have already derived Eqs. (3-19) and (3-14), which are used to determine the coordinates of the trochoid of the center of the rounding at

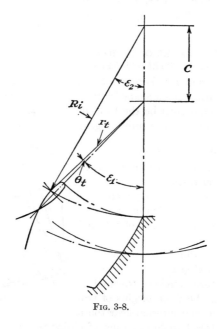

Fig. 3-8.

the tip of the pinion tooth.   We have also Eqs. (3-7) and (3-8) for the fillet form developed by the rounded corner of the pinion tooth.   Before we can use these last equations, however, we must determine values for the angle between the tangent to the trochoid and its radius vector.   Following the same method as before we have when ψ_t = angle between tangent to trochoid and its radius vector,

$$\tan \psi_t = r_t \frac{d\theta_t}{dr_t} = r_t \frac{(d\theta_t/d\epsilon_1)}{dr_t/d\epsilon_1} \tag{3-16}$$

$$\frac{d\theta_t}{d\epsilon_1} = \frac{R_o(r_t{}^2 \cos \epsilon_1 + CR_o \sin^2 \epsilon_1)}{r_t{}^2(C + R_o \cos \epsilon_1)} - \left(\frac{R_1}{R_2}\right) \tag{3-20}$$

$$\frac{dr_t}{d\epsilon_1} = -\frac{CR_o \sin \epsilon_1}{r_t} \tag{3-21}$$

With these values, we then use Eqs. (3-7) and (3-8) to determine the values of the coordinates of the fillet, and Eq. (3-15) to determine the angle between the origins of the fillet and the tooth form. Equations (3-5) and (3-6) are used as before to determine the Cartesian coordinates.

TROCHOID OF CORNER OF INTERNAL-GEAR TOOTH AT ROOT OF PINION

The corner of the tooth of the internal gear must clear both the fillet of the mating spur-pinion tooth and also its tip as it sweeps into and out of mesh. For this trochoidal path we have the following:

When R_1 = pitch radius of pinion

$\quad R_2$ = pitch radius of internal gear

$\quad C$ = center distance

$\quad R_i$ = internal or inside radius of internal gear

$\quad \epsilon_1$ = angle of rotation of pinion

$\quad \epsilon_2$ = angle of rotation of internal gear

$\quad r_t$ = any radius to trochoid on pinion

$\quad \theta_0$ = vectorial angle to corner of internal gear

$\quad \theta_t$ = vectorial angle of trochoid on pinion

$\quad \delta$ = angle between origins of trochoid and pinion-tooth form

$$\epsilon_1 = (R_2/R_1)\epsilon_2 \tag{3-22}$$

Referring to Fig. 3-8, we have from the geometrical conditions shown there

$$r_t = \sqrt{C^2 + R_i{}^2 - 2CR_i \cos \epsilon_2} \tag{3-23}$$

$$\sin (\epsilon_1 + \theta_t) = R_i \sin \epsilon_2/r_t$$

$$\theta_t = \sin^{-1} (R_i \sin \epsilon_2/r_t) - \epsilon_1 \tag{3-24}$$

Examples of these trochoids and fillets will be given later when we determine the forms of various gear teeth.

CHAPTER 4

THE INVOLUTE CURVE AND ITS PROPERTIES

At the present time the involute curve is used almost exclusively for spur-gear-tooth profiles that are employed to transmit power. It meets all the requirements for a gear-tooth profile and, in addition, has so many unique and valuable properties that it stands in a class by itself. These unique properties free it from many of the restrictions of other gear-tooth curves. In order to appreciate its many valuable features, it is best to study it by itself rather than as one of a group of gear-tooth curves.

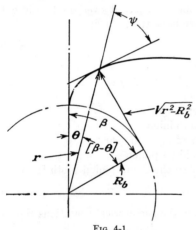

FIG. 4-1.

The involute is the curve that is described by the end of a line that is unwound from the circumference of a circle, as shown in Fig. 4-1. The circle from which the string is unwound is called the *base circle*. The equation of the involute is as follows:

Let R_b = radius of base circle

r = radius to any point of involute

θ = vectorial angle

β = angle through which line has been unwound

We have from the geometrical conditions shown in Fig. 4-1

$$\theta = \beta - \tan^{-1} \frac{\sqrt{r^2 - R_b^2}}{R_b}$$

The length of the generating line $\sqrt{r^2 - R_b^2}$ is also the length, of the circumference of the base circle, that is subtended by the angle β. Hence

$$\sqrt{r^2 - R_b^2} = R_b \beta \qquad \text{or} \qquad \beta = \frac{\sqrt{r^2 - R_b^2}}{R_b}$$

whence

$$\theta = \frac{\sqrt{r^2 - R_b^2}}{R_b} - \tan^{-1} \frac{\sqrt{r^2 - R_b^2}}{R_b} \qquad (4\text{-}1)$$

This is the polar equation of the involute curve.

When ψ = angle between tangent to curve and radius vector,

$$\tan \psi = \frac{r \, d\theta}{dr} = \frac{\sqrt{r^2 - R_b{}^2}}{R_b} \tag{4-2}$$

But this last value is also the tangent of the angle $(\beta - \theta)$, so that this tangent to the involute curve is parallel to the radial line at the start of the generating line of the involute. Hence the tangent to the involute curve is perpendicular to the generating line, or conversely, the generating line is the normal to the involute curve.

When R_c = radius of curvature of involute curve,

$$\frac{dr}{d\theta} = \frac{rR_b}{\sqrt{r^2 - R_b{}^2}} \qquad \frac{d^2r}{d\theta^2} = \frac{-rR_b{}^4}{(r^2 - R_b{}^2)^2}$$

Substituting these values into Eq. (1-34), combining, and simplifying, we obtain

$$R_c = \sqrt{r^2 - R_b{}^2} \tag{4-3}$$

But this value of R_c is the length of the generating line from its point of tangency with the base circle to the involute curve. Hence the radius of curvature of the involute curve at any point is the length of the generating line to that point.

Involute Curve as a Uniform-rise Cam. A simple conception of the involute curve is that of a uniform-rise cam, where the rise per revolution along a line tangent to the base circle of radius R_b is equal to the circumference of the base circle. Such a cam is shown in Fig. 4-2. If this cam revolves at a uniform rate of speed in the direction shown by the arrow, the roll follower will rise at a uniform rate of speed also. If the cam revolves in the reverse direction, the follower will fall accordingly.

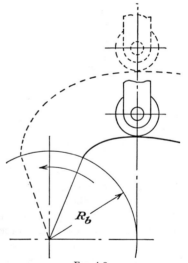

Fig. 4-2.

The path of contact between the roll on the follower and the involute cam is a straight line that is tangent to the base circle. Being a straight line, it is symmetrical in reference to any point on this straight line. This is one of the unique properties of this involute curve that set it apart from all other gear-tooth curves.

Action of One Involute against Another. If, instead of acting against a cam roll, the involute acts against another involute, we have the conditions shown in Fig. 4-3. The point of contact between the two involutes is that point where the tangents to the two curves coincide. The tangents to both involutes are always perpendicular to their generating lines. The tangents to the two involutes coincide only when the generating line of one is a continuation of the generating line of the other. Therefore the locus of points of contact between two involutes is the common tangent to the two base circles as shown in Fig. 4-3.

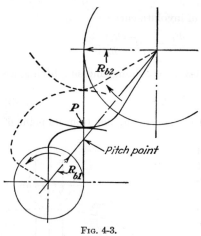

FIG. 4-3.

When one involute is revolved at a uniform rate of motion, the length of the generating line from its point of tangency to the base circle to the involute profile at point P changes uniformly. If the direction of rotation is in the direction shown by the arrow in Fig. 4-3, the length of this line increases. At the same time, the length of the generating line on the mating involute is shortened at the same uniform rate because the total length of the common tangent to the two base circles remains constant. This means that the second involute must revolve at a uniform rate in the direction shown by the arrow in Fig. 4-3.

Base Circles Determine Speed Ratio. The relative rate of motion depends only upon the relative sizes of the two base circles. No matter what the distance may be between the centers of the two base circles, when one involute acts against another, contact between them takes place only along the common tangent to the two base circles, and their relative rates of motion remain the same. If one base circle is double the size of the other, the rate of revolution of the larger involute is one-half that of the smaller. This is because the larger involute, or its base circle, revolves through only one-half the angle that the smaller one must revolve through to wind up the length of the generating line that the smaller one has unwound. The conditions are exactly the same as though two pulleys were set up and connected by a crossed belt. Hence the relative rates of the two mating involutes that act against each other are in inverse proportion to the sizes of their base circles.

The relative rates of the two involutes may be represented by two plain disks that drive each other by friction. Such disks are known as *pitch disks*, while their diameters are known as *pitch diameters*. An involute has no pitch diameter until it is brought into contact with another involute. This is another unique feature of the involute curve. All other gear-tooth curves must be developed from a preselected pitch circle or pitch line. The involute has no fixed pitch circle, but any diameter on it is a potential pitch diameter. This is because the path of contact is a straight line, a form that is symmetrical about any point in this line.

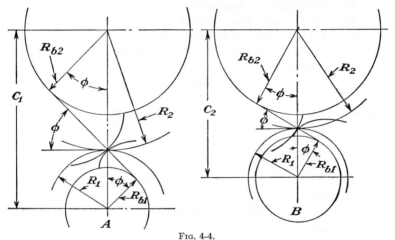

FIG. 4-4.

Furthermore, the path of contact for involute curves is also the line of action. Again, the form of the involute depends solely upon the size of the base circle.

In Fig. 4-4, two involutes are shown in contact at different center distances. The common tangent to the two base circles is both the path of contact and the line of action. We have seen before that the radii of the base circles are in inverse proportion to the rates of revolution of the involutes. The radii of the two pitch disks, which are tangent to each other at the *pitch point* and which represent the same relative rates of revolution, are directly proportional to the radii of the base circles of their respective involutes.

From the geometrical conditions shown in Fig. 4-4, we see that the intersection of the common tangent to the two base circles with the common center line of the two involutes establishes the pitch point and the radii of the two pitch circles.

The angle between the common tangent to the two base circles and a line perpendicular to their common center line is called the *pressure angle*.

This angle does not exist until two involutes are brought into contact with each other.   There is a definite relation between the pitch diameter and pressure angle of any given involute.   Thus for any established pitch diameter there is a corresponding pressure angle.

Thus both the sizes of the pitch circles and the pressure angle of a pair of mating involutes depend solely upon the sizes of their base circles and the distance between their centers.

When C = center distance

R_1 = pitch radius of first involute

R_2 = pitch radius of second involute

R_{b1} = radius of base circle of first involute

R_{b2} = radius of base circle of second involute

ϕ = pressure angle

$$C = R_1 + R_2 \tag{4-4}$$
$$R_1/R_2 = R_{b1}/R_{b2}$$

whence

$$R_1 = R_2 R_{b1}/R_{b2}$$
$$C = (R_2 R_{b1}/R_{b2}) + R_2 = R_2(R_{b1} + R_{b2})/R_{b2}$$

whence

$$R_2 = R_{b2} C/(R_{b1} + R_{b2}) \tag{4-5}$$

In like manner

$$R_1 = R_{b1} C/(R_{b1} + R_{b2}) \tag{4-6}$$

Referring again to Fig. 4-4, we have from the geometrical conditions shown there

$$\cos \phi = (R_{b1} + R_{b2})/C \tag{4-7}$$

We can obtain other simple and useful interrelations from the geometrical conditions shown in Fig. 4-4 as follows:

$$\cos \phi = R_{b1}/R_1 = R_{b2}/R_2 \tag{4-8}$$

whence

$$R_{b1} = R_1 \cos \phi \tag{4-9}$$

and

$$R_{b2} = R_2 \cos \phi \tag{4-10}$$

Action of Involute against a Straight Line.   When an involute acts against a straight line, we have the conditions shown in Fig. 4-5.   The straight line is the tangent to the involute curve and is always perpendicular to its line of action.   When it is constrained to move only in the direction of the line of action, it will be moved at a corresponding and uniform rate to that of the end of the generating line.

We shall now consider the motion of this straight line when it is constrained so that it can move only in the direction of the line AA'.   If we designate the distance that the line travels in the direction AA' as D_1, the distance that this line moves along the line of action, as D, and the angle between the line of action and the line AA' as ϕ_1, we have the following relationship:

$$D_1 = D/\cos \phi_1$$

As the value of D changes uniformly, and as the value of ϕ_1 is constant, the value of D_1 also changes uniformly. As cos ϕ_1 can never be greater than unity, the value of D_1 will never be smaller than D. Therefore

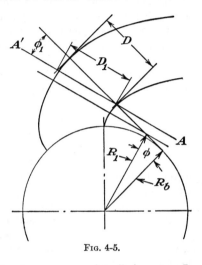

when the line against which the involute acts is constrained so that it moves only in the direction of the line AA', the distance it travels along this line will be greater than the distance along the line of action, but its rate of motion will be uniform as long as the rate of rotation of the involute is uniform.

If the involute should make one complete revolution, the value of D would become $2\pi R_b$. The value of D_1 would become $2\pi R_b/\cos \phi_1$. This last value also represents the circumference of a pitch disk that drives a straight edge by friction when this straight edge is parallel to the line AA'. The radius of

Fig. 4-5.

this pitch disk or pitch circle, R_1, thus becomes equal to $R_b/\cos \phi_1$. In Fig. 4-5, the radius of this pitch circle is established by the intersection of the line of action by a radial line, from the center of the base circle, that is perpendicular to the line AA'.

Hence the form of the basic rack of the involute is a straight line, a form that is also symmetrical in relation to any point on it. Thus any point of this basic-rack profile may be used as a pitch point without affecting its value as the basic rack of an interchangeable gear-tooth system.

SUMMARY OF INVOLUTE-CURVE PROPERTIES

It follows then that the involute curve has the following properties:

1. The shape of the involute curve is dependent only upon the size of the base circle.

2. If one involute, rotating at a uniform rate of motion, acts against another involute, it will transmit a uniform angular motion to the second regardless of the distance between the centers of the two base circles.

3. The rate of motion transmitted from one involute to another depends only upon the relative sizes of the base circles of the two involutes. This rate of motion is in inverse proportion to the sizes of the two base circles.

4. The common tangent to the two base circles is both the path of contact and the line of action. In other words, the two involutes will make contact with each other only along this common tangent to the two base circles.

5. The path of contact of an involute is a straight line. Any point on this line may therefore be taken as a pitch point, and the path of contact will remain symmetrical in relation to this pitch point.

6. The intersection of the common tangent to the two base circles with their common center line establishes the radii of the pitch circles of the mating involutes. No involute has a pitch circle until it is brought into contact with another involute, or with a straight line constrained to move in a fixed direction.

7. The pitch diameters of two involutes acting together are directly proportional to the diameters of their base circles.

8. The pressure angle of two involutes acting together is the angle between the common tangent to the two base circles and a line perpendicular to their common center line. No involute has a pressure angle until it is brought into contact with another involute, or with a straight line constrained to move in a fixed direction.

9. The form of the basic rack of the involute is a straight line. The pressure angle of an involute acting against such a rack is the angle between the line of action and a line representing the direction in which this rack moves.

10. The pitch radius of an involute acting against a straight-line rack form is the length of the radial line, perpendicular to the direction of motion of the rack, measured from the center of the base circle to its point of intersection with the line of action.

USE OF THE INVOLUTE FORM FOR GEAR-TOOTH PROFILES

When the involute form is used as a gear-tooth profile, several involute curves are developed from the same base circle to form the profiles of the several teeth. As gear teeth are generally symmetrical, at the start we shall consider but one side of the teeth.

In Fig. 4-6 is shown the development of one side of several successive teeth. Imagine a string with knots evenly spaced wound about the circumference of the base circle. As this string is unwound, each knot will describe an involute curve. The distance between these involutes, measured along any line tangent to the base circle is always the same.

This distance is equal to the length of the arc of the base circle between the origins of any two successive involutes. This is also the distance between the knots in the string. This distance is also equal to the circumference of the base circle divided by the number of teeth in the gear. It is called the *base pitch* of the involute gear. Thus when

p_b = base pitch of involute gear, in.

R_b = radius of base circle of involute, in.

N = number of teeth in gear

$$p_b = 2\pi R_b/N \qquad (4\text{-}11)$$

In a pair of mating involute gears, the base pitch must be identical on both gears to obtain smooth continuous action.

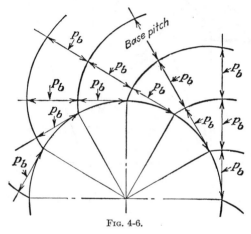

Fɪɢ. 4-6.

Rolling and Sliding Action. As pointed out before, the length of the generating line that is unwrapped from the base circle is the radius of curvature of the involute curve at any point. Figure 4-7 shows the position of this generating line at equal angular intervals. At the origin of the involute, *a*, the length of the generating line is zero. At *b* it is infinitely longer. At *c* it is twice the length that it is at *b*. At *d* it is one and a half times the length at *c*, etc. The radius of curvature of the involute thus increases rapidly in proportionate length near the base circle, and more slowly as the curve departs further from the base circle. In other words, the form near the base circle is very sensitive, but it becomes less sensitive the farther it departs from the base circle.

Sensitive curves of this type are most difficult to produce accurately whether they are on gear-tooth forms or on other types of cams, and they should always be avoided whenever possible. Thus only in cases of necessity should the active profile of an involute gear tooth extend to or very close to the base circle.

It will also be noted in Fig. 4-7 that the length of the curve *ab* is much less than the length *bc*; that *bc* is shorter than *cd*; etc.   Thus whether the involute is acting as a cam or is acting as a gear-tooth profile against another involute gear tooth, the length of the curve that must pass through the line of action for any series of equal angular movements

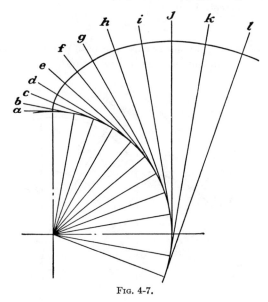

Fig. 4-7.

changes constantly.   The nearer the active part of the profile is to the base circle, the shorter is the length of this profile.

Thus when two involutes are acting against each other, a combined rolling and sliding action takes place between them because of the varying lengths of equal angular increments on the profiles.

In Fig. 4-8 are shown two equal involutes with the generating lines shown at equal angular intervals.   The part *ab* of the profile on one involute comes into contact with the profile section *gh* on the second involute.   Profile *ab* is much nearer to its base circle than is *gh*, and it is therefore much shorter.   The two profiles must slide against each other a distance equal to their difference in length to make up this difference.

The length *bc* is still much shorter than its mating section *hi*, but the amount of sliding will not be as much as with the previous sections of the mating profiles.   Spaces *cd* and *ij* are more nearly equal in length, *cd* being the shorter, so that still less sliding takes place here. The sections *de* and *jk* are almost equal in length, but the length of the profile *de* on the first involute is now slightly longer than its mating section on the second involute. Thus the small amount of sliding that takes place now acts in the opposite direction to that of the initial sliding. The remaining

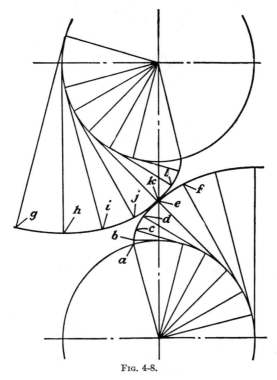

Fɪɢ. 4-8.

sections of the first involute become increasingly longer, while those on the second involute become shorter, so that the amount of sliding increases again.

It is evident that the rate of sliding between two involutes acting against each other is constantly varying. The rate of sliding starts quite high, reduces to zero at the pitch point, changes its direction, and increases again. The actual velocity of the sliding is the same for both profiles, but it is distributed over different lengths of profile.

Sliding Velocity. Equations for determining the sliding velocity at any point on a pair of involute gear teeth are derived as follows: The sliding velocity will be the difference in the speed of the ends of the generating lines of the involutes as they pass through the line of action. The angular velocity of these generating lines will be the same as the angular velocities of the gears themselves. The actual sliding velocities

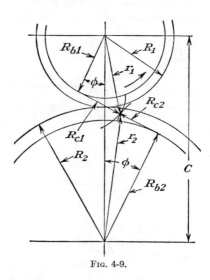

Fig. 4-9.

will be the products of these relative angular velocities and the lengths of the generating lines or radii of curvature.

Referring to Fig. 4-9, we have the following:

ω_1 = angular velocity of driving gear, radians/min

ω_2 = angular velocity of driven gear, radians/min

n = rpm of driving gear

V = pitch-line velocity of gears, ft/min

V_s = sliding velocity, ft/min

R_1 = pitch radius of driving gear, ft/min

R_2 = pitch radius of driven gear, ft/min

C = center distance, in.

R_{b1} = radius of base circle of driving gear, in.

R_{b2} = radius of base circle of driven gear, in.

ϕ = pressure angle

R_{c1} = radius of curvature of driving gear at r_1, in.

R_{c2} = radius of curvature of driven gear at r_2, in.

r_1 = any radius of driving gear-tooth profile, in.

r_2 = mating radius of driven gear-tooth profile, in.

$$V = 2\pi R_1 n/12 = R_1 \omega_1/12 \qquad (4\text{-}12)$$

whence

$$\omega_1 = 12V/R_1$$
$$V_s = (R_{c1}\omega_1 - R_{c2}\omega_2)/12$$
$$\omega_2 = R_1\omega_1/R_2$$
$$R_{c1} + R_{c2} = C \sin \phi$$
$$R_{c1} = \sqrt{r_1{}^2 - R_{b1}{}^2}$$
$$R_{c2} = \sqrt{r_2{}^2 - R_{b2}{}^2} = C \sin \phi - \sqrt{r_1{}^2 - R_{b1}{}^2}$$

Substituting these values into the equation for sliding, combining, and simplifying, we obtain

$$V_s = [V(R_1 + R_2)/R_1R_2](\sqrt{r_1{}^2 - R_{b1}{}^2} - R_1 \sin \phi) \qquad (4\text{-}13)$$

Equation (4-13) may also be written

$$V_s = V\left[\frac{1}{R_1} + \frac{1}{R_2}\right](\sqrt{r_1{}^2 - R_{b1}{}^2} - R_1 \sin \phi) \qquad (4\text{-}14)$$

When the driven member is a rack, the value of R_2 is equal to infinity, so that $1/R_2$ is equal to zero. Hence the sliding velocity between an involute gear and its rack is as follows: When $V_{sr} =$ sliding velocity on rack, feet/min, then
When the rack is driven

$$V_{sr} = V(1/R_1)(\sqrt{r_1{}^2 - R_{b1}{}^2} - R_1 \sin \phi) \qquad (4\text{-}15)$$

When the rack is the driving member

$$V_{sr} = V(1/R_2)(R_2 \sin \phi - \sqrt{r_2{}^2 - R_{b2}{}^2}) \qquad (4\text{-}16)$$

Example of Sliding Velocity. As a definite example we shall determine the relation V_s/V for a pair of gears of equal size and also for one of these gears meshing with a rack. We shall use the following values:

$$R_1 = 10.000 \qquad R_2 = 10.000 \qquad C = 20.000 \qquad \phi = 20°$$

Whence

$$R_{b1} = R_{b2} = 10 \times \cos 20° = 9.39693$$

Using a series of values for r_1 ranging from the value of R_{b1} to 11.000 in. and Eq. 4-13), we obtain the values for V_s/V plotted in Fig. 4-10.

The value of the sliding is minus on the dedendum of the driving gear and is plus on its addendum. This indicates that the direction of the sliding as the dedendum of the driving gear is in toward the center of the gear, and is out or away from the center of the gear on the addendum. The direction of sliding changes at the pitch point as its velocity passes through zero. The symbol R_{a1} used in Fig. 4-10 is for the radius to the bottom of the active profile of the driving gear. Contact does not extend down to the base circle.

Using the same series of values of r_1 as before and Eq. (4-15), we obtain the values of V_{sr}/V for the sliding between a gear and a rack. These values are plotted in Fig.

4-11. The values for this relative sliding on a rack are exactly one-half of the values for a pair of equal gears.

Duration of Contact. One of the important factors in the design of gears that are to transmit power is that the proportions of the involute

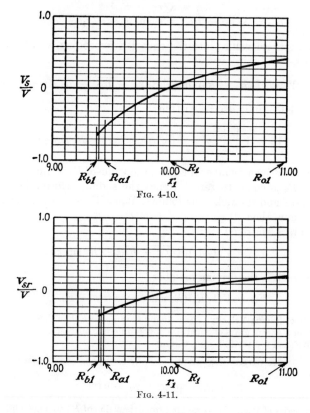

FIG. 4-10.

FIG. 4-11.

profiles must be so selected that the second pair of mating teeth will be in contact before the first pair is out of contact. The minimum amount of contact that will be adequate depends upon many conditions, and may need to be established by experience or experiment for critical cases. Except when the pitch-line velocities are high and sliding velocities are critical, a greater amount of contact than the minimum will seldom be detrimental for power drives.

The arc of action is the arc through which one tooth travels from the time it first makes contact with its mating tooth until it ceases to be in contact. The number of teeth in contact, or the *contact ratio*, is the quotient of the arc of action divided by the arc between successive teeth on the gear. Thus if an overlap of 0.60 exists, the contact ratio is 1.60.

In Fig. 4-12, that part of the line of action which is intercepted by the two outside circles of a pair of mating gears, shown as a heavy line, is the

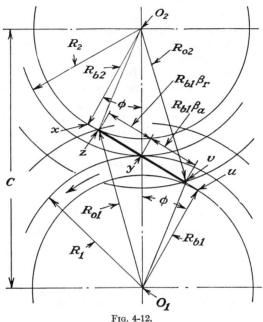

Fig. 4-12.

length of the arc of action measured at the radius of the base circle. This length divided by the length of an arc of the base circle between two successive involutes gives the contact ratio.

The arc of action is often separated into the arc of approach and the arc of recess. The arc of approach is the arc through which the tooth moves from the time it first comes into contact with its mating tooth until contact is made at the pitch point. The arc of recess is the arc through which the tooth moves from the time when contact is at the pitch point until it ceases to be in contact with its mating tooth.

Referring to Fig. 4-12, we shall let

ϕ = pressure angle
m_p = contact ratio
β_a = arc of approach
β_r = arc of recess

N_1 = number of teeth in driving gear
N_2 = number of teeth in driven gear
C = center distance, in.
R_{o1} = outside radius of driving gear, in.
R_{o2} = outside radius of driven gear, in.
R_1 = pitch radius of driving gear, in.
R_2 = pitch radius of driven gear, in.
R_{b1} = radius of base circle of driving gear, in.
R_{b2} = radius of base circle of driven gear, in.
p_b = base pitch of gears, in.

Simple equations can be derived for the values of the arc of approach, the arc of recess, and the contact ratio by solving several of the right triangles shown in Fig. 4-12. The angle of approach in circular measure or radians is found by dividing the length of the line yv by the radius of the base circle. The length of this line is equal to the length of xv minus the length xy.

$$\text{Length } xy = R_2 \sin \phi$$
$$\text{Length } xv = \sqrt{R_{o2}{}^2 - R_{b2}{}^2}$$

whence

$$\beta_a = (\sqrt{R_{o2}{}^2 - R_{b2}{}^2} - R_2 \sin \phi)/R_{b1} \qquad (4\text{-}17)$$

The arc of recess is found in a similar manner by dividing the length of the line yz by R_{b1}.

$$\beta_r = (\sqrt{R_{o1}{}^2 - R_{b1}{}^2} - R_1 \sin \phi)/R_{b1} \qquad (4\text{-}18)$$

The contact ratio is found by dividing the length of the line zv by the base pitch. The length of the line zv is equal to the sum of yz and yv.

$$m_p = (\sqrt{R_{o1}{}^2 - R_{b1}{}^2} + \sqrt{R_{o2}{}^2 - R_{b2}{}^2} - C \sin \phi)/p_b \qquad (4\text{-}19)$$

Duration of Contact with a Rack. The contact ratio for a rack and gear is determined in a similar manner. Referring to Fig. 4-13, we have the additional symbol a = addendum of rack, in.

When the gear drives the rack, we have from the geometrical conditions shown in Fig. 4-13 the following:

$$\beta_a = a/R_{b1} \sin \phi \qquad (4\text{-}20)$$
$$\beta_r = (\sqrt{R_{o1}{}^2 - R_{b1}{}^2} - R_1 \sin \phi)/R_{b1} \qquad (4\text{-}18)$$

When the rack drives the gear, the values for the arc of approach and the arc of recess are reversed.

$$m_p = \frac{(a/\sin \phi) + \sqrt{R_{o1}{}^2 - R_{b1}{}^2} - R_1 \sin \phi}{p_b} \qquad (4\text{-}21)$$

Active Profile. The active profile of a gear tooth is that portion of the tooth profile which actually comes into contact with its mating tooth along the line of action. In general, when the tooth design is such that a high rate of sliding exists, one or both active profiles will be short in relation to the length of the whole tooth profile. When the rate of sliding

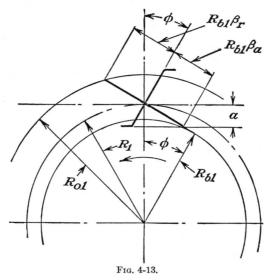

Fig. 4-13.

is low, on the other hand, the active profiles will include the greater part of the entire tooth profile.

Referring again to Fig. 4-12, the radius to the bottom of the active profile on the pinion or driving gear is equal to the length of the radial line o_1v. This line is the hypotenuse of a right triangle of which R_{b1} and the line uv are the two legs.

$$\text{Length } uv = C \sin \phi - \sqrt{R_{o2}^2 - R_{b2}^2}$$

When R_{a1} = radius to bottom of active profile on pinion, in.

R_{a2} = radius to bottom of active profile on gear, in.

$$R_{a1} = \sqrt{R_{b1}^2 + (C \sin \phi - \sqrt{R_{o2}^2 - R_{b2}^2})^2} \tag{4-22}$$

In similar manner

$$R_{a2} = \sqrt{R_{b2}^2 + (C \sin \phi - \sqrt{R_{o1}^2 - R_{b1}^2})^2} \tag{4-23}$$

In the case of a rack and pinion

$$R_{a1} = \sqrt{R_{b1}^2 + [R_1 \sin \phi - (a/\sin \phi)]^2} \tag{4-24}$$

Example of Active Profile. As a definite example we shall use the pair of equal gears and the gear and rack from the preceding example. Whence we have for the gears

$$R_{a1} = \sqrt{(9.39693)^2 + (6.84040 - 5.71820)^2} = 9.46370$$

For the example with the pinion and rack we have:

$$R_{a1} = \sqrt{(9.39693)^2 + (3.42020 - 2.92381)^2} = 9.41003$$

These values are indicated on the sliding diagrams shown in Figs. 4-10 and 4-11.

Limitation to Conjugate Action. As the involute curve starts at the base circle, no conjugate gear-tooth action can take place below it. If a

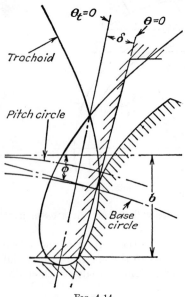

FIG. 4-14.

straight-sided rack with sharp corners acts against the involute, and these corners extend too far below the base circle, interference is present unless the tooth is undercut, as shown in Fig. 4-14. The looped curve shows the path of the sharp corner of the rack tooth as it comes into and goes out of engagement. This path not only undercuts the tooth below the base circle, but it also removes the lower part of the involute profile.

This looped path is the same trochoid that is discussed in Chap. 3 and is shown in Fig. 3-3. From there we have

R = pitch radius of gear, in.
r_t = any radius of trochoid, in.
b = distance from pitch line to corner of rack tooth, in.
θ_t = vectorial angle of trochoid

$$\theta_t = \tan^{-1} \frac{\sqrt{r_t^2 - (R - b)^2}}{R - b} - \frac{\sqrt{r_t^2 - (R - b)^2}}{R} \qquad (3\text{-}1)$$

When δ = angle between origins of trochoid and involute
ϕ = pressure angle of involute at R

$$\delta = \phi - \frac{(R - b) \tan \phi}{R} \qquad (4\text{-}25)$$

Example of Undercut. As a definite example we shall determine the form and position of the trochoid that would be formed on the gear by the corner of the following rack:

$$R = 6.000 \qquad b = 1.157 \qquad \phi = 14.500°$$
$$\cos \phi = 0.96815 \qquad \tan \phi - 0.25862$$
$$R_b = 6 \times 0.96815 = 5.80890$$
$$\delta = 0.25307 - \frac{4.843 \times 0.25862}{6.000} = 0.04432 \text{ radian}$$

From these values and the equations of the trochoid and the involute, we obtain values for the two curves that are plotted in Fig. 4-14.

Undercut with Pinion-shaped Cutter. Many gears are generated by a cutter of the form of a mating involute gear. If the sharp corner of the tooth of the pinion-shaped cutter extends too far below the base circle, undercut will be present. The form of this trochoid is also discussed in Chap. 3. For this trochoid we shall use the following symbols:

N_2 = number of teeth in gear

N_c = number of teeth in pinion-shaped cutter

R_{oc} = outside radius of pinion-shaped cutter, in.

C = center distance between axes of gear and cutter, in.

ϵ_2 = angle of rotation of gear

ϵ_c = angle of rotation of pinion-shaped cutter

θ_t = vectorial angle of trochoid

r_t = any radius to trochoid, in.

Substituting the foregoing symbols into the equations in Chap. 3, we have

From Eq. (3-12)
$$\epsilon_2 = - (N_c/N_2)\epsilon_c$$

From Eq. (3-13)
$$\cos \epsilon_c = \frac{C^2 + R_{oc}^2 - r_t^2}{2CR_{oc}}$$

From Eq. (3-14)
$$\theta_t = \sin^{-1} \frac{R_{oc} \sin \epsilon_c}{r_t} - \epsilon_2$$

When δ = angle between origins of trochoid and involute

θ_2 = vectorial angle of involute gear profile at pitch line

θ_c = vectorial angle of cutter profile at outside radius

$$\delta = (N_c/N_1)(\theta_c - \theta_2) - \theta_2 \qquad (4\text{-}26)$$

Example of Undercut with Pinion-shaped Cutter. As a definite example we shall use the same gear as before and use the following values[1] for the pinion-shaped cutter:

$$N_2 = 12 \qquad N_c = 18 \qquad R_{oc} = 10.250 \qquad C = 15.000$$
$$\theta_c = 0.06488 \qquad \theta_2 = 0.00554$$
$$\delta = (18/12)(0.06488 - 0.00554) - 0.00554 = 0.03347 \text{ radian}$$

[1] The calculations are made on the basis of 1 DP for the greatest simplicity in calculation, where DP = diametral pitch, which is the ratio of the number of teeth to the pitch diameter.

The coordinates of the involute gear profile are the same as before. Using the foregoing equations for the trochoid, we obtain the values that are plotted in Fig. 4-15.

Undercut Limit for Rack. In order to avoid this undercutting, the sharp corner of the rack or equivalent hob form, can extend below the base circle only a limited distance. Its bottom edge must not reach below the line where the line of action is tangent to the base circle, as indicated in Fig. 4-16. If the corner of the hob tooth is rounded, the position where the rounding is tangent to the flank

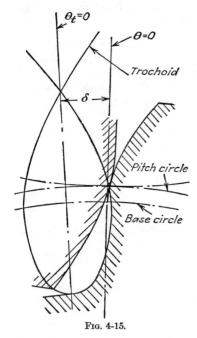

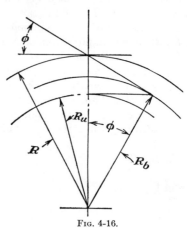

FIG. 4-15. FIG. 4-16.

of the tooth corresponds to the sharp corner.

Referring to Fig. 4-16, we shall let

R_u = radius of gear to undercut limit, in.

R_b = radius of base circle of gear, in.

R = pitch radius of gear, in.

ϕ = pressure angle of rack

We have the following from the geometrical conditions shown in Fig. 4-16:

$$R_u = R_b \cos \phi = R \cos^2 \phi \qquad (4\text{-}27)$$

Undercut or Interference Limit for Two Gears. In a similar manner, if two involute gears are acting against each other, or if a pinion-shaped cutter is used to generate another gear, their outside circles must not extend beyond the point of tangency of the line of action with the base

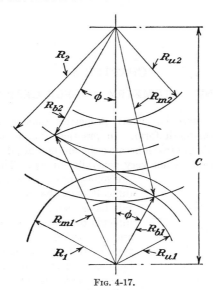

FIG. 4-17.

circle of the mating gear, as indicated in Fig. 4-17, or a similar interference will develop.

Referring to Fig. 4-17, we shall let

R_{m1} = radius of maximum addendum circle of first gear that will avoid interference, in.

R_{m2} = radius of maximum addendum circle of second gear that will avoid interference, in.

C = center distance, in.

R_{b1} = radius of base circle of first gear, in.

R_{b2} = radius of base circle of second gear, in.

R_1 = pitch radius of first gear, in.

R_2 = pitch radius of second gear, in.

R_{u1} = radius to undercut limit of first gear, in.

R_{u2} = radius to undercut limit of second gear, in.

ϕ = pressure angle of gears

We have the following from the geometrical conditions shown in Fig. 4-17:

$$R_{m1} = \sqrt{R_{b1}{}^2 + (C \sin \phi)^2} \qquad (4\text{-}28)$$

$$R_{m2} = \sqrt{R_{b2}{}^2 + (C \sin \phi)^2} \qquad (4\text{-}29)$$

$$R_{u1} = C - R_{m2} \qquad (4\text{-}30)$$

$$R_{u2} = C - R_{m1} \qquad (4\text{-}31)$$

CHAPTER 5

INVOLUTOMETRY OF SPUR GEARS

The involute curve has many properties that make it extremely valuable as a gear-tooth form. In practice, full advantage of these properties is not always taken because the method of calculating involute sizes and proportions is not very generally known. These calculations are sometimes complex, but they are not difficult once the few simple fundamentals have been mastered. It is no more difficult to calculate involute tooth sizes and proportions than it is to do the same type of thing with plane triangles. In both cases, if much of such work is to be done, a calculating machine is as necessary as the trigonometric and other tables.

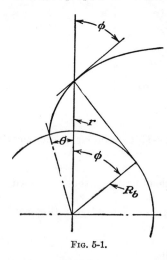

Fig. 5-1.

An involute gear-tooth form consists of two similar involute curves with a common base circle. When their relative positions are known at any radius, their relative positions at any other radius can be readily determined. In general, this is accomplished by first determining their relative positions at the base circle, and then adding or subtracting the vectorial angle of the involute for any other radius. We shall call this process of calculating involute gear-tooth relationships *involutometry*.

Equation of Involute. In Fig. 5-1 is shown the involute curve. We have derived the following equation for it in the preceding chapter. When R_b = radius of base circle, in.

r = any radius to involute form, in.

θ = vectorial angle

ϕ = pressure angle at radius r

$$\theta = \sqrt{(r^2 - R_b{}^2)}/R_b - \tan^{-1}(\sqrt{r^2 - R_b{}^2}/R_b) \qquad (4\text{-}1)$$

From the geometrical conditions shown in Fig. 5-1, we have

$$\sqrt{(r^2 - R_b{}^2)}/R_b = \tan\phi$$

78

whence

$$\theta = \tan \phi - \phi = \text{inv } \phi \tag{5-1}$$

From Fig. 5-1 we also have

$$r = R_b/\cos \phi \tag{5-2}$$

These last two equations give the fundamental relationships of the involute of the circle by means of which involute gear-tooth sizes are readily calculated. Here the origin of the gear-tooth profiles is not at the pitch point but is at the origin of the involute curve. It will be seen from Eq. (5-1) that there is a fixed relationship between the two angles ϕ and θ, which is independent of the diameters of the gears. The most convenient form for the value of θ is in circular measure or radians, and its value can be obtained by subtracting the value of the angle ϕ expressed in radians from the value of its tangent, as expressed by Eq. (5-1). This value of θ for any angle ϕ will be called the involute function of ϕ and will be expressed as inv ϕ in the equations that follow. Tables of such involute functions are available, and they enable all involute calculations to be made in a manner similar to the solution of plane triangles.

The further consideration of involutometry will be in the form of a series of problems. The ones that follow do not begin to exhaust the number of similar problems that may need a specific solution. Given the basis and the general method of operation, however, it should be possible for any one to work out the solution for any specific problem of this nature with which he is confronted.

All these problems will be solved on the basis of 1-DP gears. In most cases, such a procedure will make for the simpler solutions because then the pitch diameter is equal to the number of teeth in the gear. The first step toward the solution of any specific problem, therefore, will be to transform all given dimensions of size to the equivalent ones for a 1-DP gear. In presenting this limited selection of problems, the necessary equations will be derived first and then a specific problem will be solved numerically.

Problem 5-1. *Given the arc tooth thickness and pressure angle of an involute gear at a definite radius, to determine the coordinates of the involute profile.*

The complete profile includes both the involute profile and the trochoidal fillet. We shall start with the involute profile, and treat the fillets as separate problems.

Referring to Fig. (5-2), when

T_1 = given arc tooth thickness, in.

R_1 = given radius of profile, in.

ϕ_1 = given pressure angle at radius R_1

r = any radius of profile, in.
T = arc tooth thickness at r, in.
ϕ = pressure angle at r
R_b = radius of base circle of involute, in.
We have by transposing Eq. (5-2)

$$R_b = R_1 \cos \phi_1 \tag{5-3}$$
$$\cos \phi_1 = R_b/R_1 \tag{5-4}$$

This last equation holds true for all positions, so we can write

$$\cos \phi = R_b/r \tag{5-4}$$

As the tooth form is symmetrical, we shall deal with the half thickness

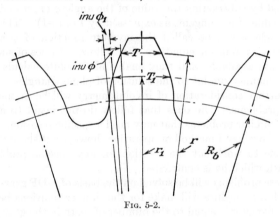

Fig. 5-2.

of the tooth. The angle of the half thickness of the tooth at R_1 in circular measure or radians is equal to $T_1/2R_1$.

The half thickness of the tooth, in radians, at the base circle is equal to $(T_1/2R_1) +$ inv ϕ_1. The value of inv ϕ_1 will be taken from a table of involute functions.

The half thickness of the tooth, in radians, at any other radius r is equal to its half thickness in radians at the base circle minus the involute function of the pressure angle at the specific radius r. Thus we have[1]

$$T/2r = (T_1/2R_1) + \text{inv } \phi_1 - \text{inv } \phi$$

Solving for T, we have

$$T = 2r[(T_1/2R_1) + \text{inv } \phi_1 - \text{inv } \phi] \tag{5-5}$$

---

[1] $T/2r$ is the angle in radians from the center of the tooth to the given point r on the involute profile.

Hence to determine the arc tooth thickness on an involute gear at any radius, Eqs. (5-3), (5-4), and (5-5) would be solved in the order given for the specific value of r that may be required.

To determine such values for the entire involute profile, we would use a series of values of r ranging from the base circle to the tip of the tooth.

Cartesian Coordinates. To obtain the Cartesian coordinates of the involute profile in reference to either the center line of the tooth or the center line of the space, we first determine the vectorial angles of the several points of the profile from the specified center line, and then use Eq. (1-5) and (1-6) as follows:

θ'' = vectorial angle of profile from specified center line

X = abscissa of profile

Y = ordinate of profile from center of gear

$$X = r \sin \theta'' \tag{1-5}$$
$$Y = r \cos \theta'' \tag{1-6}$$

When θ'' = vectorial angle of profile from center line of tooth

$$\theta'' = T/2r = (T_1/2R_1) + \text{inv } \phi_1 - \text{inv } \phi \tag{5-6}$$

When θ'' = vectorial angle of profile from center line of space

N = number of teeth in gear

$$\theta'' = (\pi/N) - (T/2r) = (\pi/N) - [(T_1/2R_1) + \text{inv } \phi_1 - \text{inv } \phi] \tag{5-7}$$

Example of Involute Profile. As a definite example we shall use a 20-tooth gear of 10 DP and 20-deg pressure angle, whose arc tooth thickness at the radius of 1 in. is equal to one-half the circular pitch of 0.31416 in. This gives the following values:

	10 DP	1 DP
Pitch radius.............................	1.000	10.000
Arc tooth thickness.....................	0.15708	1.5708
Pressure angle.........................	20°	20°

Using the 1-DP values, we have

$$R_b = 10 \times \cos 20° = 9.39693$$

From the table of involute functions we get

$$\text{inv } 20° = 0.014904$$
$$T = 2r \left(\frac{1.5708}{20.000} + 0.014904 - \text{inv } \phi \right) = 2r(0.093444 - \text{inv } \phi)$$

Using a series of values of r ranging from the radius of the base circle, 9.39693, to 11 in., we determine first the value of ϕ from Eq. (5-4), then obtain the value of inv ϕ from a table of involute functions, and then solve the foregoing equation for the several values of T. Such values have been computed and they are tabulated in Table 5-1.

For the Cartesian coordinates, we have, when the origin is at the center line of the space,

$$\theta'' = \left(\frac{3.1416}{20}\right) - \left(\frac{T}{2r}\right) = 0.15708 - \left(\frac{T}{2r}\right)$$

These coordinates have been computed. They are tabulated in Table 5-1 and are plotted in Fig. 5-3.

These tabulated values of r and T would be divided by 10 to reduce them to the original 10-DP sizes. These values give the coordinates of the involute profile only.

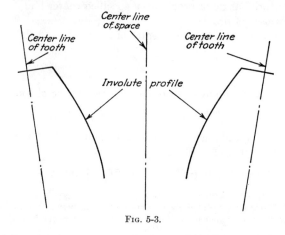

FIG. 5-3.

The tooth form includes, however, the fillet that joins the involute profile to the bottom land of the tooth space. The shape of this fillet will depend upon the form of the generating tool and the method used to produce this gear. To complete the tooth form, we must determine the coordinates of the trochoidal fillet that will be produced by the particular tool and method used to generate it. This form may be hobbed or generated by a pinion-shaped cutter. Equations for these trochoids are given in Chap. 3. The next problems will be to select and use the proper equations for the specific trochoid that will be developed. We shall start, however, with the radius to the top of the fillet when no undercut is present.

Problem 5-2. *Given the proportions of the hob and gear tooth, to determine the radius to the point of tangency of the involute profile and trochoidal fillet.*

When no undercut is present, the trochoidal fillet will be tangent to the involute profile. This point of tangency for a hob will be the point where the end of the straight-line profile of the hob or basic-rack form crosses the path of contact, as shown in Fig. 5-4.

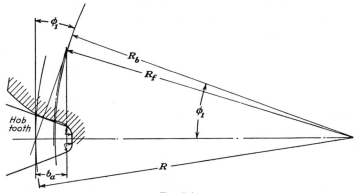

FIG. 5-4.

TABLE 5-1. COORDINATES OF INVOLUTE FORM
(Plotted in Fig. 5-3)

r, in.	ϕ, deg	inv ϕ	$T/2r$	T, in.	X, in.	Y, in.
9.39693	0.000	0.000000	0.093444	1.75617	0.59755	9.37795
9.500	8.447	0.001077	0.092367	1.75497	0.61436	9.48005
9.600	11.805	0.002966	0.090478	1.73718	0.63888	9.57869
9.700	14.359	0.005590	0.087854	1.70437	0.67085	9.67682
9.800	16.490	0.008219	0.085225	1.67041	0.70354	9.77472
9.900	18.345	0.011409	0.082035	1.62429	0.74230	9.87218
10.000	20.000	0.014904	0.078540	1.57080	0.78460	9.96920
10.100	21.504	0.018675	0.074769	1.51033	0.83032	10.06586
10.200	22.887	0.022696	0.070748	1.44326	0.87934	10.16206
10.300	24.172	0.026950	0.066494	1.36978	0.93174	10.25777
10.400	25.369	0.031399	0.062045	1.29054	0.98686	10.35310
10.500	26.499	0.036065	0.057379	1.20496	1.04517	10.44781
10.600	27.563	0.040900	0.052544	1.11393	1.10600	10.54212
10.700	28.572	0.045908	0.047536	1.01727	1.16972	10.63591
10.800	29.531	0.051074	0.042370	0.91519	1.23606	10.72904
10.900	30.447	0.056399	0.037045	0.80758	1.30517	10.82152
11.000	31.321	0.061857	0.031587	0.69491	1.37676	10.91354

When R_j = radius to top of fillet, in.
R = pitch radius of gear, in.
R_b = radius of base circle, in.
ϕ_1 = pressure angle at R
b_a = distance from pitch line of hob to point of tangency of rounded corner with straight-line form, in.

We have from the geometrical conditions shown in Fig. 5-4 the following:

$$R_f = \sqrt{[R \sin \phi_1 - (b_a/\sin \phi_1)]^2 + R_b{}^2} \tag{5-8}$$

Example of Radius to Top of Fillet.   We will use the same example as before, which gives the following 1-DP values:

$$R = 10.000 \qquad \phi_1 = 20° \qquad R_b = 9.39693 \qquad b_a = 1.000$$

Whence

$$R_1 \sin \phi_1 = 3.42020 \qquad b_a/\sin \phi_1 = 2.92380$$

Then

$$R_f = \sqrt{(0.4964)^2 + (9.39693)^2} = 9.41003 \text{ in.}$$

Problem 5-3.   *Given the proportions of the pinion-shaped cutter and the gear teeth, to determine the radius to the point of tangency of the involute profile and the trochoidal fillet.*

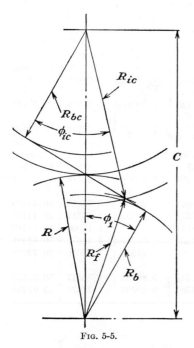

When no undercut is present, the trochoidal fillet will be tangent to the involute profile. The point of tangency for a trochoid developed by a pinion-shaped cutter will be the radius to the point where the maximum radius of the involute form of the cutter crosses the path of contact, as shown in Fig. 5-5.

When R_f = radius to top of fillet, in.

R = pitch radius of gear, in.

R_b = radius of base circle of gear, in.

ϕ_1 = pressure angle at R

R_{ic} = maximum radius of involute profile on pinion-shaped cutter, in.

R_{bc} = radius of base circle of pinion-shaped cutter, in.

C = center distance between axes of gear and cutter, in.

ϕ_{ic} = pressure angle at maximum radius of involute on cutter

Fig. 5-5.

we have the following from the geometrical conditions shown in Fig. 5-5:

$$R_f = \sqrt{(C \sin \phi_1 - R_{bc} \tan \phi_{jc})^2 + R_b{}^2} \tag{5-9}$$

Example of Radius to Top of Fillet. Using the same gear as before, and a 30-tooth pinion-shaped cutter, we have the following 1-DP values:

$$C = 25.00 \qquad R = 10.00 \qquad R_b = 9.39693 \qquad \phi_1 = 20°$$
$$R_{ic} = 16.250 \qquad R_{bc} = 14.09539$$
$$\cos \phi_{ic} = \frac{14.09539}{16.250} = 0.86741 \qquad \phi_{ic} = 29.841°$$

Whence

$$R_f = \sqrt{(0.46353)^2 + (9.39693)^2} = 9.40835 \text{ in.}$$

We shall now consider the several forms of the fillet that may be developed by different generating tools.

Problem 5-4. *Given the proportions of a hob with rounded corners and of a gear tooth, to determine the form of the trochoidal fillet.*

The form of this fillet is that whose equations have already been derived in Chap. 3. We must first determine the radius and the position of the center of the rounding at the tip of the hob tooth. Referring to Fig. 5-6, when

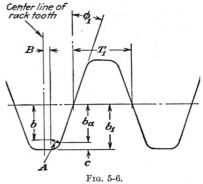

Fig. 5-6.

T_1 = arc tooth thickness of gear tooth at pitch radius R, and width of rack or hob space at pitch line, in.

R = pitch radius of gear, in.

p = circular pitch of gear, in.

ϕ_1 = pressure angle at R and one-half included angle of hob tooth

b_1 = dedendum of gear and addendum of hob, in.

b_a = distance from pitch line of hob to point of tangency of rounded corner with straight-line form, in.

A = radius of rounded corner of hob tooth, in.

b = distance from pitch line of hob to center of rounded corner, in.

B = distance from center line of rack tooth to center of rounded corner, in.

c = clearance at bottom of tooth space, in.

$$c = b_1 - b_a = A(1 - \sin \phi_1)$$

whence

$$A = \frac{b_1 - b_a}{1 - \sin \phi_1} \tag{5-10}$$

$$b = b_1 - A \tag{5-11}$$

$$B = \frac{T_1}{2} - \left(b \tan \phi_1 + \frac{A}{\cos \phi_1} \right) \tag{5-12}$$

When δ_s = angle between center line of gear-tooth space and origin of trochoid

$$\delta_s = B/R \tag{5-13}$$

When δ_t = angle between center line of gear tooth and origin of trochoid

$$\delta_t = \frac{(p/2) - B}{R} \tag{5-14}$$

For the trochoidal path of the center of the rounding, we have the following equations from Chap. 3:

When r_t = any radius of trochoid, in.

θ_t = vectorial angle of trochoid

ψ_t = angle between tangent to trochoid and radius vector and all other symbols are the same as before,

$$\theta_t = \tan^{-1} \left[\frac{\sqrt{r_t^2 - (R - b)^2}}{R - b} \right] - \frac{\sqrt{r_t^2 - (R - b)^2}}{R} \tag{3-1}$$

$$\tan \psi_t = \frac{R(R - b) - r_t^2}{R\sqrt{r_t^2 - (R - b)^2}} \tag{3-2}$$

The next step is to determine the coordinates of the actual fillet. For this we have the following from Chap. 3:

When r_f = radius to any point on actual fillet, in.

θ_f = vectorial angle of actual fillet form

and all other symbols are the same as before,

$$r_f = \sqrt{r_t^2 + A^2 - 2Ar_t \sin \psi_t} \tag{3-7}$$

$$\theta_f = \theta_t + \cos^{-1} \frac{r_t - A \sin \psi_t}{r_f} \tag{3-8}$$

Hence to determine the coordinates of the actual fillet when the corner of the hob tooth is rounded, we must first calculate the coordinates of the trochoid of the center of the rounded corner and then calculate the coordinates of the actual fillet.

Cartesian Coordinates of Fillet. To obtain the Cartesian coordinates of the fillet in reference to the center line of the space or tooth, we proceed as before. Thus when

δ = angle between origin of trochoid and center line of tooth or space

θ_f = original vectorial angle of fillet

θ''_f = vectorial angle of fillet in reference to selected center line

r_f = any radius to fillet, in.

X_f = abscissa of fillet, in.

Y_f = ordinate of fillet, in.

$$\theta''_f = \delta + \theta_f \qquad (1\text{-}4)$$
$$X_f = r_f \sin \theta''_f \qquad (1\text{-}5)$$
$$Y_f = r_f \cos \theta''_f \qquad (1\text{-}6)$$

Example of Fillet Developed from Rounded Corner of Hob Tooth. As a definite example we shall use the same 20-tooth gear of 20-deg, full-depth form as has been used in the previous examples. For this we have the following 1-DP values:

$$T_1 = 1.5708 \qquad p = 3.1416 \qquad \phi_1 = 20° \qquad b_1 = 1.157$$
$$c = 0.157 \qquad R = 10.000 \qquad b_a = 1.000$$

$$A = \frac{1.157 - 1.00}{1 - 0.34202} = 0.2386 \qquad (5\text{-}10)$$

$$b = 1.157 - 0.2386 = 0.9184 \qquad (5\text{-}11)$$

$$B = \frac{1.5708}{2} - \left(0.9184 \times 0.36397 + \frac{0.2386}{0.93969}\right) = 0.19722 \qquad (5\text{-}12)$$

Using the center line of the space as the reference line for plotting, we have

$$\delta_s = \frac{0.19722}{10} = 0.019722 \text{ radian} = 1.130° \qquad (5\text{-}13)$$

For the coordinates of the trochoidal path of the center of the rounding, we have

$$\theta_t = \tan^{-1}\left(\frac{\sqrt{r_t{}^2 - 82.475458}}{9.08160}\right)$$
$$- \frac{\sqrt{r_t{}^2 - 82.475458}}{10.000} \qquad (3\text{-}1)$$

$$\tan \psi_t = \frac{90.8160 - r_t{}^2}{10 \sqrt{r_t{}^2 - 82.475458}} \qquad (3\text{-}2)$$

For the coordinates of the actual fillet, we have

$$r_f = \sqrt{r_t{}^2 + 0.05693 - 0.47720 r_t \sin \psi_t} \qquad (3\text{-}7)$$

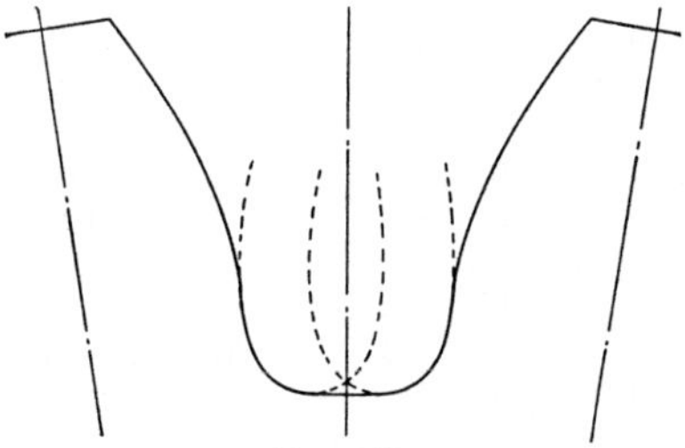

$$\theta_f = \theta_t + \cos^{-1}\frac{r_t - 0.23860 \sin \psi_t}{r_f} \qquad (3\text{-}8)$$

Fig. 5-7.

These values have been calculated. They are tabulated in Table 5-2 and plotted in Fig. 5-7.

Problem 5-5. *Given the proportions of the gear tooth and of a hob with full-rounded tips, to determine the form of the trochoidal fillet.*

On gears that must operate under highly stressed conditions, the bottom land and the fillet of the tooth space are often made in one continuously curved form so as to reduce the stress concentrations at the base of the gear tooth. Such a practice is often followed in such widely separated fields as steel-rolling-mill gears and airplane-propeller reduction gears.

One method of producing such a rounded form at the bottom of the tooth space is to use a hob with a full round or radius at the tip of the

TABLE 5-2. COORDINATES OF FILLET DEVELOPED BY ROUNDED CORNER OF HOB
TOOTH
(Plotted in Fig. 5-7)

r_t, in.	θ_t rad	ψ_t, deg	r_f, in.	θ_f rad	θ''_f rad	X_f, in.	Y_f, in.
10.000	0.01331	−12.374	10.05383	0.03649	0.05621	0.5649	10.0879
9.900	0.01533	−10.344	9.94561	0.03893	0.05865	0.5824	9.9285
9.800	0.01698	− 8.073	9.83634	0.04098	0.06070	0.5968	9.8182
9.700	0.01821	− 5.488	9.72572	0.04264	0.06236	0.6061	9.7068
9.600	0.01893	− 2.473	9.61352	0.04370	0.06342	0.6093	9.5942
9.500	0.01906	1.163	9.49815	0.04416	0.06388	0.6064	9.4788
9.400	0.01840	5.945	9.37522	0.04369	0.06341	0.5940	9.3565
9.300	0.01679	12.186	9.25258	0.04201	0.06173	0.5708	9.2349
9.200	0.01349	22.772	9.11030	0.03761	0.05733	0.5220	9.0954
9.100	0.00576	54.154	8.90769	0.02150	0.04122	0.3671	8.9001
9.0816	0.00000	90.000	8.84300	0.00000	0.01972	0.1744	8.8413
9.100	0.00576	54.154	8.90769	0.02150	−0.00178	−0.0159	8.9077
9.200	0.01349	22.772	9.11030	0.03761	−0.01789	−0.1630	9.1088
9.300	0.01679	12.186	9.25258	0.04201	−0.02229	−0.2062	9.2503
9.400	0.01840	5.945	9.37522	0.04369	−0.02397	−0.2246	9.3725
9.500	0.01906	1.163	9.49815	0.04416	−0.02444	−0.2320	9.4953
9.600	0.01893	− 2.473	9.61352	0.04370	−0.02398	−0.2305	9.6107
9.700	0.01821	− 5.488	9.72572	0.04264	−0.02292	−0.2228	9.7232
9.800	0.01698	− 8.073	9.83634	0.04098	−0.02126	−0.2091	9.8341
9.900	0.01533	−10.344	9.94561	0.03893	−0.01921	−0.1910	9.9438
10.000	0.01331	−12.374	10.05383	0.03649	−0.01677	−0.1686	10.0524

tooth as shown in Fig. 5-8. The calculation for the coordinates of the resulting fillet form is identical to that of the preceding example., For the trochoid of the center of the rounding, we have Eq. (3-1) and (3-2). For the coordinates of the actual fillet, we have Eq. (3-7) and (3-8). To obtain the Cartesian coordinates of the actual fillet, we have Eq. (1-4), (1-5), and (1-6). The only difference between this problem and the preceding one is the location of the center of the rounding, its radius,

FIG. 5-8.

and its position in reference to the involute profile of the gear tooth.

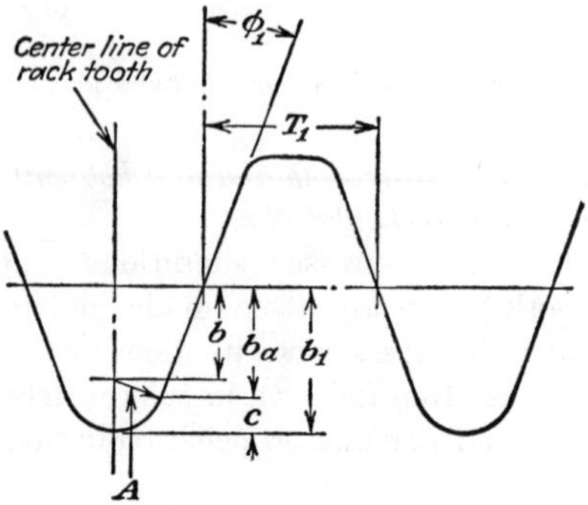

Using the same symbols here as were used in Prob. 5-4, and referring to Fig. 5-8, we have

$$B = 0 \tag{5-15}$$

$$A = \frac{(T_1/2) - b_a \tan \phi_1}{\cos \phi_1} \tag{5-16}$$

$$b = b_a - A \sin \phi_1 \tag{5-17}$$

$$b_1 = b + A \tag{5-18}$$

When δ_s = angle between center line of tooth space and origin of trochoid

$$\delta_s = 0 \tag{5-19}$$

When δ_t = angle between center line of gear tooth and origin of trochoid

$$\delta_t = \frac{p}{2R} \tag{5-20}$$

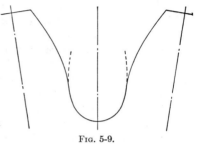

FIG. 5-9.

Example of Hobbed Full-rounded-root Form. As a definite example we shall use the same values as before.

$$T_1 = 1.5708 \qquad p = 3.1416 \qquad \phi_1 = 20° \qquad b_a = 1.000$$
$$R = 10.00 \qquad B = 0$$

$$A = \frac{0.7854 - 0.36397}{0.93969} = 0.44847 \tag{5-16}$$

$$b = 1.000 - (0.44847 \times 0.34202) = 0.84661 \tag{5-17}$$

$$b_1 = 0.84661 + 0.44847 = 1.29508 \tag{5-18}$$

Using these values in the several equations, we obtain the values tabulated in Table 5-3 and plotted in Fig. 5-9.

TABLE 5-3. COORDINATES OF FULL-ROUNDED HOBBED ROOT
(Plotted in Fig. 5-9)

r_t, in.	θ_t, rad	ψ_t, deg	r_f, in.	θ_f, rad	X_f, in.	Y_f, in.
9.15339	0.00000	90.000	8.70492	0.00000	0.0000	8.7049
9.200	0.00822	32.858	8.96460	0.05026	0.4504	8.9533
9.300	0.01332	17.048	9.17854	0.06006	0.5500	9.1593
9.400	0.01566	8.440	9.34471	0.06313	0.5896	9.3261
9.500	0.01669	2.890	9.48797	0.06390	0.6058	9.4686
9.600	0.01682	− 1.239	9.62015	0.06344	0.6099	9.6008
9.700	0.01632	− 4.552	9.74585	0.06220	0.6058	9.7270
9.800	0.01522	− 7.335	9.86727	0.06032	0.5948	9.8493
9.900	0.01368	− 9.743	9.98568	0.05796	0.5785	9.9689
10.000	0.01178	−11.873	10.10181	0.05524	0.5577	10.0864

Problem 5-6. *Given the size and form of a pinion-shaped cutter and the proportions of the gear tooth, to determine the coordinates of the trochoidal fillet.*

The form of this fillet is the trochoid whose equations have already been derived in Chap. 3. These equations are as follows:

When R_1 = pitch radius of gear, in.

$\quad R_c$ = pitch radius of cutter, in.

$\quad N_1$ = number of teeth in gear

$\quad N_c$ = number of teeth in pinion-shaped cutter

$\quad R_{oc}$ = outside radius of cutter, in.

$\quad\ \, C$ = center distance between axes of gear and cutter, in.

$\quad\ \epsilon_c$ = angle of rotation of cutter

$\quad\ \epsilon_1$ = angle of rotation of gear

$\quad\ \theta_t$ = vectorial angle of trochoid

$\quad\ r_t$ = any radius to trochoid, in.

$\quad \phi_1$ = pressure angle at pitch line of gear and cutter

$\quad \phi_{oc}$ = pressure angle at tip of cutter tooth

$\quad\ \delta_t$ = angle between origin of trochoid and center line of gear tooth

$\quad\ \delta_s$ = angle between origin of trochoid and center line of tooth space

$\quad \theta''_t$ = vectorial angle of trochoid in reference to selected center line

$\quad\ T_1$ = arc tooth thickness of gear at R_1, in.

$\quad\ T_c$ = arc tooth thickness of cutter at R_c, in.

$$\epsilon_1 = (R_c/R_1)\epsilon_c \tag{3-12}$$

$$r_t = \sqrt{C^2 + R_{oc}{}^2 - 2CR_{oc}\cos\epsilon_c} \tag{3-13}$$

$$\theta_t = \sin^{-1}[R_{oc}\sin\epsilon_c/r_t] - \epsilon_1 \tag{3-14}$$

$$\delta_s = (N_c/N_1)[(T_c/2R_c) + \text{inv }\phi_1 - \text{inv }\phi_{oc}] \tag{5-21}$$

$$\delta_t = (N_c/N_1)[(\pi/N_c) - (T_c/2R_c) - \text{inv }\phi_1 + \text{inv }\phi_{oc}] \tag{5-22}$$

$$\theta''_t = \delta \pm \theta_t \tag{1-4}$$

Example of Fillet Developed by Tip of Pinion-shaped Cutter. As a definite example we shall use the same 20-tooth gear as before, and assume the use of a 30-tooth pinion-shaped cutter. For the 1-DP values, we have the following:

$$N_1 = 20 \qquad N_c = 30 \qquad R_c = 15.000 \qquad R_{oc} = 16.250$$
$$\phi_1 = 20° \qquad T_c = 1.5708 \qquad T_1 = 1.5708$$
$$\cos\phi_{oc} = \frac{15 \times 0.93969}{16.250} = 0.86741$$
$$\phi_{oc} = 29.841° \qquad \text{inv }\phi_{oc} = 0.052832 \qquad \text{inv }\phi_1 = 0.014904 \qquad C = 25.000$$
$$\delta_s = \frac{30}{20}\left(\frac{1.5708}{30} + 0.014904 - 0.052832\right) = 0.021648 \text{ radians}$$

The coordinates for this trochoidal fillet have been calculated. They are tabulated in Table 5-4 and are plotted in Fig. 5-10.

Problem 5-7. *Given the proportions of the gear and the arc tooth thickness of the pinion-shaped cutter, to determine the radius of a full-rounded tip on the cutter and the form of the fillet produced on the gear.*

The tip of the tooth of the pinion-shaped cutter may be rounded as

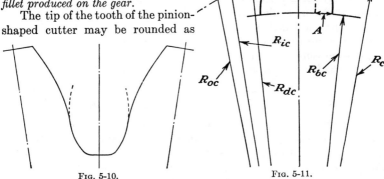

FIG. 5-10.

FIG. 5-11.

shown in Fig. 5-11 in order to obtain a continuously curved bottom land on the gear. We must first determine the radius of this rounding and the

TABLE 5-4. COORDINATES OF TROCHOID OF CORNER OF PINION-SHAPED CUTTER
(Plotted in Fig. 5-10)

ϵ_c, deg	r_t, in.	θ_t, deg	θ''_t, deg	X_t, in.	Y_t, in.
0	8.7500	0.000	1.240	0.1894	8.7480
1	8.7570	0.356	1.596	0.2439	8.7535
2	8.7783	0.704	1.944	0.2978	8.7732
3	8.8134	1.038	2.278	0.3503	8.8064
4	8.8626	1.349	2.589	0.4003	8.8535
5	8.9251	1.631	2.871	0.4471	8.9139
6	9.0008	1.878	3.118	0.4896	8.9875
7	9.0893	2.084	3.324	0.5270	9.0740
8	9.1907	2.245	3.485	0.5587	9.1736
9	9.3040	2.456	3.696	0.5990	9.2847
10	9.4289	2.415	3.655	0.6011	9.4098
11	9.5649	2.415	3.655	0.6098	9.5455
12	9.7166	2.356	3.596	0.6091	9.6925
13	9.8685	2.241	3.481	0.5992	9.8503
14	10.0346	2.064	3.304	0.5783	10.0180

position of the center of this radius.   Referring to Fig. 5-11, when

R_c = pitch radius of pinion-shaped cutter, in.

R_{ic} = radius to top of involute profile on cutter, in.

R_{oc} = outside radius of cutter, in.

R_{bc} = radius of base circle of cutter, in.

A = radius of rounding at tip of cutter tooth, in.

T_c = arc tooth thickness of cutter at pitch radius, in.

T_{ic} = arc tooth thickness of cutter at R_{ic}, in.

ϕ_1 = pressure angle at R_c

ϕ_{ic} = pressure angle at R_{ic}

R_{dc} = radius to center of rounding, in.

ϕ_{dc} = pressure angle at R_{dc}

we have the following from the geometrical conditions shown in Fig. 5-11:

$$\cos \phi_{ic} = R_{bc}/R_{ic} \tag{5-23}$$
$$T_{ic} = 2R_{ic}[(T_c/2R_c) + \operatorname{inv} \phi_1 - \operatorname{inv} \phi_{ic}] \tag{5-24}$$

An exact solution for the value of the radius A may be more complex than its importance justifies.   We can use the two preceding equations as trial solutions and then use a simple approximation for the value of A. Then with this value of A, we can determine new and exact values for R_{ic} and ϕ_{ic}.   Such an approximation is as follows:

$$A = T_{ic}/2 \cos \phi_{ic} \tag{5-25}$$
$$\operatorname{inv} \phi_{dc} = (T_c/2R_c) + \operatorname{inv} \phi_1 - (A/R_{bc}) \tag{5-26}$$
$$R_{dc} = R_{bc}/\cos \phi_{dc} \tag{5-27}$$

To determine the corrected values of R_{ic} and ϕ_{ic} for the selected value of A, we proceed as follows:

$$\tan \phi_{ic} = (R_{bc} \tan \phi_{dc} + A/R_{bc} = \tan \phi_{dc} + (A/R_{bc}) \tag{5-28}$$
$$R_{ic} = R_{bc}/\cos \phi_{ic} \tag{5-29}$$

We already have Eqs. (3-7) and (3-8) for the actual fillet form and Eqs. (3-13) and (3-14) for the trochoidal path of the center of the rounding. Equations (3-16), (3-17), and (3-18) give the value of the tangent to the trochoid.   These equations, using the preceding and following symbols, are as follows:

R_1 = pitch radius of shaped gear, in.

N_1 = number of teeth in gear

N_c = number of teeth in pinion-shaped cutter

C = center distance between axes of gear and cutter, in.

ϵ_c = angle of rotation of cutter

ϵ_1 = angle of rotation of gear

θ_t = vectorial angle of trochoid of center of rounding

r_t = any radius to trochoid of center of rounding, in.

ψ_t = angle between radius vector and tangent to trochoid

δ_t = angle between origin of trochoid and center line of gear tooth

δ_s = angle between origin of trochoid and center line of tooth space

θ_f = vectorial angle of actual fillet form

r_f = any radius to actual fillet form, in.

θ''_f = vectorial angle of actual fillet in reference to selected center line

$$\epsilon_1 = \frac{R_c}{R_1} \epsilon_c \tag{3-12}$$

$$r_t = \sqrt{C^2 + R_{dc}{}^2 - 2CR_{dc} \cos \epsilon_c} \tag{3-13}$$

$$\theta_t = \sin^{-1}\left(\frac{R_{dc} \sin \epsilon_c}{r_t}\right) - \epsilon_1 \tag{3-14}$$

$$\delta_s = 0 \tag{5-30}$$

$$\delta_t = \frac{\pi}{N_1} \tag{5-31}$$

$$\tan \psi_t = r_t \frac{d\theta_t/d\epsilon_c}{dr_t/d\epsilon_c} \tag{3-16}$$

$$\frac{d\theta_t}{d\epsilon_c} = \frac{R_{dc}(r_t{}^2 \cos \epsilon_c - CR_{dc} \sin^2 \epsilon_c)}{r_t{}^2(C - R_{dc} \cos \epsilon_c)} - \frac{R_c}{R_1} \tag{3-17}$$

$$\frac{dr_t}{d\epsilon_c} = \frac{CR_{dc} \sin \epsilon_c}{r_t} \tag{3-18}$$

$$r_f = \sqrt{r_t{}^2 + A^2 - 2Ar_t \sin \psi_t} \tag{3-7}$$

$$\theta_f = \theta_t + \cos^{-1} \frac{r_t - A \sin \psi_t}{r_f} \tag{3-8}$$

$$\theta''_f = \delta \pm \theta_f \tag{1-4}$$

$$X_f = r_f \sin \theta''_f \tag{1-5}$$

$$Y_f = r_f \cos \theta''_f \tag{1-6}$$

Example of Fillet Produced by Full-rounded Pinion-shaped Cutter. As a definite example we shall use the same gear and cutter as before. For this we have the following values:

$$N_1 = 20 \qquad N_c = 30 \qquad R_1 = 10.000 \qquad R_c = 15.000 \qquad \phi_1 = 20°$$
$$T_c = 1.5708 \qquad R_{ic} \text{ (trial)} = 16.250 \qquad R_{bc} = 14.09539 \qquad \phi_{ic} = 29.841°$$
$$\text{inv } \phi_1 = 0.014904 \qquad \text{inv } \phi_{ic} = 0.052832 \qquad \cos \phi_{ic} = 0.86741$$
$$T_{ic} = 32.50 \left(\frac{1.5708}{30} + 0.014904 - 0.052832\right) = 0.46904 \tag{5-24}$$

Trial Solution

$$A = \frac{0.46904}{2 \times 0.86741} = 0.27037 \tag{5-25}$$

$$\text{inv } \phi_{dc} = \frac{0.46904}{30} + 0.014904 - \frac{0.27037}{14.09539} = 0.048083 \tag{5-26}$$

$$\phi_{dc} = 28.985° \qquad \cos \phi_{dc} = 0.87475 \qquad \tan \phi_{dc} = 0.55397$$

$$R_{dc} = \frac{14.09539}{0.87475} = 16.11362 \tag{5-27}$$

We will use these trial values and determine the corrected values for ϕ_{ic} and R_{ic}.

$$\tan \phi_{ic} = 0.55397 + \frac{0.27037}{14.09539} = 0.57315 \qquad (5\text{-}28)$$

$$\phi_{ic} = 29.819° \qquad \cos \phi_{ic} = 0.86760$$

$$R_{ic} = \frac{14.09539}{0.86760} = 16.24641 \qquad (5\text{-}29)$$

The coordinates of the form of this fillet have been calculated. They are tabuated in Table 5-5 and plotted in Fig. 5-12.

Problem 5-8. *Given the arc tooth thickness and pressure angle of an involute gear at one radius, to determine the radius where the tooth becomes pointed.*

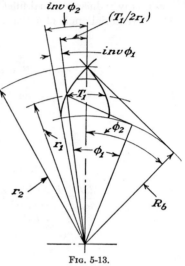

Fig. 5-12.   Fig. 5-13.

Referring to figure (5-13), let

r_1 = given radius of gear, in.

ϕ_1 = pressure angle at r_1

R_b = radius of base circle of gear, in.

T_1 = arc tooth thickness at r_1, in.

r_2 = radius where tooth becomes pointed, in.

ϕ_2 = pressure angle at r_2

In this problem, the arc tooth thickness at r_2 will be equal to zero. Hence

$$\text{inv } \phi_2 = (T_1/2r_1) + \text{inv } \phi_1 \qquad (5\text{-}32)$$

$$R_b = r_1 \cos \phi_1 \qquad (5\text{-}3)$$

$$r_2 = R_b/\cos \phi_2 = r_1 \cos \phi_1/\cos \phi_2 \qquad (5\text{-}2)$$

Example of Radius to Pointed Tip of Tooth. As a definite example we shall use the following values:

$$T_1 = 1.5708 \qquad r_1 = 9.000 \qquad \phi_1 = 20° \qquad \text{inv } \phi_1 = 0.014904$$

$$\text{inv } \phi_2 = \frac{1.5708}{18.00} + 0.014904 = 0.102171$$

$$\phi_2 = 36.422° \qquad \cos \phi_2 = 0.80467$$

$$R_b = 9 \times 0.93969 = 8.45723$$

$$r_2 = \frac{8.45723}{0.80467} = 10.51018 \text{ in.}$$

TABLE 5-5. COORDINATES OF FILLET FORM FROM FULL-ROUNDED TIP OF PINION-
SHAPED CUTTER
(Plotted in Fig. 5-12)

ϵ_c, deg	r_t, in.	θ_t, deg	ψ_t, deg	r_f, in.	θ_f, deg	X_f, in.	Y_f, in.
0	8.8864	0.000	90.000	8.6160	0.000	0.0000	8.6160
1	8.8936	0.312	59.920	8.6603	1.210	0.1829	8.6583
2	8.9140	0.617	39.119	8.7459	1.992	0.3040	8.7406
3	8.9483	0.908	25.966	8.8332	2.483	0.3827	8.8249
4	8.9963	1.178	17.086	8.9206	2.838	0.4417	8.9097
5	9.0575	1.420	10.563	9.0118	3.110	0.4889	8.9985
6	9.1314	1.630	5.432	9.1098	3.322	0.5279	9.0945
7	9.2179	1.800	1.182	9.2163	3.505	0.5634	9.1991
8	9.3170	1.927	− 2.473	9.3326	3.630	0.5908	9.3139
9	9.4279	2.007	− 5.701	9.4586	3.707	0.6116	9.4388
10	9.5502	2.037	− 8.607	9.5944	3.734	0.6248	9.5740
11	9.6834	2.013	−11.258	9.7398	3.707	0.6298	9.7194
12	9.8271	1.932	−13.703	9.8946	3.623	0.6252	9.8749
13	9.9809	1.795	−15.974	10.0586	3.482	0.6109	10.0400
14	10.1438	1.600	−18.095	10.2310	3.283	0.5859	10.2142

Problem 5-9. *Given the arc tooth thickness, radii, and pressure angle of a pair of
mating involute gears, to determine the center
distance at which they will mesh tightly.*

Referring to Fig. 5-14, let

R_1 = radius of first gear where thickness
is known, in.

T_1 = arc tooth thickness of first gear at
R_1, in

ϕ_1 = pressure angle at R_1 and R_2

N_1 = number of teeth in first gear

R_2 = radius of second gear where thickness is known, in.

T_2 = arc tooth thickness of second gear
at R_2, in.

N_2 = number of teeth in second gear

r_1 = pitch radius of first gear when
tightly meshed, in.

t_1 = arc tooth thickness of first gear at
r_1, in.

ϕ_2 = pressure angle at r_1 and r_2

r_2 = pitch radius of second gear when tightly meshed, in.

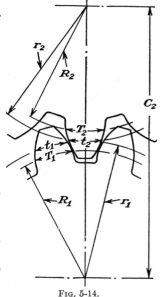

FIG. 5-14.

t_2 = arc tooth thickness of second gear at r_2
C_1 = center distance for pressure angle of ϕ_1, in.
C_2 = center distance when tightly meshed with pressure angle of ϕ_2, in.

From Eq. (5-5) we have

$$t_1 = 2r_1 \left(\frac{T_1}{2R_1} + \text{inv } \phi_1 - \text{inv } \phi_2 \right)$$

$$t_2 = 2r_2 \left(\frac{T_2}{2R_2} + \text{inv } \phi_1 - \text{inv } \phi_2 \right)$$

The sum of these two tooth thicknesses must be equal to the circular pitch at the meshing pitch line. This circular pitch is equal to the quotient of the circumference of either gear pitch circle divided by its number of teeth. Hence

$$t_1 + t_2 = \frac{2\pi r_1}{N_1} = \frac{2\pi r_2}{N_2}$$

We also know that the pitch diameters of two mating gears are directly proportional to their numbers of teeth. Whence we have

$$R_2 = \frac{N_2}{N_1} R_1 \qquad \text{and} \qquad r_2 = \frac{N_2}{N_1} r_1$$

Substituting these values into the equation for t_2, we have

$$t_2 = 2 \frac{N_2}{N_1} r_1 \left(\frac{T_2 N_1}{2N_2 R_1} + \text{inv } \phi_1 - \text{inv } \phi_2 \right)$$

whence

$$t_1 + t_2 = \frac{2\pi r_1}{N_1}$$

$$= 2r_1 \left[\frac{T_1}{2R_1} + \text{inv } \phi_1 - \text{inv } \phi_2 + \frac{N_2}{N_1} \left(\frac{T_2 N_1}{2N_2 R_1} + \text{inv } \phi_1 - \text{inv } \phi_2 \right) \right]$$

Combining terms, simplifying, and solving for inv ϕ_2, we obtain

$$\text{inv } \phi_2 = \frac{N_1(T_1 + T_2) - 2\pi R_1}{2R_1(N_1 + N_2)} + \text{inv } \phi_1 \qquad (5\text{-}33)$$

When the values are changed to the 1-DP values, then

$$2R_1 = N_1$$

and

$$\text{inv } \phi_2 = \frac{T_1 + T_2 - \pi}{N_1 + N_2} + \text{inv } \phi_1 \qquad (5\text{-}34)$$

We already have derived

$$r_1 = R_1 \frac{\cos \phi_1}{\cos \phi_2} \quad \text{and} \quad r_2 = R_2 \frac{\cos \phi_1}{\cos \phi_2}$$

Also we know that

$$R_1 + R_2 = C_1 \quad \text{and} \quad r_1 + r_2 = C_2$$

whence

$$C_2 = C_1 \frac{\cos \phi_1}{\cos \phi_2} \tag{5-35}$$

Example of Center Distance for Two Given Gears. As a definite example we shall use a pair of 6-DP gears, 20-deg nominal pressure angle, of 24 and 36 teeth, with the following given values:

	6 DP	1 DP		6 DP	1 DP
R_1	2.000	12.000	R_2	3.000	18.000
T_1	0.285	1.710	T_2	0.270	1.620
C_1	5.000	30.000			

$$\phi_1 = 20°$$
$$N_1 = 24 \qquad N_2 = 36$$
$$\text{inv } \phi_2 = \frac{1.710 + 1.620 - 3.1416}{60} + 0.014904 = 0.018044$$
$$\phi_2 = 21.268° \qquad \cos \phi_2 = 0.93189$$
$$C_2 = \frac{30 \times 0.93969}{0.93189} = 30.25110 \qquad \text{for 1-DP value}$$
$$C_2 = \frac{30.25110}{6} = 5.04185 \qquad \text{for 6-DP value}$$

Problem 5-10. *Given the arc tooth thickness, radius, and pressure angle of a gear and the proportions of a pinion-shaped cutter, to determine the generating center distance and the root radius of the gear.*

This problem is very similar to Prob. 5-9. We need only introduce symbols for values of the pinion-shaped cutter in place of those for the gear. Thus let

R_1 = radius of gear where arc tooth thickness is known, in.

R_{r1} = root radius of gear, in.

N_1 = number of teeth in gear

ϕ_1 = pressure angle at R_1 and at R_c

R_c = radius of pinion-shaped cutter where arc tooth thickness is known, in.

R_{oc} = outside radius of pinion-shaped cutter, in.

T_c = arc tooth thickness of cutter at R_c, in.

T_1 = arc tooth thickness of gear at R_1, in.

N_c = number of teeth in pinion-shaped cutter
ϕ_2 = generating pressure angle
C_1 = center distance for pressure angle of ϕ_1, in.
C_2 = generating center distance with pressure angle of ϕ_2, in.

From Eq. (5-33) we have

$$\text{inv } \phi_2 = \{[N_1(T_1 + T_c) - 2\pi R_1]/2R_1(N_1 + N_c)\} + \text{inv } \phi_1 \quad (5\text{-}36)$$

Reduced to 1-DP values, this equation becomes

$$\text{inv } \phi_2 = (T_1 + T_c - \pi)/(N_1 + N_c) + \text{inv } \phi_1 \quad (5\text{-}37)$$
$$C_2 = C_1 \cos \phi_1/\cos \phi_2 \quad (5\text{-}35)$$
$$R_{r1} = C_2 - R_{oc} \quad (5\text{-}38)$$

Example of Setting for Pinion-shaped Cutter. As a definite example we shall use the 6-DP, 24-tooth gear from the preceding example, and a 3-in. diameter, 18-tooth, pinion-shaped cutter, which gives the following values:

	6 DP	1 DP		6 DP	1 DP
R_1	2.000	12.000	R_c	1.500	9.000
T_1	0.285	1.710	T_c	0.2618	1.5708
C_1	3.500	21.000	R_{oc}	1.7083	10.250

$$N_1 = 24 \qquad N_c = 18$$
$$\phi_1 = 20°$$
$$\text{inv } \phi_2 = \frac{1.710 + 1.5708 - 3.1416}{42} + 0.014904 = 0.018218$$
$$\phi_2 = 21.334° \qquad \cos \phi_2 = 0.93147$$
$$C_2 = \frac{21 \times 0.93969}{0.93147} = 21.18532 \qquad \text{for 1-DP value}$$
$$C_2 = \frac{21.18532}{6} = 3.53089 \qquad \text{for 6-DP value}$$
$$R_{r1} = 21.18532 - 10.250 = 10.93532 \qquad \text{for 1-DP value}$$
$$R_{r1} = \frac{10.93532}{6} = 1.82255 \qquad \text{for 6-DP value}$$

Problem 5-11. *Given the proportions of a gear and its mating rack, to determine the position of the rack when meshed tightly.*

Let ϕ = pressure angle of rack
p = circular pitch of rack, in.
N_1 = number of teeth in gear
R_1 = radius of gear where pressure angle is ϕ, in.
T_1 = arc tooth thickness of gear at R_1, in.
T_r = thickness of rack tooth at R_1, in.
H = distance from center of gear to nominal pitch line of rack, in.
x = distance between R_1 and nominal pitch line of rack, in.

The nominal pitch line of the rack is the line where the rack-tooth thickness is equal to one-half the circular pitch of the rack.

We know that the circular pitch of the gear at R_1 must be the same as the circular pitch of the rack because the pressure angle at R_1 is the same as that for the rack. We also know that the sum of the arc tooth thickness of the gear at R_1 and the thickness of the rack tooth at that same position must be equal to the circular pitch of the rack. Whence

$$T_r = p - T_1$$
$$x = \frac{(p/2) - T_r}{2 \tan \phi}$$
$$H = R_1 + x$$

Combining these terms, we have

$$H = R_1 + \frac{(T_1/2) - (p/4)}{\tan \phi} \tag{5-39}$$

When this equation is reduced to the 1-DP values, it becomes

$$H = \frac{N_1}{2} + \frac{(T_1/2) - 0.7854}{\tan \phi} \tag{5-40}$$

Example of Position of Mating Rack. As a definite example we shall use the following: a 5-DP, 20-deg rack and an 18-tooth gear with an arc tooth thickness of 0.325 in. at 1.800 in. radius. This gives the following values:

	5 DP	1 DP
p	0.62832	3.1416
R_1	1.800	9.000
T_1	0.325	1.625

$$N_1 = 18$$
$$\phi = 20° \qquad \tan \phi = 0.36397$$

Using the 1-DP values, we obtain

$$H = 9 + \frac{0.8125 - 0.7854}{0.36397} = 9.07446 \qquad \text{for 1-DP value}$$

$$H = \frac{9.07446}{5} = 1.81489 \qquad \text{for 5-DP value}$$

Problem 5-12. *Given the proportions of an involute gear, to determine the proportions of a mating rack of different circular pitch.*

Let R_1 = given radius of gear, in.

ϕ_1 = pressure angle of gear at R_1

T_1 = arc tooth thickness of gear at R_1, in.

R_o = outside radius of gear, in.

R_r = root radius of gear, in.

R_b = radius of base circle of gear, in.

c = clearance, in.

p_1 = circular pitch of gear at R_1, in.

p_2 = circular pitch of rack, in.

ϕ_2 = pressure angle of rack

p_b = base pitch of gear and rack, in.

H = distance from center of gear to nominal pitch line of rack, in.

a_r = addendum of rack, in. (from nominal pitch line)

b_r = dedendum of rack, in. (from nominal pitch line)

We know that the base pitch of a pair of mating involute gears, or of a mating involute gear and rack, must be identical. Hence

$$p_b = p_1 \cos \phi_1 = p_2 \cos \phi_2$$

whence

$$\cos \phi_2 = \frac{p_1 \cos \phi_1}{p_2} \tag{5-41}$$

We already have

$$T_2 = 2R_2 \left(\frac{T_1}{2R_1} + \text{inv } \phi_1 - \text{inv } \phi_2 \right) \tag{5-5}$$

$$R_2 = \frac{R_1 \cos \phi_1}{\cos \phi_2} = \frac{R_b}{\cos \phi_2} \tag{5-2}$$

where T_2 = arc tooth thickness of gear at pressure angle of ϕ_2, in.

R_2 = radius of gear where pressure angle is ϕ_2, in.

If we let

T_{r2} = tooth thickness of rack at radius B_2, in.

x = distance between R_2 and nominal pitch line of rack, in.

then

$$T_{r2} = p_2 - T_2$$

$$x = \frac{(p_2/2) - T_{r2}}{2 \tan \phi_2}$$

Combining these expressions, we obtain

$$H = R_2 + \frac{R_2[(T_1/2R_1) + \text{inv } \phi_1 - \text{inv } \phi_2] - (p_2/4)}{\tan \phi_2} \tag{5-42}$$

$$a_r = H - R_r - c \tag{5-43}$$

$$b_r = R_0 + c - H \tag{5-44}$$

Example of Special-pitch Rack. As a definite example, we shall use a standard 12-DP, 24-tooth gear of 20-deg full-depth form, which must mesh with a rack whose circular pitch is 0.250 in. This gives the following values:

	12 DP	1 DP		12 DP	1 DP
R_1	1.000	12.000	c	0.0131	0.157
T_1	0.1309	1.5708	p_1	0.2618	3.1416
R_o	1.0833	13.000	p_2	0.250	3.0000
R_r	0.9036	10.8430			

$$\phi_1 = 20°$$
$$N_1 = 24$$

Using the 1-DP values, we obtain

$$\cos \phi_2 = \frac{3.1416 \times 0.93969}{3.000} = 0.98404$$

$$\phi_2 = 10.250° \quad \text{inv } \phi_2 = 0.001933 \quad \tan \phi_2 = 0.18083$$

$$R_2 = \frac{12 \times 0.93969}{0.98404} = 11.45917$$

$$H = 11.45917 + \frac{(11.45917 \times 0.078421) - 0.7854}{0.18083} = 12.08539 \text{ for 1 DP}$$

$$a_r = 12.08539 - 10.843 - 0.157 = 1.08539 \qquad \text{for 1 DP}$$
$$b_r = 13.000 + 0.157 - 12.08539 = 1.07161 \qquad \text{for 1 DP}$$

Problem 5-13. *Given the center distance and numbers of teeth of a pair of mating involute gears, and the tooth proportions of the hob (basic-rack form), to determine the tooth proportions of the gears.*

When N_1 = number of teeth in first gear

N_2 = number of teeth in second gear

R_1 = pitch radius of first gear, in.

R_2 = pitch radius of second gear, in.

R_{o1} = outside radius of first gear, in.

R_{o2} = outside radius of second gear, in.

R_{r1} = root radius of first gear, in.

R_{r2} = root radius of second gear, in.

b_1 = dedendum of first gear, in.

b_2 = dedendum of second gear, in.

h_t = whole depth of tooth, in.

c = clearance, in.

ϕ_1 = pressure angle of hob

ϕ_2 = pressure angle of mating gears

p = circular pitch of hob, in.

a_h = nominal addendum of hob, in.

C_1 = center distance of gears with pressure angle of ϕ_1, in.

C_2 = meshing center distance of gears, pressure angle ϕ_2, in.

P = diametral pitch of hob

we know that

$$C_1 = (N_1 + N_2)/2P \qquad (5\text{-}45)$$
$$C_2 = C_1 \cos \phi_1/\cos \phi_2 \qquad (5\text{-}35)$$

Whence

$$\cos \phi_2 = C_1 \cos \phi_1/C_2 \qquad (5\text{-}46)$$

When T_1 = arc tooth thickness of first gear where pressure angle is ϕ_1, in.

T_2 = arc tooth thickness of second gear where pressure angle is ϕ_1, in.

t_1 = arc tooth thickness of first gear where pressure angle is ϕ_2, in.

t_2 = arc tooth thickness of second gear where pressure angle is ϕ_2, in.

we know that

$$t_1 + t_2 = 2\pi C_2/(N_1 + N_2)$$

and

$$T_1 + T_2 = 2C_1\{[(t_1 + t_2)/2C_2] + \text{inv } \phi_2 - \text{inv } \phi_1\}$$

We also know that the sum of the tooth thicknesses of the hob teeth at the points where they mesh with the gears while they are being generated is equal to $2p - (T_1 + T_2)$.

When x = sum of the distances from the generating pitch circles to the nominal pitch line of the hob, in.

$$x = \frac{p - [2p - (T_1 + T_2)]}{2 \tan \phi_1} = \frac{T_1 + T_2 - p}{2 \tan \phi_1}$$

whence

$$R_{r1} + R_{r2} = C_1 + x - 2a_h$$

Introducing the values of x and $T_1 + T_2$ into the foregoing equation, we obtain

$$R_{r1} + R_{r2} = C_1 - 2a_h + \frac{2C_1\{[\pi/(N_1 + N_2)] + \text{inv } \phi_2 - \text{inv } \phi_1\} - p}{2 \tan \phi_1}$$

$$(5\text{-}47)$$

Reducing this equation to the 1-DP values, we obtain

$$R_{r1} + R_{r2} = C_1 - 2a_h + \frac{C_1(\text{inv } \phi_2 - \text{inv } \phi_1)}{\tan \phi_1} \qquad (5\text{-}48)$$

It is thus apparent that the sum of the root radii of these meshing gears is a constant, whatever size we may wish to make either radius. The next step is to select values for these radii, which involves a choice of tooth proportions.

Under some circumstances the outside radius of one gear may need to be some specific value. In such a case, we first determine the whole depth of tooth and subtract it from the fixed outside radius. This gives the value for one root radius. The root radius of the other gear would then be the remainder of the sum of the two. It is necessary in such cases to check for conditions of undercut.

In most cases, however, there will be no definite restrictions on the sizes of the gears. Under these conditions, the tooth proportions should be established in relation to the base circles rather than in relation to the actual pitch circles. The dedendum of the smaller gear should be less than the dedendum of the larger gear. When the numbers of teeth are large—about 40 or more for the smallest gear with a pressure angle of 14½ deg and 30 or more with a pressure angle of 20 deg—then the dedenda of the two gears may be the same. When smaller numbers of teeth are involved, the following equation may be used for the values of the dedenda:

$$b_1 = \frac{C_2 - (R_{r1} + R_{r2})}{1 + \sqrt{N_2/N_1}} \tag{5-49}$$

$$b_2 = C_2 - (R_{r1} + R_{r2}) - b_1 = \frac{C_2 - (R_{r1} + R_{r2})}{1 + \sqrt{N_1/N_2}} \tag{5-50}$$

We must next establish the value for the whole depth of tooth. The expression $C_2 - (R_{r1} + R_{r2})$ is equal to the sum of the dedenda of the mating gears, which is also equal to the whole depth plus the clearance. Whence

$$C_2 - (R_{r1} + R_{r2}) = h_t + c$$

It is generally best to make the clearance proportionately the same as that of the standard tooth form, which is represented by the form of the hob teeth. Thus when

h_{t1} = nominal or standard whole depth of tooth, in.

c_1 = nominal or standard clearance, in.

$$h_{t1}/(h_{t1} + c_1) = h_t/(h_t + c)$$
$$h_t = [h_{t1}/(h_{t1} + c_1)](h_t + c)$$

Substituting the value of $(h_t + c)$ into this last equation, we get

$$h_t = [h_{t1}/(h_{t1} + c_1)][C_2 - (R_{r1} + R_{r2})] \tag{5-51}$$

As the actual pitch radii are directly proportional to the numbers of teeth in the gears, we have

$$R_1 = N_1 C_2/(N_1 + N_2) \tag{5-52}$$
$$R_2 = N_2 C_2/(N_1 + N_2) \tag{5-53}$$

The following relationships should be self-evident:

$$R_{r1} = R_1 - b_1 \qquad (5\text{-}54)$$
$$R_{r2} = R_2 - b_2 \qquad (5\text{-}55)$$
$$R_{o1} = R_{r1} + h_t \qquad (5\text{-}56)$$
$$R_{o2} = R_{r2} + h_t \qquad (5\text{-}57)$$

Example of Hobbed-gear Design. As a definite example we shall use the following: 8-DP hob, 14½-deg pressure angle, gears with 18 and 30 teeth, to run at a center distance of 3.100 in. This gives the following values:

	8 DP	1 DP		8 DP	1 DP
p	0.3927	3.1416	C_2	3.100	24.800
a_h	0.1446	1.1570	C_1	3.000	24.000
h_{t1}	0.2696	2.1570	c_1	0.0196	0.1570

$$N_1 = 18 \qquad N_2 = 30$$
$$\phi_1 = 14.500° \qquad \text{inv } \phi_1 = 0.005545 \qquad \cos \phi_1 = 0.96815 \qquad \tan \phi_1 = 0.25862$$

Using the 1-DP values, we obtain

$$\cos \phi_2 = \frac{24 \times 0.96815}{24.800} = 0.93692$$

$$\phi_2 = 20.460° \qquad \text{inv } \phi_2 = 0.015995$$

$$R_{r1} + R_{r2} = 24.00 - 2.314 + \frac{24(0.015995 - 0.005545)}{0.25862} = 22.65576$$

$$\frac{h_{t1}}{h_{t1} + c_1} = \frac{2.157}{2.314} = 0.93215$$

The following values are for 1 DP:

$$h_t = 0.93215(24.800 - 22.65576) = 1.99875$$
$$b_1 = \frac{24.800 - 22.65576}{1 + \sqrt{30/18}} = \frac{2.14424}{1 + 1.2910} = 0.93594$$
$$b_2 = 2.14424 - 0.93594 = 1.20830$$
$$R_1 = \frac{18 \times 24.800}{48} = 9.300$$
$$R_2 = \frac{30 \times 24.800}{48} = 15.500$$
$$R_{r1} = 9.300 - 0.93594 = 8.36406$$
$$R_{r2} = 15.50 - 1.2083 = 14.29170$$
$$R_{o1} = 8.36406 + 1.99875 = 10.36281$$
$$R_{o2} = 14.2917 + 1.99875 = 16.29045$$

These values would be divided by 8 to obtain the values for 8 DP. The coordinates of these tooth profiles have been calculated. They are plotted in Fig. 5-15. These tooth proportions keep the active profiles of the teeth of both gears in the space between the two base circles.

Problem 5-14. *Given the center distance and numbers of teeth of a pair of mating involute gears and the proportions of a pinion-shaped cutter, to determine the tooth proportions of the gears.*

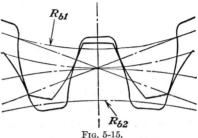

Fig. 5-15.

The best solution of this problem is in four steps, as follows:

1. Determine the tooth proportions as though the gears were hobbed from the basic-rack form of the pinion-shaped cutter.

2. Determine the arc tooth thickness of the gears at the initial pressure angle (pressure angle of basic rack and pinion-shaped cutter).

3. Determine the center distances of generation of the two gears with the pinion-shaped cutter.

4. Determine the tooth heights and clearances and the final proportions of the gears.

First Step. For the first step we have the following from Prob. 5-13:

N_1 = number of teeth in first gear

N_2 = number of teeth in second gear

R_1 = pitch radius of first gear, pressure angle of ϕ_2, in.

R_2 = pitch radius of second gear, pressure angle of ϕ_2, in.

R_{r1} = hobbed root radius of first gear, in.

R_{r2} = hobbed root radius of second gear, in.

b_1 = hobbed dedendum of first gear, in.

b_2 = hobbed dedendum of second gear, in.

ϕ_1 = pressure angle of hob and of pinion-shaped cutter

ϕ_2 = pressure angle of mating gears

a_h = nominal addendum of hob and of pinion-shaped cutter, in.

C_1 = center distance of gears with pressure angle of ϕ_1, in.

C_2 = center distance of meshing gears with pressure angle of ϕ_2, in.

P = diametral pitch of hob and of pinion-shaped cutter

$$C_1 = \frac{N_1 + N_2}{2P} \tag{5-45}$$

$$\cos \phi_2 = \frac{C_1 \cos \phi_1}{C_2} \tag{5-35}$$

$$R_{r1} + R_{r2} = C_1 - 2a_h + \frac{C_1(\text{inv } \phi_2 - \text{inv } \phi_1)}{\tan \phi_1} \tag{5-48}$$

$$b_1 = \frac{C_2 - (R_{r1} + R_{r2})}{1 + \sqrt{N_2/N_1}} \tag{5-49}$$

$$b_2 = C_2 - (R_{r1} + R_{r2}) - b_1 = \frac{C_2 - (R_{r1} + R_{r2})}{1 + \sqrt{N_1/N_2}} \qquad (5\text{-}50)$$

$$R_1 = \frac{N_1 C_2}{N_1 + N_2} \qquad (5\text{-}52)$$

$$R_2 = \frac{N_2 C_2}{N_1 + N_2} \qquad (5\text{-}53)$$

$$R_{r1} = R_1 - b_1 \qquad (5\text{-}54)$$

$$R_{r2} = R_2 - b_2 \qquad (5\text{-}55)$$

We use here the 1-DP values for calculation.

Second Step. For the second step, in order to determine the arc tooth thicknesses of the gear teeth at their generating radii in relation to the hob, we proceed as follows:

When R_{g1} = hob-generating radius of first gear, in.

R_{g2} = hob-generating radius of second gear, in.

T_1 = arc tooth thickness of first gear at R_{g1}, in.

T_2 = arc tooth thickness of second gear at R_{g1}, in.

p = circular pitch of hob, in.

and all other symbols are the same as before,

$$R_{g1} = N_1/2P \qquad R_{g2} = N_2/2P$$

$$T_1 = (p/2) + 2 \tan \phi_1 (R_{r1} + a_h - R_{g1}) \qquad (5\text{-}58)$$

$$T_2 = (p/2) + 2 \tan \phi_1 (R_{r2} + a_h - R_{g2}) \qquad (5\text{-}59)$$

Third Step. For the third step, to determine the generating center distances between each gear and the pinion-shaped cutter, we have from Prob. 5-10 the following:

When R_{oc} = outside radius of pinion-shaped cutter, in.

R_c = pitch radius of pinion-shaped cutter where pressure angle is ϕ_1, in.

T_c = arc tooth thickness of pinion-shaped cutter at R_c, in.

N_c = number of teeth in pinion-shaped cutter

ϕ_3 = generating pressure angle for first gear

ϕ_4 = generating pressure angle for second gear

C_3 = center distance for first gear and cutter with pressure angle of ϕ_1, in.

C_4 = center distance for second gear and cutter with pressure angle of ϕ_1, in.

C_{g1} = generating center distance of first gear and cutter with pressure angle of ϕ_3, in.

C_{g2} = generating center distance of second gear and cutter with pressure angle of ϕ_4, in.

R'_{r1} = root radius of first gear from pinion-shaped cutter, in.
R'_{r2} = root radius of second gear from pinion-shaped cutter, in.
and all other symbols are the same as before, we have, using the 1-DP values,

$$\text{inv } \phi_3 = (T_1 + T_c - \pi)/(N_1 + N_c) + \text{inv } \phi_1 \tag{5-37}$$
$$C_{g1} = C_3 \cos \phi_1/\cos \phi_3 \tag{5-35}$$
$$R'_{r1} = C_{g1} - R_{oc} \tag{5-38}$$
$$\text{inv } \phi_4 = (T_2 + T_c - \pi)/(N_2 + N_c) + \text{inv } \phi_1 \tag{5-37}$$
$$C_{g2} = C_4 \cos \phi_1/\cos \phi_4 \tag{5-35}$$
$$R'_{r2} = C_{g2} - R_{oc} \tag{5-38}$$

Fourth Step. For the fourth step, to determine the whole depth of tooth, we proceed as in Prob. 5-13 as follows:
When h_{t1} = nominal or standard whole depth of tooth, in.
c_1 = nominal or standard clearance, in.

$$h'_t = [h_{t1}/(h_{t1} + c_1)][C_2 - (R'_{r1} + R'_{r2})] \tag{5-51}$$

and when
R'_{o1} = outside radius of first gear, in.
R'_{o2} = outside radius of second gear, in.

$$R'_{o1} = R'_{r1} + h'_t \tag{5-56}$$
$$R'_{o2} = R'_{r2} + h'_t \tag{5-56}$$

Example of Shaped-gear Design. As a definite example we shall use the following: 10-DP, gears of 20-deg full-depth form, with 15 and 40 teeth, cut with 3-in. diameter pinion-shaped cutter, to run at a center distance of 2.800 in. This gives the following values:

	10 DP	1 DP		10 DP	1 DP
p	0.31416	3.1416	C_1	2.750	27.500
a_h	0.125	1.250	R_c	1.500	15.000
h_{t1}	0.225	2.250	R_{oc}	1.625	16.250
c_1	0.025	0.250	T_c	0.15708	1.5708
C_2	2.800	28.000			

$$N_1 = 15 \qquad N_2 = 40 \qquad N_c = 30$$
$$\phi_1 = 20° \qquad \cos \phi_1 = 0.93969 \qquad \text{inv } \phi_1 = 0.014904 \qquad \tan \phi_1 = 0.36397$$

Using the 1-DP values, we have

$$\cos \phi_2 = \frac{27.500 \times 0.93969}{28.00} = 0.92291$$
$$\phi_2 = 22.645° \qquad \text{inv } \phi_2 = 0.021952$$

$$R_{r1} + R_{r2} = 27.50 - 2.50 + \frac{27.50(0.021952 - 0.014904)}{0.36397} = 25.53251$$

$$b_1 = \frac{28.00 - 25.53251}{1 + \sqrt{40/15}} = \frac{2.46749}{1 + 1.63299} = 0.93714$$

$$b_2 = 2.46749 - 0.93714 = 1.53035$$

$$R_1 = \frac{15 \times 28}{55} = 7.63636$$

$$R_2 = \frac{40 \times 28}{55} = 20.36364$$

$$T_1 = 1.5708 + (2 \times 0.36397)(6.69922 + 1.250 - 7.50) = 1.89781$$

$$T_2 = 1.5708 + (2 \times 0.36397)(18.83329 + 1.250 - 20.0) = 1.63143$$

$$\text{inv } \phi_3 = \frac{1.89781 + 1.5708 - 3.1416}{45} + 0.014904 = 0.022170$$

$$\phi_3 = 22.717° \qquad \cos \phi_3 = 0.92243 \qquad c_3 = {}^{45}/_2 = 22.500$$

$$C_{\varrho 1} = \frac{22.500 \times 0.93969}{0.92243} = 22.92125$$

$$R'_{r1} = 22.92125 - 16.250 = 6.67125$$

$$\text{inv } \phi_4 = \frac{1.63143 + 1.5708 - 3.1416}{70} = 0.014904 = 0.015770$$

$$\phi_4 = 20.367° \qquad \cos \phi_4 = 0.93749 \qquad C_4 = {}^{70}/_2 = 35.000$$

$$C_{\varrho 2} = \frac{35 \times 0.93969}{0.93749} = 35.08213$$

$$R'_{r2} = 35.08213 - 16.250 = 18.83213$$

$$h'_t = \frac{2.25}{2.50}[28 - (6.67125 + 18.83213)] = 2.24696$$

$$R'_{o1} = 6.67125 + 2.24696 = 8.91821$$

$$R'_{o2} = 18.83213 + 2.24696 = 21.07909$$

These are the 1-DP values. For the 10-DP values, these dimensions would be divided by 10. The coordinates of these gear-tooth profiles have been calculated. They are plotted in Fig. 5-16.

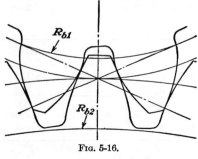

FIG. 5-16.

Problem 5-15. *Given the arc tooth thickness and pressure angle of an involute gear at a specified radius, to determine the position of a wire or roll placed in the tooth space.*

This analysis was originally made by Ernest Wildhaber. Referring to Fig. 5-17, let

R_b = radius of base circle, in.
R_1 = radius at which tooth thickness is known, in.
ϕ_1 = pressure angle at R_1
T_1 = arc tooth thickness at R_1, in.
W = radius of measuring wire or roll, in.

r_2 = radius to center of roll, in.
N = number of teeth in gear
ϕ_2 = pressure angle at r_2

The angle in radians from the center of the tooth to the center of the roll is equal to π/N.

The angle in radians from the center of the tooth to the origin of the involute tooth profile is equal to $(T_1/2R_1) + \text{inv } \phi_1$.

Another involute curve is shown in Fig. 5-17 as a dotted line. This dotted involute passes through the center of the measuring roll. The angle from the center of the tooth, in radians, to the origin of this dotted involute curve is equal to $(T_1/2R_1) + \text{inv } \phi_1 + (W/R_b)$.

The angle in radians from the origin of the dotted involute to the radial line of the gear that passes

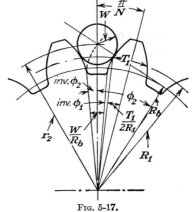

Fig. 5-17.

through the center of the roll is the involute function of ϕ_2. Whence

$$\text{inv } \phi_2 = (T_1/2R_1) + \text{inv } \phi_1 + (W/R_b) - (\pi/N) \qquad (5\text{-}60)$$
$$r_2 = R_b/\cos \phi_2 \qquad (5\text{-}2)$$

Example of Radius to Center of Measuring Roll. As a definite example we shall use the following: 6-DP, 24-tooth gear, of 20-deg pressure angle, with a tooth thickness of 0.2618 in. at 2.00 in. radius. This gives the following values:

$$R_1 = 2.000 \qquad W = 0.150 \qquad N = 24 \qquad \phi_1 = 20° \qquad T_1 = 0.2618$$
$$\text{inv } \phi_1 = 0.014904 \qquad \cos \phi_1 = 0.93969 \qquad R_b = 1.87938$$
$$\text{inv } \phi_2 = \frac{0.2618}{4.000} + 0.014904 + \frac{0.150}{1.87938} - \frac{3.1416}{24} = 0.029276$$
$$\phi_2 = 24.812° \qquad \cos \phi_2 = 0.90769$$
$$r_2 = \frac{1.87938}{0.90769} = 2.07051$$

Problem 5-16. *Given the arc tooth thickness and pressure angle of an involute gear at a definite radius, to determine the measurement over rolls placed in the tooth space.*

Here we have one of two possible conditions. When the number of teeth in the gear is even, the rolls will be diametrically opposite to each other; then the measurement over the rolls is equal to twice the sum of the radius to the center of the roll and the radius of the roll. When the number of teeth is odd, then we must determine the off-center angle of the radial lines through the centers of the rolls, and calculate accordingly.

Let M_1 = measurement over rolls, even number of teeth, in.

M_2 = measurement over rolls, odd number of teeth, in.

and all other symbols be the same as those in Prob. 5-15.

Even Number of Teeth

$$M_1 = 2(r_2 + W) \qquad (5\text{-}61)$$

Any size roll may be used provided that it makes contact on the involute profiles, and also extends beyond the outside circle of the gear.

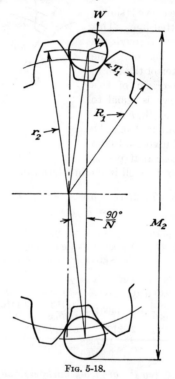

Fig. 5-18.

For the coarser pitches, it does not need to extend beyond the tips of the teeth if there is room in the tooth space for the measuring anvils of the micrometer. For gears of conventional proportions, the following equation gives a reasonable size for the measuring roll:

$$W = 0.840/P \text{ or slightly larger}$$

where P = diametral pitch.

Example of Measurement over Rolls—Even Number of Teeth. As a definite example we shall use the following: 30-tooth, 6-DP, 14½-deg tooth form of conventional proportions, where the arc tooth thickness at 2.500 in. radius is equal to 0.2618 in. This gives the following values:

$$R_1 = 2.500 \quad N = 30 \quad \phi_1 = 14.500° \quad R_b = 2.42037$$
$$T_1 = 0.2618 \quad P = 6 \quad \text{inv } \phi_1 = 0.005545 \quad \cos \phi_1 = 0.96815$$
$$W = \frac{0.84}{6} = 0.140$$

$$\text{inv } \phi_2 = \frac{0.2618}{5.000} + 0.005545 + \frac{0.140}{2.42037} - \frac{3.1416}{30} = 0.011027$$
$$\phi_2 = 18.144° \quad \cos \phi_2 = 0.95028$$
$$r_2 = \frac{2.42037}{0.95028} = 2.54701$$
$$M_1 = 2(2.54701 + 0.140) = 5.37402$$

Odd Number of Teeth. When the number of teeth is odd, the tooth spaces are not diametrically opposite to each other, as shown in Fig. 5-18. In these cases, the triangle as indicated in the figure must be solved to obtain the measurement over the rolls. Hence for odd numbers of teeth we have

$$M_2 = 2\{r_2[\cos (90°/N)] + W\} \tag{5-62}$$

Example of Measurement over Rolls—Odd Number of Teeth. As a definite example we shall use the following: 31-tooth, 6-DP, 14½ deg tooth form of conventional proportions, where the arc tooth thickness at 2.59333 in. radius is equal to 0.2618 in. This gives the following values:

$$R_1 = 2.58333 \quad N = 31 \quad \phi_1 = 14.500° \quad R_b = 2.50105 \quad \cos \phi_1 = 0.96815$$
$$T_1 = 0.2618 \quad P = 6 \quad \text{inv } \phi_1 = 0.005545 \quad (90°/N) = 2.903°$$
$$\cos (90°/N) = 0.99872 \quad W = 0.140$$
$$\text{inv } \phi_2 = \frac{0.2618}{5.16667} + 0.005545 + \frac{0.140}{2.50105} - \frac{3.1416}{31} = 0.010850$$
$$\phi_2 = 18.048° \quad \cos \phi_2 = 0.95080$$
$$r_2 = \frac{2.50105}{0.95080} = 2.63047$$
$$M_2 = 2[(2.63047 \times 0.99872) + 0.140] = 5.53420$$

When the radius of the roll and the measurement over the rolls are known, then the foregoing equations may be readily rearranged to solve for the arc tooth thickness of the gear tooth.

CHAPTER 6

INVOLUTOMETRY OF INTERNAL GEARS

We shall now turn our attention to the internal involute gears. The involute form of the gear-tooth profiles is the same as that for the spur gears, but the tooth form of the internal gear is that of the tooth space of the spur gear. In other words, contact is made on the inside or concave side of the involute curve of the internal gear instead of on the outside or convex side of this curve as on spur gears. Also the root radius of the internal gear is its largest radius, and the tips of the teeth are at its smallest radius. The fillet joins the involute at its greatest pressure angle instead of at its smallest pressure angle as is the case with spur gears. The mating spur pinion must operate inside the internal gear. This condition imposes several possible sources of interference that are not present with mating spur gears. The center distance is equal to the difference between the pitch radii instead of being equal to their sum as for spur gears.

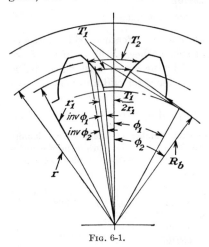

Fig. 6-1.

The further consideration of this subject will be in the form of individual problems as before.

Problem 6-1. *Given the arc tooth thickness and pressure angle of an internal involute gear at a definite radius, to determine the coordinates of the involute profile.*

Referring to Fig. 6-1, when

T_1 = given arc tooth thickness, in.

r_1 = given radius of profile, in.

ϕ_1 = given pressure angle at r_1

r_2 = any radius of profile, in.

T_2 = arc tooth thickness at r_2, in.

112

ϕ_2 = pressure angle at r_2

R_b = radius of base circle, in.

we have from the spur-gear analysis

$$R_b = r_1 \cos \phi_1 \qquad (5\text{-}3)$$
$$\cos \phi_2 = R_b/r_2 \qquad (5\text{-}4)$$

As the tooth form is symmetrical, we shall use the half thickness as before. The half thickness of the tooth at r_1, in radians, is equal to $T_1/2r_1$.

The half thickness of the tooth—extended it necessary—at the base circle is equal to $(T_1/2r_1) - \text{inv } \phi_1$.

The half thickness of the tooth, in radians, at any other radius r_2 is equal to its half thickness in radians at the base circle plus the involute function of the pressure angle at that radius. Whence we have

$$T_2/2r_2 = (T_1/2r_1) - \text{inv } \phi_1 + \text{inv } \phi_2 \qquad (6\text{-}1)$$
$$T_2 = 2r_2[(T_1/2r_1) - \text{inv } \phi_1 + \text{inv } \phi_2] \qquad (6\text{-}2)$$

Cartesian Coordinates.   To obtain the Cartesian coordinates of the involute profile in reference to either the center line of the tooth or the center line of the space, we first determine the vectorial angles of the several points of the profile from the specified center line, and then we use Eq. (1-5) and (1-6) as before.

Let θ'' = vectorial angle of profile from specified center line

X = abscissa of profile, in.

Y = ordinate of profile, in.

When θ'' is the vectorial angle from the center line of the tooth, then

$$\theta'' = T_2/2r_2 = (T_1/2r_1) - \text{inv } \phi_1 + \text{inv } \phi_2 \qquad (6\text{-}3)$$

When θ'' is the vectorial angle from the center line of the space, then

$$\theta'' = (\pi/N) - (T_2/2r_2) \qquad (6\text{-}4)$$
$$X = r_2 \sin \theta'' \qquad (1\text{-}5)$$
$$Y = r_2 \cos \theta'' \qquad (1\text{-}6)$$

Example of Internal-gear Involute Profile.   As a definite example we shall use the following: 40-tooth, 5-DP internal gear of 20-deg pressure angle, whose arc tooth thickness at a radius of 4 in. is equal to one-half the circular pitch. This gives the following values:

	5 DP	1 DP
r_1	4.000	20.000
T_1	0.31416	1.5708

$\phi_1 = 20°$ $\cos \phi_1 = 0.93969$ $\text{inv } \phi_1 = 0.014904$

Using the 1-DP values, we have

$$R_b = 20 \times 0.93969 = 18.7938$$

$$\cos \phi_2 = \frac{18.7938}{r_2}$$

$$\frac{T_2}{2r_2} = \left(\frac{1.5708}{40}\right) - 0.014904 + \text{inv } \phi_2 = 0.024366 - \text{inv } \phi_2$$

Using a series of values of r_2 ranging from 19 to 21 in. (and up to the radius where the fillet joins the involute curve), we obtain the values tabulated in Table 6-1. These values are also plotted in Fig. 6-2. In this example, the values are determined in reference to the center line of the tooth space.

The tooth form also includes the fillet that joins the involute curve to the root circle. The shape of this fillet will depend upon the size and form

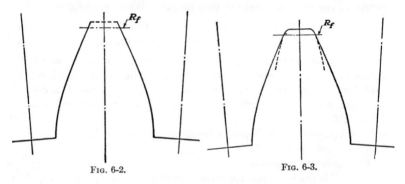

FIG. 6-2.   FIG. 6-3.

of the generating tool that is used to cut the internal gear. Pinion-shaped cutters are used almost exclusively to generate internal involute gears.

Before we determine the form of the fillet, we will first determine the radius to the point where the fillet joins the involute profile. This is the point of tangency of the two curves.

Problem 6-2. *Given the proportions of an internal involute gear and the size and position of the pinion-shaped cutter, to determine the radius to the point where the fillet joins the involute profile.*

The radius to the point where the fillet joins the involute curve will be at the point where the maximum radius of the involute profile of the pinion-shaped cutter intersects the path of contact, as shown in Fig. 6-3. When the corner of the cutter tooth is sharp, this maximum radius of the involute profile will be at the outside radius of the cutter. When the cutter tooth has a rounded corner or tip, this maximum radius of the involute profile will be at the radius where the rounding joins the involute

TABLE 6-1. COORDINATES OF INVOLUTE PROFILE OF INTERNAL GEAR
(Plotted in Fig. 6-2)

r_2, in.	ϕ_2, deg	inv ϕ_2	$T_2/2r_2$	θ'', rad	X, in.	Y, in.
18.79380	0.000	0.000000	0.024366	0.054174	1.0177	18.7662
19.000	8.449	0.001078	0.025444	0.053096	1.0083	18.9732
19.200	11.807	0.002967	0.027333	0.051207	0.9827	19.1748
19.400	14.361	0.005384	0.029750	0.048790	0.9459	19.3769
19.600	16.490	0.008219	0.032585	0.045955	0.9004	19.5792
19.800	18.345	0.011409	0.035775	0.042765	0.8464	19.7820
20.000	20.000	0.014904	0.039270	0.039270	0.7852	19.9846
20.200	21.505	0.018678	0.043044	0.035496	0.7169	20.1873
20.400	22.888	0.022699	0.047065	0.031475	0.6418	20.3900
20.600	24.172	0.026949	0.051315	0.027225	0.5607	20.5924
20.800	25.372	0.031411	0.055777	0.022063	0.4588	20.7950
21.000	26.501	0.036074	0.060440	0.018100	0.3801	20.9966
21.10627	27.072	0.038614	0.062980	0.015560	0.3286	21.1037

profile of the cutter. In all cases, the resulting fillet is tangent to the involute profile of the internal gear.

When R_{oc} = outside radius or pinion-shaped cutter, in. (or maximum radius of involute profile on cutter)

ϕ_1 = pressure angle at pitch or generating radius

C = center distance between axes of gear and cutter, in.

R_{b2} = radius of base circle of internal gear, in.

ϕ_{oc} = pressure angle at R_{oc}

R_f = radius to point where fillet joins involute, in.

$$R_f = \sqrt{(C \sin \phi_1 + R_{oc} \sin \phi_{oc})^2 + R_{b2}^2} \qquad (6-5)$$

Example of Radius to Intersection of Fillet and Involute. We shall use the values from the preceding example with a 20-tooth cutter. This gives the following values for 1 DP:

$$C = 10.000 \qquad R_{oc} = 11.250 \qquad R_{b2} = 18.7938 \qquad \phi_{oc} = 33.355°$$
$$\phi_1 = 20° \qquad \sin \phi_1 = 0.34202 \qquad \sin \phi_{oc} = 0.54982$$

Whence

$$R_f = \sqrt{(9.60367)^2 + (18.7938)^2} = 21.10627$$

The position of this point is indicated in Fig. 6-2.

Problem 6-3. *Given the proportions of a pinion-shaped cutter with sharp corners at tips of teeth and its generating center distance, to determine the coordinates of the trochoidal fillet developed on a given internal gear.*

The form of this fillet is the trochoid whose equations have already been derived in Chap. 3.

When N_2 = number of teeth in internal gear

N_c = number of teeth in pinion-shaped cutter

R_{oc} = outside radius of pinion-shaped cutter, in.

R_2 = pitch radius of internal gear, in.

R_c = pitch radius of pinion-shaped cutter, in.

C = center distance between axes of gear and cutter, in.

ϵ_c = angle of rotation of cutter

ϵ_2 = angle of rotation of internal gear

θ_t = vectorial angle of trochoid

r_t = radius to trochoid, in.

ϕ_1 = pressure angle at pitch line of cutter and internal gear

T_c = arc tooth thickness of cutter at R_c, in.

ϕ_{oc} = pressure angle at tip of cutter tooth

T_{oc} = arc tooth thickness of cutter at R_{oc}, in.

R_{bc} = radius of base circle of cutter, in.

δ_t = angle between origin of trochoid and center line of gear tooth

δ_s = angle between origin of trochoid and center line of tooth space

θ''_t = vectorial angle of trochoid in reference to selected center line

X_t = abscissa of trochoid, in.

Y_t = ordinate of trochoid, in.

$$\epsilon_2 = (R_c/R_2)\epsilon_c \tag{3-12}$$

$$r = \sqrt{C^2 + R_{oc}{}^2 + 2CR_{oc}\cos\epsilon_c} \tag{3-19}$$

$$\theta_t = \sin^{-1}[(R_{oc}\sin\epsilon_c)/r - \epsilon_2] \tag{3-14}$$

$$T_{oc}/2R_{oc} = (T_c/2R_c) + \text{inv } \phi_1 - \text{inv } \phi_{oc} \tag{5-6}$$

$$\delta_s = (N_c/N_2)(T_{oc}/2R_{oc}) \tag{6-6}$$

$$\delta_t = (\pi/N_2) - (N_c/N_2)(T_{oc}/2R_{oc}) \tag{6-7}$$

$$\theta''_t = \delta \pm \theta_t \tag{1-4}$$

$$X_t = r_t \sin\theta''_t \tag{1-5}$$

$$Y_t = r_t \cos\theta''_t \tag{1-6}$$

Example of Trochoidal Fillet on Internal Gear. As a definite example we shall use the same internal gear as before and a 4-in. diameter, 5-DP pinion-shaped cutter. This gives the following values:

	5 DP	1 DP		5 DP	1 DP
R_c	2.000	10.000	R_2	4.000	20.000
R_{oc}	2.250	11.250	T_c	0.31416	1.5708
R_{bc}	1.87938	9.3969	C	2.000	10.000

$$N_c = 20 \qquad N_2 = 40$$

$$\phi_1 = 20° \qquad \cos\phi_1 = 0.93969 \qquad \text{inv } \phi_1 = 0.014904$$

If we wish to use specific values of r_t, we can rearrange Eq. (3-19) to solve for ϵ_c as follows:

$$\cos \epsilon_c = \frac{(r_t^2 - C^2 - R_{oc}^2)}{2CR_{oc}} \tag{3-19}$$

Using the 1-DP values, we obtain the following:

$$\cos \phi_{oc} = \frac{9.3969}{11.250} = 0.83528$$

$$\phi_{oc} = 33.355° \qquad \text{inv } \phi_{oc} = 0.076097$$

$$\cos \epsilon_c = \frac{r_t^2 - 226.5625}{225.000}$$

$$\epsilon_2 = (^{10}\!\!/_{20})\epsilon_c = 0.50\epsilon_c$$

$$\theta_t = \sin^{-1}(11.25 \sin \epsilon_c/r_t) - \epsilon_2$$

We shall use the center of the tooth space as the reference line.

$$\delta_s = \frac{20}{40}\left(\frac{1.5708}{20} + 0.014904 - 0.076097\right) = 0.008673 \text{ radian}$$

Using the foregoing values and equations, and values of r_t ranging from 21.25 in. to 20.50 in., we obtain the values tabulated in Table 6-2. These values are plotted in Fig. 6-3 together with the coordinates of the involute profile, which are tabulated in Table 6-1.

TABLE 6-2. COORDINATES OF TROCHOID OF CORNER OF PINION-SHAPED CUTTER
(Plotted in Fig. 6-3)

r_t, in.	ϵ_c, deg	θ_t, deg	θ''_t, deg	X_t, in.	Y_t, in.
21.250	0.000	0.000	0.497	0.1842	21.2492
21.200	7.867	0.232	0.729	0.2697	21.1983
21.150	11.141	0.329	0.826	0.3048	21.1479
21.10627	13.359	0.395	0.892	0.3286	21.1037
21.100	13.647	0.404	0.901	0.3317	21.0973
21.050	15.672	0.467	0.964	0.3541	21.0471
21.000	17.626	0.522	1.019	0.3734	20.9966
20.950	19.312	0.573	1.070	0.3911	20.9464
20.900	20.863	0.621	1.118	0.4078	20.8960
20.850	22.308	0.664	1.161	0.4224	20.8456
20.800	23.666	0.706	1.203	0.4366	20.7954
20.750	24.949	0.746	1.243	0.4501	20.7450
20.700	26.174	0.783	1.280	0.4624	20.6948
20.650	27.344	0.820	1.317	0.4745	20.6444
20.600	28.466	0.855	1.352	0.4860	20.5942
20.550	29.547	0.889	1.386	0.4963	20.5440
20.500	30.590	0.922	1.419	0.5076	20.4936

Problem 6-4. *Given the proportions of a pinion-shaped cutter with a full-rounded tip and its generating center distance, to determine the coordinates of the fillet developed on a given internal gear.*

The form of this fillet is one whose equations have already been derived in Chap. 3.

When R_c = pitch radius of pinion-shaped cutter, in.

R_{ic} = radius to top of involute profile on cutter, in.

R_{oc} = outside radius of cutter, in.

R_{bc} = radius of base circle of cutter, in.

A = radius of rounding at tip, in.

T_c = arc tooth thickness of cutter at R_c, in.

T_{ic} = arc tooth thickness of cutter at R_{ic}, in.

ϕ_{ic} = pressure angle at R_{ic}

ϕ_1 = pressure angle at R_c

ϕ_{dc} = pressure angle at R_{dc}

R_{dc} = radius to center of rounding of tip, in.

$$\cos \phi_{ic} = R_{bc}/R_{ic} \qquad (5\text{-}23)$$
$$T_{ic} = 2R_{ic}[(T_c/2R_c) + \text{inv } \phi_1 - \text{inv } \phi_{ic}] \qquad (5\text{-}24)$$

For the trial solution we have

$$A = T_{ic}/2 \cos \phi_{ic} \qquad (5\text{-}25)$$
$$\text{inv } \phi_{dc} = (T_c/2R_c) + \text{inv } \phi_1 - (A/R_{bc}) \qquad (5\text{-}26)[1]$$
$$R_{dc} = R_{bc}/\cos \phi_{dc} \qquad (5\text{-}27)$$

To determine the corrected value of R_{ic} for the selected value of A, we proceed as follows:

$$\tan \phi_{ic} = \tan \phi_{dc} + (A/R_{bc}) \qquad (5\text{-}28)$$
$$R_{ic} = R_{bc}/\cos\phi_{ic} \qquad (5\text{-}29)$$

Trochoid of Center of Rounding. When

R_2 = pitch radius of internal gear, in.

N_2 = number of teeth in internal gear

N_c = number of teeth on pinion-shaped cutter

C = center distance for cutter and internal gear, in.

ϵ_c = angle of rotation of cutter

ϵ_2 = angle of rotation of internal gear

θ_t = vectorial angle of trochoid of center of rounding

r_t = any radius to trochoid of center of rounding, in.

ψ_t = angle between radius vector and tangent to trochoid

δ_t = angle between origin of trochoid and center line of gear tooth

δ_s = angle between origin of trochoid and center line of space

---

[1] We may use the value of A as determined from Eq. (5-25), or we may round this figure off to a value slightly smaller.

θ_f = vectorial angle of actual fillet

r_f = radius to actual fillet, in.

θ''_f = vectorial angle of actual fillet in reference to selected center line

and all other symbols are the same as before

$$\epsilon_2 = \frac{R_c}{R_2} \epsilon_c \tag{3-12}$$

$$r_t = \sqrt{C^2 + R_{dc}^2 + 2CR_{dc} \cos \epsilon_c} \tag{3-19}$$

$$\theta = \sin^{-1}[(R_{dc} \sin \epsilon_c)/r_t] - \epsilon_2 \tag{3-14}$$

$$\delta_s = 0 \tag{6-8}$$

$$\delta_t = \frac{\pi}{N_2} \tag{6-9}$$

$$\tan \psi_t = \frac{r_t \, d\theta_t/d\epsilon_c}{dr_t/d\epsilon_c} \tag{3-16}$$

$$\frac{d\theta_t}{d\epsilon_c} = \frac{R_{dc}(r_t^2 \cos \epsilon_c + CR_{dc} \sin^2 \epsilon_c)}{r_t^2(C + R_{dc} \cos \epsilon_c)} - \frac{R_c}{R_2} \tag{3-20}$$

$$\frac{dr_t}{d\epsilon_c} = -\frac{CR_{dc} \sin \epsilon_c}{r_t} \tag{3-21}$$

ACTUAL FILLET FORM

$$r_f = \sqrt{r_t^2 + A^2 - 2Ar_t \sin \psi_t} \tag{3-7}$$

$$\theta_f = \theta_t + \cos^{-1}[(r_t - A \sin \psi_t)/r_f] \tag{3-8}$$

$$\theta''_f = \delta \pm \theta_f \tag{1-4}$$

$$X_f = r_f \sin \theta''_f \tag{1-5}$$

$$Y_f = r_f \cos \theta''_f \tag{1-6}$$

Example of Fillet of Rounded Cutter Tip on Internal Gear. As a definite example we shall use the 1-DP values from the preceding example. This gives the following values:

$$R_c = 10.000 \qquad R_{ic} \text{ (trial value)} = 11.250 \qquad R_{bc} = 9.3969 \qquad T_c = 1.5708$$

$$\phi_1 = 20° \qquad \text{inv } \phi_1 = 0.014904 \qquad \cos \phi_1 = 0.93969 \qquad C = 10.000 \qquad N_2 = 40$$

$$N_c = 20$$

For the trial solution for the radius of the rounding, we have

$$\cos \phi_{ic} = \frac{9.3969}{11.250} = 0.83528$$

$$\phi_{ic} = 33.355° \qquad \text{inv } \phi_{ic} = 0.076097$$

$$T_{ic} = 22.50(0.07854 + 0.014904 - 0.076097) = 0.39031$$

$$A = \frac{0.39031}{2 \times 0.83528} = 0.23364$$

We shall use the value $A = 0.230$. Then

$$\text{inv } \phi_{dc} = 0.07854 + 0.014904 - \frac{0.230}{9,3969} = 0.068968$$

$$\phi_{dc} = 32.374° \qquad \cos \phi_{dc} = 0.84457 \qquad \tan \phi_{dc} = 0.63398$$

$$R_{dc} = \frac{9.3969}{0.84457} = 11.12625$$

Using these values, we shall recalculate the value of R_{ic}.

$$\tan \phi_{ic} = 0.63398 + \frac{0.230}{9.3969} = 0.65846$$

$$\phi_{ic} = 33.363° \qquad \cos \phi_{ic} = 0.83520$$

$$R_{ic} = \frac{9.3969}{0.83520} = 11.25107$$

$$R_{oc} = 11.12625 + 0.230 = 11.35625$$

Trochoid of Center of Rounding. We shall rearrange Eq. (3-19) to use specific values of r_t.

$$\cos \epsilon_c = \frac{r_t{}^2 - C^2 - R_{dc}{}^2}{2CR_{dc}} \tag{3-19}$$

Whence

$$\cos \epsilon_c = \frac{r_t{}^2 - 223.79344}{222.525}$$

$$\theta_t = \sin^{-1}\left(\frac{11.12625 \sin \epsilon_c}{r_t}\right) - \epsilon_2$$

$$\delta_s = 0$$

$$\frac{d\theta_t}{d\epsilon_c} = \frac{11.12625(r_t{}^2 \cos \epsilon_c + 111.2625 \sin^2 \epsilon_c)}{r_t{}^2(10 + 11.12625 \cos \epsilon_c)} - 0.500$$

$$\frac{dr_t}{d\epsilon_c} = -111.2625 \sin \frac{\epsilon_c}{r_t}$$

$$\tan \psi_t = \frac{r_t \, (d\theta_t/d\epsilon_c}{dr_t/d\epsilon_c} \tag{3-16}$$

Using the foregoing values and equations, we obtain the values tabulated in Table 6-3.

TABLE 6-3. COORDINATES OF FILLET FROM ROUNDED TIP OF PINION-SHAPED CUTTER
(Plotted in Fig. 6-4)

r_t, in.	ϵ_c, deg	θ_t, deg	ψ_t, deg	r_f, in.	θ_f, deg	X_f, in.	Y_f, in.
21.12625	0.000	0.000	−90.000	21.35625	0.000	0.0000	21.35625
21.100	5.722	0.153	−47.001	21.2688	0.578	0.2150	21.2677
21.050	9.752	0.270	−32.227	21.1735	0.795	0.2937	21.1714
21.000	12.552	0.336	−26.191	21.1025	0.896	0.3298	21.1000
20.950	14.834	0.398	−22.654	21.0397	0.973	0.3573	21.0367
20.900	16.810	0.451	−20.291	20.9809	1.039	0.3804	20.9775
20.850	18.578	0.499	−18.552	20.9243	1.097	0.4007	20.9205
20.800	20.194	0.544	−17.210	20.8692	1.152	0.4195	20.8650
20.750	21.690	0.585	−16.135	20.8151	1.198	0.4352	20.8105
20.700	23.091	0.624	−15.249	20.7617	1.240	0.4493	20.7569
20.650	24.413	0.661	−14.550	20.7088	1.279	0.4622	20.7036
20.600	25.667	0.696	−13.867	20.6563	1.316	0.4743	20.6510
20.550	26.865	0.729	−13.312	20.6042	1.351	0.4856	20.5984
20.500	28.012	0.762	−12.826	20.5523	1.388	0.4978	20.5463

Coordinates of Actual Fillet

$$r_f = \sqrt{r_t{}^2 + 0.0529 - 0.46 r_t \sin \psi_t}$$
$$\theta_f = \theta_t + \cos^{-1} (r_t - 0.23 \sin \psi_t)/r_f$$
$$\theta''_f = \theta_f$$

The values for these coordinates have been calculated. They are also tabulated in Table 6-3 and are plotted with the involute profile in Fig. 6-4.

Problem 6-5. *Given the proportions of an internal gear and mating spur pinion, and the center distance, to determine the minimum inside radius of the internal gear that will avoid involute interference.*

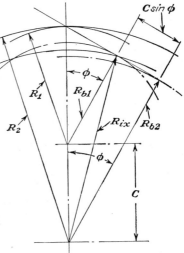

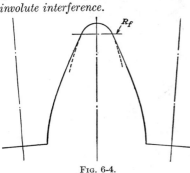

FIG. 6-4.   FIG. 6-5.

If the profile of the internal involute gear extends below the point of tangency of the path of contact with the base circle of the mating spur pinion, or pinion-shaped cutter that is used to generate it, involute interference will exist. To avoid such interference, the inside radius of the internal gear must be made large enough to be outside this point.

Referring to Fig. 6-5, when

R_1 = pitch radius of spur pinion, in.

R_2 = pitch radius of internal gear, in.

C = center distance, in.

ϕ = pressure angle at R_1 and R_2

R_{b2} = radius of base circle of internal gear, in.

R_{ix} = minimum inside radius of internal gear that will avoid involute interference, in.

we have the following from the geometrical conditions shown in Fig. 6-5:

$$C = R_2 - R_1 \tag{6-10}$$
$$R_{b2} = R_2 \cos \phi \tag{5-3}$$
$$R_{ix} = \sqrt{R_{b2}{}^2 + (C \sin \phi)^2} \tag{6-11}$$

Example of Minimum Inside Radius. As a definite example we shall determine the minimum inside radius for the 40-tooth internal gear used in the preceding example that will avoid involute interference with the 20-tooth pinion-shaped cutter. Using the 1-DP values, we have the following:

$$R_1 = 10.000 \qquad R_2 = 20.000 \qquad C = 10.000. \qquad R_{b2} = 18.7938$$
$$\phi = 20° \qquad \sin 20° = 0.34202$$

whence

$$R_{ix} = \sqrt{(18.7938)^2 + (3.4202)^2} = 19.10247$$

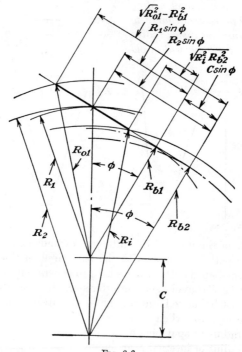

Fɪɢ. 6-6.

Problem 6-6. *Given the proportions of the internal gear and spur pinion, and the center distance, to determine the contact ratio for an internal-gear drive.*

The contact ratio of an internal-gear drive is obtained by dividing the length of the path of contact between its intersection with the inside circle of the internal gear and its intersection with the outside circle of the spur pinion by the base pitch of the series of involute curves that form the similar profiles of successive gear teeth. This condition holds true only when the inside circle of the internal gear is large enough to avoid involute interference, with both the mating spur pinion and the pinion-shaped cutter that is used to generate it.

The intersected part of the path of contact is indicated by the heavy line in Fig. 6-6. Referring to Fig. 6-6, when

R_1 = pitch radius of spur pinion, in.

R_2 = pitch radius of internal gear, in.

R_{b1} = radius of base circle of spur pinion, in.

R_{b2} = radius of base circle of internal gear, in.

R_{o1} = outside radius of spur pinion, in.

R_i = inside radius of internal gear, in.

p = circular pitch at R_1 and R_2, in.

p_b = base pitch of involute profiles, in.

C = center distance, in.

ϕ = pressure angle at R_1 and R_2

m_p = contact ratio

we know that

$$p_b = p \cos \phi \qquad R_{b1} = R_1 \cos \phi \qquad R_{b2} = R_2 \cos \phi$$

$$m_p = \frac{\sqrt{R_{o1}^2 - R_{b1}^2} + C \sin \phi - \sqrt{R_i^2 - R_{b2}^2}}{p_b} \qquad (6\text{-}12)$$

If, however, the value of $C \sin \phi$ is greater than the value of $\sqrt{R_i^2 - R_{b2}^2}$, involute interference will be present. In such cases, the maximum possible value of m_p will be something less than

$$\sqrt{R_{o1}^2 - R_{b1}^2}/p_b$$

Example of Contact Ratio for Internal-gear Drive. As a definite example we shall use the following: 10-DP, 20-tooth pinion. 40-tooth internal gear, 20-deg pressure angle, 1.000-in. center distance, with the following proportions:

	10 DP	1 DP		10 DP	1 DP
R_1	1.000	10.000	R_2	2.000	20.000
R_{b1}	0.93969	9.3969	R_{b2}	1.87938	18.7938
R_{o1}	1.125	11.250	R_i	1.940	19.400
p	0.31416	3.1416	p_b	0.29521	2.9521
C	1.000	10.000			

Using the 1-DP values, we obtain

$$m_p = \frac{\sqrt{(11.25)^2 - (9.3969)^2} + 3.4202 - \sqrt{(19.40)^2 - (18.7938)^2}}{2.95213}$$

$$m_p = \frac{6.18552 + 3.4202 - 4.81176}{2.95213} = 1.623 \text{ tooth intervals}$$

Problem 6-7. *Given the proportions of an internal-gear drive, to determine the arcs of approach and recess.*

Referring to Fig. 6-6, let

R_1 = pitch radius of spur pinion, in.

R_{o1} = outside radius of spur pinion, in.

R_{b1} = radius of base circle of spur pinion, in.

R_2 = pitch radius of internal gear, in.

R_i = inside radius of internal gear, in.

R_{b2} = radius of base circle of internal gear, in.

ϕ = pressure angle at R_1 and R_2

β_a = arc of approach of driving member

β_r = arc of recess of driving member

When the spur pinion is the driving member,

$$\beta_a = (R_2 \sin \phi - \sqrt{R_i{}^2 - R_{b2}{}^2})/R_{b1} \qquad (6\text{-}13)$$

$$\beta_r = (\sqrt{R_{o1}{}^2 - R_{b1}{}^2} - R_1 \sin \phi)/R_{b1} \qquad (6\text{-}14)$$

When the internal gear is the driving member,

$$\beta_a = (\sqrt{R_{o1}{}^2 - R_{b1}{}^2} - R_1 \sin \phi)/R_{b2} \qquad (6\text{-}15)$$

$$\beta_r = (R_2 \sin \phi - \sqrt{R_i{}^2 - R_{b2}{}^2})/R_{b2} \qquad (6\text{-}16)$$

Example of Approach and Recess on Internal Drive. As a definite example we shall use the 1-DP values from the preceding problem and determine the values when the spur pinion is the driving member. This gives the following values:

$$R_1 = 10.000 \qquad R_{o1} = 11.250 \qquad R_{b1} = 9.3969 \qquad R_2 = 20.000$$
$$R_i = 19.400 \qquad R_{b2} = 18.7938 \qquad \phi = 20° \qquad \sin 20° = 0.34202$$

$$\beta_a = \frac{6.8404 - \sqrt{(19.40)^2 - (18.7938)^2}}{9.3969} = 0.21589 \text{ radian}$$

$$\beta_r = \frac{\sqrt{(11.25)^2 - (9.3969)^2} - 3.4202}{9.3969} = 0.20427 \text{ radian}$$

Problem 6-8. *Given the proportions of an internal-gear drive, to determine the sliding velocity between the mating gear teeth.*

As with spur gears, the sliding velocity between the teeth of an internal-gear drive will be the difference in the speeds of the ends of the generating lines of the two mating involute curves as they pass through the path of contact. The angular velocities of these generating lines will be the same as the angular velocities of the gears themselves. The actual sliding velocities will be the differences between the products of these angular velocities and the lengths of the respective generating lines, or radii of curvatures. Referring to Fig. 6-7, let

ω_1 = angular velocity of driving member, radians/min

ω_2 = angular velocity of driven member, radians/min

n = rpm of driving member

V = pitch-line velocity of gears, ft/min

V_s = sliding velocity, ft/min

R_1 = pitch radius of spur pinion, in.

R_2 = pitch radius of internal gear, in.

C = center distance, in.

R_{b1} = radius of base circle of spur pinion, in.

R_{b2} = radius of base circle of internal gear, in.

ϕ = pressure angle at R_1 and R_2

R_{c1} = radius of curvature of pinion at r_1, in.

R_{c2} = radius of curvature of internal gear at r_2, in.

r_1 = any radius of pinion profile, in.

r_2 = mating radius of r_1 on internal-gear profile, in.

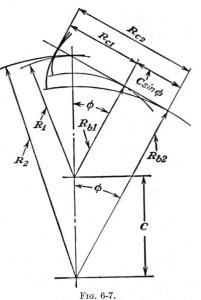

FIG. 6-7.

Spur Pinion as Driving Member. When the spur pinion is the driving member, then

$$V = (2\pi R_1 n)/12 = R_1\omega_1/12 \qquad (4\text{-}12)$$
$$\omega_1 = 12V/R_1$$
$$V_s = (\tfrac{1}{12})(R_{c1}\omega_1 - R_{c2}\omega_2)$$
$$\omega_2 = R_1\omega_1/R_2$$

Whence

$$V_s = (\omega_1/12)[R_{c1} - (R_1 R_{c2}/R_2)]$$
$$R_{c2} - R_{c1} = C \sin \phi$$
$$R_{c1} = \sqrt{r_1{}^2 - R_{b1}{}^2}$$
$$R_{c2} = \sqrt{r_2{}^2 - R_{b2}{}^2} = C \sin \phi + \sqrt{r_1{}^2 - R_{b1}{}^2}$$

Substituting these values into the previous equation for the sliding velocity, we obtain

$$V_s = (V/R_1 R_2)[(R_2 - R_1)\sqrt{r_1{}^2 - R_{b1}{}^2} - R_1 C \sin \phi]$$

But

$$C = R_2 - R_1$$

Whence

$$V_s = [V(R_2 - R_1)/R_1 R_2](\sqrt{r_1{}^2 - R_{b1}{}^2} - R_1 \sin \phi) \qquad (6\text{-}17)$$

Equation (6-17) may also be written

$$V_s = V[(1/R_1) - (1/R_2)](\sqrt{r_1^2 - R_{b1}^2} - R_1 \sin \phi) \qquad (6\text{-}18)$$

Internal Gear as Driving Member. When the internal gear is the driving member, then

$$V = 2\pi R_2 n/12 = R_2 \omega_1/12$$

And in a similar manner we obtain

$$V_s = V[(1/R_1) - (1/R_2)](R_1 \sin \phi - \sqrt{r_1^2 - R_{b1}^2}) \qquad (6\text{-}19)$$

A comparison of the two foregoing equations for sliding velocity makes it apparent that the value of this sliding velocity is the same for a given

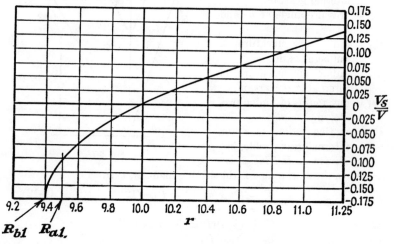

FIG. 6-8.

point of contact regardless of which member of the pair is driving, but that the sign or direction of the sliding, when the internal gear is driving, is the reverse of that when the spur pinion is driving.

Example of Sliding Velocity on an Internal-gear Drive. As a definite example we shall use the 1-DP values from the example used with Prob. 6-6. From this we have the following values:

$$R_1 = 10.00 \qquad R_2 = 20.00 \qquad R_{b1} = 9.3969 \qquad \phi = 20° \qquad \sin 20° = 0.34202$$

We shall assume the pinion to be the driving member and use values of r_1 ranging from 9.3969 to 11.250 in. Introducing the specific values into Eq. (6-18) and solving for the ratio of the sliding velocity to the pitch-line velocity, V_s/V, we obtain the following:

$$V_s/V = 0.05[\sqrt{r_1^2 - (9.3969)^2} - 3.4202]$$

These values have been calculated and they are tabulated in Table 6-4. They are also plotted in Fig. 6-8.

As with spur gears, a minus value indicates that the direction of the sliding on the driving member is toward the center of the gear, while a plus value indicates that the direction of the sliding is away from the center of the gear. The actual sliding velocity on internal gears is much less than on a pair of similar spur gears.

If this sliding velocity were plotted against the position of the contact along the path of contact, the graph would be a straight line. This position along the path of contact represents the angular movement of the gears. Therefore the velocity of sliding changes uniformly during the contact of the mating teeth. The average sliding velocity on the deden-

TABLE 6-4. SLIDING VELOCITY ON INTERNAL-GEAR DRIVE
(Plotted in Fig. 6-8)
$$V_s/V = 0.05[\sqrt{r_1{}^2 - (9.3969)^2} - 3.4202]$$

r_1, in.	R_{c1}, in.	$R_{c1} - 3.4202$	V_s/V
9.3969	0.00000	−3.4202	−0.17101
9.6000	1.96425	−1.45595	−0.07280
9.8000	2.78177	−0.63843	−0.03192
10.0000	3.4202	0.00000	0.00000
10.2000	3.96715	0.54675	0.02740
10.4000	4.45626	1.03606	0.05180
10.6000	4.90492	1.48472	0.07424
10.8000	5.32337	1.90317	0.09516
11.0000	5.71824	2.29804	0.11490
11.2500	6.18553	2.76533	0.13827

dum, for example, will be one-half the sliding velocity that exists when contact is first made. The average sliding velocity on the addendum of the driving member, on the other hand, will be one-half the sliding velocity that exists when contact is made at the tip of its tooth.

Problem 6-9. *Given the proportions of a spur pinion and an internal gear, and the center distance, to determine the radius on the pinion where contact is first made with the tip of the internal-gear tooth.*

This radius will be at the point where the inside circle of the internal gear intersects the path of contact. Referring to Fig. 6-9, when

C = center distance, in.

ϕ = pressure angle of operation

R_{b1} = radius of base circle of spur pinion, in.

R_{b2} = radius of base circle of internal gear, in.

R_{a1} = radius to bottom of active profile on spur pinion, in.

R_i = inside radius of internal gear, in.

we have from the geometrical conditions shown in Fig. 6-9 the following:

$$R_{a1} = \sqrt{(\sqrt{R_i{}^2 - R_{b2}{}^2} - C \sin \phi)^2 + R_{b1}{}^2} \qquad (6\text{-}20)$$

Example of Active Profile of Pinion with Internal Gear. As a definite example we shall use the 1-DP values from the preceding example. This gives the following values:

$$C = 10.00 \qquad \phi = 20° \qquad \sin 20° = 0.34202$$
$$R_{b1} = 9.3969 \qquad R_{b2} = 18.7938 \qquad R_i = 19.400$$

Whence

$$R_{a1} = \sqrt{(4.81372 - 3.4202)^2 + (9.3969)^2} = 9.49936$$

This radius is indicated on the sliding diagram in Fig. 6-8.

Problem 6-10. *Given the proportions of an internal-bear drive, to determine whether or not interference exists between the tips of the pinion teeth and the internal-gear teeth as the teeth come into and go out of mesh.*

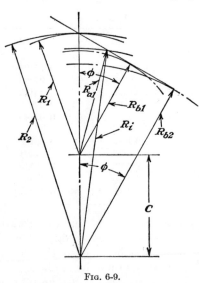

This condition of tip interference is present when the trochoid of the path of the corner of the pinion tooth intersects the involute profile of the internal-gear tooth. Thus if we solve Eqs. (3-12), (3-19), and (3-14) for θ_t, substituting the value of R_i for r_t, and establish the relationship of the origin of the trochoid in reference to the involute profile, we can compare the position of this trochoid at R_i with the position of the involute profile at the same point. If the trochoid is outside the tooth form, then no tip interference will be present. If this trochoid is inside the involute profile, then tip interference is present. In either event, we can determine the amount of clearance or the depth of interference at this inside circle of the internal gear. Thus when

Fig. 6-9.

N_1 = number of teeth in spur pinion

N_2 = number of teeth in internal gear

R_{o1} = outside radius of spur pinion, in.

R_i = inside radius of internal gear, in.

C = center distance, in.

ϵ_1 = angle of rotation of spur pinion

ϵ_2 = angle of rotation of internal gear

θ_{t1} = vectorial angle of trochoid at R_i

ϕ_1 = pressure angle at pitch line of gears

ϕ_2 = pressure angle at R_i

ϕ_{o1} = pressure angle at tip of spur-pinion tooth

R_{b1} = radius of base circle of spur pinion, in.

R_{b2} = radius of base circle of internal gear, in.

δ = angle between origins of trochoid and involute of internal gear

$$\epsilon_2 = (N_1/N_2)\epsilon_1 \qquad (3\text{-}12)$$

We have from Eq. (3-19)

$$\cos \epsilon_1 = (R_i{}^2 - R_{o1}{}^2 - C^2)/2CR_{o1}$$

and from Eq. (3-14)

$$\theta_{t1} = \sin^{-1}[R_{o1} \sin \epsilon_1/R_i] - \epsilon_2$$
$$\delta = (N_1/N_2)(\text{inv } \phi_{o1} - \text{inv } \phi_1) + \text{inv } \phi_1 \qquad (6\text{-}21)$$
$$\cos \phi_2 = R_{b2}/R_i \qquad \cos \phi_{o1} = R_{b1}/R_{o1}$$

When x_1 = angle of tip of internal-gear tooth at R_i from origin of involute, radians

x_2 = angle of trochoid at R_i from origin of involute of internal gear, radians

$$x_1 = \text{inv } \phi_2 \qquad (6\text{-}22)$$
$$x_2 = \delta - \theta_{t1} \qquad (6\text{-}23)$$

When x_1 is greater than x_2, tip interference exists. When x_2 is greater than x_1, there is clearance between the trochoid and the tip of the internal-gear tooth, and so no interference is present.

Example of Check for Tip Interference. As a definite example we shall use the following: 1-DP internal gear, of 20-deg pressure angle, with 30 teeth, and a 25-tooth spur pinion with the following values:

$$N_1 = 25 \qquad N_2 = 30 \qquad R_{o1} = 13.750 \qquad R_i = 14.400 \qquad R_{b1} = 11.74616$$
$$R_{b2} = 14.09539 \qquad C = 2.500 \qquad \phi_1 = 20° \qquad \text{inv } 20° = 0.014904$$
$$\cos \epsilon_1 = \frac{(14.40)^2 - (13.75)^2 - 2.5^2}{5 \times 13.75} = 0.17524$$
$$\epsilon_1 = 79.908° \qquad \sin \epsilon_1 = 0.98453$$
$$\epsilon_2 = {}^{25}\!/_{30} \times 79.908° = 66.590°$$
$$\theta_{t1} = 70.068° - 66.590° = 3.478° = 0.060702 \text{ radian}$$
$$\cos \phi_{o1} = \frac{11.74616}{13.75} = 0.85426$$
$$\phi_{o1} = 31.322° \qquad \text{inv } \phi_{o1} = 0.061863$$

$$\delta = (^{25}\!\!/_{30})(0.061863 - 0.014904) + 0.014904 = 0.054036 \text{ radian}$$
$$x_2 = 0.054036 - 0.060702 = -0.006666 \text{ radian}$$
$$\cos \phi_2 = \frac{14.09539}{14.40} = 0.97885$$
$$\phi_2 = 11.806° \qquad \text{inv } \phi_2 = 0.002967$$
$$x_1 = 0.002967 \text{ radian}$$

In this example, the value of x_1 is greater by 0.002967 plus 0.006666 radian, which

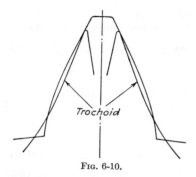

equals 0.009633 radian, than the value of x_2. Hence there is tip interference. The actual amount of this interference is equal to 0.009633 × R_i, which is equal to 0.13871 in.

The coordinates of this internal-gear tooth have been calculated, using a value of 1.3888 in. for T_2 at a radius of 15.00 in. The coordinates of the trochoidal path of the corner of the spur-pinion tooth have also been calculated. These values are tabulated in Table 6-5 and plotted in Fig. 6-10. This figure shows the trochoidal path cutting into the tip of the internal-gear tooth.

Trochoid

Fig. 6-10.

Problem 6-11. *Given the tooth proportions of a spur pinion and an internal gear, to determine the center distance at which they will mesh tightly.*

TABLE 6-5. COORDINATES OF INTERNAL-GEAR TEETH AND OF TROCHOIDAL PATH OF CORNER OF PINION TOOTH
(Plotted in Fig. 6-10)

r, in.	θ_t, rad.	X_t, in.	Y_t, in.	θ'', rad.	X, in.	Y, in.
16.250	0.00000	0.1335	16.2469	0.020499	0.3331	16.2466
16.000	0.00805	0.4373	15.9941	0.029108	0.4658	15.9949
15.800	0.01281	0.5070	15.7918	0.035675	0.5636	15.7899
15.600	0.01796	0.5809	15.5892	0.041924	0.6538	15.5863
15.400	0.02358	0.6599	15.3858	0.047824	0.7361	15.3824
15.200	0.02974	0.7450	15.1818	0.053337	0.8103	15.1784
15.000	0.03648	0.8361	14.9768	0.058427	0.8760	14.9744
14.800	0.04384	0.9337	14.7705	0.063023	0.9324	14.7705
14.600	0.05191	1.0385	14.5631	0.067044	0.9784	14.5672
14.400	0.06070	1.1506	14.3539	0.070364	1.0125	14.3644

This problem is very similar to Prob. 5-9, and will be solved in the same manner.

Let R_1 = radius of spur pinion where arc tooth thickness is known, in.

T_1 = arc tooth thickness of pinion at R_1, in.

N_1 = number of teeth in spur pinion

ϕ_1 = pressure angle at R_1 and R_2

R_2 = radius of internal gear where arc tooth thickness is known, in.

T_2 = arc tooth thickness of internal gear at R_2, in.

N_2 = number of teeth in internal gear

r_1 = pitch radius of spur pinion when tightly meshed, in.

t_1 = arc tooth thickness of pinion at r_1, in.

ϕ_2 = pressure angle at r_1 and r_2

r_2 = pitch radius of internal gear when tightly meshed, in.

t_2 = arc tooth thickness of internal gear at r_2, in.

C_1 = center distance for pressure angle of ϕ_1, in.

C_2 = center distance when tightly meshed at pressure angle of ϕ_2, in.

From Prob. 5-9 we have

$$t_1 = 2r_1[(T_1/2R_1) + \text{inv } \phi_1 - \text{inv } \phi_2]$$

From Eq. (6-2) we have

$$t_2 = 2r_2[(T_2/2R_2) - \text{inv } \phi_1 + \text{inv } \phi_2]$$

We know that

$$t_1 + t_2 = 2\pi r_1/N_1 = 2\pi r_2/N_2$$

and

$$R_2 = (N_2/N_1)R_1 \qquad \text{and} \qquad r_2 = (N_2/N_1)r_1$$

Substituting these values into the equation for t_2, combining terms, simplifying, and solving for inv ϕ_2, we obtain the following:

$$\text{inv } \phi_2 = [2\pi R_1 - N_1(T_1 + T_2)/2R_1(N_2 - N_1)] + \text{inv } \phi_1 \quad \text{(6-24)}$$

When this equation is reduced to 1-DP values, it becomes

$$\text{inv } \phi_2 = \left[\frac{\pi - (T_1 + T_2)}{N_2 - N_1}\right] + \text{inv } \phi_1 \quad \text{(6-25)}$$

We know that

$$r_1 = R_1 \cos \phi_1/\cos \phi_2$$

and

$$r_2 = R_2 \cos \phi_1/\cos \phi_2 \qquad \text{and} \qquad R_2 - R_1 = C_1$$

and

$$r_2 - r_1 = C_2$$

Whence

$$C_2 = C_1 \cos \phi_1/\cos \phi_2 \quad \text{(5-35)}$$

Example of Center Distance for Special Internal-gear Drive. As a definite example, we shall use 10-DP, 20-deg gears with a 25-tooth spur pinion and a 45-tooth internal gear with the following given values:

	10 DP	1 DP		10 DP	1 DP
R_1	1.250	12.500	R_2	2.250	22.500
T_1	0.170	1.700	T_2	0.138	1.380
C_1	1.000	10.000			

$$N_1 = 25 \qquad N_2 = 45$$
$$\phi_1 = 20° \qquad \text{inv } 20° = 0.014904 \qquad \cos 20° = 0.93969$$

Using the 1-DP values and Eqs. (6-24) and (5-35), we obtain the following:

$$\text{inv } \phi_2 = \frac{3.1416 - (1.700 + 1.380)}{20} + 0.014904 = 0.017984$$

$$\phi_2 = 21.246° \qquad \cos \phi_2 = 0.93203$$

$$C_2 = \frac{10 \times 0.93969}{0.93203} = 10.08218 \qquad \text{for 1 DP}$$

Problem 6-12. *Given the tooth proportions of an internal gear, to deter-mine the cutting or generating position of a pinion-shaped cutter.*

This problem is identical to the preceding one. With the substitution of the proper symbols into Eq. (6-25), we shall have the solution. This problem also includes the determination of the root radius of the internal gear. Thus when

R_c = pitch radius of pinion-shaped cutter, in.
T_c = arc tooth thickness of cutter at R_c, in.
N_c = number of teeth in pinion-shaped cutter
R_{oc} = outside radius of pinion-shaped cutter, in.
R_2 = pitch radius of internal gear, in.
T_2 = arc tooth thickness of internal gear at R_2, in.
N_2 = number of teeth in internal gear
R_{r2} = root radius of internal gear, in.
ϕ_1 = pressure angle at R_c and R_2
ϕ_2 = generating pressure angle
C_1 = center distance for pressure angle of ϕ_1, in.
C_2 = generating center distance with pressure angle of ϕ_2, in.

we have from Eq. (6-24)

$$\text{inv } \phi_2 = [2\pi R_c - N_c(T_c + T_2)/2R_c(N_2 - N_c)] + \text{inv } \phi_1 \qquad (6\text{-}24)$$

Reduced to the 1-DP values, this equation becomes

$$\text{inv } \phi_2 = [\pi - (T_c + T_2)]/(N_1 - N_c) + \text{inv } \phi_1 \qquad (6\text{-}25)$$

$$C_2 = C_1 \cos \phi_1 / \cos \phi_2 \qquad (5\text{-}35)$$

$$R_{r2} = C_2 + R_{oc} \qquad (6\text{-}26)$$

Example of Position of Cutter for Internal Gear. As a definite example we shall use the internal gear from the preceding problem and assume the use of a 3-in.-diameter pinion-shaped cutter with 30 teeth. This gives the following values:

	10 DP	1 DP		10 DP	1 DP
R_c	1.500	15.000	R_2	2.250	22.500
T_c	0.15708	1.5708	T_2	0.138	1.380
R_{oc}	1.625	16.250	C_1	0.750	7.500

$$N_c = 30 \qquad N_2 = 45$$
$$\phi_1 = 20° \qquad \cos 20° = 0.93969 \qquad \text{inv } 20° = 0.014904$$

Using the 1-DP values, we obtain

$$\text{inv } \phi_2 = \frac{3.1416 - (1.5708 + 1.380)}{15} + 0.014904 = 0.027264 \text{ radian}$$
$$\phi_2 = 24.362° \qquad \cos \phi_2 = 0.91096$$
$$C_2 = \frac{7.50 \times 0.93969}{0.91096} = 7.73653 \qquad \text{for 1 DP}$$
$$R_{r2} = 7.73653 + 16.250 = 23.98653 \qquad \text{for 1 DP}$$

Problem 6-13. *Given the center distance and numbers of teeth for a spur pinion and an internal gear, and the proportions of a pinion-shaped cutter, to determine the proportions of the spur pinion and the internal gear.*

This problem must be solved in five successive steps, as follows:

1. Determine the minimum inside radius for the internal gear that will avoid involute interference with either the mating pinion or the pinion-shaped cutter, whichever is the smaller.

2. Determine the position of the pinion-shaped cutter when generating the internal gear, and the arc tooth thickness of the internal gear at its nominal pitch line.

3. Determine the proportions of the teeth of the mating spur pinion. These values depend, in part, upon the method used to generate the pinion.

4. Determine the cutting data for the spur pinion.

5. Determine the clearances and the final tooth proportions of the gears.

In addition, if the difference between the numbers of teeth in the spur pinion and internal gear is small, less than about 6 or 7, for example, we must also check to be sure that the tips of the teeth will not interfere with each other as they come into and go out of mesh.

For the first step, we have Prob. 6-5 with Eqs. (6–10) and (6–11). If the pinion-shaped cutter has fewer teeth than the spur pinion, the first step would be made using the nominal pressure angle of the pinion-shaped cutter. If the spur pinion has fewer teeth than the cutter, the first step would be made using this pinion and the operating pressure angle.

If the start is made with the pinion-shaped cutter and the inside radius of the internal gear must be increased to avoid involute interference with the cutter, the generating center distance would be increased also, so that little, if any, of the depth of the tooth in the internal gear would be lost. In all cases, the inside radius that is used should be somewhat larger than the minimum inside radius.

For the second step, we have Prob. 6-12, modified as necessary to meet the particular need.

For the third step, we must rearrange some of the existing equations to solve for the values needed.

The fourth step is similar to this same type of problem on spur gears, and simple equations may be set up to solve for the values needed.

The fifth step is an over-all check. The whole depth of the tooth on the pinion must not be greater than the nominal whole depth of tooth.

Let R_{o1} = outside radius of spur pinion, in.

R_1 = nominal pitch radius of spur pinion, inches

R_{r1} = root radius of spur pinion, in.

T_1 = arc tooth thickness of spur pinion at R_1, in.

R_{b1} = radius of base circle of spur pinion, in.

N_1 = number of teeth in spur pinion

R_{oc} = outside radius of pinion-shaped cutter, in.

R_c = nominal pitch radius of pinion-shaped cutter, in.

T_c = arc tooth thickness of cutter at R_c, in.

R_{bc} = radius of base circle of pinion-shaped cutter, in.

N_c = number of teeth in pinion-shaped cutter

R_{r2} = root radius of internal gear, in.

R_2 = nominal pitch radius of internal gear, in.

R_i = inside radius of internal gear, in.

R_{b2} = radius of base circle of internal gear, in.

T_2 = arc tooth thickness of internal gear at R_2, in.

N_2 = number of teeth in internal gear

C_1 = center distance for spur pinion and internal gear for pressure angle of ϕ_1, in.

C_2 = center distance of operation of spur pinion and internal gear with pressure angle of ϕ_2, in.

C_{g1} = center distance for cutter and internal gear for pressure angle of ϕ_1, in.

C_{g2} = generating center distance for cutter and internal gear with pressure angle of ϕ_{g2}, in.

P = diametral pitch of pinion-shaped cutter

ϕ_1 = nominal pressure angle of pinion-shaped cutter

ϕ_2 = operating pressure angle of spur pinion and internal gear

ϕ_{g2} = generating pressure angle of pinion-shaped cutter and internal gear

First Step

When

$$C_1 = (N_2 - N_1)/2P$$

then

$$(N_2 - N_1)/2P = C_2$$

Otherwise

$$C_1 = C_2 \quad \text{and} \quad \phi_1 = \phi_2$$

$$\cos \phi_2 = C_1 \cos \phi_1/C_2 \tag{5-35}$$

To determine the minimum inside radius of the internal gear when the spur pinion is smaller than the pinion-shaped cutter, we have from Prob. 6-5

$$R_{ix} = \sqrt{R_{b2}^2 + (C_2 \sin \phi_2)^2} \tag{6-27}$$

When the pinion-shaped cutter has fewer teeth than the spur pinion, we will not know the generating center distance until some of the other values have been established. Here we may need to make a trial solution for this inside radius first and choose a value for it somewhat larger than the minimum. Later, such a selected value should be rechecked to be sure that involute interference has been avoided. For such a trial solution, we shall use

$$R_{ix} = \sqrt{R_{b2}^2 + (C_{g1} \sin \phi_1)^2} \tag{6-28}$$

When the pinion-shaped cutter is smaller than the spur pinion, it is a good plan to solve both of the two equations (6-27) and (6-28) and choose a value for R_i somewhat larger than the larger value of the two.

Second Step. After a value for R_i has been selected, the next step is to determine the generating center distance for the pinion-shaped cutter and the internal gear. For this we may use, when

h_{t1} = nominal whole depth of tooth, in.

$$C_{g2} = R_i + h_{t1} - R_{oc} \tag{6-29}$$

$$\cos \phi_{g2} = C_{g1} \cos \phi_1/C_{g2} \tag{5-35}$$

We must next determine the value of T_2. For this we can transpose Eq. (6-24) or (6-25) to solve for T_2 as follows:

$$\text{inv } \phi_{g2} = \{[2\pi R_c - N_c(T_c + T_2)]/2R_c(N_2 - N_c)\} + \text{inv } \phi_1 \tag{6-24}$$

Transposing this equation to solve for T_2, we obtain

$$T_2 = \{[2R_c(N_2 - N_c)(\text{inv } \phi_1 - \text{inv } \phi_{g2}) + 2\pi R_c]/N_c\} - T_c \tag{6-30}$$

Reduced to 1-DP values, this equation becomes

$$T_2 = (N_2 - N_c)(\text{inv } \phi_1 - \text{inv } \phi_{g2}) + \pi - T_c \qquad (6\text{-}31)$$
$$R_{r2} = C_{g2} + R_{oc} \qquad (6\text{-}26)$$

Third Step. We now have all the values for the internal gear. The next steps are to determine values for the spur pinion, starting with the value of T_1. This is obtained by transposing Eq. (6-24) to solve for T_1. When it is thus transposed, we have

$$T_1 = \{[2R_1(N_2 - N_1)(\text{inv } \phi_1 - \text{inv } \phi_2) + 2\pi R_1]/N_1\} - T_2 \qquad (6\text{-}32)$$

Reduced to 1-DP values, this equation becomes

$$T_1 = (N_2 - N_1)(\text{inv } \phi_1 - \text{inv } \phi_2) + \pi - T_2 \qquad (6\text{-}33)$$

Fourth Step. After the value of T_1 has been established, we turn our attention to the generation of the spur pinion. This may be done by a hob or by a pinion-shaped cutter.

SPUR PINION GENERATED BY HOB. When the spur pinion is generated by a hob, we have the equations of Prob. 5-11, where we substitute the hob proportions for those of the rack, to determine the root radius of the spur pinion. For this we have, when

p = circular pitch of hob, in.

ϕ_1 = pressure angle of hob and nominal pressure angle of gear

H = distance from center of spur pinion to nominal pitch line of hob, in.

a_h = nominal addendum of hob, including clearance, in.

$$H = R_1 + \frac{(T_1/2) - (p/4)}{\tan \phi_1} \qquad (5\text{-}39)$$
$$R_{r1} = H - a_h \qquad (6\text{-}34)$$

When c_1 = clearance at tip of pinion-tooth, in.

$$R_{o1} = R_{r2} - C_2 - c_1 \qquad (6\text{-}35)$$

SPUR PINION GENERATED BY PINION-SHAPED CUTTER. When the pinion is generated by a pinion-shaped cutter, we have the equations from Prob. 5-10 to determine the root radius. This pinion-shaped cutter may be the same that is used to cut the internal gear, or it may be of another number of teeth. In either case, the specific values for the cutter actually used should be substituted into the equations. Thus when

ϕ_{g3} = generating pressure angle of cutter and spur pinion

$$\text{inv } \phi_{g3} = \frac{N_1(T_1 + T_c) - 2\pi R_1}{2R_1(N_1 + N_c)} + \text{inv } \phi_1 \qquad (5\text{-}36)$$

Reduced to 1-DP values, this equation becomes

$$\text{inv } \phi_{g3} = \frac{T_1 + T_c - \pi}{N_1 + N_c} + \text{inv } \phi_1 \qquad (5\text{-}37)$$

When C_{g3} = generating center distance of cutter and spur pinion, in.

$$C_{g3} = [(N_1 + N_c)/2P] \frac{\cos \phi_1}{\cos \phi_{g3}} \qquad (6\text{-}36)$$

$$R_{r1} = C_{g3} - R_{oc} \qquad (6\text{-}37)$$

$$R_{o1} = R_{r2} - C_2 - c_1 \qquad (6\text{-}35)$$

The value of c_1 may generally be taken as that of the nominal clearance of the gear-tooth system that is used.

Fifth Step. The last step is to check the resulting whole depth of the pinion tooth. If this value is greater than the nominal depth of tooth, the outside radius of the spur pinion should be reduced accordingly. It may also be necessary to check for tip interference and for the duration of contact.

If tip interference is present, it can be reduced or eliminated by increasing the inside radius of the internal gear and cutting its teeth deeper; *i.e.*, increasing the root radius by the same amount that the inside radius has been increased. This will reduce the contact ratio, and so this feature must be checked.

Example of Internal-gear-drive Design. As a definite example we shall use the following: 8-DP, 20-deg internal-gear drive with a 20-tooth spur pinion and a 40-tooth internal gear operating at a center distance of 1.300 in. Both the spur pinion and the internal gear will be generated with a 3-in.-diameter, 24-tooth, pinion-shaped cutter. This gives the following values:

	8 DP	1 DP		8 DP	1 DP
R_1	1.250	10.000	R_{b2}	2.34922	18.7938
R_{b1}	1.17461	9.3969	C_2	1.300	10.400
C_1	1.250	10.000	R_{oc}	1.65625	13.250
R_c	1.500	12.000	T_c	0.19635	1.5708
R_{bc}	1.40936	11.27628	h_{t1}	0.28125	2.250
C_{g1}	1.000	8.000	c_1	0.03125	0.250
R_2	2.500	20.000			

$$N_1 = 20 \qquad N_c = 24 \qquad N_2 = 40$$
$$\phi_1 = 20^\circ \qquad \text{inv } 20^\circ = 0.014904 \qquad \cos 20^\circ = 0.93969$$

Using the 1-DP values, we have

$$\cos \phi_2 = \frac{10 \times 0.93969}{10.400} = 0.90355$$

$$\phi_2 = 25.369^\circ \qquad \text{inv } \phi_2 = 0.031399 \qquad \sin \phi_2 = 0.42845$$
$$R_{ix} = \sqrt{(18.7938)^2 + (10.40 \times 0.42845)^2} = 19.31481$$

We will select the value $R_i = 19.3200$, whence

$$C_{g2} = 19.3200 + 2.250 - 13.250 = 8.320$$
$$\cos \phi_{g2} = \frac{8.00 \times 0.93969}{8.320} = 0.90355*$$
$$\phi_{g2} = 25.369° \qquad \text{inv } \phi_{g2} = 0.031399$$
$$T_2 = (40 - 24)(0.014904 - 0.031399) + 3.1416 - 1.5708 = 1.30688$$
$$R_{r2} = 8.3200 + 13.250 = 21.5700$$
$$T_1 = (40 - 20)(0.014904 - 0.031399) + 3.1416 - 1.30688 = 1.50482$$
$$\text{inv } \phi_{g3} = \frac{1.50482 + 1.5708 - 3.1416}{2\theta + 24} + 0.014904 = 0.012155$$
$$\phi_{g3} = 18.725° \qquad \cos \phi_{g3} = 0.94707$$
$$C_{g3} = \frac{44\frac{1}{2} \times 0.93969}{0.94707} = 21.81801$$
$$R_{r1} = 21.81801 - 13.250 = 8.56801$$

This root radius is too far below the base circle. We shall therefore increase the inside radius of the internal gear to 19.500 in., and recompute the several values. When

$$R_i = 19.500$$
$$C_{g2} = 19.500 + 2.250 - 13.250 = 8.500$$
$$\cos \phi_{g2} = \frac{8.00 \times 0.93969}{8.500} = 0.88441$$
$$\phi_{g2} = 27.821° \qquad \text{inv } \phi_{g2} = 0.042141$$
$$T_2 = (40 - 24)(0.014904 - 0.042141) + 3.1416 - 1.5708 = 1.13501$$
$$R_{r2} = 8.500 + 13.250 = 21.750$$
$$T_1 = (40 - 20)(0.014904 - 0.031399) + 3.1416 - 1.13501 = 1.67669$$
$$\text{inv } \phi_{g3} = \frac{1.67669 + 1.5708 - 3.1416}{20 + 24} + 0.014904 = 0.017310$$
$$\phi_{g3} = 20.986° \qquad \cos \phi_{g3} = 0.93367$$
$$C_{g3} = \frac{44\frac{1}{2} \times 0.93969}{0.93367} = 22.14185$$
$$R_{r1} = 22.14185 - 13.250 = 8.89185$$
$$R_{o1} = 21.750 - 10.400 - 0.250 = 11.100$$

We shall now check the tooth height on the spur pinion and the clearance.

$$\text{Tooth height} = 11.100 - 8.89185 = 2.20815$$

This is less than the nominal whole depth of 2.250 in., and is satisfactory.

$$\text{Clearance at root of pinion} = R_i - C_2 - R_{r1}$$
$$= 19.500 - 10.400 - 8.89185 = 0.20815$$

This clearance is less than the nominal value of 0.250 in. and should be increased. We shall obtain the additional clearance by enlarging the inside radius of the internal gear by 0.042 in., and leaving all other values the same as before. The corrected value for the inside radius thus becomes

$$R_i = 19.542$$

* It is only a coincidence that the values of ϕ_2 and ϕ_{g2} are identical. In each case here, the center distance has been increased exactly 4 per cent.

The coordinates of the spur pinion and internal gear of this drive have been calculated, and they are plotted in Fig. 6-11.

Problem 6-14. *Given the radius, arc tooth thickness, and pressure angle of an internal gear, to determine the measurement between rolls placed in the tooth spaces.*

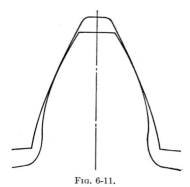

Fig. 6-11.

This problem is similar to Probs. 5-15 and 5-16 for spur gears. Referring to Fig. 6-12, when

R_{b2} = radius of base circle of internal gear, in.

R_2 = radius where tooth thickness is known, in.

ϕ_1 = pressure angle at R_2

T_2 = arc tooth thickness at R_2, in.

W = radius of wire or roll, in.

r_2 = radius from center of gear to center of roll, in.

ϕ_2 = pressure angle at r_2

N_2 = number of teeth in internal gear

M_1 = measurement between rolls, even number of teeth, in.

M_2 = measurement between rolls, odd number of teeth, in.

we have from the geometrical conditions shown in Fig. 6-12 the following:

$$\text{inv } \phi_2 = (\pi/N_2) + \text{inv } \phi_1 - (T_2/2R_2) - (W/R_{b2}) \qquad (6\text{-}38)$$
$$r_2 = R_{b2}/\cos \phi_2 \qquad (5\text{-}2)$$

Even Number of Teeth. When the number of teeth in the internal gear is even, the tooth spaces will be opposite each other; hence

$$M_1 = 2(r_2 - W) \qquad (6\text{-}39)$$

Odd Number of Teeth. When the number of teeth is odd, then the tooth spaces are not opposite to each other. In such cases,

$$M_2 = 2[r_2 \cos (90°/N_2) - W] \qquad (6\text{-}40)$$

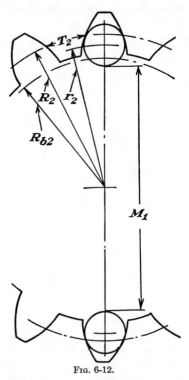

Fig. 6-12.

Example of Roll Measurement of Internal Gear. As a definite example we shall use the 40-tooth internal gear from the preceding example. This gives the following 1-DP values:

$$T_2 = 1.13501 \qquad R_2 = 20.000 \qquad R_{b2} = 18.7938 \qquad N_2 = 40$$
$$\phi_1 = 20° \qquad \text{inv } 20° = 0.014904$$

We will let $W = 0.900$.

$$\text{inv } \phi_2 = \frac{3.1416}{40} + 0.014904 - \frac{1.13501}{40} - \frac{0.900}{18.7938} = 0.017368$$
$$\phi_2 = 21.009° \qquad \cos \phi_2 = 0.93352$$
$$r_2 = \frac{18.7938}{0.93352} = 20.13219$$
$$M_1 = 2(20.13219 - 0.900) = 38.46438 \qquad \text{for 1 DP}$$

When the measurement between the rolls is known, the arc tooth thickness can be determined by rearranging the foregoing equations to solve for the arc tooth thickness.

CHAPTER 7

CONJUGATE ACTION ON HELICAL GEARS

This analysis of helical gears will be restricted to those whose teeth have a uniform axial lead. As with spur gears, for every pair of conjugate helical-gear teeth, there is a definite basic-rack form. The horizontal elements of this basic-rack form will be straight lines, but these straight lines will be at some angle other than a right angle to the side of the basic rack that lies in a plane of rotation of the mating helical gears. The form of the basic rack normal to its horizontal elements is known as the *normal basic-rack form*. Its form in the plane of rotation of the gears is known as the *basic-rack form in the plane of rotation*, or more commonly as its *basic-rack form*. In general, when any element or value of a helical gear is not specifically designated as *normal*, it refers to the conditions in the plane of rotation of the gears.

The angle of departure of the straight-line horizontal elements of the normal basic-rack form from the perpendicular to the plane of rotation is called the *helix angle*. This is also the helix angle on the pitch cylinders of the conjugate helical gears.

The *normal plane* for a pair of helical gears is the plane normal to the straight-line horizontal elements of the normal basic rack. It is not a normal plane to the forms of the teeth on the helical gears. There is no normal plane to helical-gear teeth. The helix angle of the gear changes with its diameter, and any surface that is normal to the helical-gear-tooth elements is a warped surface, and it is useless as a basis for any exact analysis.

The analysis of conjugate gear-tooth action on spur gears can be carried through completely on a single plane. That of helical gears sometimes requires the use of two or three planes. In other words, the kinematics of spur gears is a two-dimensional problem. That of helical gears is a three-dimensional problem. We must therefore develop the faculty of thinking and working in three dimensions if we are to master this subject.

The line of tangency between the pitch surfaces of any pair of gears is the locus of the pitch points, through which the normals from all points of contact of the conjugate tooth profiles must pass. From any point of contact between the teeth of two mating helical gears, or between a gear

141

and its basic rack, two normals may be drawn to the locus of the pitch points. These normals are the lines of action. One line is normal to the tooth surface at the point of contact. This is the *normal line of action.* The other line is normal to the tooth profiles in the plane of rotation and is commonly called the *line of action.*

As conjugate gear-tooth profiles act together, the points of contact will travel along definite paths, which are called *paths of contact.* Here also we have the *normal path of contact,* which is the path in the normal plane, and the *path of contact in the plane of rotation,* or more simply, the path of contact.

The tooth form of the helical gear in the plane of rotation is of constant form, and its conjugate action may be studied here as though it were a spur gear. The tooth form of the helical gear is represented by its intersection with the normal plane is of changing form as the gear rotates— only the normal basic-rack form remains constant here—so that little can be accomplished by attempting to study its action and its limitations in this normal plane. When need arises, as in the determination of the actual contact line across the faces of mating helical gears, such a study can best be made by using the basic-rack form of the system. Except for the duration of contact, the conditions between a helical gear and its basic rack are identical to those existing between a pair of mating helical gears.

All the analysis of spur gears can be applied directly to the study and design of helical gears by using this spur-gear analysis for the conditions in the plane of rotation of helical gears. But with this, we must also consider the additional conditions introduced by the helical form. In almost every case, we can use the same hob, for example, to generate both spur and helical gears. When this is done, however, the form of the helical gear in its plane of rotation is different from the form of the spur gear produced by the same hob. In effect, the form of the basic rack from which the hob is developed becomes the form of the normal basic rack of the two helical gears.

There is, in effect, a double driving action between helical gears. The one is the conjugate gear-tooth action in the plane of rotation, where the tooth forms act together as cams to transmit uniform rotary motion. The other is the rocking or rolling action between mating helices, which also acts to transmit uniform rotary motion from one shaft to another. When the face widths of the gears are large enough, the duration of contact in the plane of rotation becomes of secondary importance. Its major influence on such helical-gear drives is to control the area of the contacting tooth surfaces that are available to carry the load.

Helical gears are inherently smoother running than spur gears when sufficient face width is available to give a continuous helical action as the load is transferred from one pair or group of teeth to another. This is largely because the helix is a form of large and uniform radius of curvature, while the conjugate gear-tooth forms, which must do all the work alone on spur gears, have relatively small and rapidly changing radii of curvature. On the other hand, helical gears introduce an end thrust that is absent on spur gears, an end thrust that may or may not be detrimental to the service required of them.

Double helical gears, with the two halves of opposite hand of helix angle, are often employed where the inherent advantages of helical gears are desired without the detrimental effects of external end thrusts. Such double helical gears are known as *herringbone gears;* the end thrust of one half is counterbalanced by the end thrust in the opposite direction of the other half. Each half of a herringbone gear can be studied as a simple helical gear.

We shall start our analysis of helical gears with a given normal basic-rack form, and determine the conditions in the plane of rotation of the conjugate gear-tooth form.

Normal Basic-rack Form Given. Given the form of the normal basic rack, with the pitch point in the plane of rotation as the origin of the coordinate system, and referring to Fig. 7-1, we have the following:
When x = abscissa of basic-rack profile in plane of rotation

x_n = abscissa of normal basic-rack form

y = ordinate of basic-rack form and of path of contact, both normal and in the plane of rotation

x_p = abscissa of path of contact in plane of rotation

x_{np} = abscissa of path of contact on normal plane

ψ = helix angle at pitch line and angle of basic-rack elements

dx/dy = tan ϕ = pressure angle in plane of rotation

dx_n/dy = tan ϕ_n = normal pressure angle

$$x_p = -y/\tan \phi \tag{1-1}$$

We have the following from the geometrical conditions shown in Fig. 7-1:

$$x_n = x \cos \psi \tag{7-1}$$

and conversely

$$x = x_n/\cos \psi \tag{7-2}$$

$$dx_n/dy = \tan \phi_n = (dx/dy) \cos \psi = \tan \phi \cos \psi \tag{7-3}$$

$$x_{np} = x_p \cos \psi = -y \cos \psi/\tan \phi = -y \left[\frac{\cos^2 \psi}{\tan \phi_n}\right] \tag{7-4}$$

$$x_p = x_{np}/\cos \psi \tag{7-5}$$

The values of x_p and x_{np} will be negative when the values of x and x_n are positive, because the two sets of points will always be on opposite sides of the origin.

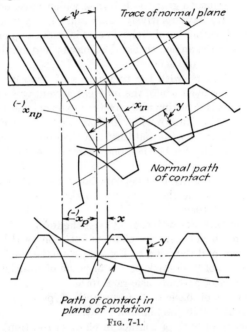

FIG. 7-1.

In order to plot the paths of contact, a series of points on the basic-rack profile together with the values of their tangents are used. From these values, the corresponding points on the paths of contact are determined by means of Eq. (1-1) and (7-4). These values can be plotted to any desired scale.

Example of Change from Normal Plane to Plane of Rotation. As a definite example we shall use the full-cycloidal rack profile as the form of the normal profile of the basic rack. The equation of this form is as follows: For the addendum of the basic rack, we have, when

a = radius of rolling circle

ϵ = angle of rotation of rolling circle

$$x_n = a(\epsilon - \sin \epsilon) \tag{1-46}$$
$$y = a(1 - \cos \epsilon) \tag{1-47}$$

For the dedendum of the basic rack, we have

$$x_n = -a(\epsilon - \sin \epsilon) \tag{7-6}$$
$$y = -a(1 - \cos \epsilon) \tag{7-7}$$

In other words, this form is symmetrical about the pitch point, and the coordinates of the addendum and of the dedendum are alike except for the change in sign. For the addendum, we have

$$\tan \phi_n = dx_n/dy = y/\sqrt{2ay - y^2} \qquad (1\text{-}49)$$
$$x_{np} = -y/\tan \phi_n = -\sqrt{2ay - y^2} \qquad (1\text{-}51)$$

Equation (1-51) is the equation of the rolling circle of the addendum at its starting position.

We will use a value of 0.500 for a. From this value and with the use of the foregoing equations, using values of ϵ from zero to 180 deg, varying by 15-deg increments, we obtain the values tabulated in Table 7-1. These values are also plotted in Fig. 7-2. That part of the path of contact which is included in the tabulated values is shown as a full line. The other half of this path of contact, when ϵ varies from 180 to

TABLE 7-1. COORDINATES OF CYCLOIDAL NORMAL BASIC RACK AND PATH OF CONTACT
(Plotted in Fig. 7-2)

ϵ, deg	x_n, in.	y, in.	$\tan \phi_n$	x_{np}, in.
180	1.57080	1.00000	∞	−0.00000
165	1.31049	0.98297	7.59575	−0.12938
150	1.05900	0.93301	3.73205	−0.25000
135	0.82455	0.85355	2.41421	−0.35356
120	0.61419	0.75000	1.73205	−0.43301
105	0.43333	0.62941	1.30323	−0.48296
90	0.28504	0.50000	1.00000	−0.50000
75	0.17154	0.37059	0.76733	−0.48296
60	0.09059	0.25000	0.57735	−0.43301
45	0.03915	0.14650	0.41421	−0.35356
30	0.01180	0.06699	0.26795	−0.25000
15	0.00149	0.01704	0.13165	−0.12938
0	0.00000	0.00000	0.00000	0.00000
− 15	−0.00149	−0.01704	0.13165	0.12938
− 30	−0.01180	−0.06699	0.26795	0.25000
− 45	−0.03915	−0.14650	0.41421	0.35356
− 60	−0.09059	−0.25000	0.57735	0.43301
− 75	−0.17154	−0.37059	0.76733	0.48296
− 90	−0.28540	−0.50000	1.00000	0.50000
−105	−0.43333	−0.62941	1.30323	0.48296
−120	−0.61419	−0.75000	1.73205	0.43301
−135	−0.82455	−0.85355	2.41421	0.35356
−150	−1.05900	−0.93301	3.73205	0.25000
−165	−1.31049	−0.98297	7.59575	0.12938
−180	−1.57080	−1.00000	∞	0.00000

360 deg is shown in dotted lines. These values are for the form of the basic rack in the normal plane and for its path of contact in that same plane.

Basic-rack Profile and Path of Contact in Plane of Rotation. We shall next determine the values of the coordinates of the basic-rack profile and its path of contact in the plane of rotation. We have for the addendum of this example

$$x = (a/\cos \psi)(\epsilon - \sin \epsilon)$$
$$y = a(1 - \cos \epsilon)$$
$$\tan \phi = \tan \phi_n/\cos \psi = y/\cos \psi \sqrt{2ay - y^2}$$
$$x_p = x_{np}/\cos \psi = -\sqrt{2ay - y^2}/\cos \psi$$

(1-47)

For this example we shall use the following values:

$$\psi = 30° \qquad \cos \psi = 0.86603$$

TABLE 7-2. COORDINATES IN PLANE OF ROTATION OF HELICAL CYCLOIDAL BASIC RACK AND PATH OF CONTACT

(Plotted in Fig. 7-3)

ϵ, deg	y, in.	x, in.	$\tan \phi$	x_p, in.
180	1.00000	1.81380	∞	0.00000
165	0.98297	1.51322	8.77081	−0.11204
150	0.93301	1.22283	4.30940	−0.21651
135	0.85355	0.95211	2.78769	−0.30619
120	0.75000	0.70921	2.00000	−0.37500
105	0.62941	0.50037	1.50484	−0.41826
90	0.50000	0.32955	1.15470	−0.43301
75	0.37057	0.19808	0.88604	−0.41826
60	0.25000	0.10460	0.66667	−0.37500
45	0.14650	0.04521	0.47829	−0.30619
30	0.06699	0.01363	0.30940	−0.21651
15	0.01704	0.00172	0.15202	−0.11204
0	0.00000	0.00000	0.00000	0.00000
− 15	−0.01704	−0.00172	0.15202	0.11204
− 30	−0.06699	−0.01363	0.30940	0.21651
− 45	−0.14650	−0.04521	0.47829	0.30619
− 60	−0.25000	−0.10460	0.66667	0.37500
− 75	−0.37057	−0.19808	0.88604	0.41826
− 90	−0.50000	−0.32955	1.15470	0.43301
−105	−0.62941	−0.50037	1.50484	0.41826
−120	−0.75000	−0.70921	2.00000	0.37500
−135	−0.85355	−0.95211	2.78769	0.30619
−150	−0.93301	−1.22283	4.30940	0.21651
−165	−0.98297	−1.51322	8.77081	0.11204
−180	−1.00000	−1.81380	∞	0.00000

From these equations and the values tabulated in Table 7-1, we obtain the values tabulated in Table 7-2. These values are also plotted in Fig. 7-3. As before, the values for the dedendum of the basic-rack form are the same as those for the addendum except for the change in sign.

The path of contact in the plane of rotation is no longer two tangent circles. This form has become two tangent ellipses, with a width on the minor axis of 0.86603 and a length on the major axis of 1.000. The resulting conjugate tooth forms are no longer cycloidal ones.

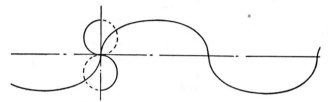

Fɪɢ. 7-2.   Normal basic rack of cycloid.

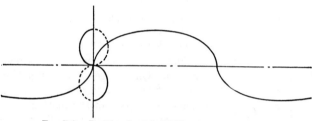

Fɪɢ. 7-3.   Basic rack of cycloid in plane of rotation.

Conjugate Gear-tooth Form in Plane of Rotation. The conjugate gear-tooth form can be determined only in the plane of rotation. All the following symbols are for values in the plane of rotation and are also the same as those for spur gears:

x = abscissa of basic-rack form

y = ordinate of basic-rack form and of path of contact

ϕ = pressure angle

x_p = abscissa of path of contact

θ = vectorial angle of conjugate gear-tooth form

r = radius to conjugate gear-tooth form

R = pitch radius of gear

$$r = \sqrt{(R - y)^2 + x_p^2} \qquad (1\text{-}2)$$
$$\theta = [(x - x_p)/R] + \tan^{-1}[x_p/(R - y)] \qquad (1\text{-}3)$$

Here we use the same coordinate system as was used for spur gears. The vectorial angle is equal to zero at the pitch point. Its values are minus for the addendum of the conjugate gear-tooth form and are plus for its dedendum.

Example of Conjugate Gear-tooth Form. As a definite example, we shall continue with the previous one and determine the tooth form of a 4-lobed helical rotor whose basic-rack form is given in Table 7-2.

When N = number of teeth in gear

p_n = normal circular pitch, in.

p = circular pitch in plane of rotation, in.

ψ = helix angle at pitch line

$$p = p_n/\cos \psi \qquad (7\text{-}8)$$
$$p_n = p \cos \psi \qquad (7\text{-}9)$$

For this example, we have the values of the coordinates of the basic rack and its path of contact that are tabulated in Table 7-2. We have also the following values:

$$N = 4 \qquad \psi = 30° \qquad \cos \psi = 0.86603 \qquad p_n = 6.2832 \qquad a = 0.500$$

$$p = \frac{6.2832}{0.86603} = 7.25521$$

$$R = pN/2\pi = \frac{4 \times 7.25521}{6.2832} = 4.61880$$

The coordinates of the conjugate gear-tooth profile in the plane of rotation have

FIG. 7-4.

been calculated from these values. They are tabulated in Table 7-3 and plotted in Fig. 7-4.

Contact Line between Helical-gear-tooth Surfaces. On spur gears the contact between meshing teeth is a straight line, which is parallel to the axes of the mating gears. As the gears revolve, this contact line travels from the bottom of the active profile of the driving member to the top of the active profile. On the driven member, this contact line travels from the top of its active profile to its bottom. It is the same contact line, but the two gears are revolving in opposite directions.

On helical gears, except for the involute form, the contact between mating teeth is a curved line that will be in a generally diagonal direction across the face width of the gears. It will reach from the top to the bottom of the active profile of the teeth when the face width is great enough to permit it to do so. With wide face widths, this contact line will be repeated on several pairs of mating teeth. As these gears revolve, this contact line will travel in an axial direction across the face width of the mating gears. A point on these contact lines, such as the point where the contact line intersects a definite plane of rotation, will travel vertically over the active profile as in the case of a spur gear, but the contact line as a whole will travel in an axial direction.

In order to determine the form of the projection of these contact lines, either on a plane parallel to the pitch plane of the basic rack or on a

TABLE 7-3. COORDINATES OF TOOTH PROFILE OF 4-LOBED ROTOR IN PLANE OF
ROTATION
(Plotted in Fig. 7-4)

ϵ, deg	y, in.	r, in.	θ, rad
180	1.00000	3.61880	0.39270
165	0.98297	3.63756	0.32107
150	0.93301	3.69214	0.25295
135	0.85355	3.77768	0.19121
120	0.75000	3.88693	0.13811
105	0.62941	4.01126	0.09443
90	0.50000	4.14150	0.06035
75	0.37057	4.26877	0.03530
60	0.25000	4.38486	0.01820
45	0.14645	4.48282	0.00772
30	0.06699	4.55696	0.00230
15	0.01704	4.60312	0.00029
0	0.00000	4.61880	0.00000
− 15	−0.01704	4.63819	−0.00047
− 30	−0.06699	4.69079	−0.00366
− 45	−0.14645	4.77508	−0.01191
− 60	−0.25000	4.88322	−0.02696
− 75	−0.37057	5.00687	−0.04981
− 90	−0.50000	5.13708	−0.08071
−105	−0.62941	5.26485	−0.11936
−120	−0.75000	5.38188	−0.16501
−135	−0.85355	5.48091	−0.21651
−150	−0.93301	5.55603	−0.27264
−165	−0.98297	5.60289	−0.33188
−180	−1.00000	5.61880	−0.39270

plane parallel to the one that contains the axes of the gears, we must
know the coordinates of the tooth profile of the basic rack and of its path
of contact in the plane of rotation, as well as the value of the helix angle.
Thus when

z = abscissa of projection of contact line on plane parallel to pitch
plane of basic rack and on plane containing axes of gears

y_1 = ordinate of projection of contact line on plane containing axes of
gears

y_2 = ordinate of projection of contact line on plane parallel to pitch
plane of basic rack

x_p = ordinate of path of contact in plane of rotation

x = abscissa of basic-rack profile in plane of rotation
y = ordinate of basic-rack profile and of path of contact
ψ = helix angle of gears at pitch line

$$z = -(x - x_p)/\tan \psi \qquad (7\text{-}10)^1$$
$$y_1 = y \qquad (7\text{-}11)$$
$$y_2 = x_p \qquad (7\text{-}12)$$

Example of Contact Line. As a definite example, the coordinates of the two projections of the contact line between two similar rotors of the size and form developed in the preceding example have been calculated. The start of this contact line is at the

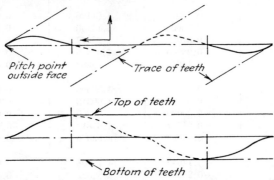

Pitch point outside face

Trace of teeth

Top of teeth

Bottom of teeth

Fig. 7-5. (Top) Projection of contact on pitch plane of basic rack. Contact on driving face in solid line; contact on nondriving face in dotted line. (Bottom) Projection of contact on plane containing axes of gears.

pitch point in the plane of rotation at one side of the face. The face widths of these rotors have been made great enough so that full contact exists around the entire rotor tooth form. In other words, one-half of the contact in this example is between nondriving sides of the mating teeth or lobes. This is accomplished in this example by making the face width equal to one-fourth of the lead of the helix. Thus when

L = lead of helix, in.
R = pitch radius, in.
ψ = helix angle at pitch line

$$L = 2\pi R/\tan \psi \qquad (7\text{-}13)$$

In this example

$$R = 4.61880 \qquad \psi = 30° \qquad \tan \psi = 0.57735$$
$$L = \frac{6.2832 \times 4.61880}{0.57735} = 50.2656$$

The face-width or length of these rotors is, therefore,

$$\frac{50.2656}{4} = 12.5664$$

---

[1] The sign of $x - x_p$ in this equation will be plus or minus depending upon the direction of the angle on the driving side of the basic-rack profile. The signs are given here to match the conditions shown in Fig. 7-1.

The calculated values of these coordinates are tabulated in Table 7-4 and plotted in Fig. 7-5.

Forms of Fillets, Internal Helical Gears, etc. The forms of the trochoidal fillets, the limitations to conjugate gear-tooth action, internal-gear problems, etc., must all be studied in the plane of rotation.   When the conditions of the drive in the plane of rotation are known, all further analyses are identical to those for spur gears.   If the normal profile of the basic rack is the starting point for the analysis, then its form and its path of contact in the plane of rotation must be established first.   After that is accomplished, it is treated as a spur gear.   If conditions of contact or any other features relating to its helical form are required, direct reference to the required conditions on the basic rack will lead to the simplest and most direct solution.

TABLE 7-4.   COORDINATES OF PROJECTIONS OF CONTACT LINE
(Plotted in Fig. 7-5)

$x - x_p$, in.	z, in.	y_1, in.	y_2, in.
0.00000	0.00000	0.00000	0.00000
0.23014	0.39861	0.06699	0.21651
0.47960	0.83069	0.25000	0.37500
0.76256	1.32078	0.50000	0.43301
1.08421	1.87791	0.75000	0.37500
1.43934	2.49301	0.93301	0.21651
1.81380	3.14160	1.00000	0.00000
2.18826	3.79018	0.93301	−0.21651
2.54339	4.40528	0.75000	−0.37500
2.86504	4.96239	0.50000	−0.43301
3.14800	5.45249	0.25000	−0.37500
3.39746	5.88457	0.06699	−0.21651
3.62760	6.28320	0.00000	0.00000
3.85774	6.68180	−0.06699	0.21651
4.10720	7.11388	−0.25000	0.37500
4.39016	7.43077	−0.50000	0.43301
4.71181	8.16109	−0.75000	0.37500
5.06694	8.77620	−0.93301	0.21651
5.44140	9.42480	−1.00000	0.00000
5.81586	10.07337	0.03301	0.21651
6.17099	10.68847	−0.75000	−0.37500
6.49264	11.24559	−0.50000	−0.43301
6.77560	11.73569	−0.25000	−0.37500
7.02506	12.16777	−0.06699	−0.21651
7.25520	12.56640	0.00000	0.00000

CHAPTER 8

INVOLUTOMETRY OF HELICAL GEARS

The involute helical gear stands in a class by itself. It retains all the unique properties of the involute spur gear, which do not need to be repeated here. In addition, when the helical action is adequate, its contact ratio in the plane of rotation is of secondary importance; hence many of the limitations to spur-gear design do not exist here. Undercut tooth profiles are about the only limitations to helical involute gear-tooth design.

It will assist to a better understanding and to a greater appreciation of the properties of the involute helical gears if we have a clear and simple

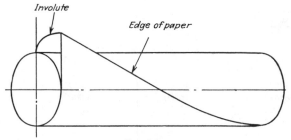

Involute

Edge of paper

Fig. 8-1.

mental picture of the physical shape or nature of the development of this involute helicoid. Thus if we take a piece of paper with square edges and wrap it tightly about a cylinder with the outside edge parallel to the axis of the cylinder, and then unwind this paper, the outside edge of it as it sweeps through space generates the surface of an involute spur gear. Here the cylinder is the base cylinder of the spur gear. Now if we cut the outside edge of this paper at an angle as indicated in Fig. 8-1, this angular edge becomes a helix of uniform lead when this paper is wrapped tightly on the cylinder. As we unwind this paper with the angular edge, each point on this edge describes an involute curve, but each point starts from a different angular position on the cylinder. This cylinder is also the base cylinder of the involute helicoid. The surface described by the angular edge of this paper is the surface of an involute helicoid, or that of a helical involute gear.

The helix angle of the angular edge of the paper when it is wrapped tightly around the cylinder is the helix angle of the involute helicoid on its base cylinder. Thus there are only two fixed or constant values of the involute helicoid: first, the size of the base cylinder, and second, the helix angle on the base cylinder. All other proportions and values are variables that are dependent upon these two fixed values. Many otherwise perplexing problems of involute helical gears can be solved simply and

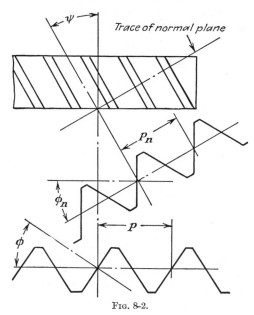

Fig. 8-2.

directly by referring back to this simple fundamental development of the involute helicoid. Thus when

R_b = radius of base cylinder, in.
ψ_b = helix angle on base cylinder
L = lead of helix, in.

$$L = 2\pi R_b/\tan \psi_b \qquad (8\text{-}1)$$
$$\tan \psi_b = 2\pi R_b/L \qquad (8\text{-}2)$$
$$R_b = L \tan \psi_b/2\pi \qquad (8\text{-}3)$$

Basic Relationships of Helical Involute Gear Elements. We will now develop some of the basic relationships of the helical involute gear. Here, as with spur gears, we will start with the basic rack of the system. Referring to Fig. 8-2, we have the following:

ϕ = pressure angle in plane of rotation
ϕ_n = normal pressure angle of basic rack
N = number of teeth in gear
p = circular pitch, plane of rotation, in.
ψ = helix angle of basic rack and at R
p_n = normal circular pitch, in.
R = pitch radius of gear, in.
P = diametral pitch, plane of rotation
P_n = normal diametral pitch

$$\tan \psi = 2\pi R/L \tag{8-4}$$
$$L = 2\pi R/\tan \psi = 2\pi R_b/\tan \psi_b \tag{7-13}$$
$$p = p_n/\cos \psi \tag{7-8}$$
$$p_n = p \cos \psi \tag{7-8}$$
$$\tan \phi_n = \tan \phi \cos \psi \tag{7-3}$$
$$\tan \phi = \tan \phi_n/\cos \psi \tag{8-5}$$
$$R_b = R \cos \phi \tag{4-6}$$

Solving Eq. (7-13) for $\tan \psi_b$, we have

$$\tan \psi_b = (R_b/R) \tan \psi$$

But

$$R_b/R = \cos \phi \tag{4-8}$$

Whence

$$\tan \psi_b = \tan \psi \cos \phi \tag{8-6}$$

This last relationship is a very important and useful one. The product of the cosine of the pressure angle of an involute helicoid at any radius and the tangent of the helix angle at that same radius is equal to a constant for any given helicoid, and that constant is the value of the tangent of the helix angle on the base cylinder.

Generating Helical Involute Gears. There are two different methods widely used for generating helical involute gears. One method uses a generating tool that represents the form of the normal basic rack either in the form of a hob or else in the form of a rack-shaped planing tool. The other method uses a helical pinion-shaped cutter that has a definite lead. This cutter and the gears it generates act together as a pair of helical gears; hence the lead of the generated gear must be proportional in the ratio of the tooth numbers of the gear and the cutter to the lead on the cutter. In other words, these helical pinion-shaped cutters are made to produce gears of definite nominal helix angles. Special helical pinion-shaped cutters must be made to produce other helix angles or leads.

With the first method, where standard basic-rack forms are used as the basis of the helical-gear design, this basic-rack form is the normal basic rack of the helical gears. Where helical pinion-shaped cutters are used, and the forms of these cutters are based on standard basic-rack forms, the basic-rack form here is the basic rack in the plane of rotation for the helical-gear design. In this case, the design is exactly the same as that for spur involute gears. Hence in the development of relationships for use in helical-gear design, we must deal with both conditions. The further consideration of helical involute gear design will be in the form of individual problems.

Problem 8-1. *Given the normal diametral pitch, numbers of teeth in the gears, and the center distance, to determine the leads and the helix angle at the pitch line.*

When N_1 = number of teeth in pinion
N_2 = number of teeth in gear
P_n = normal diametral pitch
C = center distance, in.
ψ = helix angle of basic rack
L_1 = lead of pinion, in.
L_2 = lead of gear, in.

$$\cos \psi = (N_1 + N_2)/2P_nC \tag{8-7}$$
$$L_1 = \pi N_1/P_n \sin \psi \tag{8-8}$$
$$L_2 = \pi N_2/P_n \sin \psi \tag{8-9}$$

It is apparent from an inspection of Eq. (8-8) and (8-9) that the leads are directly proportional to the numbers of teeth in the gears. Therefore whenever any approximation or rounding off of decimals is used for the values of the leads of mating helical gears, care should be taken to ensure that the leads actually used are directly proportional to the numbers of teeth in the gears.

Example of Leads and Helix Angle. As a definite example we shall use a pair of helical involute gears with the following given values:

$$P_n = 8 \qquad N_1 = 20 \qquad N_2 = 55 \qquad C = 5.250$$
$$\cos \psi = \frac{20 + 55}{2 \times 8 \times 5.250} = 0.89286$$
$$\psi = 26.765° \qquad \sin \psi = 0.45033$$
$$L_1 = \frac{3.1416 \times 20}{8 \times 0.45033} = 17.44054$$

$$L_2 = \frac{3.1416 \times 55}{8 \times 0.45033} = 47.96149$$

If we wish to round off these values of the leads, we can use

$$L_1 = 17.440 \text{ in.}$$

Then

$$L_2 = {}^{55}\!/_{20} \times 17.440 = 47.960 \text{ in.}$$

Problem 8-2. *Given the helix angle at the pitch radii, the normal diametral pitch, and the numbers of teeth in a pair of helical gears, to determine the center distance.*

When all symbols are the same as before, we have

$$C = (N_1 + N_2)/(2P_n \cos \psi) \qquad (8\text{-}10)$$

Example of Center Distance. As a definite example we will use a pair of helical gears with the following given values:

$$\psi = 25° \qquad \cos \psi = 0.90631 \qquad P_n = 12 \qquad N_1 = 30 \qquad N_2 = 60$$

$$C = \frac{30 + 60}{2 \times 12 \times 0.90631} = 4.13765$$

Problem 8-3. *Given the number of teeth in a helical gear, the helix angle of the basic rack, and the proportions of the normal basic rack, to determine the pitch radius, the radius of the base cylinder, the lead of the helix, and the helix angle on the base cylinder.*

When N = number of teeth in gear

ψ = helix angle of basic rack

P_n = normal diametral pitch of basic rack

R = pitch radius of gear, in.

ϕ_n = normal pressure angle of basic rack

ϕ = pressure angle of basic rack in plane of rotation

R_b = radius of base cylinder, in.

L = lead of helix, in.

ψ_b = helix angle on base cylinder

$$R = N/(2P_n \cos \psi) \qquad (8\text{-}11)$$
$$\tan \phi = \tan \phi_n/\cos \psi \qquad (8\text{-}5)$$
$$R_b = R \cos \phi \qquad (4\text{-}9)$$
$$\tan \psi_b = \tan \psi \cos \phi \qquad (8\text{-}6)$$
$$L = \pi N/P_n \sin \psi \qquad (8\text{-}8)$$

Example. As a definite example we shall use a helical gear with the following given values:

$$N = 40 \qquad \phi_n = 14.50° \qquad \tan \phi_n = 0.25862 \qquad P_n = 10$$
$$\psi = 30° \qquad \cos \psi = 0.86603 \qquad \tan \psi = 0.57735 \qquad \sin \psi = 0.50000$$

$$R = \frac{40}{2 \times 10 \times 0.86603} = 2.30940$$

$$\tan \phi = \frac{0.25862}{0.86603} = 0.29863$$

$$\phi = 16.627° \qquad \cos \phi = 0.95819$$

$$R_b = 2.30940 \times 0.95819 = 2.21284$$

$$\tan \psi_b = 0.57735 \times 0.95819 = 0.55321$$

$$\psi_b = 28.952°$$

$$L = \frac{3.1416 \times 40}{10 \times 0.50000} = 25.13280$$

All the problems in Chap. 5 for involute spur gears may be used directly for helical involute gears when the values for the basic-rack or tooth proportions have been established in the plane of rotation. This applies only to the proportions and sizes of the involute profiles themselves and to their interrelations with each other.

Problem 8-4. *Given the center distance, numbers of teeth or reduction ratio, and the normal basic-rack proportions, to determine the tooth proportions and the hobbing data.*

There are several different solutions possible for this problem.

1. We can determine the helix angle so that the sum of the pitch radii are equal to the given center distance. This is the more common solution, but it is not always the best one. Here we can proportion the teeth in the conventional manner, or we can increase the addendum of the smaller gear to avoid undercut, for example, and decrease the addendum of the mating gear an equal amount. This solution may introduce values for the leads that may require a considerable train of change gears on the hobbing machine.

2. The numbers of teeth may be small, and excessive undercut may be present, if the first solution is followed. Here, as with spur gears, we may use a smaller diameter and helix angle for generating the gears than is required to make the sum of the nominal pitch radii equal to the specified center distance. Then we shall determine the basic-rack proportions in the plane of rotation and proceed as with spur gears as in Prob. 5-13.

3. We can follow the practice outlined in Prob. 5-13 but shall first make a trial solution for the leads, and then select leads approximating the calculated ones but leads that are easy to obtain on the hobbing machine. Then using these simple leads as a start, we shall recompute the helix angle of generation to suit them, determine the basic-rack proportions in the plane of rotation, and proceed along the same lines as those shown in Prob. 5-13. This last method is the solution that will be carried through here.

Let ϕ_{nc} = pressure angle of rack-shaped cutter or hob

P_{nc} = diametral pitch of hob

a_h = addendum of hob, in.

C_1 = center distance with pressure angle of ϕ_1 in plane of rotation, in.

C_2 = given center distance of operation, in.

N_1 = number of teeth in pinion

N_2 = number of teeth in gear

R_{o1} = outside radius of pinion, in.

R_{o2} = outside radius of gear, in.

R_{r1} = root radius of pinion, in.

R_{r2} = root radius of gear, in.
L_1 = lead of helix of pinion, in.
L_2 = lead of helix of gear, in.
R_1 = pitch radius of pinion, in.
R_2 = pitch radius of gear, in.
b_1 = dedendum of pinion, in.
b_2 = dedendum of gear, in.
ψ_1 = helix angle of generation
ψ_2 = helix angle of operation
ϕ_1 = pressure angle of generation in plane of rotation
ϕ_2 = pressure angle of operation in plane of rotation
P_1 = diametral pitch of generation in plane of rotation
h_t = whole tooth depth of gear teeth, in.

The first step is to make a trial calculation for the leads of the gears. When the tooth numbers are small and excessive undercut may be present, the center distance used for this trial solution should be a suitable amount smaller than the specified center distance of operation.

For example, if we have a pair of such gears with 20 and 30 teeth, the sum of their tooth numbers is equal to 50. The nominal center distance for such spur gears of 1 DP is 25.000 in. A good design for such a combination would use an operating center distance of 25.5931 in., or an increase in center distance of 0.5931 in. If the helical gears are of 8 DP, we would divide this difference by 8, which gives us a value of 0.0741 in. We would then use a value for C_2 in the trial calculations for the leads somewhere between 0.070 and 0.080 in. smaller than the specified value. When the tooth numbers are large and no danger of excessive undercut exists, we would use the specified value of C_2 in the trial calculations for the leads.

Trial Calculations for the Leads

$$\cos \psi_1 = (N_1 + N_2)/2P_{nc}C_2 \qquad (8\text{-}7)$$
$$L_1 = \pi N_1/P_{nc} \sin \psi_1 \qquad (8\text{-}8)$$
$$L_2 = \pi N_2/P_{nc} \sin \psi_1 \qquad (8\text{-}8)$$

When the trial values for these leads are obtained, we will select values for them that are reasonably close to the calculated ones, leads that can be readily obtained on the hobbing machines, and leads that are directly proportional to the numbers of teeth in the gears. Then we proceed as follows:

Calculations for the Generating Helix Angle

$$\sin \psi_1 = \pi N_1/P_{nc}L_1 = \pi N_2/P_{nc}L_2 \qquad (8\text{-}12)$$
$$\tan \phi_1 = \tan \phi_{nc}/\cos \psi_1 \qquad (8\text{-}5)$$
$$P_1 = P_{nc} \cos \psi_1 \qquad (8\text{-}13)$$

$$C_1 = (N_1 + N_2)/2P_1 \tag{5-45}$$
$$\cos \phi_2 = C_1 \cos \phi_1 / C_2 \tag{5-46}$$

From here on, we follow the spur-gear problem as given in Prob. 5-13.

$$R_{r1} + R_{r2} = C_1 - 2a_h + [C_1(\text{inv } \phi_2 - \text{inv } \phi_1)/\tan \phi_1] \tag{5-48}$$

Here, as with spur gears, the sum of the root radii is a constant regardless of how the teeth are proportioned. At times we must hold the outside radius of one of the two members of the pair to a definite size because of other structural conditions of the particular mechanism. In such cases we would determine the value for the whole tooth depth, subtract it from the fixed outside radius, and thus obtain the value of the root radius for that particular member of the pair. This root radius would then be subtracted from the sum of the root radii, and the remainder would be the value of the radius of the root of the mating gear. By adding the value of the whole tooth depth to this root radius, we would obtain the outside radius of this mating gear.

The flexibility in the design of helical involute gears is almost unlimited. When the helical contact is adequate, we do not need to concern ourselves with the contact ratio of the involute profiles in the plane of rotation. Our only serious limitation is undercut. This can and should always be avoided.

Calculations for Tooth Proportions. When all other symbols are the same as before and

h_{t1} = nominal whole depth of tooth, in.

c_1 = nominal clearance, in.

$$h_t = \frac{h_{t1}}{h_{t1} + C_1} [C_2 - (R_{r1} + R_{r2})] \tag{5-51}$$

When there is no restriction on the outside radius of either gear of the pair, the following equations prove effective for determining the several tooth proportions:

$$b_1 = \frac{C_2 - (R_{r1} + R_{r2})}{1 + \sqrt{N_1/N_2}} \tag{5-49}$$

$$b_2 = C_2 - (R_{r1} + R_{r2}) - b_1 \tag{5-50}$$

$$R_1 = \frac{N_1 C_2}{N_1 + N_2} \tag{5-52}$$

$$R_2 = \frac{N_2 C_2}{N_1 + N_2} = C_2 - R_1 \tag{5-53}$$

$$R_{r1} = R_1 - b_1 \tag{5-54}$$

$$R_{r2} = R_2 - b_2 \tag{5-55}$$

$$R_{o1} = R_{r1} + h_t \tag{5-56}$$

$$R_{o2} = R_{r2} + h_t \tag{5-57}$$

Examples. In order to give some indication of the great flexibility of helical involute gear design, we shall use several different examples: each of them having to meet different sets of conditions. The examples given are only samples and do not begin to exhaust the many varying sets of conditions that can be handled effectively by this method.

First Example. As the first definite example we shall use the following: Reduction ratio of three to one; outside diameter of pinion to be held to 0.750 in., with 1.250 in. center distance; some standard 14½-deg hob is to be used to generate the helical gears. This gives the following values:

$$N_2 = 3N_1 \qquad C_2 = 1.250 \qquad R_{o1} = 0.3750$$

The first step is to determine the normal diametral pitch and numbers of teeth for these gears. If they were spur gears of conventional design, then

$$R_1 = \frac{N_1 C_2}{N_1 + N_2} = \frac{1.250}{4} = 0.3125$$

$$P_1 = \frac{N_1}{0.625}$$

If we use $P_1 = 16$, then

$$N_1 = 10 \qquad \text{and} \qquad N_2 = 30$$

The sum of the tooth numbers in this example is 40. For 1-DP spur gears, the center distance should be about 20.7428 in. to avoid undercut, an increase of 0.7428 in. over the proportional center distance. Dividing this value by 16, we obtain a value of 0.0464 in. Hence we shall use as a trial value the following:

$$C_1 = 1.250 - 0.047 = 1.203$$

But we must use a hob of some standard diametral pitch. For the helical gears, it must be of a finer pitch than for a spur gear. We have the relationship

$$P_1 = P_{nc} \cos \psi_1 \qquad (8\text{-}13)$$

In this trial example, we have

$$P_1 = \frac{N_1 + N_2}{2 \times 1.203} = 16.625$$

Assuming a trial value of $\psi_1 = 30°$,

$$\cos \psi_1 = 0.86603$$

Then

$$P_{nc} = \frac{P_1}{\cos \psi_1} = \frac{16.625}{0.86603} = 19.196$$

We must use a standard diametral pitch, so in this example it must be either 18 or 20. With the small numbers of teeth involved, the better selection would be the finer pitch, or 20 DP. Now we have the following values for this problem:

$$P_{nc} = 20 \qquad N_1 = 10 \qquad N_2 = 30 \qquad C_2 = 1.250 \qquad R_{o1} = 0.375$$
$$\phi_{nc} = 14.50° \qquad \cos \phi_{nc} = 0.96815 \qquad \tan \phi_{nc} = 0.25862 \qquad a_h = 0.0579$$

TRIAL CALCULATION FOR LEAD

$$\cos \psi_1 = \frac{40}{2 \times 20 \times 1.250} = 0.80000$$

$$\psi_1 = 36.870° \qquad \sin \psi_1 = 0.60000$$

$$L_1 = \frac{3.1416 \times 10}{20 \times 0.600} = 2.6180$$

$$L_2 = \frac{3.1416 \times 30}{20 \times 0.600} = 7.8540$$

We shall select the following values for the leads:

$$L_1 = 2.600 \qquad L_2 = {}^{30}\!/_{10} \times 2.600 = 7.800$$

We shall now determine the actual helix angle of generation

$$\sin \psi_1 = \frac{3.1416 \times 10}{20 \times 2.600} = 0.60415$$

$$\psi_1 = 37.168° \qquad \cos \psi_1 = 0.79687$$

$$\tan \phi_1 = \frac{0.25862}{0.79687} = 0.32454$$

$$\phi_1 = 17.980° \qquad \cos \phi_1 = 0.95116 \qquad \text{inv } \phi_1 = 0.010724$$

$$P_1 = 20 \times 0.79687 = 15.9374$$

We shall now treat the problem the same as if it were for a pair of spur gears.

$$C_1 = \frac{40}{2 \times 15.9374} = 1.25491$$

$$\cos \phi_2 = \frac{1.25491 \times 0.95116}{1.250} = 0.95489$$

$$\phi_2 = 17.275° \qquad \text{inv } \phi_2 = 0.009481$$

$$R_{r1} + R_{r2} = 1.25491 - 0.1158 + \frac{1.25491(0.009481 - 0.010724)}{0.32454} = 1.13863$$

$$h_{t1} = \frac{2.157}{20} = 0.1079 \qquad c_1 = 0.0079$$

$$h_t = \left(\frac{0.1079}{0.1158}\right)(1.250 - 1.13863) = 0.10377$$

With the fixed value of $R_{o1} = 0.375$, we have

$$R_{r1} = 0.3750 - 0.10377 = 0.27123$$
$$R_{r2} = 1.13863 - 0.27123 = 0.86740$$
$$R_{o2} = 0.86740 + 0.10377 = 0.97117$$

The 10-tooth pinion should be checked to make sure that its root radius is above the undercut limit. If it is below, either its value must be increased or else the drive should be recalculated using a finer pitch and a greater number of teeth. For the undercut limit, we have the following from the analysis of spur gears:

When R_u = radius to undercut limit, in.

R'_1 = generating pitch radius of pinion, in.

ϕ_1 = generating pressure angle

$$R'_1 = \frac{N_1}{2P_1} = \frac{10}{2 \times 15.9374} = 0.31373$$

$$R_u = R'_1 \cos^2 \phi_1 - c_1 = 0.31373 \times (0.95116)^2 - 0.0079 = 0.27593$$

A comparison of the values of R_{r1} and R_u shows that there will be undercut on the 10-tooth pinion. The difference is small so that we can increase the root radius of the pinion and decrease the root radius of the gear the same amount. We must leave the outside radius of the pinion unchanged because it is a fixed value. Otherwise, we would increase this outside radius the same amount as the root radius. We must reduce the outside radius of the gear the same amount that we increase the root radius of the pinion so as to maintain the clearance. The minimum change would be to make the root radius of the pinion equal to the undercut radius. This change would be

$$0.27593 - 0.27123 = 0.0047 \text{ in.}$$

We shall allow a little margin, however, and choose a correction that will make the value of the root radius of the pinion an even decimal. Hence we shall make a correction of 0.00877 in. on both members. This gives the following for the final values:

$$R_{r1} = 0.27123 + 0.00877 = 0.2800$$
$$R_{o1} = 0.375 \qquad \text{Fixed value from the start}$$
$$R_{r2} = 0.86740 - 0.00877 = 0.85863$$
$$R_{o2} = 0.97117 - 0.00877 = 0.96240$$

These last values are the actual dimensions to which the gears would be made.

If we wish to determine the tooth forms of these gears in the plane of rotation, we must first determine the arc tooth thicknesses of the teeth at their generating radii, and then proceed as in Prob. 5-1.

When T_1 = arc tooth thickness of pinion at generating radius, in.

$\quad T_2$ = arc tooth thickness of gear at generating radius, in.

$\quad R'_1$ = generating radius of pinion, in.

$\quad R'_2$ = generating radius of gear, in.

$\quad p_n$ = circular pitch of hob, in.

$\quad p$ = circular pitch of basic rack (hob) in plane of rotation, in.

and all other symbols are the same as before

$$p = p_n/\cos \phi_1 \tag{7-8}$$
$$T_1 = (p/2) + 2(R_{r1} + a_h - R'_1) \tan \phi_1 \tag{8-14}$$
$$T_2 = (p/2) + 2(R_{r2} + a_h - R'_2) \tan \phi_1 \tag{8-15}$$

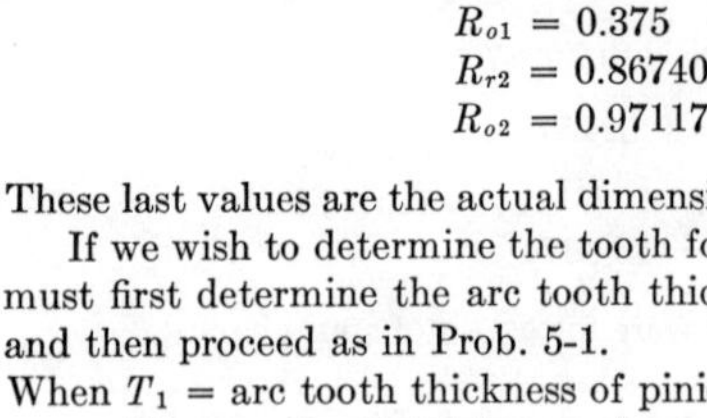

FIG. 8-3.

The coordinates of these tooth profiles have been calculated, and they are plotted in Fig. 8-3.

Second Example. For the second example we shall take one with no limitations on diameters or center distance. This example will consist of a pair of helical gears of 20 and 40 teeth, generated with a 10-DP, 14½-deg standard involute hob, with a helix angle of about 30 deg. This gives the following values for the trial calculation for the lead:

$$N_1 = 20 \qquad N_2 = 40 \qquad P_{nc} = 10 \qquad \psi_1 = 30° \qquad \cos \psi_1 = 0.86603$$
$$\sin \psi_1 = 0.50000 \qquad \phi_{nc} = 14.50° \qquad \tan \phi_{nc} = 0.25862 \qquad a_h = 0.1157$$

TRIAL CALCULATIONS FOR LEAD. Transposing Eq. (8-7) to solve for C_2, we have

$$C_2 = \frac{N_1 + N_2}{2P_{nc} \cos \psi_1} \tag{8-16}$$

$$C_2 = \frac{20 + 40}{2 \times 10 \times 0.86603} - 3.46410$$

$$L_1 = \frac{3.1416 \times 20}{10 \times 0.500} = 12.56640$$

$$L_2 = \frac{3.1416 \times 40}{10 \times 0.500} = 25.13280$$

We shall use the following values for the leads:

$$L_1 = 12.500 \qquad L_2 = 25.000$$

We shall now determine the helix angle of generation.

$$\sin \psi_1 = \frac{3.1416 \times 20}{10 \times 12.50} = 0.50266$$

$$\psi_1 = 30.176° \qquad \cos \psi_1 = 0.86449$$

$$\tan \phi_1 = \frac{0.25862}{0.86449} = 0.29916$$

$$\phi_1 = 17.407° \qquad \cos \phi_1 = 0.95420 \qquad \text{inv } \phi_1 = 0.009706$$

$$P_1 = 10 \times 0.86449 = 8.6449$$

$$C_1 = \frac{60}{2 \times 8.6449} = 3.4702$$

Comparing this with spur gears of the same tooth numbers, we find that for 1-DP gears, the center distance is increased 0.4178 in. for a total of 60 teeth in the pair. This would be 0.4178/10, which equals 0.04178 in. for 10-DP, 14½-deg gears. This increase in center distance would give

$$C_2 = 3.4702 + 0.04178 = 3.51198$$

We shall make the operating center distance an even dimension; hence we shall use

$$C_2 = 3.500 \text{ in.}$$

Whence

$$\cos \phi_2 = \frac{3.4702 \times 0.95420}{3.500} = 0.94608$$

$$\phi_2 = 18.900° \qquad \text{inv } \phi_2 = 0.012509$$

$$R_{r1} + R_{r2} = 3.4702 - 0.2314 + \frac{3.4702(0.012509 - 0.009706)}{0.29916} = 3.28221$$

$$h_{t1} = 0.2157 \qquad c_1 = 0.0157$$

$$h_t = \left(\frac{0.2157}{0.2314}\right)(3.500 - 3.28221) = 0.20301$$

$$b_1 = \frac{3.500 - 3.28221}{1 + \sqrt{40/20}} = 0.09021$$

$$b_2 = 0.21779 - 0.09021 = 0.12758$$

$$R_1 = \frac{20 \times 3.50}{60} = 1.16667$$

$$R_2 = \frac{40 \times 3.50}{60} = 2.33333$$

$$R_{r1} = 1.16667 - 0.09021 = 1.07646$$

$$R_{r2} = 2.33333 - 0.12758 = 2.20575$$

$$R_{o1} = 1.07646 + 0.20301 = 1.27947$$

$$R_{o2} = 2.20575 + 0.20301 = 2.40876$$

The coordinates of these gear-tooth profiles have been calculated, and they are plotted in Fig. 8-4.

Third Example. As a third example we shall assume that the center distance and the tooth numbers are definitely fixed, but that the pitch of the hob and the diameters of the gears are variable. For this example we shall assume the following values:

$$N_1 = 17 \qquad N_2 = 38 \qquad C = 5.250$$

together with a 14½-deg full-involute hob of some standard diametral pitch.

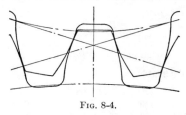

If these were spur gears of conventional design, then

$$P_1 = \frac{17 + 38}{2 \times 5.25} = 5.238$$

If $P_n = 6$, then

$$\cos \psi_1 = \frac{5.238}{6} = 0.87300$$

Fig. 8-4.

This trial calculation gives us some idea of the helix angle that would be needed with some definite diametral pitch. If the helix angle is too great, then a coarser pitch should be used. If the helix angle is too small, then a finer pitch should be used. In this example, the helix angle is slightly less than 30 deg, a value that will be considered as satisfactory. We shall therefore use a standard 6-DP hob.

Comparing this with spur gears of the same tooth numbers, we find that for a tooth-number total of 55 teeth, the center distance would be increased 0.5073 in. for 1-DP spur gears. For these 6-DP gears, this amount would be 0.5073/6, which is equal to 0.0846 in. We shall therefore reduce the value of C_2 to 5.165 in. for the purposes of the trial calculations for the leads.

TRIAL CALCULATION FOR LEADS

$$\cos \psi_1 = \frac{55}{2 \times 6 \times 5.165} = 0.88738$$

$$\psi_1 = 27.454° \qquad \sin \psi_1 = 0.46104$$

$$L_1 = \frac{3.1416 \times 17}{6 \times 0.46104} = 19.30678$$

$$L_2 = \frac{3.1416 \times 38}{6 \times 0.46104} = 43.15634$$

If we select a simple value for L_1 that is divisible by 17, then the value of L_2 will also be simple. When we divide the foregoing values of the trial leads by the respective tooth numbers (19.30678/17 for example), we obtain a factor of 1.13568. We shall use the factor 1.20, as it is a better one for the selection of change gears. Thus we shall make

$$L_1 = 17 \times 1.20 = 20.40$$
$$L_2 = 38 \times 1.20 = 45.60$$

FINAL CALCULATION FOR GEARS. We now have the following values for the further calculations:

$$N_1 = 17 \qquad N_2 = 38 \qquad C_2 = 5.250 \qquad \phi_{nc} = 14.50° \qquad P_{nc} = 6$$
$$\tan \phi_{nc} = 0.25862 \qquad L_1 = 20.40 \qquad L_2 = 45.60 \qquad a_h = 0.1928$$
$$\sin \psi_1 = \frac{3.1416 \times 17}{6 \times 20.40} = 0.43633$$
$$\psi_1 = 25.870° \qquad \cos \psi_1 = 0.89979$$

$$\tan \phi_1 = \frac{0.25862}{0.89979} = 0.28742$$

$\phi_1 = 16.036°$ $\cos \phi_1 = 0.96109$ inv $\phi_1 = 0.007544$

$$P_1 = 6 \times 0.89979 = 5.39874$$

$$C_1 = \frac{55}{2 \times 5.39874} = 5.09378$$

$$\cos \phi_2 = \frac{5.09378 \times 0.96109}{5.250} = 0.93249$$

$\phi_2 = 21.175°$ inv $\phi_2 = 0.017799$

$$R_{r1} + R_{r2} = 5.09378 - 0.3856 + \frac{5.09378(0.017799 - 0.007544)}{0.28742} = 4.88992$$

$h_{t1} = 0.3596$ $c_1 = 0.02617$

$$h_t = \left(\frac{0.3595}{0.38577}\right)(5.250 - 4.88992) = 0.33565$$

$$b_1 = \frac{0.36008}{1 + \sqrt{38/17}} = 0.14431$$

$$b_2 = 0.36008 - 0.14431 = 0.21577$$

$$R_1 = \frac{17 \times 5.25}{55} = 1.62273$$

$$R_2 = \frac{38 \times 5.25}{55} = 3.62727$$

$$R_{r1} = 1.62273 - 0.14431 = 1.47842$$
$$R_{o1} = 1.47842 + 0.33565 = 1.81407$$
$$R_{r2} = 3.62727 - 0.21577 = 3.41150$$
$$R_{o2} = 3.41150 + 0.33565 = 3.74715$$

The coordinates of these gear teeth have been calculated, and they are plotted in Fig. 8-5.

These three examples should be sufficient to indicate the flexibility inherent in the design of helical involute gears.

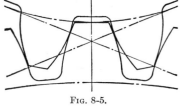

FIG. 8-5.

Problem 8-5. *Given the proportions of a pair of helical gears, to determine the face contact ratio.*

The face contact ratio is the ratio between the helical advance on the pitch cylinders of a pair of helical gears across their active face width and the circular pitch at the pitch radius in the plane of rotation. This ratio must be greater than unity to obtain continuous helical contact on a helical-gear drive. Thus when

F = active face width[1] of gears, in.

p = circular pitch at pitch radius, plane of rotation, in.

ψ = helix angle on pitch cylinder

m_f = face contact ratio

$$m_f = (F \tan \psi)/p \qquad (8\text{-}17)$$

[1] The active face width is the actual axial distance across the faces of the mating gears that are in actual contact.

Example of Face Contact Ratio. If the gears in the third example of the previous problem have an active face width of 3.00 in., we would have the values that follow. In the third example we did not determine the operating circular pitch or the helix angle on the pitch cylinders; hence we must calculate them now from the lead and the pitch radius of either gear.

$$R_1 = 1.62273 \qquad L_1 = 20.40 \qquad N_1 = 17 \qquad F = 3.000$$

$$p = \frac{2\pi R_1}{N_1} = \frac{6.2832 \times 1.62273}{17} = 0.59976$$

$$\tan \psi = \frac{2\pi R_1}{L_1} = \frac{6.2832 \times 1.62273}{20.40} = 0.49980$$

Whence

$$m_p = \frac{3 \times 0.49980}{0.59976} = 2.500$$

Problem 8-6. *Given the values of a pair of helical involute gears, to determine the projection of the contact line on a plane containing the axes of the gears and on a plane parallel to the pitch plane of the basic rack of the pair.*

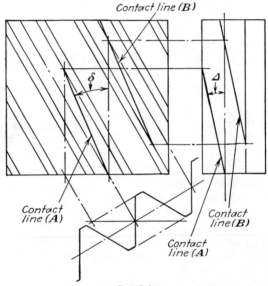

Contact line (B)

Contact line (A)

Contact line (B)

Contact line (A)

Fig. 8-6.

If we refer to Fig. 8-1, the projection of the angular edge of the paper that sweeps through space to develop the form of the involute helical-gear-tooth on either plane of reference will be the respective projections of the actual contact line between the mating gear teeth. Thus with helical involute gears, this contact line is a straight line. Referring to Fig. 8-6, when

ψ_b = helix angle on base cylinder

ϕ = pressure angle of operation in plane of rotation

ψ = helix angle on pitch cylinders

Δ = angle between projection of contact line on plane containing the axes of the gears and the trace of base or pitch cylinder on plane

δ = angle between projection of contact line on plane parallel to pitch plane of basic rack and projection of axes on plane

$$\tan \Delta = \tan \psi_b \sin \phi$$
$$\tan \zeta = \tan \psi_b \cos \phi$$

But

$$\tan \psi_b = \tan \psi \cos \phi \qquad (8\text{-}6)$$

Whence

$$\tan \Delta = \tan \psi \cos \phi \sin \phi \qquad (8\text{-}18)$$
$$\tan \delta = \tan \psi \cos^2 \phi \qquad (8\text{-}19)$$

Example of Helical Contact Line. As a definite example we shall determine the contact line for the pair of helical gears in the preceding example. From this we have the following values:

$$\phi_2 = 21.175° = \phi \qquad \sin \phi = 0.36122 \qquad \cos \phi = 0.93249$$
$$\cos^2 \phi = 0.86954 \qquad \tan \psi = 0.49980 \qquad \psi = 26.556°$$
$$\tan \Delta = 0.49980 \times 0.93249 \times 0.36122 = 0.16835 \qquad \Delta = 9.556°$$
$$\tan \delta = 0.49980 \times 0.86954 = 0.43460 \qquad \delta = 23.490°$$

Problem 8-7. *Given the proportions of a helical involute gear and the setting of the hob of given proportions, to determine the minimum distance of the center of the hob from the face of the gear before the hob starts to cut.*

Referring to Fig. 8-7, when

R_{o1} = outside radius of gear, in.

R_{oh} = outside radius of hob, in.

C = center distance between centers of hob and gear, in.

λ = angle of axis of hob with face of gear

ϵ_1 = angle on gear to point on intersection line

ϵ_h = angle on hob to point on intersection line

We shall first determine the projection, of the form of the intersection of the outside cylinder of the hob with the outside cylinder of the gear, on a plane parallel to the axes of the gear and of the hob. We shall use the intersection of these axes on the reference plane as the origin of the coordinate system, as shown in Fig. 8-7.

When x = abscissa of projection of intersection form, in.

y = ordinate of projection of intersection form, in.

we have the following from the conditions shown in Fig. 8-7:

$$x = R_{o1} \sin \epsilon_1$$

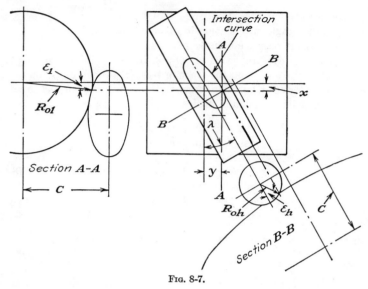

FIG. 8-7.

whence

$$\sin \epsilon_1 = x/R_{o1}$$
$$y = x \tan \lambda + (R_{oh} \sin \epsilon_h / \cos \lambda)$$
$$C = R_{o1} \cos \epsilon_1 + R_{oh} \cos \epsilon_h$$

Rearranging and combining these equations so as to solve for values of y in terms of x and the other known values, we obtain

$$\cos \epsilon_1 = \sqrt{1 - \sin^2 \epsilon_1} = \sqrt{R_{o1}^2 - x^2}/R_{o1}$$
$$C = R_{oh} \cos \epsilon_h + \sqrt{R_{o1}^2 - x^2}$$

Whence

$$\cos \epsilon_h = (C - \sqrt{R_{o1}^2 - x^2})/R_{oh}$$
$$\sin \epsilon_h = \sqrt{1 - \cos^2 \epsilon_h} = \sqrt{R_{oh}^2 - (C - \sqrt{R_{o1}^2 - x^2})^2}/R_{oh}$$

Substituting this value of $\sin \epsilon_h$ into the equation for y, we obtain

$$y = x \tan \lambda + \sqrt{R_{oh}^2 - (C - \sqrt{R_{o1}^2 - x^2})^2}/\cos \lambda \qquad (8\text{-}20)$$

This equation will give the coordinates of the projection of the inter-section of two cylinders on a plane parallel to their axes. The maximum value of y, *i.e.*, y_m, will be the minimum distance of the center of the hob from the face of the gear before the hob starts to cut, unless the hob is too short to cover the full intersection.

An examination of Eq. (8-20) shows that the maximum value of x, *i.e.*, x_m, will be reached when

$$C - \sqrt{R_{o1}^2 - x_m^2} = R_{oh}$$

because the expression under the radical in Eq. (8-20) becomes minus, and imaginary, when the value of x is greater than that given by this relationship. Solving this expression for x_m, we have

$$x_m = \sqrt{R_{o1}^2 - (C - R_{oh})^2} \qquad (8\text{-}21)$$

The value of y when x is equal to zero, *i.e.*, y_0, is given by the following:

$$y_0 = \sqrt{R_{oh}^2 - (C - R_{o1})^2}/\cos \lambda \qquad (8\text{-}22)$$

The value of x when the value of y is a maximum, *i.e.*, x_1, is given very closely by the following equation:

$$x_1 = (x_m^2 \tan \lambda)/\sqrt{y_0^2 + x_m^2 \tan^2 \lambda} \qquad (8\text{-}23)$$

Then

$$y_m = x_1 \tan \lambda + [\sqrt{R_{oh}^2 - (C - \sqrt{R_{o1}^2 - x_1^2})^2}/\cos \lambda] \qquad (8\text{-}24)$$

If the length of the hob is shorter than the distance to x_1, then the value of x_1 for use in Eq. (8-24) will be given very closely by the following: When x_h = extension of end of hob, in.

$$x_1 = x_h \cos \lambda \qquad (8\text{-}25)$$

Example of Intersection Curve of Two Cylinders. As a definite example we shall use the following values:

$$R_{o1} = 10.00 \qquad R_{oh} = 1.500 \qquad C = 11.230$$
$$\lambda = 30° \qquad \tan \lambda = 0.57735 \qquad \cos \lambda = 0.86603$$
$$x_m = \sqrt{100 - (9.73)^2} = 2.30805$$
$$y_0 = \frac{\sqrt{2.25 - (1.23)^2}}{0.86603} = 0.99135$$
$$x_1 = \frac{(2.30805)^2 \times 0.57735}{\sqrt{(0.99135)^2 + (2.30805 \times 0.57735)^2}} = 1.85180$$
$$y_m = 1.8518 \times 0.57735 + \frac{\sqrt{2.25 - (11.23 - \sqrt{100 - 3.42916})^2}}{0.86603} = 1.68199$$

If we set this hob in the machine so that the entering side has only a short extension such that $x_h = 1.500$ in., then

$$x_1 = 1.50 \times 0.86603 = 1.29904$$
$$y_m = 1.29904 \times 0.57735 + \frac{\sqrt{2.25 - (11.23 - \sqrt{100 - 1.6875})^2}}{0.86603} = 1.58382$$

TABLE 8-1. COORDINATES ON INTERSECTION CURVE OF HOB AND GEAR
(Plotted in Fig. 8-8)

x, in.	y, in.	x, in.	y, in.
0.000	0.99135	−0.000	0.99135
0.100	1.04827	−0.100	0.93280
0.200	1.10351	−0.200	0.87257
0.300	1.15708	−0.300	0.81067
0.400	1.20893	−0.400	0.74706
0.500	1.25901	−0.500	0.68167
0.600	1.30727	−0.600	0.61455
0.700	1.35364	−0.700	0.54535
0.800	1.39800	−0.800	0.47423
0.900	1.44021	−0.900	0.40098
1.000	1.48014	−1.000	0.32544
1.100	1.51761	−1.100	0.24744
1.200	1.55237	−1.200	0.16673
1.300	1.58413	−1.300	0.08302
1.400	1.61256	−1.400	−0.00402
1.500	1.63714	−1.500	−0.09490
1.600	1.65732	−1.600	−0.19020
1.700	1.67222	−1.700	−0.29076
1.800	1.68072	−1.800	−0.39774
1.900	1.68075	−1.900	−0.51318
2.000	1.67015	−2.000	−0.63925
2.100	1.64265	−2.100	−0.78222
2.200	1.58508	−2.200	−0.95526
2.30805	1.33255	−2.30805	−1.33255

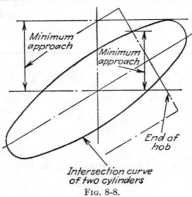

FIG. 8-8.

The coordinates for the full intersection curve of these two cylinders have been computed. They are tabulated in Table 8-1 and plotted in Fig. 8-8. The projection of the end of the hob is also shown. It will be noted that this last value of y_m is slightly greater than the true value.

Problem 8-8. *Given the proportions of a helical involute gear and of the hob, to determine the amount of overtravel needed to complete the generation of the gear.*

On hobbed herringbone gears with a clearance groove between the two sections of opposite hand of helix, it is necessary to have some measure both of the approach of the hob before it begins to cut and also of the amount of overtravel needed to complete the generation of the gear. The minimum width of groove would be the sum of the two foregoing factors.

The generating action of a hob when cutting a gear is exactly the same as the conjugate action between two helical gears whose axes are not parallel. A single-thread hob is a 1-tooth helical gear. No attempt will be made here to show the derivation of the equations because the first step towards such a derivation is the study of the conjugate gear-tooth action of such gears. This will be covered in Chap. 9 on spiral gears. Hence only the derived equations will be given here, because they are needed to complete the design of herringbone gears. Thus when

ϕ_n = normal pressure angle of hob and normal basic rack

λ_h = lead angle of hob at R_h

λ_s = angular setting of hob from face of gear blank

C = center distance between axes of gear and hob, in.

R_h = generating pitch radius of hob, in.

R_2 = generating pitch radius of gear, in.

p_n = normal circular pitch of hob, in.

ψ_2 = helix angle of gear at R_2

R_{oh} = outside radius of hob, in.

R_{o2} = outside radius of gear, in.

X = distance between pitch planes of hob and gear, in.

y_0 = overtravel of hob required to complete generation of gear tooth, in.

$$X = C - (R_h + R_2) \tag{8-26}$$
$$\sin \lambda_h = p_n/2\pi R_h \tag{8-27}$$

When the hob and gear are of the same hand, then

$$\lambda_s = \psi_2 - \lambda_h \tag{8-28}$$

When the hob and gear are of opposite hand, then

$$\lambda_s = \psi_2 + \lambda_h \tag{8-29}$$
$$U = \sin^2 \lambda_h + \tan^2 \phi_n \tag{8-30}$$

$$V = \sin^2 \psi_2 + \tan^2 \phi_n \qquad (8\text{-}31)$$
$$B_1 = (\sqrt{UR_{oh}^2 - R_h^2 \sin^2 \lambda_h} - R_h \tan \phi_n)/U \qquad (8\text{-}32)$$
$$B_2 = (\sqrt{VR_{o2}^2 - R_2^2 \sin^2 \psi_2} - R_2 \tan \phi_n)/V \qquad (8\text{-}33)$$
$$y_o = B_1 \sin \psi_2 \quad \text{or} \quad y_0 = \sin \psi_2 [B_2 - (X/\tan \phi_n)] \quad (8\text{-}34)$$

whichever is the larger.

Example of Overtravel of Hob. As a definite example we shall use substantially the same values as were used in Prob. 8-7. We shall also use a helical gear and hob of the same hand. This gives us the following values:

$$\phi_n = 14.500° \qquad \cos \phi_n = 0.96815 \qquad \tan \phi_n = 0.25862 \qquad \psi_2 = 30°$$
$$R_{oh} = 1.500 \qquad R_h = 1.3554 \qquad R_{o2} = 10.000 \qquad R_2 = 9.875$$
$$p_n = 0.3927 \qquad C = 11.2304 \qquad X = 0.000$$
$$\sin \lambda_h = \frac{0.3927}{6.2832 \times 1.3554} = 0.04611$$
$$\lambda_h = 2.643° \qquad \lambda_s = 30° - 2.643° = 27.357°$$
$$U = 0.002134 + 0.066884 = 0.069018$$
$$V = 0.250000 + 0.066884 = 0.316884$$
$$B_1 = \frac{\sqrt{0.151385} - 0.35053}{0.069018} = 0.55855$$
$$B_2 = \frac{\sqrt{7.309494} - 2.55387}{0.316884} = 0.47254$$

As the value of B_1 is the larger of the two, we have

$$y_o = 0.55855 \times 0.500 = 0.27927$$

The minimum value for the clearance groove for a herringbone gear will be $y_m + y_o$. Using the value y_m equal to 1.430 in., we have

Minimum width of groove $= 1.430 + 0.280 = 1.710$ in.

Problem 8-9. *Given the lead of a helical pinion-shaped cutter, to determine the lead of the generated gear.*

The leads of all mating helical gears operating on parallel axes must be directly proportional to the numbers of teeth in the gears, and must also be of opposite hand of helix. The pinion-shaped cutter, when generating a helical gear, operates exactly the same as a mating pinion to the gear being generated.

When N_c = number of teeth in helical pinion-shaped cutter

N_1 = number of teeth in generated helical gear

L_c = lead[1] of helical pinion-shaped cutter, in.

L_1 = lead[1] of generated gear, in.

$$L_1 = (N_1/N_c)L_c \qquad (8\text{-}35)$$

---

[1] The leads of the helical pinion-shaped cutters and of the generated gears are always of opposite hand.

Example of Lead of Gear Generated from Pinion-shaped Cutter. As a definite example we shall determine the lead of a 48-tooth helical gear generated by a 28-tooth helical pinion-shaped cutter with a lead of 25.904 in. This gives the following values:

$$N_c = 28 \quad N_1 = 48 \quad L_c = 25.904$$
$$L_1 = {}^{48}\!\!/_{28} \times 25.904 = 44.40685 \text{ in.}$$

Helical Pinion-shaped Cutters. The conventional helical pinion-shaped cutters are made to standard diametral pitches and pressure angles in the plane of rotation. The 20-deg stub tooth form is the one most commonly used. The tooth forms and proportions of these cutters in their planes of rotation are identical to those for spur gears; hence all gears to be produced by these cutters are designed, as far as tooth forms and proportions are involved, as spur gears.

Helical internal gears are generated almost exclusively by these cutters, and the design of these internal gears is also covered by the material in Chap. 6 on internal gears. The only difference is that with adequate helical contact, the contact ratio in the plane of rotation becomes of secondary importance.

Helical Internal Drives. The following tooth proportions, based on the 20-deg-stub tooth form, for pinions of 16 teeth and over, and for internal gears of 28 teeth and over, cut with helical pinion-shaped cutters of 16 teeth and larger, will avoid all interference conditions. These proportions are the same as those used for spur internal-gear drives except for the tooth heights.

When R_{o1} = outside radius of pinion, in.

$\quad R_1$ = pitch radius of pinion, in.

$\quad N_1$ = number of teeth in pinion

$\quad R_2$ = pitch radius of internal gear, in.

$\quad R_i$ = inside radius of internal gear, in.

$\quad R_{r2}$ = root radius of internal gear, in.

$\quad R_{r1}$ = root radius of pinion, in.

$\quad C$ = center distance, in.

$\quad P$ = diametral pitch, plane of rotation

$\quad T_1$ = arc tooth thickness of pinion at R_1, plane of rotation, in.

$\quad T_2$ = arc tooth thickness of internal gear at R_2, plane of rotation, in.

$$R_1 = N_1/2P \quad R_2 = N_2/2P$$
$$R_{o1} = (N_1 + 2.100)/2P = R_1 + (1.050/P) \tag{8-36}$$
$$R_i = (N_2 - 0.900)/2P = R_2 - (0.450/P) \tag{8-37}$$
$$C = (N_2 - N_1)/2P = R_2 - R_1$$
$$T_1 = 1.7528/P$$
$$T_2 = 1.3888/P$$

The difference between the tooth heights of the shaped full-depth form and the stub tooth form (1-DP values) is equal to 0.250.

Problem 8-10. *Given the measurement over a pair of rolls in the opposite tooth spaces of a helical gear, to determine the arc tooth thickness of the teeth at a given radius where the helix angle and pressure angle are known.*

Measurements over rolls on helical gears are very difficult to make with any great degree of accuracy unless definite precautions are taken. In many cases, a pair of calibrated wedges, or rack teeth, make a much more reliable measurement for tooth thickness than do rolls. However rolls are often available when needed, while the special calibrated rack-tooth wedges may not be at hand. The measurement over rolls should be made between parallel flat surfaces and not with a micrometer alone. When the rolls are held in position on the gear by two parallels, the two rolls will be on opposite sides of the gear, or diametrically opposite to each other, whether the number of teeth in the gear is odd or even. With odd numbers of teeth, one roll may make contact near one edge of the gear while the other roll makes contact near the opposite edge of the face width. If an attempt is made to measure odd numbers of teeth over the rolls directly with a micrometer, one or both rolls will be tipped away from the correct plane of measurement, and any measured values so obtained are useless for any purpose.

Ball-point micrometers may be used, but here the two balls must be definitely aligned in respect to the face of the gear blank. For example, the gear blank may be laid flat on a surface plate, and the two ball points may be held against this same surface plate. Where balls are used, when odd numbers of teeth are involved, the calculation of the actual chordal measurement must include the offset condition or position in exactly the same way as the calculations are made for spur gears with odd numbers of teeth.

The calculation for the radius to the center of the ball on helical gears is identical to the calculations for the radius to the center of a roll on the same gear. For these calculations we have the following: Referring to Fig. 8-9, when

r_2 = radius to center of roll, in.

R_b = radius of base cylinder of gear, in.

R_1 = radius at which tooth thickness is required, in.

ϕ_1 = pressure angle, plane of rotation, at R_1

ϕ_2 = pressure angle, plane of rotation, at r_2

W = radius of roll or wire, in.

ψ_1 = helix angle of gear at R_1

ψ_b = helix angle of gear on base cylinder at R_b

T_n = normal arc thickness at R_1, in.

T_1 = arc tooth thickness in plane of rotation at R_1, in.

N = number of teeth in gear

M = measurement between two flat parallel plates that hold the rolls in contact with the gear teeth, in.

we have from Eq. (5-61)

$$r_2 = (M - 2W)/2$$

We have from the conditions shown in Fig. 8-9

$$\cos \phi_2 = R_b/r_2 \tag{5-4}$$

Considering again the generation of the helical tooth surfaces of the gear by the angular edge of the sheet of paper wound about the base

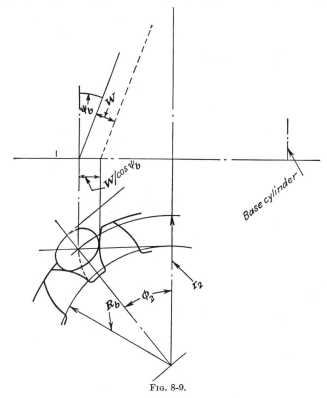

Fig. 8-9.

cylinder, the point of tangency of the roll or ball and the tooth surface will be at a distance equal to the radius of the roll or ball and normal to the angular edge of the paper when the angular edge of this paper sweeps through the center of the ball in any plane perpendicular to the axis of the helical gear.

This normal distance is equal to W. In the plane of rotation, where all involute calculations must be made, the distance from the edge of the paper to the line parallel to the edge at a distance W from it will be equal to $W/\cos \psi_b$. Whence

$$T_1 = 2R_1[(\pi/N) + \text{inv } \phi_2 - \text{inv } \phi_1 - (W/R_b \cos \psi_b)] \quad (8\text{-}38)$$

$$\tan \psi_b = (R_b/R_1) \tan \psi_1 = \cos \phi_1 \tan \psi_1 \quad (8\text{-}6)$$

$$T_n = T_1 \cos \psi_1 \quad (8\text{-}39)$$

Example of Roll Measurement. As a definite example we shall use a 30-tooth, 6-DP helical gear with the following values:

$$M = 6.1500 \qquad W = 0.140 \qquad R_1 = 2.88673 \qquad N = 30$$

$$\psi_1 = 30° \qquad \cos \psi_1 = 0.86603 \qquad \tan \psi_1 = 0.57735$$

$$\phi_1 = 16.637° \qquad \cos \phi_1 = 0.95814 \qquad \text{inv } \phi_1 = 0.008446$$

$$r_2 = \frac{6.1500 - 0.280}{2} = 2.9350$$

$$R_b = R_1 \cos \phi_1 = 2.88673 \times 0.95814 = 2.76589$$

$$\tan \psi_b = 0.95814 \times 0.57735 = 0.55318$$

$$\psi_b = 28.951° \qquad \cos \psi_b = 0.87505$$

$$\cos \phi_2 = \frac{2.76589}{2.9350} = 0.94238$$

$$\phi_2 = 19.545° \qquad \text{inv } \phi_2 = 0.013878$$

$$T_1 = 5.77346 \left(\frac{3.1416}{30} + 0.013878 - 0.008446 - \frac{0.140}{2.42029}\right) = 0.30200$$

$$T_n = 0.30200 \times 0.86603 = 0.26154$$

When the arc tooth thickness is known, the foregoing equations can be rearranged to solve for the measurement over the rolls.

CHAPTER 9

INVOLUTOMETRY OF SPIRAL GEARS

Screw Gearing. Screw gearing includes various types of gears used to drive nonparallel and nonintersecting shafts where the teeth of one or both members of the pair are of screw or helicoidal form. In these gears, the driving action is predominantly a screwing or wedging action between the contacting tooth surfaces. Sometimes conjugate gear-tooth action is present, and sometimes it is not. Spiral gears are one type of screw gears. A spiral-gear drive consists of a pair of helical gears that drive each other when mounted on nonparallel and nonintersecting shafts.

When helical gears are mounted on parallel shafts, the contact between them is line contact, and the mating gears are of opposite hand of helix. When helical gears are used to drive nonparallel shafts, the contact between them is point contact, and the mating gears are generally of the same hand of helix.

Spiral-gear Action. The exact nature of the action between a pair of spiral gears is not generally understood. Practically no present text on the subject of mechanical design gives a complete or correct statement of this action. This action must be studied in three dimensions. It is more complex than the study of the action between a pair of helical gears on parallel shafts because it cannot be shown completely on a single plane. On spiral-gear drives, the unique condition exists where each member of the pair has two distinct pitch surfaces: one is a pitch cylinder in its own plane of rotation, and the other is a pitch plane whose trace is in the plane of rotation of the mating gear.

When the axes are at right angles to each other, the pitch plane of each gear travels in the direction of its own axis. When the axes are not at right angles to each other, the pitch plane of each gear travels in the direction of rotation of the mating gear.

In addition, or, rather, complementary to the two pitch surfaces, each gear of a spiral-gear drive has two circular pitches: one is a circular pitch in its own plane of rotation and controls the size of its pitch cylinder, and the other is the axial pitch or the circular pitch of its pitch plane whose value is controlled by the axial pitch of the gear and the angle between the axes of the pair. When the axes are at right angles, the circular pitch of the pitch plane is the axial pitch of the gear and is equal to the lead of

177

the helix divided by the number of teeth in the gear. The circular pitch on the pitch cylinder of one gear is the same as the circular pitch on the pitch plane of the mating gear. When the angular position of the axes is fixed, the circular pitch of the pitch plane of each gear is the same at all distances from the axis of the gear, while the circular pitch on the pitch cylinder, which is the circular pitch of the gear in its plane of rotation, is constrained to a fixed diameter.

These pitch surfaces are shown in Fig. 9-1. As most commonly designed, the pitch planes of mating gears lie in the same plane but travel in different directions. This direction depends upon the angle between the axes of the gears.

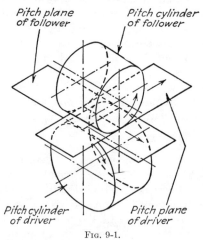

FIG. 9-1.

The driving member of a pair of spiral gears is commonly called the *driver*, while the driven member is called the *follower*. Referring to Fig. 9-1, when the lower gear is the driver and the upper gear is the follower, the pitch plane of the driver travels in the direction of rotation of the pitch cylinder of the follower as it rolls upon it and is screwed along by the helix of the driver. The direction or sense of the motion of the pitch plane is controlled by the direction of the helix on the driver. The example shown in Fig. 9-1 is for a pair of left-handed helical gears. If they were right-handed, the pitch plane of the driver and the pitch cylinder of the follower would travel in the opposite direction to that shown.

The pitch cylinder of the follower is driven by the pitch plane of the driver, while its own pitch plane is screwed along and engages the pitch cylinder of the driver, as indicated by the arrows in Fig. 9-1.

Thus we have a closed circuit of action on these spiral gears. To summarize, we have the following: The helix of the driver screws its pitch plane along its line of travel. This pitch plane is always tangent to the pitch cylinder of the follower and causes it to rotate. The rotation of the follower screws its pitch plane along its path, and this pitch plane is always tangent to the pitch cylinder of the driver. The rate of travel of this pitch plane must be the same as that of the circumference of the pitch cylinder of the driver. This completes the closed circuit.

The forms of the basic racks of the two helical gears match each other as indicated in Fig. 9-2. When the basic rack of the driver is moved in the direction of motion of the pitch plane of the driver, it acts as a wedge or cam on the basic rack of the follower, and forces it

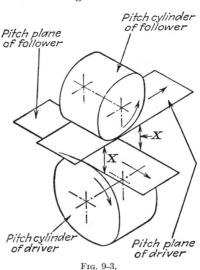

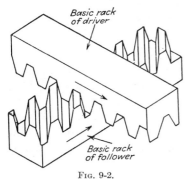

Fig. 9-2.

Fig. 9-3.

to move in the direction of motion of the pitch plane of the follower, as indicated in Fig. 9-2.

When the distance between the axes of the gears is increased, the sizes of the pitch cylinders remain unchanged, and the pitch plane of the driver remains tangent to the pitch cylinder of the follower. The pitch plane of the follower likewise remains tangent to the pitch cylinder of the driver. Under these conditions, with helical involute gears, the conjugate gear-tooth action remains correct, but the two pitch planes no longer lie in the same plane, and the two pitch cylinders no longer touch each other but are separated as indicated in Fig. 9-3.

When the distance between the axes is decreased from that shown in Fig. 9-1, the sizes of the pitch cylinders again remain unchanged and the pitch plane of each gear remains tangent to the pitch cylinder of the mating gear, thus separating the pitch planes as before, but the pitch cylinders now intersect each other. The conjugate gear-tooth action, however, remains theoretically correct. The only restriction to the reduction of the center distance is the limit of undercut on one or both of the gear-tooth profiles. The only restriction to the increase of center distance is the reduction of the contact ratio, which must be always greater than unity, and it can often be held to two or slightly more.

When the angular setting of the axes of the two gears is changed, the circular pitch in the plane of rotation of both gears is also changed. It becomes larger or smaller; the result depends upon the direction of the angular shifting of the axes and the direction of the helices on the gears. When the center distance is unchanged with this change in angular position, the position of the pitch planes shifts with the change in the size of the pitch cylinders so that they no longer lie in the same plane. The conjugate gear-tooth action here also remains theoretically correct.

Hence with the point contact between a pair of helical involute gears operating on nonparallel axes, we have the condition where neither a shift in the center distance nor a change in the angular relationship of the axes will result in the loss of theoretically correct conjugate gear-tooth action. This condition is the primary reason why a spiral-gear drive of adequate tooth design that is not loaded above its limited capacity is generally the quietest and most satisfactory of all the different types of gear drives. No small error in alignment of shafts or in the center distance between them has any detrimental effect on their action together.

Except with gears of the same helix angle, the diameters of the pitch cylinders of a pair of spiral gears are not directly proportional to their numbers of teeth. In these drives, the speed or reduction ratio is dependent upon the numbers of teeth alone.

Only helical involute gears will be considered in this study of spiral gears. Other forms may be used, but they do not have the same freedom and versatility as the helical involute gears. The individual gears of these spiral-gear drives are helical involute gears. All geometrical relationships of these gears are identical to those of helical involute gears as given in the preceding chapter.

Conjugate Action of Spiral-gear Drives. The conjugate gear-tooth action of a spiral-gear drive can be studied in the plane of rotation of each gear of the pair. The projection of the path of contact on a plane parallel to the plane of rotation of each gear is the same as the path of contact of the trace of the basic rack on this plane and the gear-tooth profile in its plane of rotation. This condition is shown in Fig. 9-4.

These are helical involute gears in this example, and the helix angle is 45 deg on both gears. Both pitch planes lie in the same plane for this example. The projection of the actual path of contact in the plan view will be normal to the elements of the basic racks. In this example, the projection of this path of contact is at 45 deg, as shown. When the pitch planes coincide, this projection of the path of contact will pass through the intersection of the projections of the two axes. The length *oa* of the path of contact is projected up from the driver, and the length *ob* is projected over from the follower into the plan view. The actual length of this path of contact is obtained by projecting this path of contact from the plan view to the common normal basic-rack section for the two gears

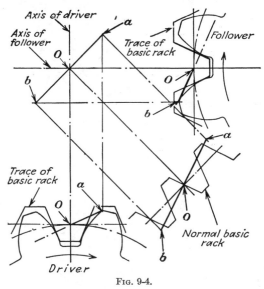

Fig. 9-4.

as shown in Fig. 9-4. This actual length is the combination of the sections *oa* and *ob* when projected to the path of contact on the normal basic-rack form.

When the pitch planes are separated by an increase in the center distance, we have the conditions shown in Fig. 9-5. The length o_1a of the projection of the path of contact in the plan view is projected from the plane of rotation of the driver as before, and the length o_2b is projected from the plane of rotation of the follower. The projection of this path of contact on the plan view no longer goes through the intersection of the projection of the two axes on this plan view, and the lengths o_1a and o_2b overlap each other, as indicated in Fig. 9-5. When this path of contact is projected on the normal basic-rack form, the projection o_1a overlaps the projection o_2b by the length of the line o_1o_2. This overlap represents the amount of action on the basic rack between the two pitch

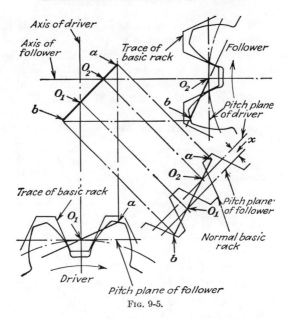

Fig. 9-5.

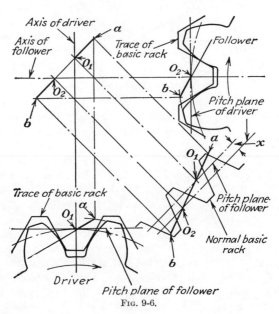

Fig. 9-6.

planes. In this case, the total length of the actual path of contact is less by the distance o_1o_2 than the sum of the lengths o_1a and o_2b.

When the pitch planes are separated because of a decrease in the center distance, we have the conditions shown in Fig. 9-6. As before, the lengths of the projections o_1a and o_2b are projected from the planes of rotation of the driver and follower, respectively, and from there to the normal section of the common basic-rack form. In this case also, these projections in the plan view do not pass through the intersection of the projections of the two axes. Also, these two projections do not meet each other but are separated by the distance o_1o_2. Conjugate action exists, however, along this distance o_1o_2, and again it represents the action that exists between the two pitch planes. In this case, in the normal plane of the basic rack, the actual length of the line ab, which is the total length of the actual path of contact, is equal to the sum of o_1a and o_2b plus the length o_1o_2.

Thus an increase in the center distance, which separates the two pitch planes, tends to decrease both the actual length of the path of contact and the contact ratio, while a decrease in the center distance, which also separates the pitch planes, but in the opposite direction, tends to increase the contact ratio of the spiral-gear drive.

The further consideration of the subject of spiral gears will be in the form of specific problems.

Problem 9-1. *Given the proportions of a pair of spiral gears and the center distance, to determine the contact ratio.*

Referring again to Figs. 9-4, 9-5, and 9-6, the value of the contact ratio is obtained by dividing the length of the path of contact on the normal basic rack (length ab) by the normal base pitch of the helical gears used in the drive. This length, as noted before, is affected by the distance that may lie between the pitch planes. Thus when

m_p = contact ratio

R_{o1} = outside radius of driver, in.

R_{o2} = outside radius of follower, in.

R_1 = radius of pitch cylinder of driver, in.

R_2 = radius of pitch cylinder of follower, in.

R_{b1} = radius of base cylinder of driver, in.

R_{b2} = radius of base cylinder of follower, in.

C = center distance, in.

ϕ_1 = pressure angle of driver at R_1 in plane of rotation

ϕ_2 = pressure angle of follower at R_2 in plane of rotation

ϕ_n = pressure angle of normal basic rack form

ψ_1 = helix angle of driver at R_1

ψ_2 = helix angle of follower at R_2

p_n = normal circular pitch of basic rack, in.

p_{bn} = normal base pitch of helical gears, in.

X = distance between the pitch planes of driver and follower, in.

$$X = C - (R_1 + R_2).$$ (9-1)[1]

For the purpose of deriving the equations for the contact ratio, we shall introduce the following symbols: Referring to Figs. 9-4, 9-5, and 9-6, we have

A_1 = length o_1a of projection of path of contact in plane of rotation of driver, in.

B_1 = length o_1a of projection of path of contact in plan view, in.

C_1 = length o_1a of path of contact on normal basic-rack section, in.

A_2 = length o_2b of projection of path of contact in plane of rotation of follower, in.

B_2 = length o_2b of projection of path of contact in plan view, in.

C_2 = length o_2b of path of contact on normal basic-rack section, in.

$$A_1 = \sqrt{R_{o1}^2 - R_{b1}^2} - R_1 \sin \phi_1$$

$$A_2 = \sqrt{R_{o2}^2 - R_{b2}^2} - R_2 \sin \phi_2$$

$$B_1 = \frac{A_1 \cos \phi_1}{\cos \psi_1}$$

$$B_2 = \frac{A_2 \cos \phi_2}{\cos \psi_2}$$

$$C_1 = \frac{B_1}{\cos \phi_n} = \frac{A_1 \cos \phi_1}{\cos \psi_1 \cos \phi_n}$$

$$C_2 = \frac{B_2}{\cos \phi_n} = \frac{A_2 \cos \phi_2}{\cos \psi_2 \cos \phi_n}$$

$$m_p = \frac{C_1 + C_2 - (X/\sin \phi_n)}{p_{bn}}$$

$$p_{bn} = p_n \cos \phi_n$$ (9-2)

As the known values are usually the normal pressure angle, the helix angles, and the outside and pitch radii, it will be convenient to transform these equations so that only these known values are required for the solution. We have to start

$$\tan \phi_n = \tan \phi_1 \cos \psi_1 = \tan \phi_2 \cos \psi_2$$

[1] The value of X is plus when the center distance is increased so that the pitch cylinders do not touch each other, and is minus when the center distance is decreased so that the pitch cylinders intersect each other.

Whence

$$\tan \phi_1 = \frac{\tan \phi_n}{\cos \psi_1}$$

$$\sin \phi_1 = \frac{\tan \phi_n}{\sqrt{\cos^2 \psi_1 + \tan^2 \phi_n}}$$

$$\cos \phi_1 = \frac{\cos \psi_1}{\sqrt{\cos^2 \psi_1 + \tan^2 \phi_n}}$$

$$R_{b1} = R_1 \cos \phi_1 = \frac{R_1 \cos \psi_1}{\sqrt{\cos^2 \psi_1 + \tan^2 \phi_n}}$$

Introducing these values into the equation for C_1, combining, and simplifying, we obtain

$$C_1 = \frac{\sqrt{(\cos^2 \psi_1 + \tan^2 \phi_n)R_{o1}^2 - R_1^2 \cos^2 \psi_1} - R_1 \tan \phi_n}{\cos \phi_n(\cos^2 \psi_1 + \tan^2 \phi_n)}$$

As one expression is repeated, to simplify the writing of the equations, we shall let

$$U = (\cos^2 \psi_1 + \tan^2 \phi_n) \tag{9-3}$$

Whence

$$C_1 = \frac{\sqrt{UR_{o1}^2 - R_1^2 \cos^2 \psi_1} - R_1 \tan \phi_n}{U \cos \phi_n}$$

In a similar manner, we obtain

$$V = (\cos^2 \psi_2 + \tan^2 \phi_n) \tag{9-4}$$

$$C_2 = \frac{\sqrt{VR_{o2}^2 - R_2^2 \cos^2 \psi_2} - R_1 \tan \phi_n}{V \cos \phi_n}$$

$$m_p = \frac{1}{p_{bn}} \left(\frac{\sqrt{UR_{o1}^2 - R_1^2 \cos^2 \psi_1} - R_1 \tan \phi_n}{U \cos \phi_n} \right.$$
$$\left. + \frac{\sqrt{VR_{o2}^2 - R_2^2 \cos^2 \psi_2} - R_2 \tan \phi_n}{V \cos \phi_n} - \frac{X}{\sin \phi_n} \right)$$

If we bring the value $1/\cos \phi_n$ outside of the parentheses, and substitute the value of p_{bn} from Eq. (9-2), we shall have as another factor

$$T = \frac{1}{p_n \cos^2 \phi_n} \tag{9-5}$$

The final equation then becomes

$$m_p = T \left(\frac{\sqrt{UR_{o1}^2 - R_1^2 \cos^2 \psi_1} - R_1 \tan \phi_n}{U} \right.$$
$$\left. + \frac{\sqrt{VR_{o2}^2 - R_2^2 \cos^2 \psi_2} - R_2 \tan \phi_n}{V} - \frac{X}{\tan \phi_n} \right) \tag{9-6}$$

Example of Contact Ratio on Spiral Gears.   As a definite example we shall use the following values:

$$R_{o1} = 2.580 \qquad R_1 = 2.480 \qquad \psi_1 = 60.500° \qquad \cos \psi_1 = 0.49242$$
$$R_{o2} = 2.620 \qquad R_2 = 2.520 \qquad \psi_2 = 29.500° \qquad \cos \psi_2 = 0.87036$$
$$\phi_n = 14.500° \qquad \cos \phi_n = 0.96815 \qquad \tan \phi_n = 0.25682 \qquad X = 0.02202$$
$$p_n = 0.31416$$

$$T = \frac{1}{0.31416 \, (0.96815)^2} = 3.39597$$

$$U = (0.49242)^2 + (0.25862)^2 = 0.30936$$

$$V = (0.87036)^2 + (0.25862)^2 = 0.82441$$

$$m_p = 3.39597 \left(\frac{\sqrt{2.059224 - 1.491329} - 0.64138}{0.30936} \right.$$
$$\left. + \frac{\sqrt{5.659080 - 4.810608} - 0.64719}{0.82441} - \frac{0.02202}{0.25862} \right) = 2.052$$

In this example the contact ratio is slightly over two, which is satisfactory.

Problem 9-2.   *To determine the form of the thread on a worm that is milled with a straight-sided thread-milling cutter with its axis parallel to the axis of the worm.*

When a cone-shaped milling cutter or grinding wheel of any diameter is used to finish a worm, and the axis of the milling cutter or grinding wheel is set parallel to the axis of the worm, the form of the threads or teeth on the finished worm will be that of a helical involute gear. Such a form is called the *involute helicoid.* This form is the limiting form of many other types of helicoids, and it may be produced in a great variety of ways. For example, the form of rolled screw threads produced by flat rolling dies, and the form of screw threads produced by hobbing with an annular thread-milling hob are both this same involute helicoid. This mathematical form has been patented over and over again as "the form produced by said method," etc. The method may be new, but the form produced is probably much older than the Patent Office itself. Such involute helicoids are suitable driving members for spur and helical involute gears when the normal base pitches of a given pair are identical. Involute helicoids produced in this manner are generally limited to single threads.

The mathematical proof of the foregoing statement about the form produced by a cone-shaped cutter or grinding wheel, originally developed by Ernest Wildhaber, is as follows: The form of the helicoid produced by a cone-shaped rotating tool whose axis is parallel to the axis of the helicoid ûis the locus of points of tangency between the cone-shaped tool and the surface of the helicoid in all the relative operating positions of the two members.

To determine this locus of points of tangency as the helicoid is screwed axially in relation to the position of the cone-shaped tool, we will establish the equations needed to locate any point of tangency. Such equations will then be general ones for all points of tangency as we change the values to define other points of tangency.

When curved or warped surfaces are tangent to each other with either point or line contact, as the case may be, there will be one and only one tangent plane that contains the point or line contact. Hence if we can locate this tangent plane in relation to the two members, we can soon determine whether line or point contact exists and where it is.

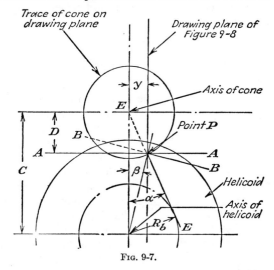

Fɪɢ. 9-7.

Referring to Fig. 9-7, the line *AA* is the trace of a plane, parallel to the axes of the cone and helicoid, on the drawing plane. The point *P* is any point of tangency between the cone and helicoid and is in the drawing plane. This point *P* is also on the line *AA*. The line *EE* is the projection of an element of the cone that also contains the point *P*. The line *BB* is the projection on the drawing plane of a tangent line to the helicoid that contains the point *P*. This line is perpendicular to a radial line of the helicoid. The tangent plane of the two members must contain the point *P* and the two lines *BB* and *EE*.

The intersection curve of the cone with the intersecting plane *AA* is a hyperbola, and its equation is given by the following expression:
When γ = one-half included angle of cone-shaped tool

D = distance of intersecting plane *AA* from axis of cone

$$x = \tan \gamma \sqrt{D^2 + y^2} \qquad (9\text{-}7)$$
$$dx/dy = \tan \phi = y \tan \gamma / \sqrt{D^2 + y^2} \qquad (9\text{-}8)$$

The tangent to this hyperbola that passes through the point P will lie in the plane AA and also in the tangent plane of the two surfaces. In other words, the tangent to the hyperbola, on plane AA, that passes through point P is the trace of the tangent plane of the two members on plane AA.

When λ = angle of line BB with the drawing plane, which is also the lead angle of the helicoid at point P

R = radius on helicoid to point P

C = distance between axes of cone and helicoid

L = lead of helicoid

$$R = \sqrt{(C - D)^2 + y^2}$$
$$\tan \lambda = L/2\pi R = L/2\pi \sqrt{(C - D)^2 + y^2} \tag{9-9}$$

As noted before, this line BB, whose projection is shown on the drawing plane, is this tangent to the helicoid at point P, and it must also lie in

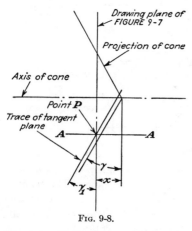

Drawing plane of FIGURE 9-7

Projection of cone

Axis of cone

Point P

Trace of tangent plane

A —— *A*

γ

x

λ_1

Fig. 9-8.

the tangent plane to the surfaces of the two members. When this line BB is revolved about point P in the tangent plane of the two members, it must coincide with the line tangent at point P to the hyperbola of the cone member when it reaches the plane AA. When this line BB is revolved further about point P until it meets a plane through point P that is parallel to and at a distance y from the axes of the cone and helicoid, it will be the trace of the tangent plane of the two members on this new plane. This new plane will be the drawing plane of Fig. 9-8. Referring to Fig. 9-8, this trace of the tangent plane of the two surfaces will be at some angle γ_1 to the drawing plane of Fig. 9-7. Thus when

γ_1 = angle of trace of tangent plane with drawing plane of Fig. 9-7

λ_1 = angle of tangent line BB with drawing plane of Fig. 9-7 when revolved to intersecting plane AA

α = angle between projection of cone element EE^* on drawing plane of Fig. 9-7 and center line of cone and helicoid

---

* This element of the cone, which passes through point P, lies also in the tangent plane of the two surfaces.

β = angle between radial line of helicoid to point P and center line of cone and helicoid

we have the following from the geometrical conditions described:

$$\cos \alpha = D/\sqrt{D^2 + y^2} \qquad (9\text{-}10)$$
$$\sin \alpha = y/\sqrt{D^2 + y^2} \qquad (9\text{-}11)$$
$$\tan \gamma_1 = (\tan \gamma \sqrt{D^2 + y^2} - y \tan \phi)/D$$
$$\tan \phi = y \tan \gamma / \sqrt{D^2 + y^2} \qquad (9\text{-}8)$$

Substituting this value into the preceding equation, combining, and simplifying, we obtain

$$\tan \gamma_1 = D \tan \gamma / \sqrt{D^2 + y^2} \qquad (9\text{-}12)$$

This last equation gives the value of the angle of the trace of the tangent plane on the drawing plane of Fig. 9-8. In order to obtain the value of the angle of the trace of this tangent plane on the intersecting plane AA, we proceed as follows:

$$\sin \beta = \frac{y}{\sqrt{(C - D)^2 + y^2}} \qquad (9\text{-}13)$$

$$\cos \beta = \frac{C - D}{\sqrt{(C - D)^2 + y^2}} \qquad (9\text{-}14)$$

$$\tan \lambda_1 = \frac{\tan \lambda - \sin \beta \tan \gamma_1}{\cos \beta}$$

Substituting the values of the angles into this last equation, combining, and simplifying, we obtain

$$\tan \lambda_1 = \frac{L \sqrt{D^2 + y^2} - 2\pi D y \tan \gamma}{2\pi(C - D) \sqrt{D^2 + y^2}} \qquad (9\text{-}15)$$

As noted before

$$\tan \phi = \tan \lambda_1$$

whence

$$\frac{y \tan \gamma}{\sqrt{D^2 + y^2}} = \frac{L \sqrt{D^2 + y^2} - 2\pi D y \tan \gamma}{2\pi(C - D) \sqrt{D^2 + y^2}}$$

Solving this equation for y, we obtain

$$y = \frac{LD}{\sqrt{(2\pi C \tan \gamma)^2 - L^2}} \qquad (9\text{-}16)$$

It is apparent from an inspection of Eq. (9-16) that the value of y will always be directly proportional to the value of D for any given values of L, C, and γ. Therefore the locus of points of tangency of the cone and helicoid when they are in contact with each other will lie in a straight line, and this straight line must be the element of the cone, EE. This line will be tangent to a cylinder concentric with the axis of the helicoid, a cylinder that we will call the *base cylinder*. Thus when

R_b = radius of base cylinder of helicoid

λ_b = lead angle at the base cylinder of the helicoid

$$R_b = C \sin \alpha = Cy/\sqrt{D^2 + y^2}$$

Substituting the value of y from Eq. (9-16), combining, and simplifying, we obtain

$$R_b = L/2\pi \tan \gamma \qquad (9\text{-}17)$$

The value of R_b is therefore independent of the value of C or D. Its value depends only upon the value of the factors L and γ. Therefore, regardless of the diameter of the cone-shaped tool and the center distance between the tool and the helicoid, the same form of helicoid will be produced as long as the lead of the helicoid and the angle of the cone remain unchanged.

$$\tan \lambda_b = L/2\pi R_b \qquad (9\text{-}18)$$

Transposing Eq. (9-17) to solve for $\tan \gamma$, we obtain

$$\tan \gamma = L/2\pi R_b \qquad (9\text{-}19)$$

Therefore

$$\gamma = \lambda_b \qquad (9\text{-}20)$$

In other words, the cone angle of the cone-shaped tool is the lead angle of the helicoid upon its base cylinder. If we consider this angle as that of the angular edge of a sheet of paper that is wound around the base cylinder, it is apparent that the form of this helicoid is the same as that of an involute helical gear, or an involute helicoid.

Therefore to produce an involute helicoid worm to mesh with a spur gear with shafts at right angles to each other, or to mesh with any helical involute gear where the axis of the worm is at right angles to the normal basic rack of the helical gear, we can mill or grind this worm with a cone-shaped tool with its axis parallel to that of the worm. The lead of this worm would be equal to the normal circular pitch of the basic rack, and the half angle of the cone would be equal to the normal pressure angle of the basic rack. This type of worm is generally restricted to single-thread worms, or single-tooth helical gears.

Problem 9-3. *Given the shaft angle and helix angles, with the diameters and speed of rotation, to determine the peripheral sliding velocity between them.*

The greatest part of the sliding action on a spiral-gear drive is that which is represented by the sliding of the two basic racks of the system on

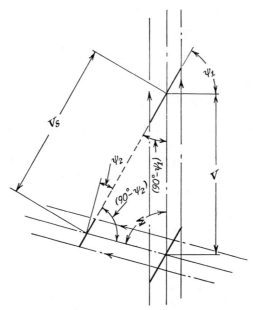

FIG. 9-9.

each other. Such conditions are shown in Fig. 9-9. Thus when

R_1 = radius of pitch cylinder of driver, in.

R_2 = radius of pitch cylinder of follower, in.

n = number of rpm of driver

V = pitch-line velocity of driver, ft/min

V_s = sliding velocity between basic tacks, ft/min

ψ_1 = helix angle of driver at R_1

ψ_2 = helix angle of follower at R_2

Σ = shaft angle

$$\Sigma = \psi_1 + \psi_2$$
$$V = 2\pi R_1 n/12 = 0.5236 R_1 n \tag{4-12}$$

Figure 9-9 gives a graphic solution of this problem. Whence we have

$$V \sin \Sigma = V_s \sin (90° - \psi_2) = V_s \cos \psi_2$$

Substituting the value of V into this equation, we obtain

$$V_s = 0.5236 R_1 n \sin \Sigma/\cos \psi_2 \tag{9-21}$$

When the shafts are at right angles to each other, then

$$\Sigma = 90° \qquad \text{and} \qquad \sin \Sigma = 1.000$$

and Eq. (9-21) becomes

$$V_s = 0.5236R_1n/\cos \psi_2 \qquad (9\text{-}22)$$

These equations apply to the conditions that exist when the pitch planes of the two gears coincide. When these pitch planes are separated, there may be some small variations from these values. The equations are still correct, however, for the velocity of sliding on the basic racks of the system.

In the case of an involute helicoid worm driving a spur gear, the value of R_1 is infinite. The actual basic racks, however, are acting at the pitch radius of the spur gear; hence the distance from the axis of the driver to its pitch plane, R_3, would be used instead of the infinite value of R_1.

$$R_3 = C - R_2 \qquad (9\text{-}23)$$

When the values of ψ_1 and ψ_2 are equal, the sliding conditions are the most favorable. For one thing, with a fixed ratio and a given center distance, the sliding velocities are least under these conditions. For another, the actual amount of sliding on both gears will be alike under such conditions. In other words, under such conditions, the sliding will be evenly distributed over the same distance on the tooth surfaces of both gears. In other cases, as when the helix angle of the driver is much greater than the helix angle of the follower, the amount of sliding is, of course, the same on both gears, but it is distributed over a longer distance on the driver, and over a shorter distance on the follower.

Example of Sliding Velocity on Spiral Gears. As a definite example we shall use the following values:

$$R_1 = 2.500 \qquad n = 600 \qquad \Sigma = 85° \qquad \sin \Sigma = 0.99619$$
$$\psi_1 = 50° \qquad \psi_2 = 35° \qquad \cos \psi_2 = 0.81915$$
$$V_s = \frac{0.5236 \times 2.5 \times 600 \times 0.99619}{0.81915} = 955 \text{ ft/min}$$

CHAPTER 10

HELICOID SECTIONS

Conjugate Action on Worm-gear Drives. In many respects, a worm-gear drive is a development of a spiral-gear drive such that one member of the pair has been made to envelop the other. This construction will introduce line contact between the mating members in place of the point contact that exists between a pair of spiral gears. Starting from a spiral-gear drive, either member of the pair may be made to envelop the other. A worm-gear drive can be designed as a substitute or replacement for any spiral-gear drive. The conventional design of a worm-gear drive is mounted on nonintersecting axes with planes of rotation at right angles to each other. Worm-gear drives can be made, however, to drive shafts that are not at right angles to each other.

The conjugate gear-tooth action between a worm and a worm gear is identical to that of a spur gear and a rack. As the worm revolves, the thread form on the worm advances along its axis, and the worm gear is rotated a corresponding amount. The pitch surfaces of such a drive consist of a pitch plane for the worm or rack member and a pitch cylinder for the worm gear.

In this analysis, the worm will always be the member with a uniform axial lead, whether it is the driver or follower, or whether it is the smaller or larger member of the pair.

When one member of a spiral-gear drive is made to envelop the other, the enveloping member loses its uniform axial pitch, and with it, its pitch plane. It retains only its pitch cylinder. The mating member, with a uniform axial lead, retains its pitch plane to match the pitch cylinder of the enveloping member, but loses its pitch cylinder because it has no matching pitch surface on the enveloping member to act against. Thus the worm has no true pitch diameter or pitch cylinder. The radial distance from the axis of the worm to its pitch plane is commonly called the *pitch radius* of the worm.

The conjugate gear-tooth action between a worm and a worm gear is the same whether the worm is revolved to screw the thread form along its axis or whether the worm is moved axially without revolving.

The basic-rack form of the worm gear is the form of that section of the worm thread which actually engages with the worm-gear teeth. This

193

form changes across the face of the worm gear. When these forms are established for any given planes of rotation of the worm gear, conjugate gear-tooth forms and trochoidal fillets of the worm gear are determined for these planes of rotation in exactly the same manner as for spur gears. When the contact on a series of planes of rotation of the worm gear has been determined, the position and the projection of the actual contact line between the worm and worm gear can also be established.

The first step toward the study of the nature and the amount of the contact on such drives and of the forms of the conjugate gear teeth on the worm gear is the determination of the form of the basic rack of the worm gear in any desired plane of rotation of the worm gear.

Helicoid Sections. In order to determine the basic-rack forms on various planes of rotation of the worm gear, we must determine the equations of the intersection profiles of the worm or helicoid with these planes of rotation.

The exact thread form used for the worm, within certain limits, as for all spur-gear and rack forms, is of small importance. The essential requirement is that the thread form of the worm and that of the hob or other tool used to generate the worm gear be as nearly identical as possible. The actual form of the thread on the worm depends entirely upon the type, form, and size of the thread-cutting tool, and upon the type of process used to finish the worm thread. With tools of identical form of cutting profile, threads chased in a lathe or milled on a thread-milling machine or ground with a conical wheel of appreciable diameter will all have different thread forms. We must therefore determine the intersection profiles of several different types of helicoids. These will be as follows:

Convolute Helicoid. This type of helicoid has its straight-line generatrix tangent to a cylinder of any diameter that is concentric with the axis of the helicoid. The concentric cylinder will be called the *base cylinder.* The inclination of the generatrix, measured from any plane of rotation of the helicoid, is in the same direction as the inclination of the helix of the thread. This form is an approximation to that of a milled or ground worm thread. It is a general form of helicoid, of which the screw helicoid and the involute helicoid are limiting or specific types.

Screw Helicoid. This type of helicoid has its straight-line generatrix passing through the axis of the helicoid. It is a convolute helicoid with its base cylinder reduced to a zero diameter. It can be produced on a lathe with the cutting edges of a straight-sided cutting tool set in a plane that contains the axis of the helicoid. This is the common screw-thread form.

Involute Helicoid. This type of helicoid has its straight-line generatrix tangent to a concentric cylinder of such diameter that the lead angle on this base cylinder is the same as the angle of the generatrix with a plane perpendicular to the axis of the helicoid. Such a plane is a plane of rotation of the helicoid. This is another specific type of the convolute helicoid. This is also the form of a helical involute gear. It is also the form of a rolled and of a hobbed screw thread.

Chased Helicoid. This type of helicoid has its straight-line generatrix tangent to a cylinder concentric with the axis of the helicoid. The inclination of this generatrix, measured from a plane of rotation of the helicoid, is in the opposite direction to the inclination of the helix of the thread. This is the form produced by a straight-sided cutting tool that is tipped to the helix of the thread.

Milled Helicoid. This type of helicoid is the form produced by a cone-shaped milling cutter or grinding wheel that is tipped to the helix of the thread. The generatrix in this case is not a straight line but is a slightly curved one; its exact form depends upon the cone angle, the diameter of the rotating tool, and the diameter and lead of the worm.

We shall now determine the equations of the intersection profiles of these different types of helicoids, starting with the convolute helicoid.

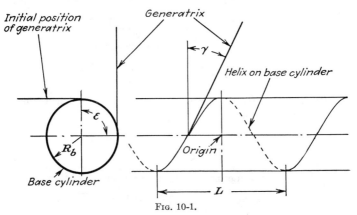

Fig. 10-1.

Convolute Helicoid. Referring to Fig. 10-1, which represents the convolute helicoid, we have the following symbols:

L = lead of generatrix, in.

r = any radius to helicoidal surface, in.

γ = angle between generatrix and plane perpendicular to axis

θ = vectorial angle

ψ = angle between tangent to intersection curve and radius vector

ϕ = angle between tangent to intersection curve and trace of plane perpendicular to axis

ϵ = angle of rotation of generatrix

R_b = radius of base cylinder, in.

D = distance from axis to intersecting plane, in.

x_a = abscissa of intersection curve, origin at initial point of tangency of generatrix and base cylinder, in.

y_a = ordinate of intersection curve, origin at axis of helicoid, in.

Intersection Curve with Axial Plane. We shall first determine the equation of the intersection curve of this convolute helicoid with the axial

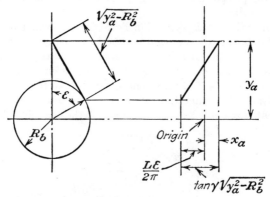

Fig. 10-2.

plane. Referring to Fig. 10-2, we have from the geometrical conditions shown there the following:

$$x_a = \tan \gamma \sqrt{y_a^2 - R_b^2} - (L\epsilon/2\pi)$$
$$\tan \epsilon = \sqrt{y_a^2 - R_b^2}/R_b$$

Substituting this last value into the first equation, we obtain

$$x_a = \tan \gamma \sqrt{y_a^2 - R_b^2} - (L/2\pi) \tan^{-1} (\sqrt{y_a^2 - R_b^2}/R_b) \quad (10\text{-}1)$$

For the tangent to this curve, we have the following:

$$\tan \phi = dx_a/dy_a = (2\pi y_a^2 \tan \gamma - LR_b)/2\pi y_a \sqrt{y_a^2 - R_b^2} \quad (10\text{-}2)$$

The form of this intersection curve is plotted in Fig. 10-3. The corresponding conic section would be two straight lines, the trace of the cone on its axial plane. These same straight lines are the asymptotes of the hyperbola or the intersection curve of the cone with a plane parallel to its axis.

These curves representing the intersection of the convolute helicoid with its axial plane repeat themselves for every turn or thread of the helicoid. These intersection curves also have their asymptotes. An

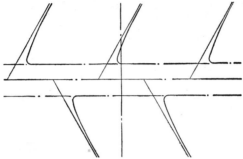

FIG. 10-3. Axial section of convolute helicoid.

inspection of Eq. (10-1) gives the following equation for the asymptotes of this intersection curve:

$$x = y \tan \gamma - (L/2\pi) \tan^{-1} \infty \qquad (10\text{-}3)$$

The value of an arc when its tangent is equal to infinity is $\pi/2$, $3\pi/2$, $5\pi/2$, etc. These asymptotes are also shown in Fig. 10-3.

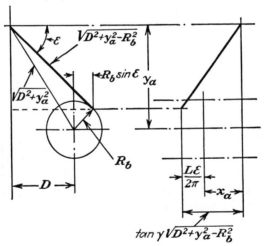

FIG. 10-4.

Intersection Curve of Convolute Helicoid with a Plane Parallel to the Axis and at a Distance D from the Axis. Referring to Fig. 10-4, we have the following from the geometrical conditions shown there:

$$x_a = \tan \gamma \sqrt{D^2 + y_a{}^2 - R_b{}^2} - \frac{L\epsilon}{2\pi}$$

$$\cos \epsilon = \frac{D + R_b \sin \epsilon}{\sqrt{D^2 + y_a{}^2 - R_b{}^2}} = \frac{D + R_b \sqrt{1 - \cos^2 \epsilon}}{\sqrt{D^2 + y_a{}^2 - R_b{}^2}}$$

Solving this last equation for $\cos \epsilon$, we obtain

$$\cos \epsilon = \frac{R_b y_a + D \sqrt{D^2 + y_a{}^2 - R_b{}^2}}{D^2 + y_a{}^2}$$

Substituting this last value into the first equation, we obtain

$$x_a = \tan \gamma \sqrt{D^2 + y_a{}^2 - R_b{}^2}$$

$$- \frac{L}{2\pi} \cos^{-1} \frac{R_b y_a + D \sqrt{D^2 + y_a{}^2 - R_b{}^2}}{D^2 + y_a{}^2} \quad (10\text{-}4)$$

$\tan \phi$

$$= \frac{dx_a}{dy_a} = \frac{2\pi y_a \tan \gamma (D^2 + y_a{}^2) - L(R_b y_a + D \sqrt{D^2 + y_a{}^2 - R_b{}^2})}{2\pi (D^2 + y_a{}^2) \sqrt{D^2 + y_a{}^2 - R_b{}^2}} \quad (10\text{-}5)$$

The form of this intersection curve is shown in Fig. 10-5. The corresponding conic section is a hyperbola. This intersection curve also

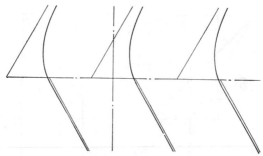

FIG. 10-5.   Off-center section of convolute helicoid.

has its asymptotes. An inspection of Eq. (10-4) gives the following equation for these asymptotes:

$$x = y \tan \gamma - \left(\frac{L}{2\pi}\right) \cos^{-1} (0) \quad (10\text{-}6)$$

The value of $\cos^{-1} (0)$ will be $\pi/2$, $3\pi/2$, etc. Hence this equation is identical to Eq. (10-3). In other words, the asymptotes to these off-center sections are identical to those of the axial section. These asymptotes are also shown in Fig. 10-5.

Intersection Curve of the Convolute Helicoid with a Plane Perpendicular to Its Axis (plane of Rotation). Referring to Fig. 10-6, we have the following from the geometrical conditions shown there:

$$r^2 = \left(\frac{L\epsilon}{2\pi \tan \gamma}\right)^2 + R_b{}^2$$

Whence

$$\epsilon = \left(\frac{2\pi \tan \gamma}{L}\right)\sqrt{r^2 - R_b{}^2}$$

$$\tan \delta = \sqrt{\frac{r^2 - R_b{}^2}{R_b{}^2}}$$

$$\theta = \epsilon - \delta = \left(\frac{2\pi \tan \gamma}{L}\right)\sqrt{r^2 - R_b{}^2} - \tan^{-1}\sqrt{\frac{r^2 - R_b{}^2}{R_b{}^2}} \quad (10\text{-}7)$$

$$\frac{d\theta}{dr} = \frac{2\pi r^2 \tan \gamma - R_b L}{rL \sqrt{r^2 - R_b{}^2}}$$

$$\tan \psi = \frac{r \, d\theta}{dr} = \frac{2\pi r^2 \tan \gamma - R_b L}{L \sqrt{r^2 - R_b{}^2}} \quad (10\text{-}8)$$

The form of this intersection curve is shown in Fig. 10-7. The corresponding conic section is a circle. This curve is a spiral, starting at the radius of the base cylinder.

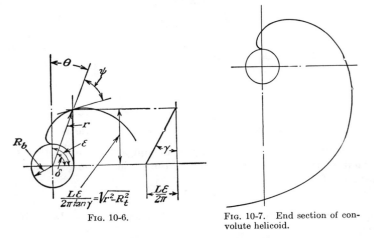

FIG. 10-6.

FIG. 10-7. End section of convolute helicoid.

Limits of Conjugate Gear-tooth Action on Convolute Helicoid. The extreme limit of conjugate gear-tooth action on these helicoids is reached when the pressure angle ϕ of the intersection curves with planes parallel to the axis of the helicoid is equal to zero. This same condition exists when the tangent to the spiral intersection curve in the plane of rotation is perpendicular to the axis of the mating worm gear.   Referring to Fig. 10-8, we have from the geometrical conditions shown there the following:

$$\theta + \psi = 90°$$

$$\tan \theta = 1/\tan \psi = L \sqrt{r^2 - R_b{}^2}/2\pi r^2 \tan \gamma - R_b L \qquad (10\text{-}9)$$

The form of this curve is shown in Fig. 10-9.   Conjugate gear-tooth action is possible only in the region indicated as the "field of conjugate action" on this figure.

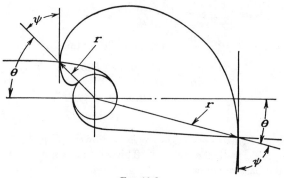

Fig. 10-8.

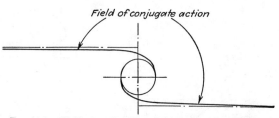

Fig. 10-9.   Field of conjugate action on convolute helicoid.

This curve, which shows the limits of conjugate action, approaches an asymptote that is parallel to the axis of the mating worm gear and is at a distance of

$$y = L/2\pi \tan \gamma \qquad (10\text{-}10)$$

from the axis of the helicoid or worm.   This asymptote is also plotted in Fig. 10-9.

Screw Helicoid.   The screw helicoid has a straight-line generatrix that passes through the axis of the helicoid.   Thus it is one limiting example of the convolute helicoid with the size of its base cylinder reduced to zero.   Hence the equations for the several intersection curves may be established by using the equations of the convolute helicoid and making the value of R_b equal to zero.

Intersection Curve of Screw Helicoid with Axial Plane. Introducing the value of $R_b = 0$ into Eq. (10-1), we obtain the following:

$$x_a = y_a \tan \gamma - (L/2\pi) \tan^{-1} \infty \qquad (10\text{-}11)$$

This equation is the same as that for the asymptote to the convolute helicoid, Eq. (10-3). Hence the axial-section curve of the screw helicoid is the asymptote of the intersection curve of any similar convolute helicoid with a plane parallel to its axis.

$$\tan \phi = (dx_a/dy_a) = \tan \gamma \qquad (10\text{-}12)$$

The form of this intersection curve is shown in Fig. 10-10.

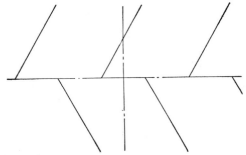

FIG. 10-10. Axial section of screw helicoid.

Intersection Curve of Screw Helicoid with a Plane Parallel to the Axis and at a Distance D from the Axis. Introducing the value of $R_b = 0$ into Eq. (10-4), we obtain

$$x_a = \tan \gamma \sqrt{D^2 + y_a^2} - \left(\frac{L}{2\pi}\right) \cos^{-1} \frac{D}{\sqrt{D^2 + y_a^2}}$$

This equation can be simplified by introducing the value of $\tan^{-1}$ in place of $\cos^{-1}$, as follows:

$$\cos \epsilon = \frac{D}{\sqrt{D^2 + y_a^2}}$$

$$\tan \epsilon = \sqrt{\frac{1 - \cos^2 \epsilon}{\cos \epsilon}} = \frac{y_a}{D}$$

Whence

$$x_a = \tan \gamma \sqrt{D^2 + y_a^2} - \left(\frac{L}{2\pi}\right) \tan^{-1} \frac{y_a}{D} \qquad (10\text{-}13)$$

$$\tan \phi = \frac{dx_a}{dy_a} = \frac{2\pi y_a \tan \gamma \sqrt{D^2 + y_a^2} - LD}{2\pi(D^2 + y_a^2)} \qquad (10\text{-}14)$$

For the equation of the asymptote we have

$$x = y \tan \gamma - \left(\frac{L}{2\pi}\right) \tan^{-1} \infty \qquad (10\text{-}3)$$

The form of this intersection curve and its asymptotes are plotted in Fig. 10-11.

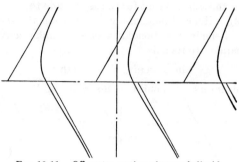

FIG. 10-11.   Off-center section of screw helicoid.

Intersection Curve of Screw Helicoid with a Plane Perpendicular to its Axis.   Referring again to Fig. 10-6, when $R_b = 0$, then $\epsilon = \theta$, whence

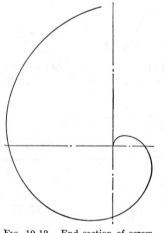

FIG. 10-12.   End section of screw helicoid.

$$r = L\epsilon/2\pi \tan \gamma$$

and

$$\theta = 2\pi r \tan \gamma/L \qquad (10\text{-}15)$$

This curve is an Archimedean spiral with a uniform rise. Its rise per revolution is equal to $L/\tan \gamma$. The equation of its tangent is as follows:

$$\tan \psi = (r\, d\theta/dr) = 2\pi r \tan \gamma/L \qquad (10\text{-}16)$$

But

$$2\pi r/L = 1/\tan \lambda$$

where λ is the lead angle at radius r. Hence

$$\tan \psi = \tan \gamma/\tan \lambda \qquad (10\text{-}17)$$

The form of this intersection curve is shown in Fig. 10-12.

Limits of Conjugate Gear-tooth Action on Screw Helicoid.   Substituting the value $R_b = 0$ into Eq. (10-9), we obtain

$$\theta = \tan^{-1} (L/2\pi r \tan \gamma) \qquad (10\text{-}18)$$

This curve, which represents the limits of conjugate gear-tooth action, approaches an asymptote that is parallel to the axis of the mating worm gear and is at a distance of

$$y = L/2\pi \tan \gamma \tag{10-10}$$

from the axis of the helicoid. This asymptote is the same as that for the convolute helicoid. This curve and its asymptotes are plotted in Fig. 10-13.

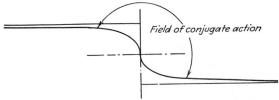

Field of conjugate action

FIG. 10-13. Field of conjugate action on screw helicoid.

Involute Helicoid. The involute helicoid has its straight-line generatrix tangent to a base cylinder of such a diameter that the lead angle of the helix on this base cylinder is the same as the angle of the generatrix with the plane of rotation. This helicoid is another specific or limiting case of the convolute helicoid, where

$$\tan \gamma = \tan \lambda_b = L/2\pi R_b \tag{10-19}$$

and λ_b is the lead angle on base cylinder.

Intersection Curve of Involute Helicoid with Axial Plane. Substituting the value $\tan \gamma = L/2\pi R_b$ into Eq. (10-1), we obtain

$$x_a = \frac{L}{2\pi R_b} \sqrt{y_a{}^2 - R_b{}^2} - \left(\frac{L}{2\pi}\right) \tan^{-1} \frac{\sqrt{y_a{}^2 - R_b{}^2}}{R_b}$$

$$x_a = \frac{L}{2\pi} \left(\frac{\sqrt{y_a{}^2 - R_b{}^2}}{R_b} - \tan^{-1} \frac{\sqrt{y_a{}^2 - R_b{}^2}}{R_b} \right) \tag{10-20}$$

$$\tan \phi = \frac{dx_a}{dy_a} = \frac{L \sqrt{y_a{}^2 - R_b{}^2}}{2\pi R_b y_a} \tag{10-21}$$

The equation of the asymptotes to this intersection curve is as follows:

$$x = y \tan \gamma - \left(\frac{L}{2\pi}\right) \tan^{-1} \infty = \frac{L}{2\pi} \left(\frac{y_a}{R_b} - \tan^{-1} \infty\right) \tag{10-22}$$

These asymptotes are the same as those for the convolute helicoid. The form of this intersection curve and its asymptotes are plotted in Figure (10-14).

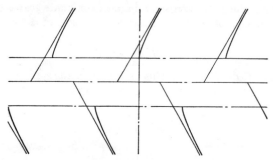

FIG. 10-14. Axial section of involute helicoid.

Intersection Curve of the Involute Helicoid with a Plane Parallel to Its Axis and at a Distance D from Its Axis. Substituting the value of $\tan \gamma = L/2\pi R_b$ into Eq. (10-4), we obtain

$$x_a = \frac{L}{2\pi R_b} \sqrt{D^2 + y_a^2 - R_b^2} - \frac{L}{2\pi} \cos^{-1} \frac{R_b y_a + D \sqrt{D^2 + y_a^2 - R_b^2}}{D^2 + y_a^2}$$

$$x_a = \frac{L}{2\pi} \left(\frac{\sqrt{D^2 + y_a^2 - R_b^2}}{R_b} - \cos^{-1} \frac{R_b y_a + D \sqrt{D^2 + y_a^2 - R_b^2}}{D^2 + y_a^2} \right)$$

$$(10\text{-}23)$$

$$\tan \phi = \frac{dx_a}{dy_a} = \frac{L}{2\pi} \left[\frac{y_a \sqrt{D^2 + y_a^2 - R_b^2} - D R_b}{R_b(D^2 + y_a^2)} \right] \qquad (10\text{-}24)$$

The equation of the asymptotes to this intersection curve is as follows:

$$x = \frac{L}{2\pi} \left(\frac{y_a}{R_b} - \tan^{-1} \infty \right) \qquad (10\text{-}22)$$

These asymptotes are the same as those for the convolute helicoid. The form of this intersection curve and its asymptotes are plotted in Fig. 10-15.

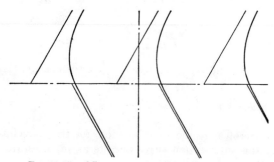

FIG. 10-15. Off-center section of involute helicoid.

Intersection Curve of the Involute Helicoid with a Plane Perpendicular to Its Axis.

Substituting the value of $\tan \gamma = L/2\pi R_b$ into Eq. (10-7), we obtain

$$\theta = \sqrt{r^2 - R_b{}^2}/R_b - \tan^{-1} (\sqrt{r^2 - R_b{}^2}/R_b) \qquad (4\text{-}1)$$

This equation is the polar equation of the involute curve, which was derived in Chap. 4. For the equation of the tangent we have

$$\tan \psi = r\, d\theta/dr = \sqrt{r^2 - R_b{}^2}/R_b \qquad (4\text{-}2)$$

The form of this intersection curve is plotted in Fig. 10-16. This involute is a uniform-rise spiral along a line tangent to the base cylinder. The rise per revolution is equal to the circumference of the base circle.

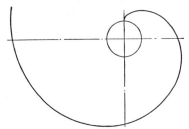

FIG. 10-16.   End section of involute helicoid.

Limits of Conjugate Gear-tooth Action on Involute Helicoid. Substituting the value of $\tan \gamma = L/2\pi R_b$ into Eq. (10-9), we obtain

$$\theta = \tan^{-1} R_b/\sqrt{r^2 - R_b{}^2}$$

$$\tan \theta = R_b/\sqrt{r^2 - R_b{}^2} \qquad \sin \theta = \tan \theta/\sqrt{1 + \tan^2 \theta} = R_b/r$$

If the equation for this curve, which represents the limits of conjugate gear-tooth action, is given in Cartesian coordinates, then

$$y = r \sin \theta = R_b \qquad (10\text{-}25)$$

This curve is a straight line. But

$$R_b = L/2\pi \tan \gamma$$

Whence

$$y = L/2\pi \tan \gamma \qquad (10\text{-}10)$$

This last equation is that of the asymptote to the curves representing the limits of conjugate gear-tooth action for both of the previous helicoids, in fact, for all varieties of helicoids with straight-line generatrices. Hence the limit of conjugate gear-tooth action for the involute helicoid is the asymptotic value of this limit for all other types of helicoids with straight-line generatrices. This limit of conjugate gear-tooth action for the involute helicoid is shown in Fig. 10-17.

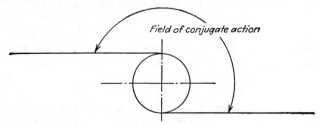

FIG. 10-17.　Field of conjugate action on involute helicoid.

Chased Helicoid. The chased helicoid is the form produced by a straight-sided threading tool set at an angle to the axis of the helicoid as shown in Fig. 10-18.

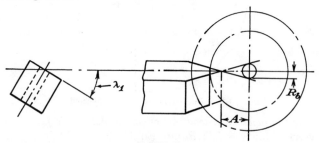

FIG. 10-18.

When γ_n = one-half included angle of threading tool

λ_1 = angular setting of threading tool

A = distance between sharp point (extended) of tool and axis of helicoid, in.

γ = angle between generatrix (cutting edge of tool) and plane perpendicular to axis of helicoid

R_b = radius of cylinder to which generatrix is tangent, in.

we have the following from the geometrical conditions shown in Fig. 10-18:

$$R_b = \frac{A \tan \gamma_n \sin \lambda_1}{\sqrt{1 + \tan^2 \gamma_n \sin^2 \lambda_1}} \tag{10-26}$$

$$\tan \gamma = \frac{\tan \gamma_r \cos \lambda_1}{\sqrt{1 + \tan^2 \gamma_n \sin^2 \lambda_1}} = \frac{R_b}{A \tan \lambda_1} \tag{10-27}$$

This helicoid is similar to the convolute helicoid except that the direction of the inclination of the generatrix is in the opposite direction to that of the helices of the helicoid. Referring to Fig. 10-1, the helix on the base cylinder would run in the opposite direction for this chased helicoid from that shown in Fig. 10-1, while all other elements would remain unchanged. Hence for the equations of the several intersection curves, the distance $L\epsilon/2\pi$, which the generatrix travels as the helicoids revolves, would be added instead of subtracted to obtain the distance through which a given intersection point moves axially as it rotates.

Intersection Curve of Chased Helicoid with Axial Plane. Introducing this change of direction into Eq. (10-1), we obtain

$$x_a = \tan \gamma \sqrt{y_a^2 - R_b^2} + (L/2\pi) \tan^{-1} (\sqrt{y_a^2 - R_b^2}/R_b) \quad (10\text{-}28)$$

For the equation of the tangent to this curve we have

$$\tan \phi = dx_a/dy_a = (2\pi y_a^2 \tan \gamma + R_b L)/2\pi y_a \sqrt{y_a^2 - R_b^2} \quad (10\text{-}29)$$

For the equation of the asymptotes we have

$$x = y \tan \gamma + (L/2\pi) \tan^{-1} \infty \quad (10\text{-}30)$$

The form of this intersection curve and its asymptotes are plotted in Fig. 10-19.

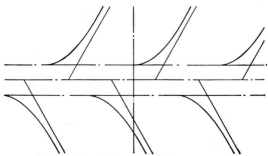

FIG. 10-19. Axial section of chased helicoid.

Intersection Curve of Chased Helicoid with a Plane Parallel to the Axis and at a Distance D from the Axis. Introducing the change of direction of the helices into Eq. (10-4), we obtain

$$x_a = \tan \gamma \sqrt{D^2 + y_a^2 - R_b^2} + \frac{L}{2\pi} \cos^{-1} \frac{R_b y_a + D \sqrt{D^2 + y_a^2 - R_b^2}}{D^2 + y_a^2} \quad (10\text{-}31)$$

The equation of the tangent to this curve is as follows:

$$\tan \phi = \frac{dx_a}{dy_a} = \frac{2\pi y_a \tan \gamma \, (D^2 + y_a^2) + L(R_b y_a + D \sqrt{D^2 + y_a^2 - R_b^2})}{2\pi (D^2 + y_a^2) \sqrt{D^2 + y_a^2 - R_b^2}} \quad (10\text{-}32)$$

For the equation of the asymptotes we have

$$x = y \tan \gamma + \left(\frac{L}{2\pi}\right) \tan^{-1} \infty \qquad (10\text{-}30)$$

The form of this intersection curve and its asymptotes are plotted in Fig. 10-20.

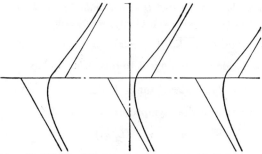

FIG. 10-20.   Off-center section of chased helicoid.

Intersection Curve of Chased Helicoid with a Plane Perpendicular to Its Axis.   Introducing the change of direction of the helices into Eq. (10-7), we obtain

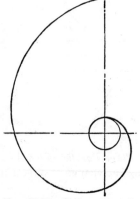

FIG. 10-21.   End section of chased helicoid.

$$\theta = (2\pi \tan \gamma/L) \sqrt{r^2 - R_b{}^2} \\ + \tan^{-1} (\sqrt{r^2 - R_b{}^2}/R_b) \quad (10\text{-}33)$$

For the equation of the tangent to this curve we have

$$\tan \psi = r\, d\theta/dr = (2\pi r^2 \tan \gamma \\ + R_b L)/L \sqrt{r^2 - R_b{}^2} \quad (10\text{-}34)$$

The form of this intersection curve is plotted in Fig. 10-21.

Limits of Conjugate Gear-tooth Action on Chased Helicoid.   Introducing the change of direction of the helices into Eq. (10-9), we obtain

$$\theta = \tan^{-1} [L \sqrt{r^2 - R_b{}^2}/(2\pi r^2 \tan \gamma + R_b L)] \qquad (10\text{-}35)$$

This curve approaches an asymptote that is parallel to the axis of the mating worm gear, and is at a distance of

$$y = L/2\pi \tan \gamma \qquad (10\text{-}10)$$

from the axis of the helicoid.   This curve and its asymptotes are plotted in Fig. 10-22.

Milled Helicoid. The milled helicoid is produced by a cone-shaped milling cutter or grinding wheel that is set at an angle to the axis of the helicoid, as shown in Fig. 10-23. The general method used to derive the equation of its intersection curve is similar to the one used in Prob. 9-2.

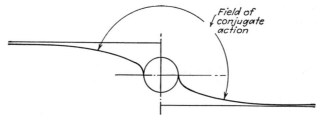

FIG. 10-22. Field of conjugate action on chased helicoid.

As in Prob. 9-2, the form of the helicoid produced is the locus of all points of tangency between the cone-shaped cutter and the helicoid in all the relative operating positions of the two members.

The contact between the helicoid and the cone-shaped tool set at an angle to the axis of the helicoid does not remain on the same element of the cone. The problem is complex, and the derivation of these equations

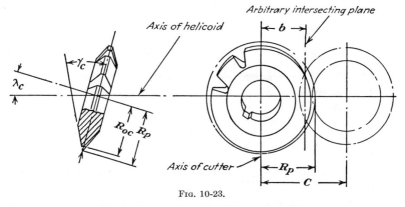

FIG. 10-23.

would use up so much space that it has been omitted. Only the series of equations needed for the solution of a definite example is given.

Referring to Fig. 10-23, let

C = distance between axes of helicoid and cone-shaped tool, in.

γ_c = one-half included angle of cutting edges of tool

λ_c = angular setting of cone-shaped tool

R_{oc} = outside radius of cone-shaped tool, in.

R_p = radius to sharp point (extended) of tool, in.

b = distance between axis of cone-shaped tool and any arbitrary intersecting plane,[1] in.

f = width of flat at outside radius of cone-shaped tool, in.

Other arbitrary symbols will be introduced as needed to simplify the equations and the calculations.

We have the following from the geometrical conditions shown in Fig. 10-23:

$$R_p = R_{oc} + f/2 \tan \gamma_c \qquad (10\text{-}36)$$

Intersection Curve of Milled Helicoid with Axial Plane. The following equations must be solved in the order as given to determine the coordinates of this intersection curve:

$$E = \tan \gamma_c (L \tan \lambda_c + 2\pi C) \qquad (10\text{-}37)$$

$$F = 2\pi R_p \tan^2 \gamma_c \tan \lambda_c \qquad (10\text{-}38)$$

$$G = L - 2\pi C \tan \lambda_c + 2\pi b \tan \lambda_c (1 + \tan^2 \gamma_c) \qquad (10\text{-}39)$$

$$Y = \frac{G \sqrt{F^2 + E^2 - G^2} - EF}{E^2 - G^2} \qquad (10\text{-}40)$$

$$R_a = b \sqrt{Y^2 + 1} \qquad (10\text{-}41)$$

$$\tan \Delta = \frac{bY \cos \lambda_c + (R_p - R_a) \tan \gamma_c \sin \lambda_c}{C - b} \qquad (10\text{-}42)$$

$$y_a = \frac{C - b}{\cos \Delta} \qquad (10\text{-}43)$$

$$x_a = (R_p - R_a) \tan \gamma_c \cos \lambda_c - bY \sin \lambda_c + \frac{\Delta L}{2\pi} \qquad (10\text{-}44)$$

These last two equations give the values of the coordinates of the intersection curve of this milled helicoid with an axial plane. The origin of the system of coordinates is on the axis of the helicoid and at the center of the tooth space produced by the cone-shaped tool.

For the tangent to this intersection curve we have the following equations:

$$H = R_a \tan \gamma_c \cos \lambda_c - \left(\frac{L}{2\pi y_a}\right) \cos \Delta (y_a \sin \Delta - R_p \tan \gamma_c \sin \lambda_c) \qquad (10\text{-}45)$$

$$J = b \left[Y \sin \lambda_c - \left(\frac{L}{2\pi y_a}\right) \sin \Delta \right] \qquad (10\text{-}46)$$

$$K = \sin \Delta (y_a \sin \Delta - R_p \tan \gamma_c \sin \lambda_c) \qquad (10\text{-}47)$$

$$M = b \left(\frac{1}{\cos \Delta} - \sin \Delta \tan \Delta \right) = b \cos \Delta \qquad (10\text{-}48)$$

$$\tan \phi = \frac{dx_a}{dy_a} = \frac{H + J}{M - K} \qquad (10\text{-}49)$$

---

[1] These intersecting planes are parallel to both the axis of the tool and the axis of the helicoid.

When the diameter of the cutter is small in relation to the diameter of the helicoid, and when the lead angle of the helicoid is large, this form will be concave and approach the form of a chased helicoid. As the relative diameter of the cutter increases, this form will be partly concave and partly convex. With a further increase in the diameter, or as the setting angle becomes smaller, the form will be convex and will approach that of a convolute helicoid. With a cutter of infinite diameter, or when the setting angle is reduced to zero, the form will be that of an involute helicoid.

Intersection Curve of the Milled Helicoid with a Plane Parallel to the Axis and at a Distance D from the Axis. For this intersection curve we shall derive general equations, which can be used for any type of helicoid when the coordinates of the intersection curve with the axial plane are known.

When x_a and y_a = coordinates of intersection curve with axial plane

$\tan \phi = dx_a/dy_a$ = tangent to intersection curve with axial plane

x_2 and y_2 = coordinates of intersection curve with plane parallel to axis at a distance D from axis

$\tan \phi_2 = dx_2/dy_2$ = tangent to intersection curve with off-center plane

$$x_2 = x_a - (L/2\pi) \sin^{-1} (D/y_a) \qquad (10\text{-}50)$$
$$y_2 = \sqrt{y_a{}^2 - D^2} \qquad (10\text{-}51)$$
$$\tan \phi_2 = (y_2/y_a) \tan \phi + (LD/2\pi y_a{}^2) \qquad (10\text{-}52)$$

Intersection Curve of Milled Helicoid with Plane Perpendicular to the Axis. For this intersection curve we shall also use a general equation, which gives us the values in terms of the coordinates of the form in the axial section. For the equations of this intersection curve we have

$$\theta = 2\pi x_a/L \qquad (10\text{-}53)$$
$$r = y_a \qquad (10\text{-}54)$$

For the equation of the tangent to this intersection curve we have

$$\tan \psi = r \, d\theta/dr = 2\pi y_a \tan \phi/L \qquad (10\text{-}55)$$

This completes the analysis of this group of helicoid sections. We shall next consider their application to the analysis of the contact and to the determination of the position of the actual contact lines on worm-gear drives.

CHAPTER 11

CONTACT ON WORM-GEAR DRIVES

Path of Contact and Conjugate Gear-tooth Profiles. When the coordinates of the intersection curves of the worm or helicoid with planes parallel to the axis of the worm are known, we can use the equations derived in Chap. 1 to determine the paths of contact and the conjugate gear-tooth profiles of the mating worm gear in various planes of rotation of the worm gear. The origin of the coordinate system of the helicoids is on the axis of the worm, while the origin of the coordinate system used for the path of contact and for the conjugate gear-tooth profiles is at the pitch point. We must, therefore, first shift the origin of the coordinates of the helicoid sections to the pitch point. For this we have the following:
When x_a and y_a = coordinates of helicoid sections with origin on axis of helicoid
$\quad\quad x$ and y = coordinates of helicoid sections with origin at the pitch point, in.
$\quad\quad R_1$ = radius to pitch plane of worm, in.
$\quad\quad x'_a$ = value of x_a when y_a is equal to R_1, in.
then

$$y = y_a - R_1 \tag{11-1}$$
$$x = x_a - x'_a \tag{11-2}$$

Path of Contact. We have from Chap. 1 the following equations for the path of contact:
When y = ordinate of basic-rack profile and of path of contact, in.
$\quad\quad x_p$ = abscissa of path of contact, in.

$$\tan \phi = dx/dy = dx_a/dy_a \tag{11-3}$$
$$x_p = -y/\tan \phi \tag{1-1}$$

Conjugate Gear-tooth Profile. We also have from Chap. 1 the following equations for the conjugate gear-tooth profile:
When θ = vectorial angle of conjugate gear-tooth form
$\quad\quad r$ = length of radius vector of conjugate tooth profile, in.
$\quad\quad R_2$ = radius of pitch cylinder of worm gear, in.

$$r = \sqrt{(R_2 - y)^2 + x_p{}^2} \tag{1-2}$$
$$\theta = [(x - x_p)/R_2] + \tan^{-1}[x_p/(R_2 - y)] \tag{1-3}$$

Trochoidal Fillet. We also have from Chap. 3 the following equations for the form of the trochoidal fillet at the root of the worm gear-teeth: When R_2 = pitch radius of worm gear, in.

b = distance from pitch line to corner of basic-rack tooth or to center of rounded corner, in.

r_t = any radius of trochoid, in.

θ_t = vectorial angle of trochoid

δ = angle between origins of gear-tooth profile and trochoid

x_t = abscissa of corner of rack tooth or center of rounding, measured from the pitch point, in.

$$\theta_t = \tan^{-1}\left[\frac{\sqrt{r_t^2 - (R_2 - b)^2}}{R_2 - b}\right] - \frac{\sqrt{r_t^2 - (R_2 - b)^2}}{R_2} \quad (3\text{-}1)$$

$$\text{arc } \delta = \frac{x_t}{R_2} \quad (3\text{-}3)$$

Example of Path of Contact and Conjugate Gear-tooth Profile on Worm Gear. As a definite example we shall determine the basic-rack forms (helicoid sections), paths of contact, conjugate gear-tooth profiles, and forms of trochoidal fillets on the following worm-gear drive. The worm will be a screw helicoid.

VALUES FOR WORM

Axial pitch..	1.000 in.
Number of starts or threads.........................	6
Lead of thread.....................................	6.000 in.
Radius to pitch plane..............................	1.750 in.
Outside radius.....................................	2.036 in.
Root radius..	1.413 in.
One-half included angle of thread.....................	14.500°

VALUES FOR WORM GEAR

Number of teeth....................................	36
Pitch radius.......................................	5.7294 in.
Throat radius......................................	6.0154 in.
Outside radius.....................................	6.1970 in.
Face width..	2.6000 in.

An analysis will be made on five sections across the face of the gear as indicated in Fig. 11-1. The nonactive profile of the rack form in section 1 will be the same as the active profile in section 5. Their relative positions will be determined from the known thread thickness of the worm at any convenient diameter, such as the outside diameter, for example. In a similar manner, the nonactive profile in section 2 will be the same as the active profile in section 4. Section 3 is an axial section, and the basic-rack form here will be symmetrical. In this case of a screw helicoid, this section is formed of straight lines.

For this screw helicoid we have the following values:

$$\gamma = 14.500° \qquad \tan \gamma = 0.25862 \qquad L = 6.000 \qquad R_1 = 1.750$$

We shall use the following values for the distances of the intersecting planes from the axis of the helicoid:

Section	Value of D, in.
1	1.250
2	0.625
3	0.000
4	−0.625
5	−1.250

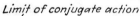

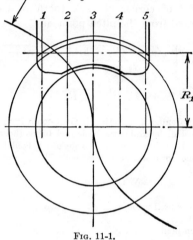

Fig. 11-1.

Sections 1 and 5. Substituting the foregoing values into Eq. (10-13) we obtain
For section 1

$$x_a = 0.25862 \sqrt{1.5625 + y_a^2} - 0.95493 \tan^{-1} \frac{y_a}{1.250}$$

For section 5

$$x_a = 0.25862 \sqrt{1.5626 + y_a^2} - 0.95493 \tan^{-1} \frac{-y_a}{1.250}$$

Substituting the values into Eq. (10-14), we obtain
For section 1

$$\tan \phi \; \frac{1.62496 y_a \sqrt{1.5625 + y_a^2} - 7.500}{6.28319(1.5625 + y_a^2)}$$

For section 5

$$\tan \phi = \frac{1.62496 y_a \sqrt{1.5625 + y_a^2} + 7.500}{6.28319(1.5625 + y_a^2)}$$

Solving these equations for a series of values of y_a ranging from 0.630 in. to 1.670 in., which covers the depth of thread, and for 1.750 in., which gives the value on the pitch line, we obtain the values tabulated in Table 11-1.

Substituting the values from Table 11-1 into Eqs. (11-1) and (11-2), we obtain

$$y = y_a - 1.750$$

For section 1
$$x = x_a + 0.35152$$
For section 5
$$x = x_a + 1.53612$$

These values of x and y are tabulated in Table 11-2. Substituting the values from Tables 11-1 and 11-2 into Eq. (1-1), we obtain the values of x_p, which are also tabulated in Table 11-2.

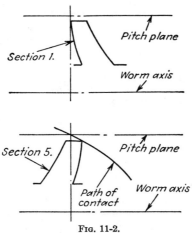

Fig. 11-2.

There is no path of contact in section 1 because all the values of tan ϕ are minus, which indicates that this part of the helicoid form is outside of the field of conjugate action.

These values in Table 11-2 are plotted in Fig. 11-2.

TABLE 11-1. COORDINATES OF BASIC-RACK PROFILE IN SECTIONS 1 AND 5
(Plotted in Fig. 11-2)

y_a, in.	Section 1		Section 5	
	x_a, in.	tan ϕ	x_a, in.	tan ϕ
0.630	−0.08379	−0.49280	−2.19219	0.72559
0.760	−0.14334	−0.42340	−2.10000	0.69211
0.890	−0.19400	−0.35694	−2.01232	0.65694
1.020	−0.23632	−0.29508	−1.92920	0.62209
1.150	−0.27096	−0.23864	−1.85850	0.58884
1.280	−0.29863	−0.18788	−1.77579	0.55794
1.410	−0.32005	−0.14266	−1.70533	0.52950
1.540	−0.33593	−0.10263	−1.63815	0.50419
1.670	−0.34693	−0.06727	−1.57411	0.48136
1.750	−0.35152	−0.04764	−1.53612	0.46853

Using the values from the preceding tables in Eqs. (1-2) and (1-3), we obtain the values of the coordinates of the conjugate gear-tooth profiles of section 5, which are tabulated in Table 11-3. The form of the worm-gear-tooth profile in section 1 is that of the trochoidal path of the corner of the hob tooth.

TABLE 11-2. COORDINATES OF BASIC-RACK PROFILE AND PATH OF CONTACT, SECTIONS 1 AND 5

(Plotted in Fig. 11-2)

y_a, in.	Section 1		Section 5		
	y, in.	x, in.	y, in.	x, in.	x_p, in.
0.630	−1.120	0.26773	−1.120	−0.65607	1.54357
0.760	−0.990	0.20818	−0.990	−0.56388	1.43040
0.890	−0.860	0.15752	−0.860	−0.47620	1.30910
1.020	−0.730	0.11520	−0.730	−0.39308	1.17346
1.150	−0.600	0.08056	−0.600	−0.31438	1.01895
1.280	−0.470	0.05289	−0.470	−0.23985	0.84238
1.410	−0.340	0.03147	−0.340	−0.16921	0.64187
1.540	−0.210	0.01559	−0.210	−0.10203	0.41650
1.670	−0.080	0.00459	−0.080	−0.03799	0.16619
1.750	0.000	0.00000	0.000	0.00000	0.00000

TABLE 11-3. COORDINATES OF CONJUGATE GEAR-TOOTH FORM, SECTION 5

(Plotted in Fig. 11-3)

y_a, in.	r, in.	θ, rad	$r\theta$, in.
0.630	7.021	−0.16226	−1.146
0.890	6.870	−0.13833	−0.950
1.020	6.565	−0.09372	−0.615
1.150	6.411	−0.07310	−0.468
1.280	6.256	−0.05384	−0.337
1.410	6.103	−0.03621	−0.221
1.540	5.954	−0.02049	−0.122
1.670	5.811	−0.00704	−0.041
1.750	5.7294	0.00000	0.000

Using the value of $b = -0.080$ in Eq. (3-1), we obtain the values of the coordinates of the trochoidal fillet for sections 1 and 5 that are tabulated in Table 11-4. The only difference in the trochoids of these two sections at equal distances on opposite sides of the axis of the worm is in their relative locations from the origins of their respective conjugate gear-tooth forms, or from the pitch points of the profiles.

For section 1, the value of $x_t = 0.00459$, whence

$$\text{arc } \delta = \frac{0.00459}{5.7294} = 0.00080 \text{ radian}$$

For section 5, the value of $x_t = -0.03799$, whence

$$\text{arc } \delta = -\frac{0.03799}{5.7294} = -0.00663 \text{ radian}$$

The values tabulated in Tables 11-3 and 11-4 are plotted in Fig. 11-3.

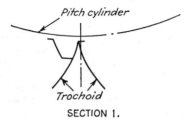

SECTION 1.

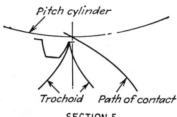

SECTION 5.
FIG. 11-3.

TABLE 11-4. COORDINATES OF TROCHOIDAL FILLET, SECTIONS 1 AND 5
(Plotted in Fig. 11-3)

r_t, in.	θ_t, rad	$r_t\theta_t$, in.
5.8094	0.00000	0.0000
5.9000	−0.00602	−0.0355
6.0000	−0.00913	−0.0547
6.1000	−0.01478	−0.0901
6.2000	−0.02117	−0.1312
6.3000	−0.02816	−0.1774
6.4000	−0.03570	−0.2284
6.5000	−0.04373	−0.2842
6.6000	−0.05221	−0.3445

Sections 2 and 4. The values of the coordinates of the several curves in sections 2 and 4 are calculated in the same manner as were those for sections 1 and 5. These values are tabulated in Tables 11-5, 11-6, 11-7, and 11-8. The angles between the origins of the conjugate gear-tooth profiles and the trochoidal fillets are as follows:
For section 2

$$\text{arc } \delta = \frac{0.02752}{5.7294} = 0.00480 \text{ radian}$$

TABLE 11-5.   COORDINATES OF BASIC-RACK PROFILE, SECTIONS 2 AND 4
(Plotted in Fig. 11-4)

y_a, in.	Section 2		Section 5	
	x_a, in.	tan ϕ	x_a, in.	tan ϕ
1.267	−0.69701	−0.06710	−1.57224	0.53095
1.350	−0.70123	−0.03500	−1.53130	0.50435
1.450	−0.70301	−0.00191	−1.48028	0.47687
1.550	−0.70176	0.02616	−1.43379	0.45351
1.650	−0.69792	0.05012	−1.38945	0.43355
1.750	−0.69185	0.07070	−1.34698	0.41637
1.850	−0.68389	0.08848	−1.30642	0.40152
1.990	−0.66433	0.10460	−1.25670	0.37896

TABLE 11-6.   COORDINATES OF BASIC RACK AND PATH OF CONTACT, SECTIONS 2 AND 4
(Plotted in Fig. 11-4)

y_a, in.	Section 2			Section 4		
	y, in.	x, in.	x_p, in.	y, in.	x, in.	x_p, in.
1.267	−0.483	−0.00516	....	−0.483	−0.22526	0.90969
1.350	−0.400	−0.00938	....	−0.400	−0.18432	0.79310
1.450	−0.300	−0.01116	....	−0.300	−0.13330	0.62910
1.550	−0.200	−0.00991	7.64526	−0.200	−0.08681	0.44100
1.650	−0.100	−0.00607	1.99521	−0.100	−0.04272	0.23065
1.750	0.000	0.00000	0.00000	0.000	0.00000	0.00000
1.850	0.100	0.00796	−1.13020	0.100	0.04056	−0.24905
1.990	0.240	0.02752	−2.29445	0.240	0.09028	−0.63331

TABLE 11-7.   COORDINATES OF CONJUGATE GEAR-TOOTH PROFILES, SECTIONS 2 AND 4
(Plotted in Fig. 11-5)

y_a, in.	Section 2		Section 4	
	r, in.	$r\theta$, in.	r, in.	$r\theta$, in.
1.267	....	....	6.2392	−0.3288
1.350	....	....	6.1804	−0.2590
1.450	....	....	6.0621	−0.1764
1.550	9.67511	−4.1118	5.9457	−0.1063
1.650	6.16139	−0.1204	5.8339	−0.0474
1.750	5.72940	0.0000	5.7294	0.0000
1.850	5.74173	....	5.6349	0.0426
1.990	5.9490	....	5.5258	0.0631

TABLE 11-8. COORDINATES OF TROCHOIDAL FILLET, SECTIONS 2 AND 4
(Plotted in Fig. 11-5)

r_t, in.	$r_t\theta_t$, in.
5.4894	0.00000
5.6000	0.03237
5.7000	0.02702
5.8000	0.01113
5.9000	−0.01286
6.0000	−0.04320
6.1000	−0.07960
6.2000	−0.12127

For section 4

$$\text{arc } \delta = \frac{0.09028}{5.7294} = 0.01575 \text{ radian}$$

The basic-rack forms and paths of contact for these two sections are plotted in Fig. 11-4. The conjugate gear-tooth profiles and trochoidal fillets are plotted in Fig. 11-5.

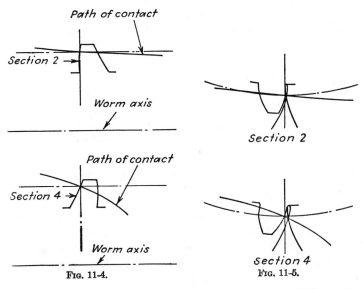

FIG. 11-4.   FIG. 11-5.

In section 2, the lower part of the basic-rack form is beyond the field of conjugate action. The pressure angle of this form is so low that the conjugate gear-tooth profile is undercut. There is also a cusp above the pitch line because the values of r increase again. Such cusps and undercut are always found together.

Section 3. Section 3 is the axial section of the screw helicoid. The basic-rack form here is a straight-line profile, and the conjugate gear-tooth profile is the involute of a circle. The coordinates of these intersection curves are tabulated in Table 11-9. The profile of this basic-rack form and its path of contact are plotted in Fig. 11-6.

TABLE 11-9. COORDINATES OF PROFILES IN SECTION 3
(Plotted in Fig. 11-7)
tan φ = 0.25862

y_a, in.	x_a, in.	y, in.	x, in.	x_p, in.	r, in.	$r\theta$, in.
1.413	−1.13457	−0.337	−0.08715	1.30308	6.2043	−0.1927
1.550	−1.09914	−0.200	−0.05172	0.77334	5.9786	−0.0855
1.650	−1.07328	−0.100	−0.02586	0.38667	5.8422	−0.0337
1.750	−1.04742	0.000	0.00000	0.0000	5.7294	0.0000
1.850	−1.02156	0.100	0.02586	−0.38667	5.6426	0.0193
1.950	−0.99570	0.200	0.05172	−0.77334	5.5832	0.0281
2.086	−0.96052	0.336	0.08690	−1.29921	5.5476	0.0308

The location of the origin of the trochoidal fillet in reference to the pitch point of the gear-tooth profile is as follows:

$$\text{arc } \delta = \frac{0.08690}{5.7924} = 0.01500 \text{ radian}$$

The values of the coordinates of the trochoid are tabulated in Table 11-10. The coordinates of the conjugate gear-tooth profile and of the trochoid are plotted in Fig. 11-7.

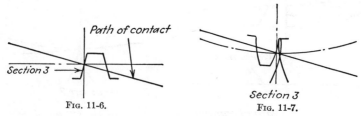

FIG. 11-6.   FIG. 11-7.

Field of Contact and Contact Lines. We are now in a position to determine the projection of the field of actual contact and the actual position of the line contact between the worm and the worm gear. The example used here is a very poor design and was selected purposely to make apparent the nature of the conditions outside of the field of conjugate action, and also those of undercut. The angle of the generatrix

TABLE 11-10. COORDINATES OF TROCHOIDAL FILLET, SECTION 3
(Plotted in Fig. 11-7)

r_t, in.	$r_t\theta_t$, in.
5.3934	0.0000
5.5000	0.0502
5.6000	0.0529
5.7000	0.0433
5.8000	0.0250
6.0000	−0.0319
6.2000	−0.1113

of the helicoid must not be much less than the lead angle of the worm if such conditions are to be avoided. Also the worm diameter must be large enough to bring the thread contour into the field of conjugate action.

The projection of the field of contact is determined graphically as follows: To develop this field of contact and the projections of the actual contact lines between the worm and the gear, we need three views of the worm: the end view, the plan view, and the side view. In the side view, we first plot the paths of contact of the several sections. Through these paths of contact we then draw straight lines, parallel to the pitch plane, which represent the trace of the outside cylinder of the worm with the several intersection planes of the different sections. The intersection of this line for section 5, for example, with the path of contact for section 5 is projected into the plan view of the worm. Where this projection line crosses the line representing section 5 is one point of the boundary of the projection of the field of contact in the plan view of the worm. This process is repeated for all sections, and gives a series of points through which a curve is drawn. This curve represents the end of the recess action between the worm and the gear.

To determine the boundary at the beginning of the approach action, circles are drawn from the center of the worm gear that represent the trace of the outside of the worm-gear blank with the several intersection planes. The intersections of these circles with their respective paths of contact establish the several points on the boundary of the field of contact at the beginning of mesh. These points are projected into the plan view of the worm as before, and so this part of the form of the field of contact is established. Consideration must be given here to the form of the outside of the worm-gear blank, as these projected points generally belong on two intersecting curves or forms, and do not lie in one continuous curve.

The positions of the sides of the field of contact are controlled by the intersection of the two sides or faces of the worm-gear blank with the threads of the worm. These are commonly straight lines that represent the edges of the worm-gear blank.

When undercut is present, the boundaries will then be inside the full potential field of contact. The end of the recess action, in such cases, will be the point where a circle of the worm gear that passes through the intersection of the trochoid and the conjugate gear-tooth profile crosses the path of contact for the particular section. This gives another line or curve at the end or side of the field of contact.

These operations have been carried out in Fig. 11-8. The paths of contact, straight lines, and circles are numbered, corresponding to the number of the section to which they belong.

The determination of the projection of the actual contact lines between the two members of the drive is made in a similar manner. The several intersection profiles of the worm (helicoid sections) are drawn, using the original coordinates x_a and y_a, with the y_a axis passing through the pitch point. The several paths of contact are drawn as before. The point where a given path of contact crosses the corresponding basic-rack form is one point on one contact line. This point is projected to the line in the plan view that represents the given section. The height of this point

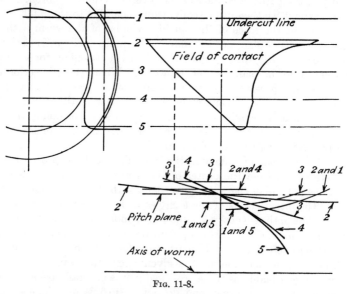

Fig. 11-8.

from the axis of the worm is transferred to the corresponding section in the end view of the worm. When all these points for a given position of the basic-rack profiles are obtained, curves are drawn through them. This gives the projection of one contact line in the plan view and in the end view of the worm.

For the other contact lines, the basic-rack profiles are moved together in a direction parallel to the axis of the worm for a distance equal to the axial pitch of the worm. Points on this second contact line are established and projected as before. This process is repeated until all the contact lines within the field of contact have been established. Points outside the field of contact are useful to establish more definitely the forms of some of these contact lines, particularly those which lie near the ends of the field of contact.

These operations have been carried out in Fig. 11-9. Generally, the operations described can be carried out on the same layout. They are separated here to avoid possible confusion.

Such an analysis may be made of any worm-gear drive. It gives a very illuminating picture of the nature and amount of the contact on worm-gear drives. Often it will help to explain seemingly paradoxical conditions, which are often met with on such drives. Many actual analyses and layouts must be made to obtain a full comprehension of this phase of the subject.

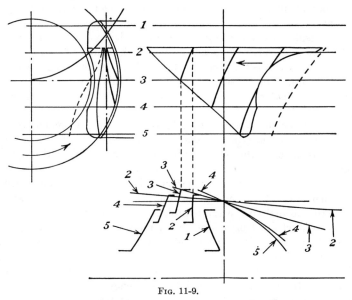

Fɪɢ. 11-9.

Direct Analysis of Contact Lines. It is possible to determine the positions of the contact lines between a worm and a worm gear without first determining the sectional basic-rack profiles of the worm and their paths of contact. A direct analysis of the contact lines is the simpler procedure, but it gives little indication of the possible conditions of undercut. Therefore until one has mastered the characteristics of helicoid sections and their application to the analysis of worm-drive contact, and has become fully aware of the influences of the thread angle and the lead angle on the off-center sections of the worm, the full analysis will always be the more valuable one.

We shall now proceed to study the contact lines on a helicoid used as a worm. This direct method of analysis was originated by Ernest

Wildhaber. In order to simplify the problem, we shall at first ignore the diameter of the worm gear and also all problems of interference and undercutting, and other similar limitations to the field of contact. Eventually these problems must be solved in exactly the same manner here as is used for the solution of similar problems with spur gears. The diameter of the worm gear has no influence on the forms or positions of these contact lines: the larger worm gear only uses more of them.

The contact lines between a worm and a worm gear are determined in the same manner as would be used to determine the contact lines between a rack of varying form of profile across its face and a spur gear. At any moment, an infinitely small part of the motion between the two members can be considered as a turning motion about the contact line between the pitch cylinder of the worm gear and the pitch plane of the worm. This contact line, which is the locus of all the pitch points of the mating conjugate profiles, when considered as the axis of this turning motion, will be called the *momentary center of the motion*, or more briefly, the *momentary axis*.

In order for the worm to be able to turn about this momentary axis relative to the worm gear, the normal to any contact point between the worm thread and the gear-tooth surface must pass through this momentary axis. Otherwise it would be impossible for the worm to turn, or rock slightly, about the momentary axis in both directions. All this is another way of expressing the fact that for conjugate gear-tooth action, the normal to the tooth profiles of the mating members at the point of contact must pass through the pitch point. Here, however, we have a three-dimensional problem instead of a two-dimensional one.

The normal to the surface of a helicoid must be perpendicular to its straight-line generatrix. It must also be perpendicular to the tangent of the helix of the helicoid at its point of tangency to the helicoid surface. It must also be perpendicular to the tangent of any intersection profile of the helicoid with any plane that contains the point of contact.

We shall now study the screw helicoid and determine its contact lines.

Contact Lines on Screw Helicoid. Referring to Fig. 11-10, the position of the surface of the helicoid relative to the drawing plane can be determined from the location of its intersection curve with the drawing plane. In this case, the drawing plane is perpendicular to the axis of the helicoid. The intersection curve for the screw helicoid is an Archimedean spiral whose equation is given by Eq. (10-15), which is as follows:

$$\theta = 2\pi r \tan \gamma / L \qquad (10\text{-}15)$$

This intersection curve can be located in any angular position that may be desired. In this example we shall start with the origin revolved one-half a revolution from its initial position as used for the original derivation of the equations in helicoid sections.

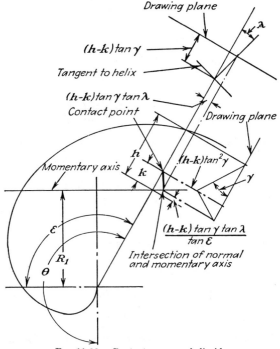

Fig. 11-20.   Contact on screw helicoid.

Referring to Fig. 11-10, let

L = lead of generatrix, in.

γ = angle of generatrix

θ = angle of rotation of generatrix

h = length along radial line from momentary axis to intersection curve of screw helicoid, in.

R_1 = distance of pitch plane from axis of helicoid, in.

k = distance to projection of contact point on radial line from momentary axis, in.

r = length of radius vector to intersection curve, in.

ϵ = angle of rotation of generatrix from horizontal position

From the geometrical conditions shown in Fig. 11-10 we have the following:

$$h = r - (R_1/\sin \epsilon) \tag{11-4}$$

The intersection of the normal to the helicoid surface with the momentary axis is at a distance $(h - k) \tan \gamma \tan \lambda$ from the projection of the generatrix on the end view of the worm as shown in Fig. 11-10, where

λ = lead angle at point of contact
r_1 = radius to point of contact

$$\tan \lambda = L/2\pi r_1$$
$$r_1 = (R_1/\sin \epsilon) + k$$

whence

$$\tan \lambda = L \sin \epsilon/2\pi(R_1 + k \sin \epsilon)$$
$$k = (h - k) \tan^2 \gamma - [(h - k) \tan \gamma \tan \lambda/\tan \epsilon]*$$

Substituting the value of $\tan \lambda$, combining, and simplifying, we obtain

$$k = (h - k)\{\tan^2 \gamma - [L \tan \gamma \cos \epsilon/2\pi(k \sin \epsilon + R_1)]\} \quad (11\text{-}5)$$

When Eq. 11-5 is solved for k, we obtain an extended quadratic equation. For simplification when solving, we will let

$A = 2\pi \sin \epsilon$
$B = 2\pi R_1 - L \sin \gamma \cos \gamma \cos \epsilon - 2\pi h \sin^2 \gamma \sin \epsilon$
$C = h(2\pi R_1 \sin^2 \gamma - L \sin \gamma \cos \gamma \cos \epsilon)$

Then

$$k = (-B + \sqrt{B^2 + 4AC})/2A \quad (11\text{-}6)$$

When $\epsilon = 90°$, Eq. (11-6) reduces to the following:

$$k = h \sin^2 \gamma \quad (11\text{-}7)$$

The value of h is plus when the intersection curve of the helicoid is above the pitch plane. When the value of h is plus, the value of k is plus also. When the value of h is minus, the value of k is minus also.

A study of Fig. 11-10 will make it apparent that the contact line will always lie between the momentary axis and the intersection curve of the helicoid with the drawing plane that contains the momentary axis. The form and position of this contact line will depend upon the lead of the helicoid, the angle of the generatrix, and the position of the pitch plane. The smaller the lead and the lower the angle of the generatrix, the closer the contact line will be to the momentary axis, which is the intersection of the pitch plane with the drawing plane.

To obtain the contour of the projection of any contact line on the end view of the worm for any given position of the helicoid in relation to the momentary axis, the positions of a series of contact points must be

---

* The sign here is minus instead of plus because the value of ϵ as shown is in the second quadrant where the value of its tangent is minus.

established. This is done by solving Eq. (11-6) for several different positions of the radial plane, or for different values of ϵ.

To obtain the projections of these contact lines on the plan view of the worm, we proceed as follows:
When x = distance of contact point from momentary axis in an axial direction, in.
then

$$x = (h - k) \tan \gamma \qquad (11\text{-}8)$$

The value of the coordinate y in this plan view, or the distance of the projection of the contact point from the projection of the axis of the worm on the plan view, is determined graphically by projecting the position of the desired point from the end view into the plan view.

To determine the change in the form and position of the contact lines as the worm is revolved into other positions, the preceding equations are solved for successive positions of the momentary axis in relation to the helicoid. This is done by determining new values for h corresponding to the change in position of the helicoid in relation to the momentary axis. For example, if the helicoid is revolved one-quarter of a revolution, all the original values of h will be altered an amount equal to $L/4 \tan \gamma$.
When N = number of threads or starts on worm
    h_1 = length of h on first thread from original position, in.
    h_2 = length of h on second thread from original position, in.
then to obtain values for successive contact lines that exist simultaneously on successive threads of the worm

$$h_1 = h + (L/N \tan \gamma) \qquad (11\text{-}9)$$
$$h_2 = h_1 + (L/N \tan \gamma) = h + 2(L/N \tan \gamma) \qquad (11\text{-}10)$$
etc.

The actual duration of contact can be determined by establishing the turning angle of the helicoid that will carry the contact lines across the face of the worm gear. This, of course, will also be influenced by the diameter of the worm gear. The duration of contact will be greater when worm gears of greater diameter (and numbers of teeth) are used. These conditions will be considered later when we study the field of contact.

Contact Lines in Involute Helicoids. The form and position of the contact lines on an involute helicoid are easy to determine. The nature of an involute helicoid is such that every normal to its surface is tangent to the base cylinder. Thus if we use for our analysis a series of planes that are tangent to the base cylinder, these planes will contain both the generatrix and the normal to the helicoidal surface. This leads to a very simple solution for the positions of the contact points. Referring to Fig. 11-11, when

L = lead of generatrix, in.

γ = angle of generatrix

ϵ = angle of plane tangent to base cylinder with pitch plane

R_b = radius of base cylinder, in.

h = length along line tangent to base cylinder from the momentary axis to intersection curve of involute helicoid, in.

k = distance to projection of contact point on line tangent to base circle from momentary axis, in.

R_1 = distance of pitch plane from axis of helicoid, in.

ϵ_1 = angle of rotation of generatrix from origin of involute

we have from the geometrical conditions shown in Fig. 11-11 the following:

$$h = R_b\epsilon_1 - [(R_1 - R_b \cos \epsilon)/\sin \epsilon] \qquad (11\text{-}11)$$
$$k = h \sin^2 \gamma \qquad (11\text{-}12)$$
$$R_b = L/2\pi \tan \gamma \qquad (11\text{-}13)$$

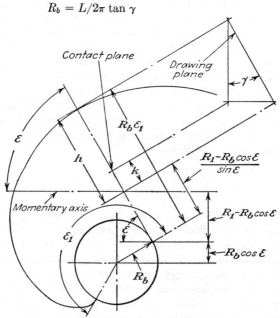

FIG. 11-11.   Contact on involute helicoid.

Equation (11-12) is the same for all positions of the intersecting tangent planes.

In order to determine the change in the projection of the form and position of other contact lines as the helicoid is revolved to different positions, the changed length of h is multiplied by the constant value of $\sin^2 \gamma$. The length of h changes uniformly on all tangent planes. For a turning movement of one revolution, it changes an amount equal to the circumference of the base cylinder. For all other movements, it changes in direct proportion to the fraction of a revolution that may be made.

To determine the projection of these contact points on the plan view of the worm, we proceed exactly as before as in the case of a screw helicoid. For this purpose we have Eq. (11-8).

Examples of Contact Lines. Before considering the method of establishing the projections of the field of contact, we shall solve a few definite examples to determine the difference, if any, in the nature of the contact lines between similar worms of the form of screw helicoids and of involute helicoids. These two, as noted before, are limiting types of the convolute helicoid. The range between them covers practically all the types of worms commonly used. We will also study the influence of the position of the pitch plane on the form and nature of these contact lines.

First Example: Single-thread Screw Helicoid. The first example will be a single-thread screw helicoid of the following proportions:

Outside radius.. 2.500 in.
Root radius... 2.000 in.
Angle of generatrix................................... 30°
Lead.. 1.000 in.
Distance to pitch plane............................... 2.250 in.

For the initial position of the helicoid in respect to the monetary axis, we shall make the intersection curve of this helicoid in the drawing plane pass through the point where $r = 2.250$ in. on the radial plane where $\epsilon = 90°$. Under these conditions

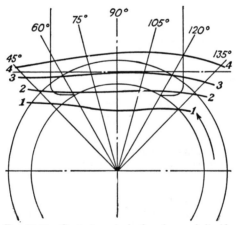

Fig. 11-12. Contact on single-thread screw helicoid.

we obtain the values tabulated in Table 11-11. With a single-thread worm, it must be revolved a full revolution to determine the contact line on the adjacent thread or tooth. Several contact lines, 1 thread apart, have been determined. The values of their coordinates are tabulated in Table 11-11. These values are also plotted in Fig. 11-12.

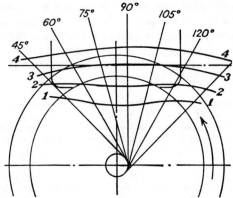

Fig. 11-13. Contact on single-thread involute helicoid.

Second Example: Single-thread Involute Helicoid. As a second example, we shall use an involute helicoid with the same values as were used for the screw helicoid, and make the intersection curve of this helicoid with the drawing plane pass through the same point as before. For this example, we get the values tabulated in Table 11-12 and plotted in Fig. 11-13.

TABLE 11-11. COORDINATES OF CONTACT LINES ON SINGLE-THREAD SCREW HELICOID
(Plotted in Fig. 11-12)

Position of worm	Values of ϵ						
	45°	60°	75°	90°	105°	120°	135°
1. Turned back two revolutions							
h	−4.6171	−3.9565	−3.6156	−3.4641	−3.4713	−3.6678	−4.5065
k	−1.0329	−0.8773	−0.8692	−0.8660	−0.9010	−0.9831	−1.1476
2. Turned back one revolution							
h	−2.8850	−2.2244	−1.8836	−1.7320	−1.7392	−1.9358	−2.4475
k	−0.6605	−0.5235	−0.4568	−0.4330	−0.4439	−0.5111	−0.6607
3. Original position							
h	−1.1485	−0.4924	−0.1515	0.0000	−0.0072	−0.2037	−0.7154
k	−0.2662	−0.1171	−0.0369	0.0000	−0.0018	−0.0554	−0.2165
4. Turned ahead one revolution							
h	0.5836	1.2397	1.5805	1.7320	1.7249	1.5283	1.0166
k	0.1366	0.2970	0.3870	0.4330	0.4398	0.3972	0.2691

A comparison of Fig. 11-12 and 11-13 shows some minor differences in detail, but on the whole they are substantially alike. The arrows in the figures indicate the direction of rotation of the worm. The lower contact lines are those at the beginning of mesh, while the contact lines above the trace of the pitch plane are those of the recess action at the end of mesh. Except where undercut is present, the size of the worm gear has no influence on the amount of recess action. A larger worm gear will give more approach action.

TABLE 11-12. COORDINATES OF CONTACT LINES ON SINGLE-THREAD INVOLUTE HELICOID

(Plotted in Fig. 11-13)

Position of worm	Values of ϵ_1						
	45°	60°	75°	90°	105°	120°	135°
1. Turned back two revolutions							
h	-4.3199	-3.7803	-3.5248	-3.4771	-3.5282	-3.8100	-4.4382
k	-1.0800	-0.9451	-0.8812	-0.8693	-0.8820	-0.9525	-1.1095
2. Turned back one revolution							
h	-2.5879	-2.0483	-1.7927	-1.7150	-1.7961	-2.0779	-2.7062
k	-0.6470	-0.5121	-0.4482	-0.4288	-0.4490	-0.5195	-0.6765
3. Original position							
h	-0.8558	-0.3162	-0.0607	0.0170	-0.0641	-0.3459	-0.9741
k	-0.2139	-0.0791	-0.0152	-0.0042	-0.0160	-0.0865	-0.2435
4. Turned ahead one revolution							
h	0.8762	1.4158	1.6714	1.7490	1.6680	1.3862	0.7579
k	0.2191	0.3539	0.4178	0.4373	0.4170	0.3465	0.1895

With the pitch plane at the middle of the thread form in the axial section, most of the contact is approach action, which is the less favorable type of contact on worm-gear drives. As the pitch plane is dropped toward the root circle, more of the action will be the more favorable recess action.

To distinguish the two sides of the face of the worm gear from each other, we shall call the side where the thread is entering into mesh the *entering side,* and the opposite side of the face will be called the *leaving side.* It will be noted that the projections of the contact lines on the end view of the worm tend to converge toward the pitch plane on the leaving side of the face of the worm gear.

In order to determine the effect of dropping the pitch plane of the worm, we shall determine the projections of the contact lines on both the preceding examples when the pitch plane has been dropped to the root circle of the worm.

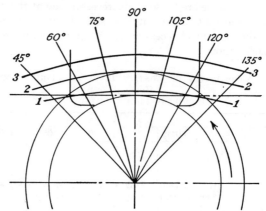

Fig. 11-14.   Contact on single-thread screw helicoid.

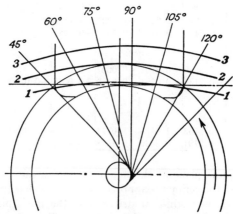

Fig. 11-15.   Contact on single-thread involute helicoid.

Single-thread Screw Helicoid with Pitch Plane Dropped.   All factors will be the same for this example as for the first example except that the value of R_1 will be reduced to 2.000 in.   The coordinates of the projection of these contact lines on the end section of the worm are tabulated in Table 11-13 and are plotted in Fig. 11-14.

Single-thread Involute Helicoid with Pitch Plane Dropped.   All factors will also be the same here as for the second example except that the value of R_1 is reduced to 2.000 in.   The coordinates of the projections of these contact lines are tabulated in Table 11-14 and plotted in Fig. 11-15.

TABLE 11-13.   COORDINATES OF CONTACT LINES ON SINGLE-THREAD SCREW HELICOID
WITH PITCH PLANE DROPPED TO ROOT CIRCLE
(Plotted in Fig. 11-14)

Position of worm	Values of ϵ						
	45°	60°	75°	90°	105°	120°	135°
1. Original positions							
h	−0.7949	−0.2037	0.1073	0.2500	0.2516	0.0850	−0.3619
k	−0.1828	−0.0482	0.0261	0.0625	0.0645	0.0223	−0.0971
2. Turned ahead one revolution							
h	0.9371	1.5283	1.8393	1.9820	1.9837	1.8170	1.3701
k	0.2180	0.3648	0.4496	0.4955	0.5065	0.4735	0.3642
3. Turned ahead two revolutions							
h	2.6692	3.2604	3.5714	3.7141	3.7157	3.5491	3.1022
k	0.6265	0.7832	0.8757	0.9285	0.9459	0.9197	0.8187

TABLE 11-14.   COORDINATES OF CONTACT LINES ON SINGLE-THREAD INVOLUTE
HELICOID WITH PITCH PLANE DROPPED TO ROOT CIRCLE
(Plotted in Fig. 11-15)

Position of worm	Values of ϵ_1						
	45°	60°	75°	90°	105°	120°	135°
1. Original position							
h	−0.5023	−0.0276	0.1982	0.2670	0.1948	−0.0572	−0.6206
k	−0.1256	−0.0069	0.0495	0.0667	0.0487	−0.0143	−0.1551
2. Turned ahead one revolution							
h	1.2298	1.7045	1.9302	1.9990	1.9268	1.6748	1.1115
k	0.3074	0.4261	0.4825	0.4998	0.4817	0.4187	0.2779
3. Turned ahead two revolutions							
h	2.9618	3.4365	3.6627	3.7311	3.6589	3.4069	2.8435
k	0.7405	0.8591	0.9156	0.9328	0.9147	0.8517	0.7109

A comparison of **Fig. 11-14** and **11-15** shows that no material differences exist between the two sets of contact lines.

For further examples, we shall next determine the form and position of the contact lines on the end views of worms with multiple threads and higher lead angles.

6-thread Screw Helicoid. The next example will be a screw helicoid with the following proportions:

Outside radius	2.500 in.
Root radius	2.000 in.
Lead	6.000 in.
Number of threads	6
Angle of generatrix	30°
Distance to pitch plane	2.250 in.

For the initial position of the helicoid in respect to the momentary axis, we will make the intersection curve in the drawing plane pass through the point where $r = 2.2500$, on the intersecting plane where $\epsilon = 90°$. Under these conditions we obtain the values tabulated in Table 11-15. For successive contact lines on adjacent threads, the worm is revolved one-sixth of a revolution. The coordinates of the contact lines tabulated in Table 11-15 are plotted in Fig. 11-16.

TABLE 11-15.   COORDINATES OF CONTACT LINES ON END SECTION OF 6-THREAD SCREW
HELICOID
(Plotted in Fig. 11-16)

Position of worm	Values of ϵ						
	45°	60°	75°	90°	105°	120°	135°
1. Turned back 120°							
h	−5.6951	−4.6782	−3.9765	−3.4641	−3.1104	−2.9461	−3.0970
k	−0.6068	−0.6729	−0.7665	−0.8660	−0.9512	−1.0282	−1.1703
2. Turned back 60°							
h	−3.9630	−2.9461	−2.2444	−1.7320	−1.3784	−1.2141	−1.5161
k	−0.4587	−0.4590	−0.4790	−0.4330	−0.4330	−0.4008	−0.3924
3. Original position							
h	−2.2310	−1.1241	−0.5124	0.0000	0.3536	0.5180	0.3671
k	−0.2797	−0.2027	−0.1079	0.0000	0.1000	0.1604	0.1224
4. Turned ahead 60°							
h	−0.4989	0.5180	1.2197	1.7320	2.0857	2.2500	2.0991
k	−0.0674	0.0916	0.2641	0.4330	0.5788	0.6758	0.6772
5. Turned ahead 120°							
h	1.2331	2.2500	2.9517	3.4641	3.8177	3.9820	3.8312
k	0.1785	0.4484	0.6525	0.8660	1.0454	1.1733	1.2053

6-thread Involute Helicoid. As another example of a 6-thread worm, we will use an involute helicoid with the same values as those of the similar screw helicoid, and with the intersection curve passing through the same point on the drawing plane as on the single-thread involute helicoid used in the second example. For this we get the values tabulated in Table 11-16 and plotted in Fig. 11-17.

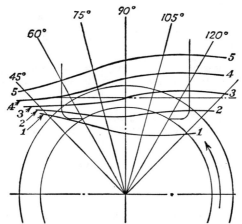

FIG. 11-16. Contact on 6-thread screw helicoid.

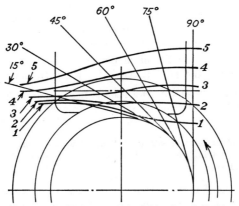

FIG. 11-17. Contact on 6-thread involute helicoid.

A comparison of Figs. 11-16 and 11-17 shows some minor differences in detail in the forms of the projections of the contact lines on the end view of the worm, but that their general nature and positions are substantially the same, particularly on those portions of the contact lines which lie in the thread section of the worm.

TABLE 11-16.   COORDINATES OF CONTACT LINES ON END SECTION OF 6-THREAD
INVOLUTE HELICOID
(Plotted in Figs. 11-17 and 11-20)

Position of worm	Values of ϵ_1						
	5°	15°	30°	45°	60°	75°	90°
1. Turned back 120°							
h	−9.9379	−5.2585	−3.9401	−3.3999	−3.0820	−2.8930	−2.8228
k	−2.4843	−1.3146	−0.9850	−0.8500	−0.7705	−0.7232	−0.7057
2. Turned back 60°							
h	−8.2052	−3.5264	−2.2080	−1.6678	−1.3499	−1.1610	−1.0908
k	−2.0513	−0.8816	−0.5520	−0.4169	−0.3375	−0.2902	−0.2727
3. Original position							
h	−6.4632	−1.7944	−0.4760	0.0642	0.3821	0.5711	0.6413
k	−1.6183	−0.4486	−0.1190	0.0161	0.0955	0.1428	0.1603
4. Turned ahead 60°							
h	−4.7411	−0.0623	1.2560	1.7963	2.1142	2.3031	2.3733
k	−1.1853	−0.0156	0.3140	0.4491	0.5285	0.5758	0.5933
5. Turned ahead 120°							
h	−3.0091	1.6697	2.9881	3.5283	3.8462	4.0352	4.1054
k	−0.7523	0.4177	0.7470	0.8821	0.9615	1.0088	1.0263

We shall again drop the pitch planes of these worms to their root radii and examine the conditions under these changed circumstances.

6-thread Screw Helicoid with Pitch Plane at Root Radius.   When the value of R_1 is reduced to 2.000 in., we obtain the values, for the coordinates of the projections of the contact lines on the end view of the worm, that are tabulated in Table 11-17 and plotted in Fig. 11-18.

6-thread Involute Helicoid with Pitch Plane at Root Radius.   When the value of R_1 is reduced to 2.000 in., we obtain the values that are tabulated in Table 11-18 and plotted in Fig. 11-19.

A comparison of Figs. 11-18 and 11-19 again shows a very close agreement between the forms and positions of the projections of the contact lines of the screw helicoid and the involute helicoid on the end sections of the worms, particularly those portions of them which lie on the thread sections of the worms.   Therefore, except for critical drives, which should always be analyzed in detail, the conditions on either type of helicoid may be used as a very close approximation to those on any other type of helicoid with a straight-line generatrix.

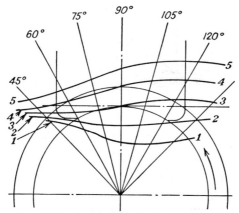

Fig. 11-18.   Contact on 6-thread screw helicoid.

TABLE 11-17.   COORDINATES OF CONTACT LINES ON END SECTION OF 6-THREAD SCREW
HELICOID WITH PITCH PLANE AT ROOT RADIUS
(Plotted in Fig. 11-18)

Position of worm	Values of ϵ						
	45°	60°	75°	90°	105°	120°	135°
1. Turned back 120°							
h	−5.3415	−4.3895	−3.7176	−3.2141	−2.8516	−2.6575	−2.7435
k	−0.4795	−0.5727	−0.6867	−0.8035	−0.8975	−0.9649	−1.0783
2. Turned back 60°							
h	−3.6095	−2.6574	−1.9856	−1.4820	−1.1196	−0.9254	−1.0114
k	−0.3581	−0.3829	−0.3912	−0.3705	−0.3315	−0.3052	−0.3618
3. Original position							
h	−1.8774	−0.9254	−0.2535	0.2500	0.6125	0.8066	0.7206
k	−0.2058	−0.1453	−0.0523	0.0625	0.1723	0.2532	0.2442
4. Turned ahead 60°							
h	−0.1454	0.8066	1.4785	1.9820	2.3445	2.5387	2.4527
k	−0.0175	0.1362	0.3156	0.4955	0.6548	0.7716	0.7993
5. Turned ahead 120°							
h	1.5867	2.5387	3.2105	3.7141	4.0766	4.2707	4.1847
k	0.2091	0.4547	0.6364	0.9285	1.1217	1.2680	1.3300

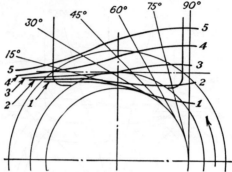

FIG. 11-19.   Contact on 6-thread involute helicoid.

TABLE 11-18. COORDINATES OF CONTACT LINES ON END SECTION OF 6-THREAD
INVOLUTE HELICOID WITH PITCH PLANE AT ROOT RADIUS
(Plotted in Figs. 11-19 and 11-21)

Position of worm	Values of ϵ_1						
	5°	15°	30°	45°	60°	75°	90°
1. Turned back 120°							
h	−7.0690	−4.2926	−3.4401	−3.0463	−2.7933	−2.6332	−2.5728
k	−1.7672	−1.0731	−0.8600	−0.7616	−0.6983	−0.6583	−0.6432
2. Turned back 60°							
h	−5.3369	−2.5605	−1.7080	−1.3142	−1.0613	−0.9011	−0.8408
k	−1.3342	−0.6401	−0.4270	−0.3286	−0.2653	−0.2259	−0.2102
3. Original position							
h	−3.6049	−0.8285	−0.0240	0.4178	0.6708	0.8309	0.8913
k	−0.9012	−0.2071	−0.0060	0.1044	0.1677	0.2077	0.2228
4. Turned ahead 60°							
h	−1.8728	0.9036	1.7560	2.1498	2.4028	2.5630	2.6233
k	−0.4682	0.2259	0 4390	0.5375	0.6007	0.6407	0.6558
5. Turned ahead 120°							
h	−0.1408	2.6356	3.4881	3.8819	4.1349	4.2950	4.3554
k	−0.0352	0.6589	0.8720	0.9705	1.0337	1.0737	1.0888

The involute helicoid is the easiest type to analyze in this manner.
The equations are simple and easy to solve.   We shall therefore use the
conditions on the involute helicoid in the further analysis of worm-gear
contact with the assurance that the results are representative of those on
other types of helicoids with straight-line generatrices.

Field of Contact. When we have the basic-rack forms (helicoid sections) and the paths of contact on several sections of the worm, we can determine graphically the forms and positions of the projections of the contact lines on the end view of the worm and on the plan view of the worm, and also the projection of the field of contact on the plan view of the worm.

On the other hand, when we have the forms and positions of the projections of the contact lines on the end view and on the plan view of the worm, we can determine the paths of contact on the several sections of the worm and the projection of the field of contact on the plan view of the worm by a similar graphical process.

For the projection of the contact lines on the plan view of the worm, we have Eq. (11-8), as follows:

$$x = (h - k) \tan \gamma \qquad (11\text{-}8)$$

Example of Contact Lines on Plan View of Worm. As definite examples, we shall use the 6-thread involute helicoid: one example with the pitch plane at the middle of the thread form at the axial section, and the other example with the pitch plane at the root radius of the worm in the axial section. For these examples, using the values previously determined, we obtain the values tabulated in Tables 11-19 and 11-20, respectively.

The values from Table 11-19 together with those from Table 11-16 have been plotted in Fig. 11-20. The y values for the projections of the contact lines in the

TABLE 11-19.   COORDINATES OF CONTACT LINES ON PLAN VIEW OF 6-THREAD INVO-
LUTE HELICOID—PITCH PLANE AT MIDDLE OF THREAD FORM
(Plotted in Fig. 11-20)

Position of worm	Values of ϵ_1					
	15°	30°	45°	60°	75°	90°
1. Turned back 120°						
$h - k$	−3.9439	−2.9551	−2.5499	−2.3115	−2.1698	−2.1171
$(h - k) \tan \gamma$	−2.2770	−1.7061	−1.4722	−1.3345	−1.2527	−1.2223
2. Turned back 60°						
$h - k$	−2.6448	−1.6560	−1.2509	−1.0124	−0.8708	−0.8181
$(h - k) \tan \gamma$	−1.5270	−0.9561	−0.7222	−0.5845	−0.5027	−0.4723
3. Original position						
$h - k$	−1.3458	−0.3570	0.0481	0.2866	0.4283	0.4810
$(h - k) \tan \gamma$	−0.7770	−0.2061	0.0278	0.1655	0.2473	0.2777
4. Turned ahead 60°						
$h - k$	−0.0467	0.9420	1.3472	1.5857	1.7273	1.7800
$(h - k) \tan \gamma$	−0.0270	0.5439	0.7778	0.9155	0.9973	1.0277
5. Turned ahead 120°						
$h - k$	1.2523	2.2411	2.6462	2.8847	3.0264	3.0791
$(h - k) \tan \gamma$	0.7230	1.2939	1.5278	1.6655	1.7473	1.7777

plan view are projected from the end view as indicated in the figure. This figure shows the contact conditions that exist when the pitch plane is at the middle of the axial thread form.

Paths of Contact. To determine graphically the paths of contact in the several sections of the worm, we proceed as follows: Lines are drawn through the end view of the worm, numbered 1 to 5, respectively, representing the several desired sections of the worm. These lines are continued through the plan view. The intersection of these lines with the projections of the contact lines in the plan view gives points on the path of contact. These intersection points are projected down into the side

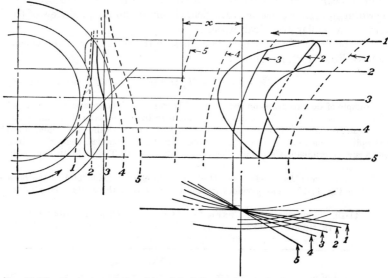

Fig. 11-20. Contact on 6-thread involute helicoid with pitch plane at middle of thread form.

view of the worm. The distance of these points on the paths of contact above the axis of the worm are transferred from the end view of the worm. These heights are those from the center line of the worm to the intersection of the lines representing the given sections with the projections of the contact lines on the end view. These distances may also be measured from the trace of the pitch plane, and be transferred to the side view at the same distance above or below the trace of the pitch plane there. Through the several points of the path of contact in the side view of the worm, the curve is drawn as indicated in the figure.

The projection of the field of contact on the plan view of the worm is then determined exactly as before. The intersection of the outside line of the worm on any section with its respective path of contact establishes the position of the end of the contact for that section. The intersection of the outside circle of the worm-gear blank for this same section with its respective path of contact establishes the point of the beginning of contact for that section. Through the several points, projected

TABLE 11-20. COORDINATES OF CONTACT LINES ON PLAN VIEW OF 6-THREAD INVO-
LUTE HELICOID—PITCH PLANE AT ROOT OF THREAD FORM
(Plotted in Fig. 11-21)

Position of worm	Values of ϵ_1					
	15°	30°	45°	60°	75°	90°
1. Turned back 120°						
$h - k$	−3.2195	−2.5801	−2.2847	−2.0950	−1.9749	−1.9296
$(h - k) \tan \gamma$	−1.8588	−1.4896	−1.3191	−1.2095	−1.1402	−1.1140
2. Turned back 60°						
$h - k$	−2.0104	−1.2810	−0.9857	−0.7960	−0.6758	−0.6303
$(h - k) \tan \gamma$	−1.1608	−0.7396	−0.5691	−0.4695	−0.3902	−0.3640
3. Original position						
$h - k$	−0.6212	−0.0180	0.3134	0.5031	0.6232	0.6685
$(h - k) \tan \gamma$	−0.3588	−0.0106	0.1809	0.2905	0.3598	0.3860
4. Turned ahead 60°						
$h - k$	0.6777	1.3170	1.6123	1.8021	1.9223	1.9675
$(h - k) \tan \gamma$	0.3912	0.7604	0.9309	1.0405	1.1098	1.1360
5. Turned ahead 120°						
$h - k$	1.9757	2.6161	2.9114	3.1012	3.2213	3.2666
$(h - k) \tan \gamma$	1.1412	1.5104	1.6809	1.7905	1.8598	1.8860

into the plan view, curves or lines are drawn that give the outline of the field of
contact on the plan view of the worm.

In these examples, the worm gear has 40 teeth with a pitch radius (R_2) equal to
6.36618 in. The values tabulated in Table 11-20 are plotted in Fig. 11-21, together
with the projection of the field of contact on the plan view of the worm.

Contact Lines on Worm Drives with Shafts at Any Angle. Thus far,
the analysis of the conjugate gear-tooth action and the position of the
contact lines on worm-gear drives has been restricted to those drives
where the axes of the two members are at right angles to each other. As
noted before, a worm-gear drive can be substituted for any spiral-gear
drive. In order to determine the conditions of contact on worm-gear
drives when the axes are not at right angles to each other, we shall first
establish the projections of the contact lines on the end and plan views of
the worm. This will be done analytically. The remainder of the analysis will be completed graphically.

Using the more common condition where the axes are at right angles
to each other as the point of departure, the worm-gear position may be
altered so that its axis is twisted either toward the direction of the helices
on the helicoid worm or else away from this direction. In the first case,
the angle between the shafts, or the shaft angle, will be less than 90 deg.
In the second case, the shaft angle will be greater than 90 deg.

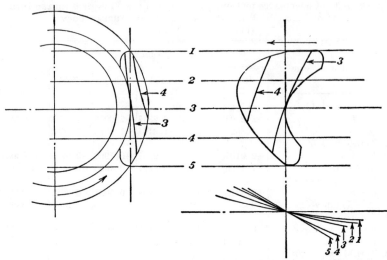

Fɪɢ. 11-21.   Contact on 6-thread involute helicoid with pitch plane at root of thread form.

In this analysis we shall use the involute helicoid as the form of the worm because of the greater simplicity of its analysis.   If necessary, the analysis of other forms of helicoids can be made in a similar, but more complicated, manner.   Referring to Fig. 11-22, let

Σ = shaft angle

L = lead of generatrix, in.

γ = angle of generatrix

ϵ = angle of plane tangent to base cylinder with pitch plane

R_b = radius of base cylinder, in.

h = length along line tangent to base cylinder from momentary axis to intersection curve of involute helicoid with plane through pitch point, in.

h_1 = length along line tangent to base cylinder from its intersection with the trace of the pitch plane with drawing plane to intersection curve of involute helicoid with drawing plane, in.

k = distance to projection of contact point on line tangent to base cylinder from trace of pitch plane with drawing plane, in.

R_1 = distance of pitch plane from axis of helicoid, in.

ϵ_1 = angle of rotation of generatrix from origin of involute curve

For the intersection of the involute helicoid with the drawing plane, we already have from Eq. (11-11) by substituting the symbol h_1 for h

$$h_1 = R_b\epsilon_1 - [(R_1 - R_b \cos \epsilon)/\sin \epsilon] \qquad (11\text{-}14)$$

In this problem, the momentary axis is at an angle to the drawing plane. The drawing plane will be located so that its trace in the plan view intersects the momentary axis where the trace of the plane $\epsilon = 90°$ also intersects the momentary axis, as shown in Fig. 11-22.

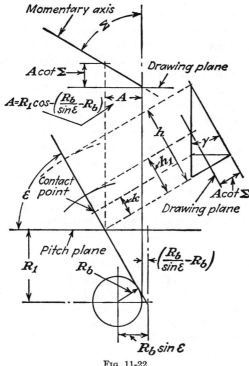

Fig. 11-22.

From the geometrical conditions in Fig. 11-22 we have the following:

$$h = h_1 + \cot \Sigma \cot \gamma [R_1 \cot \epsilon - (R_b/\sin \epsilon) + R_b] \qquad (11\text{-}15)$$
$$k = h \sin^2 \gamma \qquad (11\text{-}12)$$

In order to determine the change in the projection of the form and position of other contact lines as the helicoid is revolved to different positions, the changed length of h is multiplied by the constant value of $\sin^2 \gamma$. For a turning movement of one revolution, the value of h changes an amount equal to the circumference of the base cylinder. For all other movements, it changes in direct proportion to the fraction of a revolution that may be made.

To determine the projections of these contact lines on the plan view of the worm, we proceed as follows:

When x = distance of contact point from momentary axis in an axial direction, in.

$$x = (h - k) \tan \gamma \qquad (11\text{-}8)$$

The value of the coordinate y in the plan view, or the distance of the projection of the contact point from the projection of the axis of the worm on the plan view, is determined graphically by projecting the position of the desired point from the end view of the worm into the plan view.

Field of Contact. The projection of the field of contact on the plan view of the worm is determined graphically in a similar manner to that for worm gears with axes at 90 deg, except that the sections on which the paths of contact are determined are planes of rotation of the worm gear instead of planes parallel to the axis of the worm. Hence lines are drawn parallel to the edge of the worm gear, or perpendicular to its axis, which are traces of planes of rotation of the worm gear. The intersection of one of these lines with the projection of a contact line in the plan view is projected into the side view, or end view, of the worm gear. The height of this point above the axis of the worm is transferred from the height of this same point in the end view of the worm. The path of contact is then drawn through the several points thus determined.

The trace of the outside cylinder of the worm on these planes of rotation of the worm gear will be an ellipse and not a straight line. The origins of these ellipses will be the point, in the plan view, where the projection of the axis of the worm crosses the trace of the specific plane of rotation involved. All these ellipses will be the same, but each will start from a different origin. The equation of this ellipse is as follows:

When R_{o1} = outside radius of worm, in.

Σ = shaft angle

a = major radius of ellipse, in.

b = minor radius of ellipse, in.

x_e = abscissa of ellipse, in.

y_e = ordinate of ellipse, in.

$$a = R_{o1}/\sin \Sigma \qquad b = R_{o1}$$

For the equation of the ellipse, we have

$$(x_e/a)^2 + (y_e/b)^2 = 1$$

Whence

$$x_e = (a/b) \sqrt{b^2 - y_e^2} = \sqrt{R^o{}_1{}^2 - y_e^2}/\sin \Sigma \qquad (11\text{-}16)$$

The point where the trace of the outside cylinder of the worm crosses the path of contact for the specific section is the end of the action on that section. The point where the trace of the outside circle of the worm gear crosses the path of contact for any specific section is the beginning of action on that section. These points are projected into the plan view of the worm, and the outline of the projection of the field of contact is drawn through these points as before.

Example with Axes Twisted in Direction of Helices on Worm. As a definite example we shall use the 6-thread involute helicoid from the preceding examples with

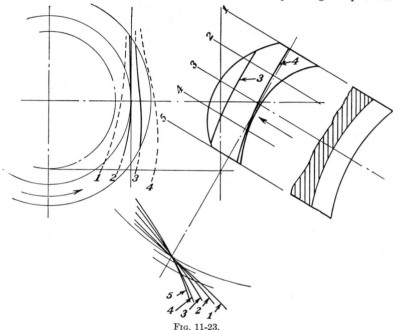

FIG. 11-23.

a 40-tooth worm gear. For the first of these examples, the axis of the worm gear will be twisted in the direction of the helices on the worm. We will use the following values for this first example:

Shaft angle.. 60°
Outside radius of worm............................. 2.500 in.
Root radius of worm................................ 2.000 in.
Distance to pitch plane............................. 2.000 in.
Angle of generatrix................................ 30°
Lead of generatrix.................................. 6.000 in.
Number of starts or threads........................ 6
Number of teeth in worm gear....................... 40
Pitch radius of worm gear........................... 5.7625 in.

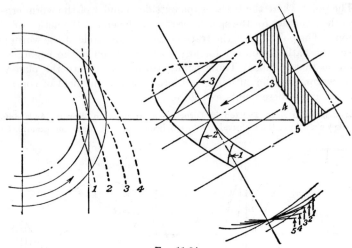

FIG. 11-24.

TABLE 11-21.   COORDINATES OF CONTACT LINES ON 6-THREAD INVOLUTE HELICOID
WITH SHAFT ANGLE OF 60 DEG
(Plotted in Fig. 11-23)

Position of worm	Values of ϵ_1						
	5°	15°	30°	45°	60°	75°	90°
1. Turned back 120°							
h	−1.5311	−1.5649	−1.6300	−1.7314	−1.8945	−2.1556	−2.5728
k	−0.3828	−0.3912	−0.4075	−0.4328	−0.4736	−0.5389	−0.6432
x	−0.6630	−0.6776	−0.7058	−0.7497	−0.8203	−0.9334	−1.1141
2. Turned back 60°							
h	0.2010	0.1672	0.1021	0.0007	−0.1624	−0.4236	−0.8408
k	0.0503	0.0418	0.0255	0.0002	−0.0406	−0.1059	−0.2102
x	0.0870	0.0724	0.0442	0.0003	−0.0703	−0.1834	−0.3641
3. Original position							
h	1.9330	1.8992	1.8341	1.7327	1.5696	1.3085	0.8913
k	0.4833	0.4748	0.4585	0.4332	0.3924	0.3271	0.2228
x	0.8370	0.8224	0.7942	0.7503	0.6797	0.5666	0.3859
4. Turned ahead 60°							
h	3.6651	3.6312	3.5662	3.4648	3.3017	3.0405	2.6233
k	0.9163	0.9078	0.8915	0.8662	0.8254	0.7601	0.6558
x	1.5870	1.5724	1.5442	1.5003	1.4297	1.3166	1.1359

With these values and the use of the preceding equations, we obtain the coordinates for the contact lines, which are tabulated in Table 11-21. These coordinates are plotted in Fig. 11-23. The graphic determination of the projection of the field of contact in the plan view of the worm has also been completed here.

Worm Gear with Axis Twisted Away from the Direction of Helices on Worm. As another example, with the axis of the worm gear twisted away from the direction of the helices on the worm, we shall use the same involute helicoid as before and the following values for the shaft angle and for the worm gear:

Shaft angle.. 120°
Number of teeth in worm gear...................... 40
Pitch radius of worm gear......................... 10.1485 in.

The coordinates for the form and position of these contact lines are tabulated in Table 11-22 and plotted in Fig. 11-24, together with the graphic solution for the field of contact.

TABLE 11-22. COORDINATES OF THE PROJECTION OF CONTACT LINES ON 6-THREAD INVOLUTE HELICOID WITH SHAFT ANGLE OF 120 DEG
(Plotted in Fig. 11-24)

Position of worm	Values of ϵ						
	5°	15°	30°	45°	60°	75°	90°
1. Original position							
h	−9.1428	−3.5562	−1.7861	−0.8971	−0.2281	0.3536	0.8913
k	−2.2857	−0.8890	−0.4465	−0.2243	−0.0570	0.0883	0.2228
x	−3.9590	−1.5404	−0.7734	−0.3885	−0.0988	0.1530	0.3859
3. Turned ahead 60°							
h	−7.4108	−1.8241	−0.0541	0.8349	1.5040	2.0854	2.6233
k	−1.8527	−0.4560	−0.0135	0.2087	0.3760	0.5214	0.6558
x	−3.2090	−0.7904	−0.0234	0.3615	0.6512	0.9030	1.1359
3. Turned ahead 120°							
h	−5.6787	−0.0921	1.6780	2.5670	3.2360	3.8175	4.3554
k	−1.4197	−0.0230	0.4195	0.6417	0.8090	0.9544	1.0888
x	−2.4590	−0.0404	0.7266	1.1115	1.4012	1.6530	1.8859
4. Turned ahead 180°							
h	−3.9467	1.6400	3.4100	4.2990	4.9681	5.5495	6.0874
k	−0.9867	0.4100	0.8525	1.0748	1.2420	1.3874	1.5219
x	−1.7090	0.7096	1.4766	1.8615	2.1512	2.4030	2.6359

CHAPTER 12

DESIGN OF WORM-GEAR DRIVES

The starting point for the design of a worm-gear drive is the worm itself, and the companion hob, which should always match the worm. Except for small-diameter multiple-thread worms where the thread form and worm-gear face come close to or overlap the limits of conjugate action, the exact type of helicoid used for the worm is of small importance. For the very small diameter worms, the chased helicoid and the screw helicoid permit conjugate action closer to the worm axis than does any other type of helicoid. The type of helicoid is determined by the method used in the shop to make the finishing cut or grind on the worm threads. If a screw helicoid is needed, then the grinding wheel must be suitably formed to produce it.

There is, however, one important feature about the design of the worm itself. This is the relation between the half thread angle, or pressure angle in the axial section, and the lead angle of the worm. As the lead angle is increased, a larger thread angle is essential to avoid excessive undercutting on the off-center sections of the worm gear on the leaving side. Hence this pressure angle should never be much less than the lead angle. For example, a pressure angle of $14\frac{1}{2}$ deg should never be used for worms with lead angles greater than about 16 deg; a pressure angle of 20 deg should never be used for worms with lead angles greater than about 25 deg; a pressure angle of 25 deg should never be used for worms with lead angles greater than about 35 deg; while a pressure angle of 30 deg can be used for worms with lead angles up to 45 deg. If a single pressure angle is to be used for all worms, it should be chosen to meet the highest lead angle that is to be used. If these lead angles cover the entire range, the 30-deg pressure angle is the best choice.

As with all other types of gear drives, there is a relatively wide range of good design, while outside this range, the conditions become more and more unfavorable. The worm-gear face should lie inside the field of conjugate action. The relationship between the pressure angle and the lead angle of the worm is another limitation to the field of good design. For another, the total number of teeth in the worm and in the worm gear should never be less than about 40, and more will be better. Because of the changing profile of the basic-rack form on the off-center sections of

248

the worm, undercut starts there on the leaving side of the worm-gear face very much sooner than it does on the axial section. This is apparent from only a casual study of helicoid sections. This is a limitation at times to the effective face width for the worm gear.

At present there is no general standard practice for the design of worms or of worm-gear drives. This condition may always be present to some degree because of the different restrictions that are imposed upon the design of the worm by the characteristics and influences of the different methods available to produce the worms. The first step toward the development of a standard design for worm drives by any organization is the standardization of the worm itself. This problem will be discussed later.

Given the definite specifications for the worm, the hob form must match the type of helicoid used or produced on the worm. Also the diameter of the hob must never be less than that of the worm. To express this in another way, the center distance used for hobbing the worm gear must never be less than the center distance of operation. The hob may be larger than the worm; the amount larger depends primarily upon the lead angle of the worm. For example, with a lead angle of 10 deg or less and with a nominal pitch radius or effective radius of about 1.500 in., the hob may be as much as $\frac{1}{4}$ in. larger in diameter than the worm without affecting the form of the off-center sections appreciably. Any such difference is small enough to be soon wiped out by the plastic flow of the material of the worm gear in operation. As the lead angle increases, however, the amount of oversize of the hob must be reduced to maintain commensurate conditions. Thus with a lead angle of about 35 deg, the hob should be held to within about 0.005 in. of the diameter of the worm. The ductility of the material of the worm gear also plays a part here. A more ductile material will permit more plastic deformation of the surface in the running-in period than will a harder material.

The tops of the teeth of all hobs should be rounded with a continuous curve, even though this will require a greater amount of clearance. The form of this curve may be an arc of a circle or any other smooth curve. This is essential, because the pitch plane of the worm generally crosses the root of the worm-gear tooth on either side of the axial section of the worm. Under such conditions, the form of the fillet at the root of the gear tooth is almost an exact duplicate of the form of the corner of the hob tooth. When this form is almost sharp, the stress concentrations here will be extremely high, and may result in fatigue cracks, which are often the cause of failure of such drives in service.

Many of the dimensions and proportions of the worm gear are directly dependent upon the dimensions of the worm. We shall therefore start the consideration of the design of worm-gear drives with the proportions of the worm gear, which will be based upon known or assumed dimensions of the worm.

DESIGN OF WORM GEAR

We shall start with the following symbols:

p_x = axial pitch of worm, in.

p_2 = circular pitch of worm gear, in.

L = lead of worm, in.

N_1 = number of starts or threads on worm

N_2 = number of teeth in worm gear

a_1 = addendum of worm, axial section, in.

a_2 = addendum of worm gear, throat section, in.

c_1 = clearance at root of worm, in.

c_2 = clearance at root of worm-gear tooth, in.

h_t = whole depth of thread of worm, in.

R_e = effective radius of worm, in. (*i.e.*, radius on worm to cylinder where thread thickness and space width are both equal to one-half the axial pitch)

R_1 = radius to pitch plane on worm, in.

R_2 = pitch radius of worm gear, in.

R_{o1} = outside radius of worm, in.

R_{r1} = root radius of worm, in.

R_{o2} = outside radius of worm gear, in.

R_t = throat radius of worm gear, in. (plane of rotation)

T_r = radius of form at throat of worm gear, in. (axial section of worm gear)

I_t = throat increment of worm gear, in.

C = center distance, in.

F_h = minimum cutting length of hob, in.

F_2 = face width of worm gear, in.

λ = lead angle of worm at R_1

Σ = shaft angle

Worm Drives with Shafts at Right Angles. When the shaft angle is equal to 90 deg, we have the following:

$$p_x = L/N_1 \tag{12-1}$$

$$p_2 = p_x \tag{12-2}$$

$$R_2 = p_2 N_2 / 2\pi \tag{12-3}$$

$$R_1 = C - R_2 \tag{12-4}$$

$$R_t = C - (R_{r1} + c_1) \tag{12-5}$$

$$\tan \lambda = L/2\pi R_1 \tag{12-6}$$

$$T_r = R_{r1} + c_1 \tag{12-7}$$
$$R_{o2} = R_t + (I_t/2) \tag{12-8}$$
$$F_h = 2\sqrt{(2R_t - h_t)h_t} \tag{12-9}$$
$$\text{Edge round} = 0.25p_x \tag{12-10}$$

It is good practice to reduce the face width of the worm gear with an increase in the lead angle of the worm because of the poorer contact conditions that exist as the distance from the axial section of the worm towards the leaving side of the worm-gear face becomes greater. These critical conditions become more acute as the lead angle of the worm increases. On critical drives, or for any others where an analysis of the actual contact conditions is made, these face widths should be limited by the conditions of the actual contact. In other cases where no definite contact analysis is made, the following equations will give good proportions: When the lead angle is 15 deg or less

$$F_2 = 2\sqrt{(2R_1 + a_1)a_1} + 0.50p_x \tag{12-11}$$
$$I_t = 0.70h_t \tag{12-12}$$

When the lead angle is greater than 15 deg

$$F_2 = 2\sqrt{(2R_1 + a_1)a_1} + 0.25p_x \tag{12-13}$$
$$I_t = 0.35h_t \tag{12-14}$$

The pitch plane of the worm may be located anywhere between the effective radius R_e and the root radius R_{r1} of the worm. At times it may even be below the outside circle of the worm gear. If the pitch plane is below the root circle of the worm, an actual contact analysis should be made to be certain that the contact is adequate. The lower on the worm thread that this pitch plane is placed, the greater will be the amount of recess action and the less will be the amount of the approach action. The recess action is by far the more favorable action on worm gears. This latitude in the position of the pitch plane of the worm also permits the selection in most cases of an even dimension for the center distance of the worm drive.

Example with Lead Angle Less than 15 Deg. As the first example we shall use a worm with a lead angle of less than 15 deg. We will use the following values: The hob will have a rounded tip on its teeth, which will give a greater clearance at the root of the worm-gear-tooth form that exists at the root of the worm thread.

$$N_1 = 1 \qquad N_2 = 60 \qquad R_e = 1.500 \qquad L = 0.500 \qquad c_1 = 0.025$$
$$c_2 = 0.060 \qquad R_{o1} = 1.6592 \qquad R_{r1} = 1.3159 \qquad h_t = 0.3433$$
$$p_x = \frac{0.500}{1} = 0.500 \qquad R_2 = \frac{0.500 \times 60}{6.2832} = 4.7446$$
$$I_t = 0.70 \times 0.3433 = 0.2404$$
$$C\,(\text{max}) = R_2 + R_e = 4.7446 + 1.500 = 6.2446$$
$$C\,(\text{min}) = R_2 + R_{r1} = 4.7446 + 1.3159 = 6.0605$$

For the center distance we may select any value, between these two values, that may be most favorable for other structural or design reasons. In this example we will select

$$C = 6.125 \text{ in.}$$

Then

$$R_1 = 6.125 - 4.7446 = 1.3804$$
$$a_1 = 1.6592 - 1.3804 = 0.2788$$
$$b_1 = 1.3804 - 1.3159 = 0.0645$$
$$R_t = 6.125 - (1.3159 + 0.025) = 4.7841$$
$$T_r = 1.3159 + 0.025 = 1.3409$$
$$\tan \lambda = \frac{0.500}{6.2832 \times 1.3804} = 0.05765$$
$$\lambda = 3.300°$$
$$R_{o2} = 4.7841 + 0.1202 = 4.9043$$
$$\text{Edge round} = 0.25 \times 0.500 = 0.125$$
$$F_2 = 2 \sqrt{(2.7608 + 0.2788) \times 0.2788} + 0.250 = 2.091$$

We shall make the face width of the worm gear an even dimension, and for this example we shall use $F_2 = 2.000$ in. These values have been plotted in Fig. 12-1.

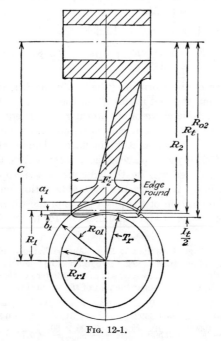

Fig. 12-1.

Example with Lead Angle Greater than 15 Deg. As a second example we shall use a multiple-thread worm having all values except the lead angle, lead, and number of threads on the worm the same as before. This gives the following values:

$$N_1 = 8 \qquad N_2 = 60 \qquad R_e = 1.500 \qquad L = 4.000 \qquad c_1 = 0.025 \qquad h_t = 0.3433$$
$$c_2 = 0.060 \qquad R_{o1} = 1.6592 \qquad R_{r1} = 1.3159 \qquad R_2 = 4.7446$$
$$p_x = \frac{4.00}{8} = 0.500$$
$$I_t = 0.35 \times 0.3433 = 0.1202$$
$$C \text{ (max)} = 6.2446 \qquad C \text{ (min)} = 6.0605$$

We shall use the same value as before, whence

$$C = 6.125 \qquad R_1 = 1.3804 \qquad a_1 = 0.2788$$
$$b_1 = 0.0645 \qquad R_t = 4.7841 \qquad T_r = 1.3409$$
$$\tan \lambda = \frac{4.00}{6.2832 \times 1.3804} = 0.46117$$
$$\lambda = 24.757°$$
$$R_{02} = 4.7841 + 0.0601 = 4.8442$$
$$F_2 = 2 \sqrt{(2.7608 + 0.2788) \times 0.2788} + (0.25 \times 0.50) = 1.966$$

We shall use $F_2 = 1.875$ in.

Worm Drives with Axes at Any Angle. When the axes of the two members are not at right angles to each other, we have the following:

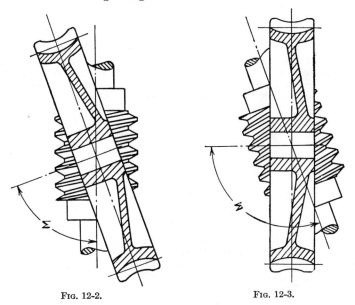

Fig. 12-2.   Fig. 12-3.

When the axis of the worm gear is twisted in the direction of the helices on the worm as shown in Fig. 12-2, the value of the shaft angle is less than 90 deg. When the axis of the worm gear is twisted away from the direction of the helices on the worm as shown in Fig. 12-3, the value of the shaft angle is greater than 90 deg.

$$p_x = L/N_1 \tag{12-1}$$
$$p_2 = p_x \cos \lambda/\sin (\Sigma + \lambda) \tag{12-15}$$
$$R_2 = p_2 N_2/2\pi \tag{12-3}$$
$$R_1 = C - R_2 \tag{12-4}$$
$$R_t = C - (R_{r1} + c_1) \tag{12-5}$$
$$\tan \lambda = L/2\pi R_1 \tag{12-6}$$
$$T_r = (R_{r1} + c_1)/\sin^2 \Sigma \tag{12-16}$$
$$I_t = 0.35 h_t \tag{12-14}$$
$$R_{o2} = R_t + (I_t/2) \tag{12-8}$$
$$F_2 = 2 \sqrt{(2R_1 + a_1)a_1} + 0.50 p_x \tag{12-13}$$
$$\text{Edge round} = 0.25 p_x \tag{12-10}$$

Example when Shaft Angle Is Less than 90 Deg. For this example we shall use the following values:

$$\Sigma = 70° \qquad R_e = 1.500 \qquad N_1 = 3 \qquad L = 1.500 \qquad h_t = 0.3433 \qquad c_1 = 0.025$$
$$N_2 = 60 \qquad c_2 = 0.060 \qquad R_{o1} = 1.6592 \qquad R_{r1} = 1.3159$$
$$p_x = \frac{1.500}{3} = 0.500$$

In this example we must select the position of the pitch plane on the worm first, and use the center distance that results. Otherwise we shall have an indeterminate equation to solve by a series of trials. We could first solve this problem as a spiral-gear drive and then choose one of the members of the pair as a worm. In this case, however, we would use a higher pressure angle for the basic-rack form than that which is used for a spiral-gear drive. This second solution would also introduce a special worm and a special hob. In this example we shall use the worm as specified and select the value

$$R_1 = 1.375 \text{ in.}$$
$$a_1 = 1.6592 - 1.375 = 0.2842$$
$$\tan \lambda = \frac{1.500}{6.2832 \times 1.375} = 0.17362$$
$$\lambda = 9.850° \qquad \cos \lambda = 0.98526$$
$$\Sigma + \lambda = 79.850° \qquad \sin (\Sigma + \lambda) = 0.98435$$
$$p_2 = \frac{0.500 \times 0.98526}{0.98435} = 0.50046$$
$$R_2 = \frac{0.50046 \times 60}{6.2832} = 4.7790$$
$$C = 4.7790 + 1.375 = 6.154$$
$$R_t = 4.7790 - (1.3159 + 0.025) = 4.8131$$
$$I_t = 0.35 \times 0.3433 = 0.1202$$
$$R_{o2} = 4.8131 + 0.0601 = 4.8732$$
$$\Sigma = 70° \qquad \sin \Sigma = 0.93969 \qquad \sin^2 \Sigma = 0.88302$$
$$T_r = \frac{1.3409}{0.88302} = 1.5185$$
$$F_2 = 2 \sqrt{(2.750 + 0.2842) \times 0.2842} + (0.50 \times 0.50) = 2.1072$$

We shall use $F_2 = 2.125$ in.

$$\text{Edge round} = 0.25 \times 0.50 = 0.125$$

This worm drive is plotted in Fig. 12-2.

Example when Shaft Angle Is Greater than 90 Deg. For this example we shall use the same values as before except for the shaft angle. Thus we have

$\Sigma = 110°$ $R_e = 1.500$ $N_1 = 3$ $L = 1.500$ $h_t = 0.3433$ $c_1 = 0.025$
$N_2 = 60$ $c_2 = 0.060$ $R_{o1} = 1.6592$ $R_{r1} = 1.3159$ $p_x = 0.500$
$R_1 = 1.375$ $a_1 = 0.2842$ $\lambda = 9.850°$ $\cos \lambda = 0.98526$
$\Sigma + \lambda = 119.850°$ $\sin (\Sigma + \lambda) = 0.86333$

$$p_2 = \frac{0.500 \times 0.98526}{0.86333} = 0.57061$$

$$R_2 = \frac{0.57061 \times 60}{6.2832} = 5.4489$$

$$C = 5.4489 + 1.375 = 6.8239$$
$$R_t = 6.8239 - 1.3409 = 5.4830$$
$$I_t = 0.1202$$
$$R_{o2} = 5.4830 + 0.0601 = 5.5431$$
$\Sigma = 110°$ $\sin \Sigma = 0.93969$ $\sin^2 \Sigma = 0.88302$
$T_r = 1.5185$ $F_2 = 2.1072$

We shall use $F_2 = 2.125$ in.

Edge round $= 0.125$

This worm-gear drive is plotted in Fig. 12-3.

When the lead angles of the worms are small, under 15 deg, for example, and when the angles of the shafts do not vary too much from 90 deg, under 30 deg, for example, then the axis of the worm gear may be twisted in either direction. As either the lead angle or the departure of the shaft angle from 90 deg becomes greater than the foregoing limits, the shaft angle should always be less than 90 deg. For critical drives of this nature, the relative diameters of the worm and of the worm gear may be determined from those of a similar spiral-gear drive. In every case of a critical or important drive, a complete contact analysis should be made to determine the contact conditions.

DESIGN OF WORM

The design of the worm is influenced by two major factors or considerations: first, the nature of the manufacturing facilities available to produce it; and, second, the conditions for which it is manufactured. For one condition, it may be a specific unit for a standard product that is manufactured in large quantities, and only one worm drive is involved. As specific tools and other equipment must be provided for its manufacture, these tools may be made as required without any consideration of the possible use of these tools for the production of worms for other uses. Here, any possible advantage gained by the use of any available standard worm is the possibility of buying the corresponding hob from the stock of some tool manufacturer.

Another condition of manufacture is that where a wide variety of different worm drives must be made. Here it becomes essential to use the same tools for as many different worm drives as possible. Here, in order to obtain the widest possible use of a given tool, and also to reduce the variety of these tools to a minimum, certain restrictions are imposed upon the design.

The major influence of the nature of the manufacturing equipment available to finish the threads of the worm is on the exact type of helicoid that this available process will produce. This factor, except for the minimum diameters of multiple-thread worms as noted before, is of minor importance.

The majority of the early uses of worm-gear drives were for large speed reductions. Here the single-thread worm was the most common one. Today, worm gears are needed and used for much smaller speed ratios, and multiple-thread worms are much more common. Hence we must consider the problems of these multiple-thread worms in any attempt to set up a standard series.

Proportions of Chased or Milled Worm Threads. We shall start the consideration of the problem of the proportions of worm threads with the conditions that exist when we must use a tool of constant form and proportions to chase or mill the threads of worms with different numbers of starts, all with the same axial pitch. The common practice is to use axial pitches of even fractions of an inch because of the change-gear problem on the machines used to finish the threads of the worm. For the purposes of this discussion, we shall use a constant axial pitch of 1 in. We shall also use a 60-deg included angle for the form of the threading tool, with the proportions shown in Fig. 12-4.

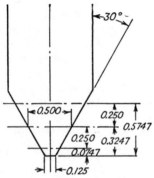

Fig. 12-4. Form of threading tool for 1-in. axial pitch.

Ignoring for the present the changes caused because of the setting of this tool to the lead angle of the worm at its effective radius, this would give us the following thread proportions: The effective radius, where the thread thickness and the thread-space width are both equal to one-half of the axial pitch, is considered here as the pitch line.

	Inches
Addendum	0.250
Dedendum	0.3247
Clearance	0.0747
Whole depth of thread	0.5747

When this tool is set to the lead angle at the effective radius, and if we cut the thread to the full depth, the width of the thread space will be increased to over one-half of the axial pitch because the value of the tool thickness at this radius now becomes the normal space width and not the axial width. This would make the thread thickness less than one-half of the axial pitch and would also reduce the width of flat at the tip of the thread. If we maintain the thread thickness here at a constant value, equal to one-half of the axial pitch, then we must cut the thread to a lesser depth. Thus when

λ_e = lead angle at effective radius

ϕ_n = one-half included angle of threading tool

a_e = nominal addendum of worm thread, in.

b_e = nominal dedendum of worm thread, in.

a_1 = actual addendum of worm thread measured from R_e, in.

b_1 = actual dedendum of worm thread measured from R_e, in.

c_1 = clearance at root of worm, in.

t_n = thickness of tool at R_e, in.

t_1 = axial width of space at R_e, in.

R_e = effective radius of worm, in.

then

$$t_n = t_1 \cos \lambda_e = 0.500 \cos \lambda_e$$
$$b_1 = b_e - (t_1 - t_n)/2 \tan \phi_n$$
$$b_1 = b_e - t_1(1 - \cos \lambda_e)/2 \tan \phi_n$$

But for the particular form given

$$t_1/2 \tan \phi_n = 4b_e/3$$

whence

$$b_1 = (b_e/3)(4 \cos \lambda_e - 1) \tag{12-17}$$

To keep the form of the same proportions, we will have

$$a_1 = (a_e/3)(4 \cos \lambda_e - 1) \tag{12-18}$$
$$c_1 = b_1 - a_1 = (c_e/3)(4 \cos \lambda_e - 1) \tag{12-19}$$

For a series of lead angles, ranging from 0 to 45 deg, we have the following changes in the thread proportions:

λ_e, deg	a_1	b_1	c_1	h_t
0	0.2500	0.3247	0.0747	0.5747
10	0.2449	0.3181	0.0732	0.5630
20	0.2299	0.2986	0.0687	0.5285
30	0.2053	0.2667	0.0614	0.4720
40	0.1720	0.2234	0.0514	0.3957
45	0.1524	0.1979	0.0455	0.3503

Thus in order to use the same threading tool for worms of all numbers of threads of the same axial pitch, we must vary the thread proportions with the changing lead angles as indicated above.

If we wish to keep the thread depths constant for all numbers of threads, and also to use the same threading tool in all cases for a given nominal pitch, then we must change the leads when we change either the diameter of the worm or the number of threads or starts on the worm. This would give a variable axial pitch and a constant normal circular pitch. Thus when

p_n = normal and nominal pitch, in.

p_x = axial pitch, in.

$$p_x = p_n/\cos \lambda$$

Of the two foregoing solutions, the first would best fit the existing processes now generally available for finishing the worm threads. The second method would be identical to the practice followed for hobbed helical gears. If the worm threads were hobbed, then this second method would be the logical one to follow. This would permit the unification of cutting practices for helical gears and worm threads, and the worm could best be made as an involute helical gear. In such an event, the standard pitches for worms should be the same standard diametral pitches as those now used for spur and helical gears.

Diameter of Worm.   Some worms are made integral with the driving shaft and others are made as separate components, which are mounted on the driving shafts. We shall call the first type of worms *integral worms,* and the second type *shell worms.*

It should be apparent that the diameters of shell worms must be larger than the diameters of integral worms because additional space must be provided on the shell worms to take care of the bore, keyways, or splines, and an adequate wall thickness between the root of the worm thread and the tops of the keyways or splines. In addition, consideration must be given to the depth of the cutting flutes on the hob. To meet these hob requirements for integral worms, the corresponding hob must also be made integral with its cutting arbor or shank. For shell worms, the corresponding hob is made with a bore and a keyway, and will be called a *shell hob.*

Integral worms are generally required for specific applications where space must be reduced to a minimum and where the most effective designs possible are essential. There is probably but little chance of standardizing such drives to any great extent, although a standard for integral worms for more general purposes might be of value.

Several possibilities are present for the development of standards for shell worms and hobs. For one, the American Gear Manufacturers' Association has considered the following equation for the effective diameters of shell worms of the lower lead angles as a possible basis for standardization:

When D_e = effective diameter of worm, in.

p_x = axial pitch of worm, in.

$$D_e = 2.4p_x + 1.10 \qquad 20)$$

The constant in this equation is provided to allow for the bore and keyway of shell worms and hobs. This association has also considered the adoption of the following axial pitches as standards in order to reduce the variety:

Inches

¼ ⁵⁄₁₆ ⅜ ½ ⅝ ¾ 1 1¼ 1½ 1¾ 2

Using Eq. (12-20) and the foregoing pitches, we obtain the values tabulated in Table 12-1. This table gives the axial pitch, the calculated effective diameter, and the lead angles at this effective diameter for worms with different numbers of threads. It also gives the number of threads, N'_1, that would give a lead angle close to 45 deg.

It should be apparent that when the lead angle reaches some maximum value, generally below 45 deg, the diameter of the worm must be increased over the original one with increasing numbers of threads in order to keep the value of the lead angle within the selected maximum one. If a fixed maximum value is used, then all worms with this maximum lead angle, and with the same axial pitch, will have diameters directly proportional to the numbers of threads. It is also apparent that such worms of the same number of threads but of different axial pitch will be geometrically similar to each other.

The efficiency of a worm drive depends largely upon the lead angle of the worm. When this angle is small, the efficiency is low. As it increases up to about 45 deg, the efficiency improves. The difference in efficiency, however, between drives where the lead angle is about 35 deg and over, and those where the lead angle is 45 deg, is very small. The worms and hobs with lead angles of 45 deg are much more difficult to make than are those with lead angles of 35 deg. Hence maximum-lead-angle values of between 35 and 40 deg are as good as any.

A study of Table 12-1 will show that until the lead angles have reached the maximum values and the diameters are changed accordingly, there is no geometrical similarity between any of the worms. This means that individual analyses must be made of each of these drives if definite contact conditions are to be studied.

TABLE 12-1. WORM DIAMETERS AND LEAD ANGLES

p_x, in.	D_e, in.	Lead angle, deg				N'_1
		$N_1 = 1$	$N_1 = 2$	$N_1 = 4$	$N_1 = 8$	
0.2500	1.700	2.680	5.348	10.605	20.530	21
0.3125	1.850	3.077	6.137	12.138	23.275	18
0.3750	2.000	3.416	6.807	13.427	25.523	16
0.5000	2.300	3.959	7.879	15.471	28.966	14
0.6250	2.600	4.376	8.701	17.017	31.472	13
0.7500	2.900	4.706	9.349	18.226	33.368	12
1.0000	3.500	5.197	10.309	19.991	36.039	11
1.2500	4.100	5.544	10.984	21.216	37.825	10
1.5000	4.700	5.801	11.485	22.114	39.101	9
1.7500	5.300	6.000	11.871	22.806	40.058	9
2.0000	5.900	6.158	12.178	23.345	40.801	9

Module System of Worms. The module of a gear is the amount of diameter of the gear for each tooth. It is the ratio of the pitch diameter of the gear divided by the number of teeth. It is the reciprocal of the diametral pitch. The practice is followed in some places of making the effective diameter of the worm some integral number of modules in diameter. For example, if a single-thread worm has an effective diameter of 12 modules, this diameter will be the same as that of a 12-tooth worm gear of the given pitch. Such a practice will introduce a greater degree of geometrical similarity between worms of different pitches than any other practice can attain. We shall therefore turn our attention to this practice.

The value of the module for any given axial pitch is obtained by dividing the axial pitch by the value 3.1416. Thus when

M = module, in.

N'_1 = number of modules for effective diameter of worm

D_e = effective diameter of worm, in.

λ_e = lead angle of worm at D_e

N_1 = number of starts or threads on worm

$$D_e = N'_1 M \qquad (12\text{-}21)$$
$$\tan \lambda_e = N_1/N'_1 \qquad (12\text{-}22)$$

One of the major objects of standardization is the reduction of variety. One move in this direction is the selection of a limited number of axial pitches for worms. Another good move would be the selection of a limited number of threads or starts for multiple-thread worms. This has been done to great advantage in several manufacturing plants. Although no general and comprehensive standards have yet been developed for worm gears, certain general practices have come into more

TABLE 12-2. TABLE OF SHELL WORMS—MODULE SYSTEM

p_x, in.	Module	N'_1	D_e, in.	Lead angle, deg			
				$N_1 = 1$	$N_1 = 3$	$N_1 = 6$	$N_1 = 12$
0.2500	0.079577	20	1.5915	2.862	8.531	16.699	30.964
0.3125	0.099471	18	1.7905	3.180	9.462	18.435	33.690
0.3750	0.119366	16	1.9099	3.576	10.620	20.556	36.870
0.5000	0.159154	14	2.2282	4.085	12.094	23.199	
0.6250	0.198943	14	2.7852	4.085	12.094	23.199	
0.7500	0.238731	14	3.3422	4.085	12.094	23.199	
1.0000	0.318309	12	3.8197	4.764	14.037	26.565	
1.2500	0.397886	12	4.7746	4.764	14.037	26.565	
1.5000	0.447463	12	5.7296	4.764	14.037	26.565	
1.7500	0.557041	12	6.6845	4.764	14.037	26.565	
2.0000	0.636618	12	7.6394	4.764	14.037	26.565	

TABLE 12-3. TABLE OF INTEGRAL WORMS—MODULE SYSTEM

p_x, in.	N'_1	D_e, in.	Lead angle, deg		
			$N_1 = 1$	$N_1 = 3$	$N_1 = 6$
0.2500	12	0.9549	4.764	14.037	26.565
0.3125	12	1.1937	4.764	14.037	26.565
0.3750	12	1.4324	4.764	14.037	26.565
0.5000	10	1.5915	5.711	16.699	30.964
0.6250	10	1.9894	5.711	16.699	30.964
0.7500	10	2.3878	5.711	16.699	30.964
1.0000	10	3.1831	5.711	16.699	30.964
1.2500	8	3.1831	7.125	20.556	36.870
1.5000	8	3.8197	7.125	20.556	36.870
1.7500	8	4.4563	7.125	20.556	36.870
2.0000	8	5.0937	7.125	20.556	36.870

or less general use over a period of many years. Some of these, unfortunately, lead to very poor designs of worm gears. For one thing, in the effort to use single-thread worms as much as possible, much coarser pitches have been used than are actually needed for the service rendered.

ANALYTICAL MECHANICS OF GEARS

For all other types of gears, the trend in practice over the past 10 to 20 years has been towards the use of finer pitches because of many obvious advantages. For worm gears, the more general practice of today uses single-thread worms of about three times as coarse a pitch as is actually

TABLE 12-4. SINGLE-THREAD WORMS

p_x, in.	L, in.	R_{o1}, in.	R_e, in.	R_{r1}, in.	h_t, in.	c_1, in.	λ_e, deg
			Shell worms				
0.2500	0.2500	0.8581	0.7957	0.7147	0.1434	0.0186	2.862
0.3125	0.3125	0.9732	0.8952	0.7940	0.1792	0.0232	3.180
0.3750	0.3750	1.0484	0.9549	0.8335	0.2149	0.0279	3.576
0.5000	0.5000	1.2387	1.1141	0.9523	0.2864	0.0372	4.085
0.6250	0.6250	1.5483	1.3926	1.1904	0.3579	0.0465	4.085
0.7500	0.7500	1.8580	1.6711	1.4284	0.4296	0.0558	4.085
1.0000	1.0000	2.1586	1.9098	1.5866	0.5720	0.0744	4.764
1.2500	1.2500	2.6984	2.3873	1.9833	0.7151	0.0929	4.764
1.5000	1.5000	3.2381	2.8648	2.3800	0.8581	0.1115	4.764
1.7500	1.7500	3.7777	3.3422	2.7766	1.0011	0.1301	4.764
2.0000	2.0000	4.3174	3.8197	3.1733	1.1441	0.1487	4.764
			Integral worms				
0.2500	0.2500	0.5396	0.4774	0.3966	0.1430	0.0186	4.764
0.3125	0.3125	0.6746	0.5968	0.4958	0.1788	0.0232	4.764
0.3750	0.3750	0.8095	0.7162	0.5950	0.2145	0.0279	4.764
0.5000	0.5000	0.9199	0.7957	0.6344	0.2855	0.0371	5.711
0.6250	0.6250	1.1499	0.9947	0.7931	0.3568	0.0464	5.711
0.7500	0.7500	1.3799	1.1936	0.9517	0.4282	0.0556	5.711
1.0000	1.0000	1.8398	1.5915	1.2689	0.5709	0.0743	5.711
1.2500	1.2500	1.9008	1.5915	1.1893	0.7110	0.0924	7.125
1.5000	1.5000	2.2809	1.9098	1.4278	0.8531	0.1109	7.125
1.7500	1.7500	2.6611	2.2281	1.6657	0.9954	0.1294	7.125
2.0000	2.0000	3.0413	2.5464	1.9037	1.1376	0.1478	7.125

needed. Many of these worm drives would be materially improved by the use of a 3-thread worm of about one-third the axial pitch of the single-thread worms now used.

The number of threads for worms could be restricted to the following which will cover a very wide range of applications:

1, 3, 6, 12, 18, 24 threads or starts

In many places, the first three will cover the great majority of applications.

Values are tabulated in Table 12-2 for such shell worms based on the module system. This table gives the axial pitch, the value of the module, the number of modules in the effective diameter of the worm, the effective diameter, and the lead angle at the effective diameter for worms with different numbers of threads. When the number of threads gives a value

TABLE 12-5. 3-THREAD WORMS

p_x, in.	L, in.	R_{o1}, in.	R_e, in.	R_{r1}, in.	h_t, in.	c_1, in.	λ_e, deg
			Shell worms				
0.2500	0.7500	0.8573	0.7957	0.7157	0.1416	0.0184	8.531
0.3125	0.9375	0.9719	0.8952	0.7956	0.1763	0.0229	9.462
0.3750	1.1250	1.0465	0.9549	0.8359	0.2106	0.0274	10.620
0.5000	1.5000	1.2354	1.1141	0.9566	0.2788	0.0362	12.094
0.6250	1.8750	1.5442	1.3926	1.1957	0.3485	0.0453	12.094
0.7500	2.2500	1.8531	1.6711	1.4348	0.4183	0.0543	12.094
1.0000	3.0000	2.1498	1.9098	1.5980	0.5518	0.0718	14.037
1.2500	3.7500	2.6874	2.3873	1.9976	0.6898	0.0896	14.037
1.5000	4.5000	3.2249	2.8648	2.3971	0.8278	0.1076	14.037
1.7500	5.2500	3.7623	3.3422	2.7966	0.9657	0.1255	14.037
2.0000	6.0000	4.2998	3.8197	3.1962	1.1036	0.1434	14.037
			Integral worms				
0.2500	0.7500	0.5374	0.4774	0.3995	0.1379	0.0179	14.037
0.3125	0.9375	0.6718	0.5968	0.4994	0.1724	0.0224	14.037
0.3750	1.1250	0.8062	0.7162	0.5993	0.2069	0.0269	14.037
0.5000	1.5000	0.9137	0.7957	0.6425	0.2712	0.0352	16.699
0.6250	1.8750	1.1422	0.9947	0.8032	0.3390	0.0440	16.699
0.7500	2.2500	1.3706	1.1936	0.9638	0.4068	0.0528	16.699
1.0000	3.0000	1.8274	1.5915	1.2851	0.5423	0.0705	16.699
1.2500	3.7500	1.8775	1.5915	1.2201	0.6574	0.0854	20.556
1.5000	4.5000	2.2530	1.9098	1.4641	0.7889	0.1025	20.556
1.7500	5.2500	2.6285	2.2281	1.7081	0.9204	0.1196	20.556
2.0000	6.0000	3.0040	2.5464	1.9521	1.0519	0.1367	20.556

for the tangent of the lead angle greater than 0.75000, the diameter of the worm is increased to maintain this maximum lead angle. Thus for all 12-thread worms with axial pitches greater than 0.375 in., the effective diameters will be equal to 16 modules. For all 18-thread worms, the effective diameters will be equal to 24 modules. For all 24-thread worms, the effective diameters will be equal to 32 modules. On all these enlarged worms, the tangent of the lead angle at the effective diameter will be equal to 0.75000. This maximum lead angle is equal to 36.870 deg.

Similar values are tabulated in Table 12-3 for integral worms based on the module system. In this case, all worms with 12, 18, and 24 threads will be enlarged to 16, 24, and 32 modules, respectively. Except for the 0.250-in. and 0.3125-in. axial pitches, these enlarged worms will be identical to the shell type of worms with the same number of threads.

TABLE 12-6. 6-THREAD WORMS

p_x, in.	L, in.	R_{o1}, in.	R_e, in.	R_{r1}, in.	h_t, in.	c_1, in.	λ_e, deg
			Shell worms				
0.2500	1.5000	0.8547	0.7957	0.7191	0.1356	0.0176	16.699
0.3125	1.8750	0.9680	0.8952	0.8007	0.1673	0.0217	18.435
0.3750	2.2500	1.0407	0.9549	0.8435	0.1972	0.0256	20.556
0.5000	3.0000	1.2256	1.1141	0.9693	0.2563	0.0333	23.199
0.6250	3.7500	1.5320	1.3926	1.2115	0.3205	0.0417	23.199
0.7500	4.5000	1.8384	1.6711	1.4538	0.3846	0.0500	23.199
1.0000	6.0000	2.1246	1.9098	1.6308	0.4938	0.0642	26.565
1.2500	7.5000	2.6558	2.3873	2.0386	0.6172	0.0802	26.565
1.5000	9.0000	3.1870	2.8648	2.4463	0.7407	0.0963	26.565
1.7500	10.5000	3.7181	3.3422	2.8540	0.8641	0.1123	26.565
2.0000	12.0000	4.2493	3.8197	3.2617	0.9876	0.1284	26.565
			Integral worms				
0.2500	1.5000	0.5311	0.4774	0.4077	0.1234	0.0164	26.565
0.3125	1.8750	0.6639	0.5968	0.5096	0.1543	0.0201	26.565
0.3750	2.2500	0.7968	0.7162	0.6116	0.1852	0.0240	26.565
0.5000	3.0000	0.8969	0.7957	0.6642	0.2327	0.0303	30.964
0.6250	3.7500	1.1213	0.9947	0.8303	0.2911	0.0378	30.964
0.7500	4.5000	1.3455	1.1936	0.9963	0.3492	0.0454	30.964
1.0000	6.0000	1.7940	1.5915	1.3285	0.4655	0.0605	30.964
1.2500	7.5000	1.8207	1.5915	1.2939	0.5268	0.0684	36.870
1.5000	9.0000	2.1848	1.9098	1.5526	0.6322	0.0822	36.870
1.7500	10.5000	2.5489	2.2281	1.8114	0.7375	0.0959	36.870
2.0000	12.0000	2.9131	2.5464	2.0702	0.8429	0.1095	36.870

Suggestion for Standard Worms. Using the tool with 60-deg included angle shown in Fig. 12-4 for all worms, and Eqs. (12-17), (12-18), and (12-19) for the tooth proportions, we obtain the values for the worms, both shell and integral types, that are tabulated in the following tables. Table 12-4 gives the values for the single-thread worms; Table 12-5 gives the 3-thread worms; Table 12-6 gives the 6-thread worms; Table 12-7 gives

the 12-thread worms; Table 12-8 gives the 18-thread worms; and Table
12-9 gives the values for the 24-thread worms.

On any detailed drawing of the worm, the thread form should always
be specified by giving the form, size, type, and setting of the tool that is
to be used to produce it. This is the only definite way in which this
information can be given. This information is essential for the maker of

TABLE 12-7.   12-THREAD WORMS

p_x, in.	L, in.	R_{o1}, in.	R_e, in.	R_{r1}, in.	h_t, in.	c_1, in.	λ_e, deg
			Shell worms				
0.2500	3.000	0.8463	0.7957	0.7301	0.1162	0.0150	30.964
0.3125	3.750	0.9558	0.8952	0.8165	0.1393	0.0181	33.690
0.3750	4.500	1.0236	0.9549	0.8656	0.1580	0.0206	36.870
0.5000	6.000	1.3649	1.2732	1.1541	0.2108	0.0274	36.870
0.6250	7.500	1.7061	1.5915	1.4427	0.2634	0.0342	36.870
0.7500	9.000	2.0473	1.9098	1.7312	0.3161	0.0411	36.870
1.0000	12.000	2.7298	2 5465	2.3084	0.4214	0.0548	36.870
1.2500	15.000	3.4123	3.1831	2.8855	0.5268	0.0684	36.870
1.5000	18.000	4.0947	3.8197	3.4625	0.6322	0.0822	36.870
1.7500	21.000	4.7771	4.4563	4.0396	0.7375	0.0959	36.870
2.0000	24.000	5.4596	5.0929	4.6167	0.8429	0.1095	36.870
			Integral worms				
0.2500	3.000	0.6824	0.6366	0.5771	0.1053	0.0137	36.870
0.3125	3.750	0.8531	0.7958	0.7214	0.1317	0.0171	36.870
0.3750	4.500	1.0236	0.9549	0.8656	0.1580	0.0206	36.870
0.5000	6.000	1.3649	1.2732	1.1541	0.2108	0.0274	36.870
0.6250	7.500	1.7061	1.5915	1.4427	0.2634	0.0342	36.870
0.7500	9.000	2.0473	1.9098	1.7312	0.3161	0.0411	36.870
1.0000	12.000	2.7298	2.5465	2.3084	0.4214	0.0548	36.870
1.2500	15.000	3.4123	3.1831	2.8855	0.5268	0.0684	36.870
1.5000	18.000	4.0947	3.8197	3.4625	0.6322	0.0822	36.870
1.7500	21.000	4.7771	4.4563	4.0396	0.7375	0.0959	36.870
2.0000	24.000	5.4596	5.0929	4.6167	0.8429	0.1095	36.870

the hob if he is to make the form to match the helicoid of the worm. Any
notation such as "normal form of thread," or "form of thread in normal
section," etc., is incorrect, incomplete, and generally misleading. The
thread form desired should be specified as, for example, "form of 4-in.-
diameter milling cutter set to lead angle of worm at its effective diameter."
The axial or normal thread thickness at this same diameter should be
given, as well as the outside and root diameters.

TABLE 12-8. 18-THREAD WORMS, SHELL AND INTEGRAL
$N'_1 = 24$ $\lambda_e = 36.870°$

p_x, in.	L, in.	R_{o1}, in.	R_e, in.	R_{r1}, in.	h_t, in.	c_1, in.
0.2500	4.500	1.0007	0.9549	0.8954	0.1053	0.0137
0.3125	5.625	1.2510	1.1937	1.1193	0.1317	0.0171
0.3750	6.750	1.5011	1.4324	1.3431	0.1580	0.0206
0.5000	9.000	2.0015	1.9098	1.7907	0.2108	0.0274
0.6250	11.250	2.5019	2.3873	2.2385	0.2634	0.0342
0.7500	13.500	3.0023	2.8648	2.6862	0.3161	0.0411
1.0000	18.000	4.0030	3.8197	3.5816	0.4214	0.0548
1.2500	22.500	5.0038	4.7746	4.4770	0.5268	0.0684
1.5000	27.000	6.0046	6.7296	5.3724	0.6322	0.0822
1.7500	31.500	7.0053	6.6845	6.2678	0.7375	0.0959
2.0000	36.000	8.0061	7.6394	7.1632	0.8429	0.1095

TABLE 12-9. 24-THREAD WORMS, SHELL AND INTEGRAL
$N'_1 = 32$ $\lambda_e = 36.870°$

p_x, in.	L, in.	R_{o1}, in.	R_e, in.	R_{r1}, in.	h_t, in.	c_1, in.
0.2500	6.000	1.3190	1.2732	1.2137	0.1053	0.0137
0.3125	7.500	1.6488	1.5915	1.5171	0.1317	0.0171
0.3750	9.000	1.9785	1.9098	1.8205	0.1580	0.0206
0.5000	12.000	2.6382	2.5465	2.4274	0.2108	0.0274
0.6250	15.000	3.2977	3.1831	3.0342	0.2634	0.0342
0.7500	18.000	3.9572	3.8197	3.6411	0.3161	0.0411
1.0000	24.000	5.2762	5.0929	4.8548	0.4214	0.0548
1.2500	30.000	6.5954	6.3662	6.0686	0.5268	0.0684
1.5000	36.000	7.9144	7.6394	7.2822	0.6322	0.0822
1.7500	42.000	9.2335	8.9127	8.4960	0.7375	0.0959
2.0000	48.000	10.5526	10.1859	9.7097	0.8429	0.1095

Examples of Contact Conditions on Worm Gears. In order to show the nature of the contact conditions that will exist on worm drives that use the foregoing series of worms, several different examples will be analyzed. As noted before, the sum of the numbers of teeth in the worm and worm gear should never be less than 40. We shall use the 0.500-in. axial-pitch worms as representative of this series.

Single-thread Worm, 42-tooth Worm Gear. For the first example we shall use a single-thread shell worm and a 42-tooth worm gear. This gives the following values:

$$p_x = 0.500 \quad L = 0.500 \quad N_1 = 1 \quad N_2 = 42 \quad h_t = 0.2864 \quad c_1 = 0.0372$$
$$R_e = 1.1141 \quad R_{o1} = 1.2387 \quad R_{r1} = 0.9523 \quad \lambda_e = 4.085°$$
$$\text{Module} = M = 0.159154$$

From these values we obtain

$$R_2 = N_2 M / 2 = 21 \times 0.159154 = 3.3422$$
$$C \text{ (max)} = 3.3422 + 1.1141 = 4.4563$$
$$C \text{ (min)} = 3.3422 + 0.9523 = 4.2945$$

We shall make the center distance an even fraction of an inch and shall use $C = 4.375$ in. Then

$$R_1 = 4.375 - 3.3422 = 1.0328$$
$$R_t = 4.375 - (0.9523 + 0.0372) = 3.3855$$

For purposes of determining the face width of the worm gear, we can use the value of the lead angle at the effective radius, the value of the effective radius, and the value of the nominal addendum of the worm-thread form. This practice will give a constant value for the face width of a worm gear that meshes with a worm of given axial pitch as long as its diameter does not change or as its lead angle does not increase beyond 15 deg. Whence

$$F_2 = 2 \sqrt{(2.2282 + 0.25) \times 0.25} + (0.50 \times 0.50) = 1.8244$$

We shall use $F_2 = 1.8125$ in.

$$T_r = 0.9523 + 0.0372 = 0.9895$$
$$I_t = 0.70 \times 0.2864 = 0.2004$$
$$R_{o2} = 3.3858 + 0.1002 = 3.4860$$
$$\text{Edge round} = 0.25 \times 0.50 = 0.125$$

The contact lines and the field of contact for this drive have been determined by the use of methods shown in Chap. 11, and they are plotted in Fig. 12-5. An examina-

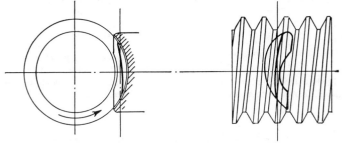

FIG. 12-5. Single thread worm and 42-tooth worm gear.

tion of this figure will show that the field of contact extends over almost $1\frac{1}{2}$ threads. This is a measure of the duration of contact on this drive.

3-thread Worm, 42-tooth Worm Gear. We shall use the same center distance as before. Many of the values will be the same as in the preceding examples. Thus we have

$$p_x = 0.500 \qquad L = 1.500 \qquad N_1 = 3 \qquad N_2 = 42 \qquad h_t = 0.2788 \qquad c_1 = 0.0362$$
$$R_e = 1.1141 \qquad R_{o1} = 1.2354 \qquad R_{r1} = 0.9566 \qquad \lambda_e = 12.094° \qquad M = 0.159154$$
$$C = 4.375 \qquad R_2 = 3.3422 \qquad F_2 = 1.8125 \qquad R_1 = 1.0328$$
$$R_t = 4.375 - (0.9566 + 0.0362) = 3.3822$$
$$T_r = 0.9566 + 0.0362 = 0.9928$$
$$I_t = 0.70 \times 0.2788 = 0.1952$$
$$R_{o2} = 3.3822 + 0.0976 = 3.4798$$

The contact lines and field of contact for this drive have been determined as before. They are plotted in Fig. 12-6. An examination of this figure will show that

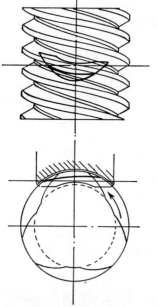

the field of contact extends over about 1.40 threads on the worm, which gives this drive about the same duration of contact as that in the preceding example.

6-*thread Worm, 42-tooth Worm Gear.* We shall use the same center distance as before. Many of the values will be the same as in the two preceding exam les. The lead angle will be greater than 15 deg; hence the face width of the worm gear will be reduced. For this example we have

$$p_x = 0.500 \qquad L = 3.000 \qquad N_1 = 6$$
$$N_2 = 42 \qquad h_t = 0.2563 \qquad c_1 = 0.0333$$
$$R_e = 1.1141 \qquad R_{o1} = 1.2256$$
$$R_{r1} = 0.9693 \qquad \lambda_e = 23.199°$$
$$M = 0.159154 \qquad C = 4.375$$
$$R_2 = 3.3422 \qquad R_1 = 1.0328$$
$$R_t = 4.375 - (0.9693 + 0.0333) = 3.3724$$
$$T_r = 0.9693 + 0.0333 = 1.0026$$
$$I_t = 0.35 \times 0.2563 = 0.0897$$
$$R_{o2} = 3.3724 + 0.0448 = 3.4172$$
$$F_2 = 2\sqrt{(2.2282 + 0.25) \times 0.25} + (0.25 \times 0.50) = 1.6994$$

We shall use $F_2 = 1.6875$ in.

FIG. 12-6.   3-thread worm and 42-tooth worm gear.

The contact lines and field of contact for this drive have been determined as before. They are plotted in Fig. 12-7. An examination of the figure will show that the field of contact extends over slightly more than 2 threads. On this drive, therefore, there will always be at least 2 threads in contact.

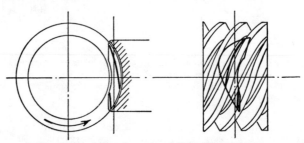

FIG. 12-7.   6-thread worm and 42-tooth worm gear.

12-*thread Worm, 36-tooth Worm Gear.* For the next example we shall use a 12-thread worm. We shall keep the sum of the numbers of teeth in the worm and worm gear equal to 48, the same sum as for the preceding example   This will give a 36-tooth worm gear. For this we have the following values:

$p_x = 0.500$ $\quad L = 6.000$ $\quad N_1 = 12$ $\quad N_2 = 36$ $\quad h_t = 0.2108$ $\quad c_1 = 0.0274$
$R_e = 1.2732$ $\quad R_{o1} = 1.3649$ $\quad R_{r1} = 1.1541$ $\quad \lambda_e = 36.870°$ $\quad M = 0.159154$

$$R_2 = 18 \times 0.159154 = 2.8648$$
$$C \text{ (max)} = 2.8648 + 1.2732 = 4.1380$$
$$C \text{ (min)} = 2.8648 + 1.1541 = 4.0189$$

We shall use $C = 4.000$.

This value for the center distance will bring the pitch plane of the worm slightly below the root radius of the worm. We shall make an analysis of the contact, however, to be sure that the duration of contact is adequate. If it is not, then we must move the pitch plane of the worm up by increasing the center distance.

$$R_1 = 4.000 - 2.8648 = 1.1352$$
$$R_t = 4.000 - (1.1541 + 0.0274) = 2.8185$$
$$T_r = 1.1541 + 0.0274 = 1.1815$$
$$I_t = 0.35 \times 0.2108 = 0.0738$$
$$R_{o2} = 2.8185 + 0.0369 = 2.8554$$
$$F_2 = 2 \sqrt{(2.5464 + 0.25) \times 0.25} + (0.25 \times 0.50) = 1.7972$$

We shall use $F_2 = 1.750$ in.

The contact lines and the field of contact for this drive have been determined. They are plotted in Fig. 12-8. An examination of this figure will show that the field

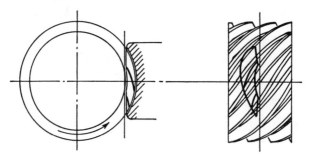

Fig. 12-8.   12-thread worm and 36-tooth worm gear.

of contact extends over about $2\frac{1}{2}$ threads of the worm. Even though the pitch plane of the worm is below the root radius of the worm, and the pitch radius of the worm gear is larger than its outside radius, the contact is adequate. In this example, all the action is recess action.

18-thread Worm, 30-tooth Worm Gear. We shall keep the sum of the numbers of teeth equal to 48. This gives a 30-tooth worm gear for this example. We have the following values:

$p_x = 0.500$ $\quad L = 9.000$ $\quad N_1 = 18$ $\quad N_2 = 30$ $\quad h_t = 0.2108$ $\quad c_1 = 0.0274$
$R_e = 1.9098$ $\quad R_{o1} = 2.0015$ $\quad R_{r1} = 1.7907$ $\quad \lambda_e = 36.870°$ $\quad M = 0.159154$

$$R_2 = 15 \times 0.159154 = 2.3873$$
$$C \text{ (max)} = 2.3873 + 1.9098 = 4.2971$$
$$C \text{ (min)} = 2.3873 + 1.7907 = 4.1780$$

We shall use $C = 4.1875$.

$$R_1 = 4.1875 - 2.3873 = 1.8002$$
$$R_t = 4.1875 - (1.7907 + 0.0274) = 2.3694$$
$$T_r = 1.7907 + 0.0274 = 1.8181$$
$$I_t = 0.35 \times 0.2108 = 0.0738$$
$$R_{o2} = 2.3694 + 0.0369 = 2.4063$$
$$F_2 = 2 \sqrt{(3.8169 + 0.25) \times 0.25} + 0.125 = 2.1422$$

We shall use $F_2 = 2.125$ in.

The contact lines and the field of contact for this drive have been determined. They are plotted in Fig. 12-9. An examination of this figure will show that the field of contact extends over $2\frac{1}{2}$ threads of the worm.

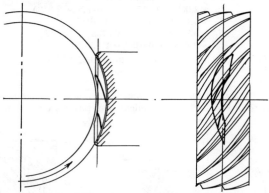

Fig. 12-9.   18-thread worm and 30-tooth worm gear.

24-thread Worm, 24-tooth Worm Gear. For the next example we shall use a 24-thread worm and a 24-tooth worm gear. This keeps the sum of the numbers of teeth in the drive equal to 48. It also gives us a one-to-one ratio. For this we have the following values:

$$p_x = 0.500 \qquad L = 12.000 \qquad N_1 = 24 \qquad N_2 = 24 \qquad h_t = 0.2108 \qquad c_1 = 0.0274$$
$$R_e = 2.5465 \qquad R_{o1} = 2.6382 \qquad R_{r1} = 2.4274 \qquad \lambda_e = 36.870° \qquad M = 0.159154$$
$$R_2 = 12 \times 0.159154 = 1.9098$$
$$C \text{ (max)} = 1.9098 + 2.5465 = 4.4563$$
$$C \text{ (min)} = 1.9098 + 2.4274 = 4.3372$$

We shall use $C = 4.375$ in.

$$R_1 = 4.375 - 1.9098 = 2.4652$$
$$R_t = 4.375 - (2.4274 + 0.0274) = 1.9202$$
$$T_r = 2.4274 + 0.0274 = 2.4548$$
$$I_t = 0.0738$$
$$R_{o2} = 1.9202 + 0.0369 = 1.9571$$
$$F_2 = 2 \sqrt{(5.0930 + 0.25) \times 0.25} + 0.125 = 2.5298$$

We shall use $F_2 = 2.500$ in.

The contact lines and the field of contact for this drive have been determined They are plotted in Fig. 12-10. An examination of this figure will show that the field of contact extends over about 2¾ threads on the worm.

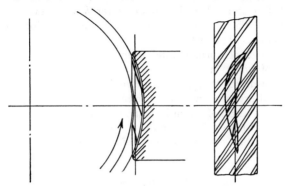

FIG. 12-10.   24-thread worm and 24-tooth worm gear.

Driving Member of Enveloping Form. As noted before, a worm-gear drive can be considered as a development from a spiral-gear drive where one member of the pair is made to envelop the other so as to obtain line contact instead of point contact between the mating teeth. Either member of the pair may be made to envelop the other. The more common practice is to make the larger member, or the member with the greater number of teeth, envelop the smaller one or the driver. There are occasions, however, when a definite advantage can be gained by reversing this practice and by making the driver to envelop the follower. This has been done in some cases. Such drives are sometimes called *hourglass worm drives,* but they are true worm-gear drives as long as one member has a uniform axial lead.

In drives of this kind, the large, multiple-thread worm or helicoid member can best be produced as a helical involute gear. In such cases, a standard diametral pitch and pressure angle could best be used. The enveloping member can be generated by a helical pinion-shaped cutter of the same number of teeth as is used for the helicoid member.

For the further examples, however, we will use the same combinations of teeth and the same center distances as before, so as to obtain a direct comparison of the conditions of contact when the opposite member is made to envelop the other. We will make the helicoid follower as a helical involute gear with a 30-deg normal basic-rack form of the same proportions as those of the threading tool used to produce the worms. The first step to this end will be to transform the values of any given drive into those of the equivalent spiral-gear drive. Here we must use the actual distance to the pitch plane on the worms of the preceding examples as the pitch radius of the enveloping member, and the actual lead angle at this radius as the helix angle of the follower. We shall use the same symbols as before and call the helicoid follower *the worm* even though it is not the driver. Then we can use the same equations for the contact analysis as before. Thus when

λ_1 = lead angle of worm at pitch radius

L_1 = lead of worm, in.

L_2 = lead of driver of spiral-gear drive, in.

ϕ_n = normal pressure angle of basic rack of worm

ϕ_1 = pressure angle of worm in plane of rotation

R_{b1} = radius of base cylinder of worm, in.

and all other symbols are the same as before, then

$$\cot \lambda_1 = L_2/2\pi R_2 \qquad (12\text{-}23)$$
$$L_1 = 2\pi R_1 \tan \lambda_1 \qquad (12\text{-}24)$$
$$\tan \phi_1 = \tan \phi_n/\sin \lambda_1 \qquad (12\text{-}25)$$
$$R_{b1} = R_1 \cos \phi_1 \qquad (5\text{-}3)$$

and all other equations are the same as before.

Examples of Enveloping Driver. *30-tooth Worm, 18-tooth Worm Gear.* As the first example of an enveloping driver we shall use an 18-tooth enveloping driving member and a 30-tooth helicoid follower or worm. From the corresponding worm-gear example we have the following values:

$$N_1 = 30 \qquad N_2 = 18 \qquad R_1 = 2.3873 \qquad R_{o1} = 2.3694 \qquad R_{r1} = 2.1586$$
$$R_2 = 1.8002 \qquad C = 4.1875 \qquad h_t = 0.2108 \qquad c_1 = 0.0274 \qquad L_2 = 9.000$$
$$\phi_n = 30° \qquad \tan \phi_n = 0.57735$$
$$\cot \lambda_1 = \frac{9.000}{6.2832 \times 1.8002} = 0.79568$$
$$\lambda_1 = 51.492° \qquad \tan \lambda_1 = 1.25663 \qquad \sin \lambda_1 = 0.78252$$
$$L_1 = 6.2832 \times 2.3873 \times 1.25663 = 18.8494$$
$$\tan \phi_1 = \frac{0.57735}{0.78252} = 0.73781$$
$$\phi_1 = 36.420° \qquad \cos \phi_1 = 0.80469$$
$$R_{b1} = 2.3873 \times 0.80469 = 1.9210$$
$$R_t = 4.1875 - (2.1586 + 0.0274) = 2.0015$$
$$T_r = 2.1586 + 0.0274 = 2.1860$$
$$I_t = 0.35 \times 0.2108 = 0.0738$$
$$R_{o2} = 2.0015 + 0.0369 = 2.0384$$
$$F_2 = 2 \sqrt{(4.7746 + 0.25) \times 0.25} + 0.125 = 2.3666$$

We shall use $F_2 = 2.375$ in.

The contact lines and the field of contact for this drive have been determined as before. They are plotted in Fig. 12-11. An examination of this figure will show that the field of contact extends over more than $2\frac{1}{2}$ threads of the worm.

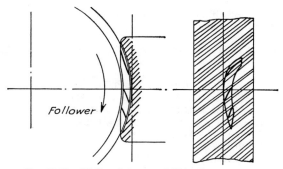

Fig. 12-11. 30-thread worm and 18-tooth worm gear.

36-*tooth Worm*, 12-*tooth Worm Gear*. As the next example of an enveloping driver we shall use a 12-tooth enveloping member and a 36-tooth helicoid follower. From the corresponding worm-gear drive we obtain the following:

$$N_1 = 36 \quad N_2 = 12 \quad R_1 = 2.8648 \quad R_{o1} = 2.8185 \quad R_{r1} = 2.6077$$
$$C = 4.000 \quad h_t = 0.2108 \quad c_1 = 0.0274 \quad R_2 = 1.1352 \quad L_2 = 6.000$$
$$\cos \lambda_1 = \frac{6.000}{6.2832 \times 1.1352} = 0.84119$$
$$\lambda_1 = 49.930° \qquad \tan \lambda_1 = 1.18880$$
$$L_1 = 6.2832 \times 2.8646 \times 1.18880 = 21.3884$$
$$R_t = 4.000 - (2.6077 + 0.0274) = 1.3649$$
$$T_r = 2.6077 + 0.0274 = 2.6351$$
$$I_t = 0.35 \times 0.2108 = 0.0738$$
$$R_{o2} = 1.3649 + 0.0369 = 1.4018$$
$$F_2 = 2 \sqrt{(5.7296 + 0.25) \times 0.25} + 0.125 = 2.5696$$

We shall use $F_2 = 2.625$ in.

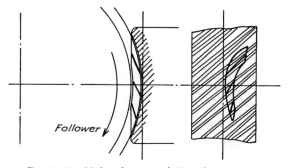

Fig. 12-12. 36-thread worm and 12-tooth worm gear.

The contact lines and field of contact for this drive are plotted in Fig. 12-12. An examination of this figure will show that the field of contact extends over about 3 threads of the worm.

42-*tooth Worm*, 6-*tooth Worm Gear*. From the corresponding worm-gear example we obtain the following values:

$$N_1 = 42 \qquad N_2 = 6 \qquad R_1 = 3.3422 \qquad R_{o1} = 3.3724 \qquad R_{r1} = 3.1161$$
$$C = 4.375 \qquad h_t = 0.2563 \qquad c_1 = 0.0333 \qquad R_2 = 1.0328 \qquad L_2 = 3.000$$

$$\cot \lambda_1 = \frac{3.000}{6.2832 \times 1.0328} = 0.46230$$

$$\lambda_1 = 65.189° \qquad \tan \lambda_1 = 2.16311$$

$$L_1 = 6.2832 \times 3.3422 \times 2.16311 = 45.4253$$

$$R_t = 4.375 - (3.1161 + 0.0333) = 1.2256$$

$$T_r = 3.1161 + 0.0°33 = 3.1494$$

$$I_t = 0.0897$$

$$R_{o2} = 1.2256 + 0.0448 = 1.2704$$

$$F_2 = 2 \sqrt{(6.6844 + 0.25) \times 0.25} + 0.125 = 2.7582$$

We shall use $F_2 = 2.750$ in.

The contact lines and the field of contact for this drive are plotted in Fig. 12-13. An examination of this figure will show that the field of contact extends over about 4 threads of the worm.

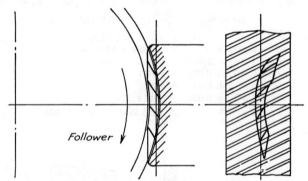

Follower

Fig. 12-13.   42-thread worm and 6-tooth worm gear.

42-*tooth Worm*, 3-*tooth Worm Gear*. From the corresponding worm-gear example we obtain the following values:

$$N_1 = 42 \qquad N_2 = 3 \qquad R_1 = 3.3422 \qquad R_{o1} = 3.3822 \qquad R_{r1} = 3.1034$$
$$C = 4.375 \qquad h_t = 0.2788 \qquad c_1 = 0.0362 \qquad R_2 = 1.0328 \qquad L_2 = 1.500$$

$$\cot \lambda_1 = \frac{1.500}{6.2832 \times 1.0328} = 0.23115$$

$$\lambda_1 = 76.985° \qquad \tan \lambda_1 = 4.32476$$

$$L_1 = 6.2832 \times 3.3422 \times 4.32476 = 90.8200$$

$$R_t = 4.375 - (3.1034 + 0.0362) = 1.2354$$

$$T_r = 3.1034 + 0.0362 = 3.1396$$

$$I_t = 0.35 \times 0.2788 = 0.0976$$

$$R_{o2} = 1.2354 + 0.0488 = 1.2842$$

$$F_2 = 2.750$$

The contact lines and the field of contact for this drive have been plotted in Fig. 12-14. An examination of the figure will show that the field of contact extends over more than 4 threads of the worm.

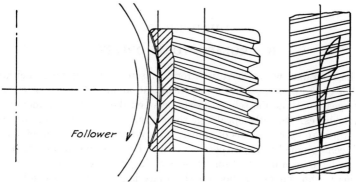

FIG. 12-14. 42-thread worm and 3-tooth worm gear.

42-tooth Worm, Single-tooth Worm Gear. As the last example of an enveloping driver, we shall use a single-tooth enveloping driving member. From the corresponding worm-gear example we obtain the following values:

$$N_1 = 42 \qquad N_2 = 1 \qquad R_1 = 3.3422 \qquad R_{o1} = 3.3855 \qquad R_{r1} = 3.0991$$
$$C = 4.375 \qquad h_t = 0.2864 \qquad c_1 = 0.0372 \qquad R_2 = 1.0328 \qquad L_2 = 0.500$$
$$\cot \lambda_1 = \frac{0.500}{6.2832 \times 1.0328} = 0.07705$$
$$\lambda_1 = 85.594° \qquad \tan \lambda_1 = 12.97850$$
$$L_1 = 6.2832 \times 3.3422 \times 12.97850 = 272.5485$$
$$R_t = 4.375 - (3.0991 + 0.0372) = 1.2387$$
$$T_r = 3.0991 + 0.0372 = 3.1363$$
$$I_t = 0.35 \times 0.2864 = 0.1002$$
$$R_{o2} = 1.2387 + 0.0501 = 1.2888$$
$$F_2 = 2.750$$

The contact lines and the field of contact for this drive have been plotted in Fig. 12-15. An examination of this figure will show that the field of contact extends over about 5 threads of the worm.

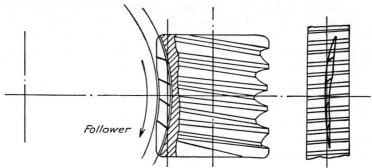

FIG. 12-15. 42-thread worm and single-tooth worm gear.

CHAPTER 13

HOURGLASS-WORM DRIVES

The hourglass-worm drive is a form of screw gearing where neither member of the pair has a uniform helical lead along its axis. True conjugate gear-tooth action seldom exists between the two mating members of such drives. Strictly speaking, these gears are not true gear drives but rather are special forms of cams. Many types of this form of drive have been designed to meet specific needs. For example, a wide variety of them have been designed for use as steering units for automobiles.

There is no general method for the analysis of the contact conditions that exist on these hourglass-worm drives. Each type must be analyzed individually. A few examples of this type of drive will be analyzed in this chapter. The contact on these drives is generally line contact. Surface contact is sometimes claimed or implied, but line contact is the most that can be obtained together with smooth continuous action.

Multiple tooth contact can be readily attained on many of these drives. However when the amount of contact is adequate to carry the loads imposed, additional contact is a liability and not an asset. If these drives are run continuously, or enough to develop any appreciable heat of operation, the thermal expansion of the gear member is generally greater than that of the worm member. As a result, the contact tends to become concentrated at the two ends of the worm member. This condition limits the amount of effective contact that can actually be used on this type of drive.

HINDLEY-WORM DRIVE

The first example we shall analyze is the Hindley-worm drive. This was developed by a Mr. Hindley in England for use in a dividing engine about 1765. It consists of an hourglass worm that is formed by a straight-sided threading tool set in the axial plane of the worm and traveling in a circular path with the center of this path at the center of the mating worm gear. The mating gear, in turn, is made to envelop the worm partially. It is generally produced by a hob of the form of the worm. This construction is indicated in Fig. 13-1.

There are several ways in which this drive may be analyzed. For one, the profiles of the worm threads may be determined on a series of planes parallel to the axial plane of the worm. These intersection profiles will show a change in form from one thread to another on the same off-center plane, and will also show a varying angular displacement in reference to the center of the gear except on the axial plane. It is obvious that all the teeth of the gear must be identical.

Another method would be to determine the trace of both the worm and the worm gear on a cylinder concentric with the gear. These traces could then be compared directly. This second method will be used here.

Figure 13-2 shows the axial and end sections of the worm. The straight-line radial elements of this worm-thread form are all tangent to

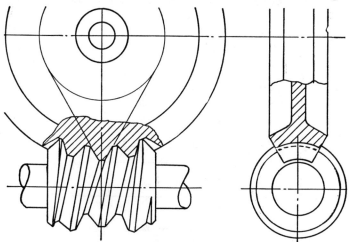

Fig. 13-1. Hindley worm-gear drive.

a base cylinder concentric with the gear. This holds true only in the axial plane of the worm, which is also a plane of rotation of the gear. Thus when

C = center distance, in.

R_b = radius of base circle, in.

R_c = radius of intersecting cylinder, in.

N_1 = number of threads or starts on worm

N_2 = number of teeth in worm gear

θ = angular position of thread element when $y = 0$

ϵ_1 = turning angle of worm from original position

ϵ_2 = turning angle of gear from original position

A = distance of intersection point from center line of gear, in.

α = angle between axial thread element and radial line of gear

ϕ = pressure angle at A when worm element is brought into axial plane

x = abscissa of trace of worm thread on intersecting cylinder, in.

y = ordinate of trace of worm thread on intersecting cylinder, in.

we have to start

$$\epsilon_2 = N_1\epsilon_1/N_2 \tag{13-1}$$

$$\sin \alpha = R_b/R_c \tag{13-2}$$

We have the following from the geometrical conditions shown in Fig. 13-2:

$$\phi = \alpha - (\theta + \epsilon_2) \tag{13-3}$$

$$A = R_b \cos \phi - (C - R_b \sin \phi) \tan \phi + r \tan \phi \tag{13-4}$$

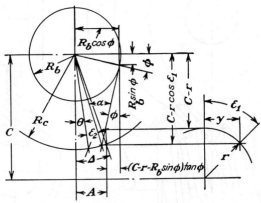

$$A = R_b \cos\phi - (C - r - R_b \sin\phi)\, \tan\phi$$
$$C - r\cos \mathcal{E}_1 = \sqrt{R_c^2 - A^2}$$
$$\phi = \alpha - (\theta + \mathcal{E}_2)$$
$$y = r \sin \mathcal{E}_1$$
$$\sin \Delta = A/R_c$$
$$x = R_c \Delta = R_c \sin^{-1}(A/R_c)$$

FIG. 13-2.

To simplify and condense this equation, we shall let

$$B = R_b \cos \phi - (C - R_b \sin \phi) \tan \phi \tag{13-5}$$

whence

$$A = B + r \tan \phi \tag{13-6}$$

But

$$A^2 = R_c^2 - (C - r \cos \epsilon_1)^2 \tag{13-7}$$

Expanding this last equation, we obtain the following:

$$A^2 = 2rC \cos \epsilon_1 - r^2 \cos^2 \epsilon_1 - (C^2 - R_c^2)$$

We shall let
$$D = C^2 - R_c{}^2$$
whence
$$A^2 = 2rC \cos \epsilon_1 - r^2 \cos^2 \epsilon_1 - D \qquad (13\text{-}8)$$

Squaring Eq. (13-6), equating it to Eq. (13-8), and solving for r, we obtain

$$r = \frac{(C \cos \epsilon_1 - B \tan \phi) - \sqrt{(C \cos \epsilon_1 - B \tan \phi)^2 - (B^2 + D)(\tan^2 \phi + \cos^2 \epsilon_1)}}{\tan^2 \phi + \cos^2 \epsilon_1} \qquad (13\text{-}9)$$

We have from Fig. 13-2

$$y = r \sin \epsilon_1 \qquad (13\text{-}10)$$

$$\sin \Delta = \frac{A}{R_c}$$

$$x = R_c \Delta = R_c \sin^{-1} \frac{A}{R_c} \qquad (13\text{-}11)$$

To determine the complete trace of the worm thread on the intersecting cylinder, we first select a series of values for θ that will include all the successive threads on the worm. Then for each value of θ, we use a series of values of ϵ_1 and ϵ_2 that will cover the face of the worm gear. For the values of ϵ_1 and ϵ_2 on one side of the axis of the coordinate system, their signs will be plus. For the opposite side, their values will be minus. The values of θ on one side of the central section where the worm diameter is a minimum will be plus. On the opposite side they are minus.

If the values of θ for opposite sides of the thread space of the worm at the central section are the same but of opposite sign, then the forms of the intersection curves of these opposite sides are alike, but inverted in their relative positions. In other words, the plus values of x and y become minus for the opposite side and the minus values become plus.

Example of Hindley Worm. As a definite example we shall use the following values:

$$N_1 = 9 \qquad N_2 = 36 \qquad R_b = 2.000 \qquad R_c = 4.000 \qquad C = 5.000$$

We shall use an initial value of $\theta = 2.50°$ for the position of the thread near the central section. We will use values of ϵ_1 in steps of 8 deg. From these values, we obtain

$$\epsilon_2 = \frac{9\epsilon_1}{36} = \frac{\epsilon_1}{4}$$

$$\sin \alpha = \frac{2.00}{4.00} = 0.50000$$

whence $\alpha = 30°$.

In this example we shall determine the coordinates of the top half of the first thread, the complete forms of the next 4 threads, and the bottom half of the sixth thread. These are the threads that engage the worm gear. Values of ϵ_1 will vary from 0 to ±40 deg in steps of 8 deg. This range will give the full form of the trace of the worm threads on the intersecting cylinder. The angular values used to compute the coordinates are tabulated in Table 13-1.

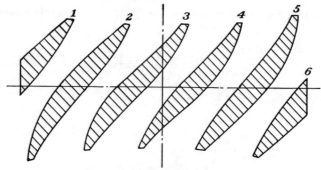

FIG. 13-3. Trace of threads of Hindley worm.

The values for the coordinates of the traces of these threads on the intersecting cylinder have been calculated. They are tabulated in Table 13-2 and plotted in Fig. 13-3. It will be noted that the values of the coordinates of the second half of the first thread are the same, but of opposite sign, as are those for the first half of the sixth thread, etc. An examination of Table 13-2 or Fig. 13-3 will show that the intersection form of the threads of the worm varies from thread to thread.

TABLE 13-1. VALUES OF ANGLES USED FOR CALCULATING COORDINATES OF INTERSECTION PROFILE OF HINDLEY WORM

Values of ϵ_1, deg	Thread					
	1st $\theta = -27.5°$	2d $\theta = -17.5°$	3d $\theta = -7.5°$	4th $\theta = 2.5°$	5th $\theta = 12.5°$	6th $\theta = 22.5°$
	Values of ϕ, deg					
40	47.5	37.5	27.5	17.5	7.5	
32	49.5	39.5	29.5	19.5	9.5	
24	51.5	41.5	31.5	21.5	11.5	
16	53.5	43.5	33.5	23.5	13.5	
8	55.5	45.5	35.5	25.5	15.5	
0	57.5	47.5	37.5	27.5	17.5	7.5
− 8	....	49.5	39.5	29.5	19.5	9.5
−16	....	51.5	41.5	31.5	21.5	11.5
−24	....	53.5	43.5	33.5	23.5	13.5
−32	....	55.5	45.5	35.5	25.5	15.5
−40	....	57.5	47.5	37.5	27.5	17.5

TABLE 13-2. COORDINATES OF TRACES OF THREADS OF HINDLEY WORM ON INTERSECTING CYLINDER
(Plotted in Fig. 13-3)

Front		Back	
x, in.	y, in.	x, in.	y, in.
First thread			
−0.92740	0.92891	−0.99358	0.94359
−1.17447	0.73184	−1.07582	0.71473
−1.39403	0.55230	−1.18187	0.52241
−1.59132	0.37632	−1.30278	0.34704
−1.76725	0.19452	−1.43354	0.17625
−1.91988	0.00000	−1.57080	0.00000
Second thread			
−0.30285	0.84872	−0.34181	0.85135
−0.52569	0.64642	−0.40568	0.63771
−0.72522	0.47442	−0.50000	0.45912
−0.90708	0.31611	−0.61128	0.30012
−1.07289	0.16064	−0.73709	0.15005
−1.22173	0.00000	−0.87266	0.00000
−1.35075	−0.17223	−1.01571	−0.15856
−1.45491	−0.36179	−1.16462	−0.33502
−1.52653	−0.57335	−1.31807	−0.54104
−1.55502	−0.81138	−1.47466	−0.79281
−1.52514	−1.08014	−1.63278	−1.11487
Third thread			
0.34174	0.85135	0.30285	0.84878
0.13781	0.62635	0.26166	0.63021
−0.04740	0.44535	0.18347	0.44710
−0.21852	0.28846	0.07980	0.28722
−0.37727	0.14304	−0.04050	0.14057
−0.52360	0.00000	−0.17453	0.00000
−0.65562	−0.17092	−0.31912	−0.14232
−0.76969	−0.35312	−0.47403	−0.29479
−0.85912	−0.56714	−0.63565	−0.46767
−0.91707	0.04203	−0.79671	−0.65888
−0.92747	−1.22650	−0.99274	−0.94196

TABLE 13-2. COORDINATES OF TRACES OF THREADS OF HINDLEY WORM ON INTER-SECTING CYLINDER. *(Continued)*
(Plotted in Fig. 13-3)

Front		Back	
x, in.	y, in.	x, in.	y, in.
Fourth thread			
0.99274	0.94196	0.92747	1.22650
0.79671	0.65888	0.91707	0.84203
0.63565	0.46767	0.85912	0.56713
0.47403	0.29479	0.76969	0.35312
0.31912	0.14232	0.65562	0.17079
0.17453	0.00000	0.52360	0.00000
0.04050	−0.14057	0.37727	−0.14304
−0.07980	−0.28722	0.21852	−0.28846
−0.18347	−0.44710	0.04740	−0.44535
−0.26166	−0.63021	−0.13781	−0.62635
−0.30285	−0.84872	−0.34174	−0.85135
Fifth thread			
1.63279	1.11487	1.52514	1.08014
1.47466	0.79281	1.55502	0.81138
1.31807	0.54104	1.52653	0.57335
1.16462	0.33502	1.45491	0.36179
1.01571	0.15856	1.35075	0.17229
0.87266	0.00000	1.22173	0.00000
0.73709	−0.15005	1.07289	−0.16064
0.61125	−0.30012	0.90708	−0.31611
0.50000	−0.45912	0.72522	−0.47442
0.40568	−0.63771	0.52569	−0.64642
0.34181	−0.85135	0.30285	−0.84872
Sixth thread			
1.57080	0.00000	1.91988	0.00000
1.43354	−0.17625	1.76725	−0.19452
1.30278	−0.34707	1.59132	−0.37632
1.18187	−0.52241	1.34903	−0.55230
1.07582	−0.71473	1.17447	−0.73184
0.99358	−0.94359	0.92740	−0.92891

Trace of Gear Teeth on Intersecting Cylinder. The intersection curves of the worm threads with a cylinder concentric with the axis of the worm gear vary from position to position. All the worm-gear teeth, however, must be identical and must not interfere with the worm threads. Hence one side, or one-half of each side, of the gear tooth will be the form produced by the hob at the central section where the lead angle is greatest, and the other part of the gear tooth will be the form produced by that end of the hob which has the highest pressure angle, and also has the lowest lead angle.

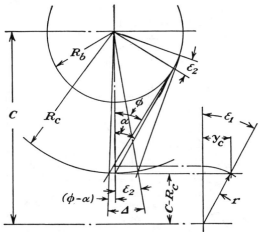

Fig. 13-4.

It should be apparent that the hob must never be shorter than the worm. It must be either of the same length or else somewhat longer. If it is made of the same length, then as the hob is resharpened, the worm must be reduced in length on one side at least by an amount that is dependent upon the amount of relief on the hob and upon the amount of metal removed from the hob in sharpening.

Thus to determine the intersection curve of the worm-gear teeth, we must analyze the conditions at the central section and those at one end.

Intersection Curve at Central Section. Referring to Fig. 13-4, let

x_c = abscissa of trace of central section of hob, in.

y_c = ordinate of trace of central section of hob, in.

and all other symbols be the same as before. Some additional symbols are also shown in Fig. 13-4.

We have as before

$$\epsilon_2 = N_1\epsilon_1/N_2 \qquad (13\text{-}1)$$
$$\sin \alpha = R_b/R_c \qquad (13\text{-}2)$$

We have the following from the geometrical conditions shown in Fig. 13-4:

$$y_c = (C - R_c) \tan \epsilon_1 \qquad (13\text{-}12)$$
$$r = (C - R_c)/\cos \epsilon_1 \qquad (13\text{-}13)$$
$$\sin \phi = R_b/(C - r) \qquad (13\text{-}14)$$
$$\Delta = \epsilon_2 + \phi - \alpha \qquad (13\text{-}15)$$
$$x_c = R_c \, \Delta \qquad (13\text{-}16)$$

The polar equation of the intersection curve of the worm at the central section with a plane perpendicular to the axis of the worm is as follows: When θ_c = vectorial angle

r_c = length of radius vector, in.

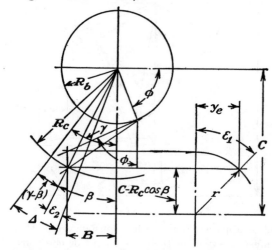

Fig. 13-5.

we have from the geometrical conditions shown in Fig. 13-4 the following:

$$r_c = C - (R_b/\sin \phi) \qquad (13\text{-}17)$$
$$\phi = \alpha - \epsilon_2 + \Delta \qquad (13\text{-}18)$$
$$\epsilon_2 = N_1\epsilon_1/N_2 \qquad (13\text{-}1)$$
$$\theta_c = \epsilon_1 = N_2\epsilon_2/N_1 \qquad (13\text{-}19)$$

Intersection Curve at End of Hob. Referring to Fig. 13-5, let us use the same symbols as before, with the addition of the following:

B = distance to end of hob from central section, in.

β = angle at radius R_c to end of hob

x_e = abscissa of trace of end section of hob, in.

y_e = ordinate of trace of end section of hob, in.

and other symbols shown in Fig. 13-5.

We have the following from the geometrical conditions shown in Fig. 13-5:

$$\sin \beta = \frac{B}{R_c} \tag{13-20}$$

$$y_e = (C - R_c \cos \beta) \tan \epsilon_1 \tag{13-21}$$

$$r = \frac{C - R_c \cos \beta}{\cos \epsilon_1} \tag{13-22}$$

$$\tan \phi = \frac{B + R_b \cos \phi}{C - r - R_b \sin \phi}$$

Solving this last equation for $\sin \phi$, we obtain

$$\sin \phi = \frac{R_b(C - r) + B \sqrt{(C - r)^2 - (R_b^2 - B^2)}}{(C - r)^2 + B^2} \tag{13-23}$$

$$\gamma = \phi - \alpha \tag{13-24}$$

$$\Delta = \epsilon_2 + \gamma - \beta \tag{13-25}$$

$$x_e = R_c \Delta \tag{13-26}$$

The polar equation of the intersection curve of the end section of the worm with a plane perpendicular to the axis is as follows:

θ_e = vectorial angle

r_e = length of radius vector, in.

we have the following from the geometrical conditions shown in Fig. 13-5:

$$r_e = C - R_b \sin \phi - \frac{B + R_b \cos \phi}{\tan \phi} \tag{13-27}$$

$$\phi = \alpha + \beta - \epsilon_2 \tag{13-28}$$

$$\epsilon_2 = \frac{N_1 \epsilon_1}{N_2} \tag{13-1}$$

$$\theta_e = \epsilon_1 = \frac{N_2 \epsilon_2}{N_1} \tag{13-29}$$

Example of Trace of Hindley-worm Gear. Using the same example as before, we shall use a length of hob and worm of 3 in. This gives a value for B of 1.50 in.

For the trace of the central section, we have

$$\alpha = 30° \qquad C = 5.000 \qquad R_c = 4.000 \qquad R_b = 2.000$$

Using these values in the several equations, we obtain the values of the coordinates that are tabulated in Table 13-3.

For the trace of the end section, using the foregoing values in the several equations, we obtain the values of the coordinates that are tabulated in Table 13-3. All these values are plotted in Fig. 13-6.

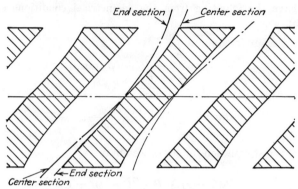

FIG. 13-6.   Trace of teeth of Hindley worm gear.

TABLE 13-3.   TRACE OF HINDLEY WORM ON INTERSECTING CYLINDER
(Plotted in Fig. 13-6)

Center section		End section	
x_c, in.	y_c, in.	x_e, in.	y_e, in.
0.89179	0.83910	1.09530	1.08402
0.66762	0.62487	0.74211	0.80726
0.47501	0.44523	0.53351	0.57518
0.30278	0.28675	0.32721	0.37045
0.14528	0.14054	0.15114	0.18156
0.00000	0.00000	0.00000	0.00000
−0.13397	−0.14054	−0.12811	−0.18156
−0.25573	−0.28675	−0.23129	−0.37045
−0.36275	−0.44523	−0.30425	−0.57518
−0.44939	−0.62487	−0.37490	−0.80726
−0.50447	−0.83910	−0.30096	−1.08402

Contact Lines on Hindley-worm Drive. The traces of the first, third, fourth, and sixth threads of the worm have been plotted with the trace of the worm-gear teeth in Fig. 13-7. With geometrically perfect conditions, the only continuous contact is along the straight-line element of the worm in its axial plane. There is an intermittent line contact, through from 25 to 30 deg of angular rotation of the worm at the central section. The form of this contact will be the form of the intersection curve at the central section with a plane perpendicular to the axis of the worm. The coordinates of this form have been calculated, and they are tabulated in Table 13-4.

When the hob and the worm are of the same length, there will also be another intermittent line contact on the end thread of the worm. The form of this line contact will be the form of the intersection curve

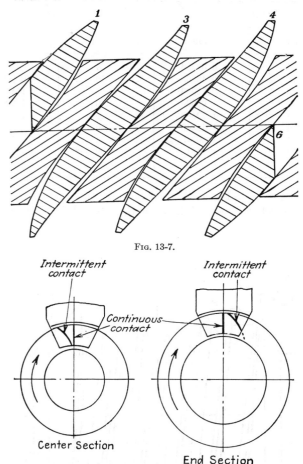

Fig. 13-7.

Fig. 13-8.   Contact of Hindley worm-gear drive.

at the end section of the worm with a plane perpendicular to its axis. The coordinates of this form are also tabulated in Table 13-4.

The contact lines are plotted in Fig. 13-8.   The straight, radial, line contact at the axial section exists on all threads of the worm.   As noted before, the curved-line contact on the central and end threads is intermittent and revolves with the worm.

TABLE 13-4. HINDLEY-WORM INTERSECTION CURVES WITH PLANES PERPENDICULAR
TO AXIS OF WORM
(Plotted in Fig. 13-8)

Center section		End section	
θ_c, deg	r_c, in.	θ_e, deg	r_e, in.
24	0.08286	24	0.77348
16	0.43765	16	0.96026
8	0.73988	8	1.13248
0	1.00000	0	1.29188
− 8	1.22585	− 8	1.43976
−16	1.42340	−16	1.57739
−24	1.59743	−24	1.70583
−32	1.75146	−32	1.82598
−40	1.88857	−40	1.93862

Referring again to Fig. 13-7, it will be noted that there is an appreciable clearance between most of the worm threads and the worm-gear teeth

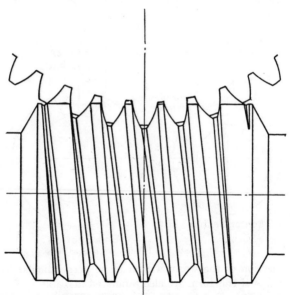

FIG. 13-9. Enveloping worm for involute spur gear.

at the sides of the worm-gear face. Many times in actual practice, considerable scraping is done on the hobbed gear so as to reduce the amount of the clearance showing at the edges of the worm gear. Under such conditions, the nature of the actual contact will be indeterminate, and the smoothness and uniformity of motion transmitted by such a scraped drive will be a matter of the circumstances existing on each individual drive.

ENVELOPING WORM FOR SPUR GEARS

As another example we will examine the contact conditions that will exist between an involute spur gear and an enveloping worm. The

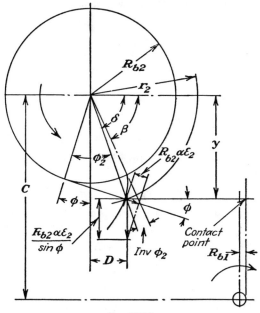

Fig. 13-10.

enveloping worm could be generated by a pinion-shaped cutter whose cutting edges represent the forms of the teeth of the spur gear. Such a drive is shown in Fig. 13-9.

The intersection profile of the spur-gear-tooth form with any plane parallel to its axis will be a straight line that is parallel to the axis of the spur gear. As the spur gear revolves, this straight line will rise or fall—depending upon the direction of rotation. The rate of motion will depend upon the position of the intersecting plane in relation to the axis of the gear, the size of the base circle of the involute spur gear, and the angular velocity of the spur gear. Referring to Fig. 13-10, when

R_{b2} = radius of base circle of spur gear, in.

D = distance of intersecting plane from axis of spur gear, in.

ϵ_1 = angle of rotation of worm

ϵ_2 = angle of rotation of gear

N_1 = number of starts on worm (generally 1)

N_2 = number of teeth in involute spur gear

ϕ = momentary pressure angle

y = distance of gear-tooth intersection from projection of axis of spur gear, in.

R_{b1} = momentary radius of base circle of worm, in.

C = center distance, in.

we have to start

$$\epsilon_1 = N_2\epsilon_2/N_1 \tag{13-30}$$

We have the following from the geometrical conditions shown in Fig. 13-10:

$$\sin \phi = (y - R_{b2} \cos \phi)/\sqrt{D^2 + y^2 - R_{b2}{}^2}$$

Solving this equation for $\sin \phi$, we obtain

$$\sin \phi = (y \sqrt{D^2 + y^2 - R_{b2}{}^2} - DR_{b2})/(D^2 + y^2) \tag{13-31}$$

As the gear is revolved in the direction shown on Fig. 13-10, we have the following relationship:

$$R_{b2} \, d\epsilon_2/\sin \phi = R_{b1} \, d\epsilon_1$$

The momentary base circle of the worm is a circle whose circumferential velocity is the same as the velocity of the contact point at y.

$$d\epsilon_1/d\epsilon_2 = N_1/N_2$$

Solving for R_{b1}, we obtain

$$R_{b1} = N_1 R_{b2}/N_2 \sin \phi \tag{13-32}$$

As indicated in Fig. 13-10, these equations give us the position of a contact point for any given value of y and D. When the intersecting plane is on the opposite side of the axis of the spur gear, the value of D will be minus. The contact point, in respect to the axis of the worm, will be at a distance $C - y$ above the worm axis and at a distance of R_{b1} to the side of it, as indicated in the figure. By using a series of values of y for a given value of D, we obtain a series of points on the path of contact of this drive on the given intersection plane. By determining the path of contact on several intersecting planes we can establish the projection of the field of contact for this drive.

Contact Lines on Teeth of Spur Gear. Using this same general method of analysis, we can determine the projections of the actual contact lines on the several teeth of the involute spur gear that may be engaged. Referring again to Fig. 13-10, we shall use the same symbols as before with the addition of the following:

β = position of origin of involute gear-tooth profile

ϕ_2 = pressure angle at radius r_2

r_2 = length of radius vector of involute gear-tooth profile, in.

δ = angular position of radius vector

We have the following from the geometrical conditions shown in Fig. 13-10:

$$\cos \phi_2 = R_{b2}/r_2 \qquad (13\text{-}33)$$
$$\delta = \beta + \text{inv } \phi_2 \qquad (13\text{-}34)$$
$$y = r_2 \sin \delta \qquad (13\text{-}35)$$
$$D = r_2 \cos \delta \qquad (13\text{-}36)$$
$$\phi = \phi_2 + \delta - (\pi/2) \qquad (13\text{-}37)$$
$$R_{b1} = N_1 R_{b2}/N_2 \sin \phi \qquad (13\text{-}32)$$

A series of values of r_2 for each tooth engaged with the enveloping worm would be used, and the corresponding values of R_{b1}, $C - y$, and D would be determined. With these values, the contact lines and the field of contact could be plotted.

TABLE 13-5

	Values of r_2, in.				
	4.750	4.875	5.000	5.125	5.250
First tooth					
R_{b1}, in.	....	....	....	1.74532	0.90458
y, in.	....	....	....	4.82555	4.96283
D, in.	....	....	....	1.72620	1.71255
Second tooth					
R_{b1}	....	1.79630	0.77215	0.52779	0.41452
y	....	4.77341	4.90395	5.03614	5.16962
D	....	0.99002	0.97545	0.95002	0.91508
Third tooth					
R_{b1}, in.	1.24811	0.53226	0.38529	0.31548	0.27322
y, in.	4.74326	4.86954	4.99665	5.12280	5.24921
D, in.	0.25256	0.23108	0.19630	0.15052	0.09529
Fourth tooth					
R_{b1}, in.	0.47231	0.31700	0.26096	0.22900	0.20758
y, in.	4.72426	4.84570	4.96535	5.08328	5.19939
D, in.	-0.49253	-0.53352	-0.58770	-0.65272	-0.72723
Fifth tooth					
R_{b1}, in.	0.29576	0.22974	0.20104	0.18327	0.17076
y, in.	4.58879	4.70262	4.81230	4.91857	5.02163
D, in.	-1.22555	-1.28500	-1.35720	-1.43987	-1.53164

Example of Contact Lines on Enveloping Worm for Spur Gear. As a definite example we shall use a 40-tooth spur gear that will mesh with a single-thread enveloping worm. This gear will have the following values:

                                                                    Inches
Outside radius of spur gear................................. 5.250
Root radius of spur gear................................... 4.7109
Radius of base circle of spur gear......................... 4.69846
Face width of spur gear.................................... 2.500
Center distance... 6.750
Length of enveloping worm................................. 5.000
Maximum outside radius of worm........................... 2.000

Using the foregoing values and the equations for the contact lines on the teeth of the spur gear, we obtain the values given in Table 13-5.

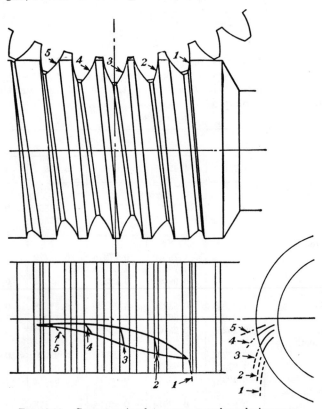

FIG. 13-11.   Contact on involute spur gear and enveloping worm.

These values, together with the projection of the field of contact on the involute spur gear, are plotted in Fig. 13-11.

WILDHABER-WORM DRIVE

As another example of an hourglass-worm drive, we shall examine the contact conditions on the Wildhaber-worm drive. This design was developed by Ernest Widhaber about 1922, primarily as an accurate index worm. It has since been used successfully as a power drive. This drive consists of an index plate or gear with straight, plane gashes or notches and a mating and enveloping worm. This design of worm and gear has the following advantages:

1. The form of the gear is extremely simple, and it can be readily and accurately produced.

2. It meshes with straight-line contact, and the contact can be extended over several teeth.

3. The production of both the worm and the gear is simple and direct.

4. No hobs or other special cutting tools are required for different sizes and designs once the initial manufacturing equipment is set up.

5. Both the worm and the gear can be ground in a simple manner. This makes possible the use of hardened steel for both members to carry heavy static and low-speed loads.

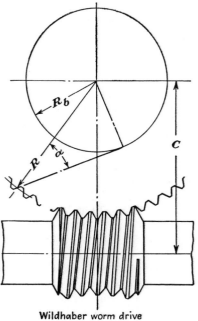

Wildhaber worm drive
Fig. 13-12.

6. The gear can be split and adjusted to take up backlash when used for accurate indexing purposes without affecting the accuracy or extent of the contact.

7. The tooth forms of both the worm and the gear can be accurately and simply measured, thus permitting the production of fully interchangeable parts to any measurable degree of accuracy.

8. The correct assembled positions of the worm and the gear can be determined by a simple visual inspection, thus ensuring proper contact conditions when the drive is assembled.

This worm-gear drive is shown in Fig. 13-12. The gear, as noted before, consists of a disk with straight notches cut into its rim. The worm is generated from a flat-sided milling cutter or grinding wheel, so positioned as to represent the side of the notch in the rim of the gear. Instead of generating the worm from a straight line, it is generated from a plane. Referring to Fig. 13-12, let

R = radius on gear to any point of tooth profile, in.

α = angle of tooth at R with radial line of gear

R_b = radius of base cylinder of gear to which plane sides of teeth are tangent, in.

The plane tooth flanks of the gear are parallel to the axis of the gear. These planes are also tangent to the base cylinder. Whence

$$R_b = R \sin \alpha \qquad (13\text{-}38)$$

With a given form for the gear teeth, the form of the worm threads are controlled by the gear-tooth forms. Here the worm is a barrel-shaped cam with the straight, plane teeth of the gear as successive followers. The cam or thread form of the worm is the surface that is generated by the different successive positions of the plane that forms the tooth surface of the gear. In other words, these cam surfaces are those which are enveloped by the successive positions of the plane tooth flanks of the gear teeth.

The worm, so developed, is not composed of helices. Its outlines are not cylindrical but follow the contour of the mating gear.

Contact on Wildhaber-worm Drive. The contact between the two members of this drive can be analyzed by the same method as has been used for the enveloping worm of a spur gear. However, as these gear-tooth surfaces are planes, a more direct analysis is possible.

The tooth surfaces of the worm are those which are enveloped by the infinitely large number of different positions of the tooth flanks of the gear. The contact line between the worm and gear is therefore that curve which belongs to any given position of the gear-tooth surface and to the infinitely close successive position of it. In other words, the contact line is the line that is common to two infinitely close successive positions of the gear-tooth flank relative to the worm, and is therefore the limiting case of the intersection of the two successive positions of the gear-tooth flanks relative to the worm.

In this case, the tooth surfaces of the gear are planes, which are tangential planes to the thread surface of the worm. The contact line is the intersection line of two infinitely close positions of these tangential planes. The intersection line of two planes is always a straight line. Therefore the contact line between this gear and its worm is always a straight line.

Referring to Fig. 13-13, let

C = center distance, in.

R_b = radius of base cylinder, in.

N_1 = number of starts on worm

N_2 = number of teeth in gear

ϕ = momentary pressure angle of gear

The contact line at position A is the projection, on a plane perpendicular to the axis of the worm, of the intersection line of this position A with the infinitely close position represented by B. As the gear has turned through an angle of $\Delta\phi$, the tangential plane has turned about the axis of the worm an angle of $(N_2/N_1)\Delta\phi$.

In order to find the intersection line of the planes in the positions A and B (called for brevity *planes A and B*), we shall first determine the

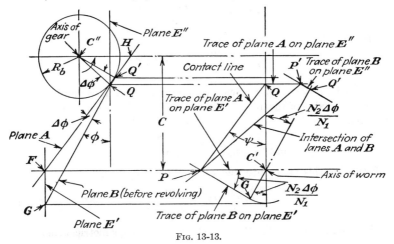

Fig. 13-13.

intersection lines of these planes with a plane E', which is perpendicular to the axis of the worm. These intersection lines are the lines PC' and PG, which intersect at P. The point P, which is a common point of the two tangential planes A and B, is therefore one point on the straight contact line between the gear and the worm.

When we determine a second point of the contact line, we shall establish the contact line by joining the two points. We shall next determine the intersection lines of the planes A and B with the plane E'', which passes through Q and which is also perpendicular to the axis of the worm. These intersection lines are the lines QP' and $Q'P'$, which intersect at P'. The point P' is therefore the second point of the line that represents the intersection of the two planes A and B. The projection of this intersection line on a plane perpendicular to the axis of the worm is the line PP'.

We shall now determine the actual position of this contact line so as to be able to determine its position for any definite example.   We shall first determine the distance PC'.   From the geometrical conditions shown in Fig. 13-13 we have

$$PC' = \frac{C'G}{\sin (N_2 \,\Delta\phi/N_1)}$$

Consider the triangle FGQ, where

$$\text{Angle } FQG = \Delta\phi$$

and

$$\text{Angle } FGQ = \phi$$

$$C'G = FG = \frac{FQ \sin \Delta\phi}{\sin \phi}$$

From these equations we get

$$PC' = \frac{FQ \sin \Delta\phi/\sin \phi}{\sin (N_2 \,\Delta\phi/N_1)}$$

Consider now the limiting case where the value of $\Delta\phi$ is infinitely small so that the value of the arc and the value of the sine are identical.   Here we have

$$\frac{\sin \Delta\phi}{\sin (N_2 \,\Delta\phi/N_1)} = \frac{\Delta\phi}{N_2 \,\Delta\phi/N_1} = \frac{N_1}{N_2}$$

Substituting this value into the equation for PC', we get

$$PC' = (FQ) \frac{N_1}{N_2 \sin \phi} \tag{13-39}$$

Referring again to Fig. 13-13, we have from the geometrical conditions there

$$FQ = FH - QH = \frac{C}{\cos \phi} - R_b \tan \phi = \frac{C - R_b \sin \phi}{\cos \phi}$$

Substituting this value into Eq. 13-39, we obtain

$$PC' = \frac{N_1(C - R_b \sin \phi)}{N_2 \sin \phi \cos \phi} = \frac{2N_1(C - R_b \sin \phi)}{N_2 \sin 2\phi} \tag{13-40}$$

We shall now determine the length of the line $P'Q$.   This length depends upon the value of $\Delta\phi$.   In the limiting case the value of $\Delta\phi$ is infinitely small; hence

$$\Delta\phi = 0 \quad \text{and} \quad P'Q = 0$$

and the point Q' coincides with the point P'.

The projection of the actual contact line is therefore the line PQ, which is at the angle ψ from the y axis. The ordinate of point Q is

$$y = QC' = C - R_b \sin \phi \tag{13-41}$$

The abscissa of the point P will be x; hence

$$x = PC' = \frac{2N_1(C - R_b \sin \phi)}{N_2 \sin 2\phi} \tag{13-40}$$

$$\tan \psi = \frac{x}{y} = \frac{2N_1}{N_2} \sin 2\phi \tag{13-42}$$

Summary of Contact Analysis. We have now established the position of the contact lines for this drive. To summarize, we have when

C = center distance, in.
R_b = radius of base cylinder of gear, in.
N_1 = number of starts on worm
N_2 = number of teeth in gear
ϕ = momentary pressure angle of gear
R = radius on gear to tooth profile, in.
α = angle of gear tooth at R with radial line
x = abscissa of contact line (when y is zero), in.
y = ordinate of contact line (when x is zero), in.
ψ = angle of contact line

$$R_b = R \sin \alpha \tag{13-38}$$
$$x = 2N_1(C - R_b \sin \phi)/N_2 \sin 2\phi \tag{13-40}$$
$$y = C - R_b \sin \phi \tag{13-41}$$
$$\tan \psi = x/y = 2N_1/N_2 \sin 2\phi \tag{13-42}$$

The contact lines on every tooth of the mesh can be determined from these equations. In every case, the contact line at the beginning of mesh where the momentary pressure angle is least should be determined. If this minimum angle is too small, the theoretical initial contact will be outside of the meshing surfaces of the worm and gear, which would introduce edge contact at the beginning of mesh. The minimum width of face for the gear is determined from a layout of the contact line at the beginning of mesh. It should always lie on the contacting surfaces.

Example of Wildhaber-worm Drive. As a definite example we shall examine the contact conditions on a Wildhaber-worm drive that corresponds to the example used for the enveloping worm for an involute spur gear. The angle of the teeth will be selected so that opposite flanks of the gear teeth at some definite number of tooth spaces apart will be parallel. This will make possible a very simple method of measuring these teeth. To accomplish this, the included angle of the gear-tooth flanks must be equal to some even multiple of the angle between successive teeth.

This drive will have the following values:

$$N_1 = 1 \qquad N_2 = 40 \qquad C = 6.750 \qquad R = 5.000$$

The angle between successive teeth in this example is equal to $^{360}\!/_{40}$ or 9 deg. Hence the included angle of the gear tooth will be some multiple of this, say, 54 deg. The angle of the tooth flank with a radial line at the middle of the gear tooth will be

$$\alpha = (54°/2) + (9°/4) = 29.25° \qquad \sin \alpha = 0.48862$$
$$R_b = 5.000 \times 0.48862 = 2.4431$$

We will use the following values:

Tooth	ϕ, deg
First	9
Second	18
Third	27
Fourth	36
Fifth	45

Using these values in the foregoing equations, we obtain the following values for the coordinates of the several contact lines:

Tooth	x, in.	y, in.
First	1.03032	6.36783
Second	0.50996	5.99503
Third	0.34862	5.64086
Fourth	0.27937	5.31397
Fifth	0.25112	5.02246

These values are plotted in Fig. 13-14.

The radius on the gear to the sharp point of the tooth will be equal to the base radius divided by the sine of one-half the included angle of the tooth. Thus when

r_1 = radius to sharp point of gear tooth, in.

$$r_1 = \frac{2.4431}{\sin 27°} = 5.381$$

The radius on the gear to the sharp root of the tooth space will be equal to the value of the base-circle radius divided by the sine of one-half the included angle of the tooth space. The value of the included angle of the tooth space is equal to the sum of the included angle of the tooth and the angle between successive teeth. Thus when

r_2 = radius to sharp root of tooth space, in.

$$r_2 = \frac{2.4431}{\sin 31.50°} = 4.675$$

We shall use a truncated form for the gear tooth such that

$$R_0 = 5.250 \qquad R_r = 4.750 \qquad \text{Clearance} = 0.050$$

From the layout in Fig. 13-14, we find that the minimum face width for the gear in this example is 2 in. The contact on one set of thread flanks is all on one side of the center line of the worm. The contact on the opposite flanks of the thread surfaces is the same but reversed in position, and is all on the opposite side of the center line of the worm. Thus a split worm gear can be used to take up backlash, and would be

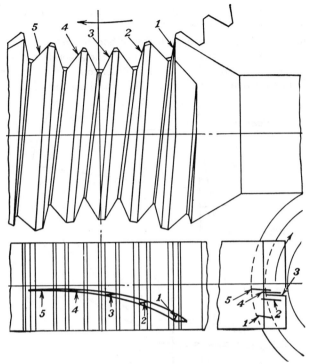

FIG. 13-14. Contact on Wildhaber worm-gear drive.

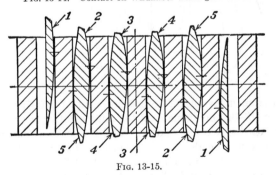

FIG. 13-15.

adjusted to the "high" side of the worm threads, or to the side that makes contact with a line that is at right angles to the projection of the worm axis.

The nature of this contact will be more apparent if we consider the traces of the worm and gear on a cylinder that is concentric with the gear and passes through the teeth of this drive. Such a concentric cylinder was used in the analysis of the contact on a Hindley-worm drive. These contact conditions on the developed surface of such a concentric cylinder are shown in Fig. 13-15.

Setting of Milling Cutter or Grinding Wheel for Worm. The milling cutter or grinding wheel that is used to finish the generation of the worm must be set off the center of the worm in order to cover the extreme contact and thus prevent interference. The actual setting of the wheel is

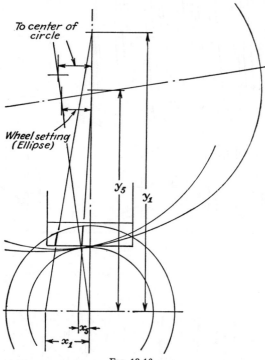

Fɪɢ. 13-16.

determined from a layout. The extreme contact lines are plotted as shown in Fig. 13-16; the diameter of the milling cutter or grinding wheel that is to be used is selected, and the wheel is located so that its circumference will cover the ends of the extreme contact lines. Generally the use of the arc of the circle is adequate, and the difference, because of the actual tipping of the wheel, is slight. When the clearance is small, however, the form of the projected ellipse of the wheel should be used. When the finishing cut is made by a grinding wheel, it is a good practice to provide an undercut root for grinding clearance. In any event, the off-center position of the wheel is established from the layout.

CHAPTER 14

CONJUGATE TOOTH ACTION ON BEVEL GEARS

When gears are used to drive shafts whose axes intersect, the pitch surfaces are cones. The point where the apexes of the two pitch cones intersect or meet is the intersection point of the two axes. This point is called the *cone center*.

The line of tangency between the two pitch cones is the locus of all pitch points of the contacting gear teeth. At any specific point along this line of tangency between the pitch cones, the diameters of the cones are directly proportional to the numbers of teeth in the gears. Referring to Fig. 14-1, when

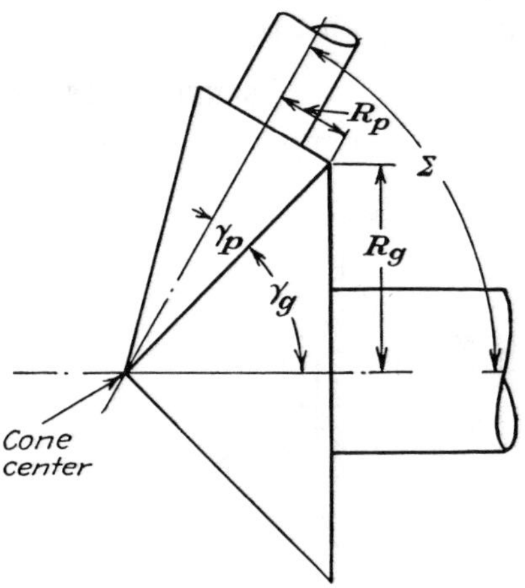

R_p = pitch radius of pinion pitch cone at large end, in.

R_g = pitch radius of gear pitch cone at large end, in.

γ_p = pitch angle of bevel pinion

γ_g = pitch angle of bevel gear

Σ = shaft angle

N_p = number of teeth in bevel pinion

N_g = number of teeth in bevel gear

Fig. 14-1.

$$\Sigma = \gamma_p + \gamma_g \tag{14-1}$$

$$\tan \gamma_p = \frac{\sin \Sigma}{(N_g/N_p) + \cos \Sigma} \tag{14-2}$$

When $\Sigma = 90°$, this equation reduces to

$$\tan \gamma_p = N_p/N_g \tag{14-3}$$

$$\tan \gamma_g = N_g/N_p \tag{14-4}$$

When P is the diametral pitch at the large end of the pitch cone,

301

$$R_p = \frac{N_p}{2P} \qquad (14\text{-}5)$$

$$R_g = \frac{N_g}{2P} \qquad (14\text{-}6)$$

There are many possible designs for bevel gears. Practically, the choice is restricted to such designs as can be made on available equipment. The forms may be symmetrical in relation to the pitch cone, or they may be formed without reference to the pitch cones. When a gear of any form is mounted on one of two intersecting shafts, the normal from the point of contact between it and its conjugate gear must pass through a pitch point that lies in the line of tangency of the two pitch cones.

LANTERN PINION AND FACE GEAR

The lantern pinion and face gear is probably the earliest form of a

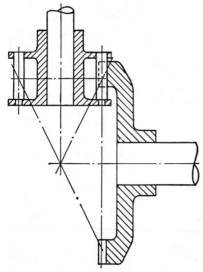

bevel-gear drive. Such a drive is shown in Fig. 14-2. The pinion consists of a number of cylindrical pins equally spaced on a circle concentric with the axis of the pinion. These pins are mounted in flanges. The gear may also have cooperating formed pins, or it may have shaped teeth to mesh with the cylindrical pins of the lantern pinion. In this last case, when the pins are of appreciable size, the teeth of the face gear can be generated by an end mill, which is moved, in relation to the motion of the face gear, in the same manner as the movement of the cylindrical pins in the lantern pinion in respect to the face gear.

We shall determine the contact conditions and the forms of

FIG. 14-2. Lantern pinion and face gear.

the teeth of the face gear for this last type of drive. Referring to Fig. 14-3, let

R_1 = radius to center of pins in lantern pinion, also nominal pitch radius of lantern pinion, in.

R_2 = nominal pitch radius of face gear, in.

A = radius of pins in lantern pinion, in.

ϕ = momentary pressure angle of operation
F = distance of pitch point from pinion nominal pitch circle, in.
G = height to pitch point and to contact point, in.
E = distance from contact point to tangent of pinion pitch circle, in.
r_2 = radius to contact point on face gear, in.
δ = angle to contact point on face gear
ϵ_1 = angle of rotation of lantern pinion
ϵ_2 = angle of rotation of face gear
γ_p = pitch-cone angle of lantern pinion
θ_2 = vectorial angle of face-gear-tooth form

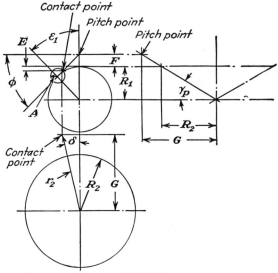

FIG. 14-3.

The pitch surfaces of this drive are pitch cones, the same as for other bevel gears. The angles of these pitch cones are determined from the tooth numbers of the gears, or from their nominal pitch radii, exactly as for other bevel gears.

In order to establish the contact points between the pins in the lantern pinion and the teeth of the face gear, we will use a series of values of E with a series of values of ϵ_1 and ϵ_2 for each value of E, as indicated in Fig. 14-3. The position of the contact point in the plan view of the lantern pinion will be the point where the trace of a plane at E intersects the circumference of the pin. The line of action will be a radial line of the pin that passes through this contact point. The point at the distance

F, where this line of action intersects the center line of the pinion, is the projection of the pitch point for this specific contact point. This pitch point is on the common tangent element of the two pitch cones at a distance G above the axis of the face gear. This tangent element of the two pitch cones is the locus of all pitch points. The line of action is perpendicular to the axis of the pin of the lantern pinion.

Referring again to Fig. 14-3, we have the following from the geometrical conditions shown there:

$$\sin \phi = [R_1(1 - \cos \epsilon_1) - E]/A \tag{14-7}$$
$$F = R_1(\cos \epsilon_1 + \sin \epsilon_1 \tan \phi - 1) \tag{14-8}$$
$$G = R_2 + F \cot \gamma_p \tag{14-9}$$
$$\tan \delta = (R_1 \sin \epsilon_1 - A \cos \phi)/G \tag{14-10}$$
$$r_2 = G/\cos \delta \tag{14-11}$$
$$\theta_2 = \delta - \epsilon_2 \tag{14-12}$$

Example of Lantern Pinion and Face Gear. As a definite example we shall determine the contact conditions and the form of the face-gear teeth for the following drive:

Number of pins in lantern pinion........................... 18
Number of teeth in face gear............................... 36
Nominal pitch radius of lantern pinion..................... 5.000
Nominal pitch radius of face gear.......................... 10.000
Radius of pins in lantern pinion........................... 0.400

Whence

$$R_1 = 5.000 \qquad R_2 = 10.000 \qquad A = 0.400$$
$$\cot \gamma_p = {}^{36}\!/_{18} = 2.000$$

Using a series of values of E, ϵ_1, and ϵ_2, we obtain from the foregoing equations the values that are tabulated in Table 14-1 These values are also plotted in Fig. 14-4.

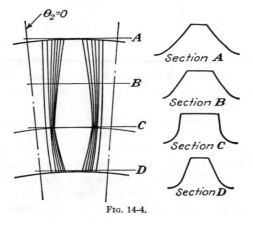

FIG. 14-4.

TABLE 14-1. COORDINATES OF TEETH OF FACE GEAR
(Plotted in Fig. 14-4)

ϵ_1, deg	ϕ, deg	r_2, in.	θ_2, deg	G, in.
		$E = -0.30$		
−10	70.030	5.1679	−6.212	5.0693
−5	52.904	8.8480	−1.856	8.8224
0	48.590	10.0035	−1.516	10.0000
5	52.904	11.1031	−1.521	11.1014
10	70.030	14.6452	−2.136	14.6269
		$E = -0.20$		
−15	67.800	3.6183	−16.043	3.3171
−10	43.541	8.2793	−3.042	8.1979
−5	33.204	9.4290	−2.157	9.3979
0	30.000	10.0060	−1.984	10.0000
5	33.204	10.5264	−1.952	10.5259
10	43.541	11.5128	−2.121	11.4983
15	67.800	16.0422	−3.414	16.0015
		$E = -0.10$		
−15	42.523	7.4569	−4.802	7.2857
−10	26.096	9.0809	−2.768	8.9975
−5	17.315	9.7273	−2.293	9.6933
0	14.477	10.0075	−2.218	10.0000
5	17.315	10.2306	−2.225	10.2305
10	26.096	10.7108	−2.275	10.6968
15	42.523	12.0743	−2.753	12.0329
		$E = 0.000$		
−15	25.206	7.6231	−5.046	7.4018
−10	10.945	9.5955	−2.551	9.5123
−5	2.730	9.9555	−2.285	9.9208
0	0.000	10.0080	−2.291	10.0000
5	2.730	10.0030	−2.321	10.0030
10	10.945	10.1950	−2.327	10.1839
15	25.206	10.9173	−2.602	10.8775
		$E = 0.100$		
−20	30.257	7.6820	−5.521	7.4018
−15	10.129	9.3510	−2.899	9.1974
−12	1.325	9.8393	−2.413	9.7334
12	1.325	9.8504	−2.281	9.8296
15	10.129	10.1611	−2.417	10.1212
20	30.257	11.4735	−3.170	11.3920
25	67.090	19.1633	−6.637	19.0631
		$E = 0.200$		
−25	42.154	5.7648	−12.203	5.2372
−20	14.707	8.7541	−3.859	8.4992
-17	2.651	9.6096	−2.669	9.4276
17	2.651	9.7563	−2.249	9.6984
20	14.707	10.3793	−2.676	10.2946
25	42.154	13.0164	−4.477	12.8890
		$E = 0.300$		
−25	24.969	7.5147	−6.736	7.0952
−20	0.222	9.6180	-2.673	9.3837
20	0.222	9.5007	−2.074	9.4101
25	24.969	11.1690	−3.484	11.0310

Figure 14-4 shows a series of outlines of the face-gear teeth at different heights. In effect, it is a contour map of the tooth. From this layout, the forms of the different sections of the gear teeth have been determined graphically.

All the sectional contours for the sections where the value of E is plus are two intersecting curves. One of these curves is that of the approach action, while the other is that of the recess action. Contact on these upper sections does not exist when the pin is in its central position, but starts at some distance on either side of this central position. The

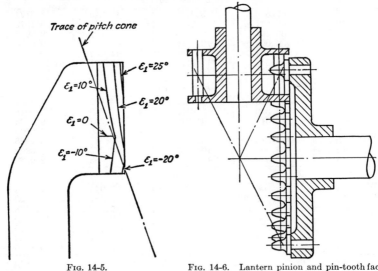

Fig. 14-5. Fig. 14-6. Lantern pinion and pin-tooth face gear.

start of this contact will be the minimum value of ϵ_1. Using the same symbols as before, we have

$$\cos \epsilon_1 \text{ (min)} = (R_1 - E)/R_1 \qquad (14\text{-}13)$$

Some of this theoretical contact on these upper sections is lost as part of the end of the approach action is cut away by the contour for the start of the recess action. The actual intersection of the two contour curves is determined graphically.

The point where the contact ceases for all positions of the several sections is given by the following equation for the maximum value of ϵ_1:

$$\cos \epsilon_1 \text{ (max)} = (R_1 - E - A)/R_1 \qquad (14\text{-}14)$$

In Fig. 14-5, the projection of the actual contact lines between the pin of the lantern pinion and the teeth of the face gear are shown for several angular positions of the lantern pinion. The coordinates of the projection of these contact lines are given by the values of G and E for the same value of ϵ_1.

LANTERN PINION AND PIN-TOOTH FACE GEAR

As noted before, the face gear that meshes with a lantern pinion may have teeth consisting of formed pins, as indicated in Fig. 14-6. The load capacity of such a drive is very small because only point contact can exist between the mating pins or teeth.

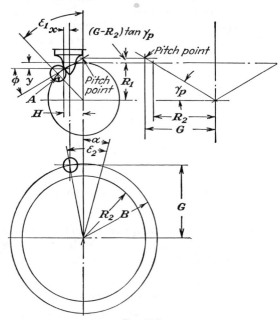

Fig. 14-7.

We shall determine the contact conditions and the form of the pins in the face gear for this type of drive. Referring to Fig. 14-7, let

R_1 = radius to center of pins in lantern pinion, also nominal pitch radius of lantern pinion, in.

R_2 = nominal pitch radius of face gear, in.

A = radius of pins in lantern pinion, in.

ϕ = momentary pressure angle

ϵ_1 = angle of rotation of lantern pinion

ϵ_2 = angle of rotation of face gear

γ_p = pitch-cone angle of lantern pinion

B = radius to center of pins in face gear, in.

α = original angular position of pin in face gear

N_1 = number of teeth or pins in lantern pinion

N_2 = number of teeth or pins in face gear

x = radius of form of pin in face gear, in.

y = height to radius x of pin in face gear, in.

The pitch surfaces of this drive are cones, just as for any other type of bevel gear. The line of action must be a radial line of the pin in the lantern pinion and be perpendicular to the axis of this pin. This line of action must also pass through the axis of the pin in the face gear. Referring again to Fig. 14-7, where several additional symbols are shown for the purpose of establishing the equations, we have the following from the geometrical conditions shown there:

$$\alpha = \frac{180°}{N_2} \qquad (14\text{-}15)$$

$$G = B \cos(\epsilon_2 - \alpha)$$

$$\tan\phi = \frac{G \tan\gamma_p - R_1 \cos\epsilon_1}{R_1 \sin\epsilon_1}$$

$$\tan\phi = \frac{B \tan\gamma_p \cos(\epsilon_2 - \alpha) - R_1 \cos\epsilon_1}{R_1 \sin\epsilon_1} \qquad (14\text{-}16)$$

$$H = R_1 \sin\epsilon_1 - A \cos\phi$$

$$x = R_1 \sin\epsilon_1 - A \cos\phi - B \sin(\epsilon_2 - \alpha) \qquad (14\text{-}17)$$

$$y = R_1 (1 - \cos\epsilon_1) - A \sin\phi \qquad (14\text{-}18)$$

We shall determine by trial whether or not anything can be gained by making the radius to the center of the pins in the face gear greater, less than, or equal to the nominal pitch radius of the face gear.

First Example of Pin-tooth Face Gear. As the first example we shall use the same values as were used in the preceding example. Thence we have

$$N_1 = 18 \qquad N_2 = 36 \qquad R_1 = 5.000 \qquad R_2 = 10.000 \qquad A = 0.400$$

$$\tan\alpha_p = \frac{18}{36} = 0.500 \qquad \alpha = \frac{180°}{36} = 5°$$

For this first example we shall use $B = 10.50$.

Using these values and the preceding equations, we obtain the values tabulated in Table 14-2. These values are also plotted in Fig. 14-8, and show the form of the pins for the face gear.

When the pins in the face gear are located outside the nominal pitch circle of the gear, and the lantern pinion is the driver, all the action is approach action. The curve for the recess action where the values of ϵ_1 are minus, lies outside the curve for the approach action. This recess curve also has a cusp. Such a drive would be better for a speed-up drive

TABLE 14-2. COORDINATES OF FACE-GEAR PIN TEETH
(Plotted in Fig. 14-8)
$B = 10.50$

ϵ_1, deg	ϕ, deg	x, in.	y, in.
	Approach action		
40	21.112	0.1232	1.0257
35	19.752	0.2188	0.7691
30	18.575	0.2975	0.5424
25	17.679	0.3614	0.3476
20	17.267	0.4233	0.1828
15	17.795	0.4552	0.0481
10	20.577	0.4938	−0.0646
5	31.503	0.5478	−0.1901
0	90.000	0.9048	−0.4000
	Recess action		
0	−90.000	0.9048	0.4000
− 5	−27.482	0.5848	0.2036
−10	−15.830	0.5702	0.1851
−15	−12.880	0.5886	0.2595
−20	−12.296	0.6167	0.3867
−25	−12.682	0.6541	0.5563
−30	−13.565	0.7024	0.7637
−35	−14.742	0.7634	1.0060
−40	−16.105	0.8393	1.2808

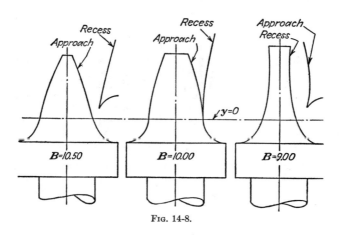

FIG. 14-8.

with the face gear driving the lantern pinion.   In such a case, all the
action would be recess action.

Second Example of Pin-tooth Face Gear.

For the second example we shall use the
same values as before except for the location of the pins in the face gear.   In this
example we will locate the pins at the nominal pitch radius of the face gear, so that
$B = 10.000$.

With these values and the preceding equations, we obtain the values tabulated in
Table 14-3 and plotted in Fig. 14-8.   In this case, when the lantern pinion is the
driving member, all the action again is approach action.   The curve of the recess
action is outside the curve for the approach action.   There is no cusp here, however.
To obtain recess action, the pins of the face gear must be located inside of the nominal
pitch circle of the face gear.

TABLE 14-3.   COORDINATES OF FACE-GEAR PIN TEETH
(Plotted in Fig. 14-8)
$B = 10.00$

ϵ_1, deg	ϕ, deg	x, in.	y, in.
Approach action			
40	17.274	0.2438	1.0510
35	15.322	0.3177	0.7985
30	13.364	0.3743	0.5774
25	11.389	0.4157	0.3895
20	9.380	0.4537	0.2364
15	7.292	0.4611	0.1196
10	4.999	0.4698	0.0411
5	1.901	0.4673	0.0058
0	90.000	0.8617	−0.4000
Recess action			
0	90.000	0.8617	−0.4000
− 5	3.155	0.4751	−0.0030
−10	0.000	0.4682	0.0759
−15	− 2.286	0.4706	0.1863
−20	− 4.387	0.4793	0.3321
−25	− 6.395	0.4965	0.5130
−30	− 8.381	0.5245	0.7278
−35	−10.348	0.5654	0.9761
−40	−12.310	0.6214	1.2551

Third Example of Pin-tooth Face Gear.

In order to have recess action, the pres-
sure angle for this part of the action must start with a plus value.   This is obtained
by locating the pins in the face gear on a circle smaller than the nominal pitch circle
of the face gear.   This brings the pitch point farther down on the pitch cones and
inside the projection of the nominal pitch circle of the pins in the lantern pinion.   We

Table 14-4. Coordinates of Face-gear Pin Teeth
(Plotted in Fig. 14-8)
$B = 9.00$

ϵ_1, deg	ϕ, deg	x, in.	y, in.
	Approach action		
40	9.129	0.4896	1.1063
35	5.925	0.5221	0.8630
30	2.325	0.5375	0.6536
25	− 1.899	0.5385	0.4817
20	− 7.185	0.5377	0.3516
15	−14.469	0.5142	0.2703
10	−26.031	0.5086	0.2515
5	−48.398	0.5578	0.3182
0	−90.000	0.7755	0.4000
	Recess action		
0	90.000	0.7755	−0.4000
− 5	50.328	0.4886	−0.2856
−10	29.559	0.3467	−0.1214
−15	18.621	0.2758	+0.0426
−20	11.623	0.2275	0.2210
−25	6.475	0.1958	0.4233
−30	2.326	0.1785	0.6536
−35	−1.233	0.1763	0.9129
−40	−4.416	0.1908	1.2006

shall therefore make the value of B equal to 9.00 for this example, with all other values the same as before.

With these values we obtain the values tabulated in Table 14-4 and plotted in Fig. 14-8. Here the contact is all recess action when the lantern pinion is the driver. In this case the approach action is outside the form of the pins in the face gear, and is therefore imaginary. The form for this approach action has a cusp.

From these examples we see that if we wish to use the more favorable recess action when the lantern pinion is the driving member, the pins in the face gear must be located inside the nominal pitch circle of the face gear. When the face gear is the driving member, then to secure recess action, these pins must be located at or outside the nominal pitch circle of the face gear.

The distance at which these pins can be located inside or outside the nominal pitch circle of the face gear is limited however. The diameter of these pins will be greatest when they are located on the nominal pitch circle of the face gear. This pin diameter is reduced as the location is shifted either way. This condition limits the distance to which they can be moved.

If we wish to have both approach and recess action on one of these drives, then we must use two sets of pins in the face gear, one set at or outside the nominal pitch circle and the second set inside the nominal pitch circle of the face gear.

THE FELLOWS FACE-GEAR DRIVE

Another design for bevel gears is the Fellows spur-pinion-and-face-gear drive. This drive consists of an involute spur pinion meshing with a face gear that is generated by a pinion-shaped cutter of substantially the same size and form as the mating spur pinion. Such a drive is shown in Fig. 14-9.

The pitch surfaces here are cones, just as for all other types of bevel-gear drives. The momentary pressure angle of the drive changes with a change in the distance from the center of the face gear. It is constant for any specific distance from the center of the face gear. It is reduced to zero at the distance from the center of the face gear where the radius of the pitch cone of the pinion is the same as the radius of the base circle of the involute spur gear. It is equal to the nominal pressure angle of the spur gear at the distance from the center of the face gear where the radius of the pitch cone of the pinion is equal to the nominal pitch radius of the involute spur gear.

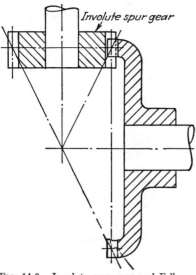

Involute spur gear

FIG. 14-9. Involute spur gear and Fellows face gear.

The active face width of the face gear cannot extend below the point where the momentary pressure angle is equal to zero. In fact, it cannot extend to this point because the trochoidal fillet of the tip of the cutter tooth will cut away the conjugate face-gear-tooth profile before this point is reached. Hence very little of the tooth inside the nominal pitch circle of the face gear is an effective part of the tooth profile.

A simple approximation to the tooth form of the face gear can be obtained by treating the face-gear tooth as a rack of varying pressure angle and circular pitch at different distances from the center of the face gear. Such an approximation will be developed first.

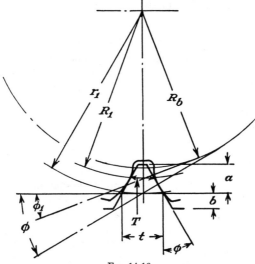

Fig. 14-10.

Approximation for Face-gear-tooth Form.   Referring to Fig. 14-10, let

N_1 = number of teeth in involute spur gear

N_2 = number of teeth in face gear

R_1 = nominal pitch radius of involute spur gear, also nominal pitch radius of bevel pinion, in.

R_2 = nominal pitch radius of face gear, in.

r_1 = momentary pitch radius of bevel pinion, in.

r_2 = momentary pitch radius of face gear, in.

R_r = root radius of involute spur gear, in.

R_b = radius of base circle of involute spur gear, in.

R_{oc} = outside radius of pinion-shaped cutter, in.

ϕ = momentary pressure angle of involute spur gear and pressure angle of rack

ϕ_1 = nominal pressure angle of involute spur gear

γ_p = pitch-cone angle of pinion

T = arc space width of involute spur gear at R_1, in.

t = arc space width of involute spur gear at r_1, also tooth thickness of rack, in.

a = addendum of rack, in.

b = dedendum of rack, in.
c = clearance, in.

Referring again to Fig. 14-10, we have

$$r_1 = r_2 \tan \gamma_p \qquad (14\text{-}19)$$
$$\cos \phi_1 = R_b/r_1 \qquad (14\text{-}20)$$

Solving for the arc space width of the pinion at r_1 by the same method that was used for the development of Eq. (5-5), we obtain

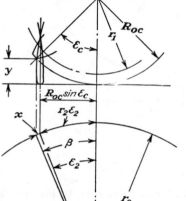

$$t = 2r_1[(T/2R_1) - \text{inv } \phi_1 + \text{inv } \phi] \qquad (14\text{-}21)$$
$$a = r_1 - R_r - c \qquad (14\text{-}22)$$
$$b = R_{oc} - r_1 \qquad (14\text{-}23)$$

Before solving any definite example, we shall first derive equations for the trochoidal fillet on the face gear that is developed by the corner of the tooth of the pinion-shaped cutter as it comes into and goes out of mesh with the face gear. These trochoidal fillets limit the extent of the active face width of the face gear on the side toward the center of the face gear. These face widths are limited in the other side by the increasing pressure angle and pointed teeth of the face gear.

Trochoid on Face Gear. We shall determine the form of this trochoid

Fig. 14-11.

on an intersecting cylinder concentric with the face gear. Referring to Fig. 14-11, when

R_{oc} = outside radius of pinion-shaped cutter, in.
r_1 = momentary pitch radius of bevel pinion, in.
r_2 = corresponding momentary pitch radius of face gear, in.
ϵ_c = angle of rotation of pinion-shaped cutter
ϵ_2 = angle of rotation of face gear
x = abscissa of trochoidal form on cylinder, in.
y = ordinate of trochoidal form on cylinder, in.

we have the following from the geometrical conditions shown in Fig. 14-11:

$$y = R_{oc}(1 - \cos \epsilon_c) \qquad (14\text{-}24)$$
$$\sin \beta = R_{oc} \sin \epsilon_c/r_2$$
$$x = r_2(\epsilon_2 \pm \beta) \qquad (14\text{-}25)$$

We must next determine the distance between the origins of the two trochoids on opposite sides of the rack tooth. Referring to Fig. 14-12, when

Δ = angle of tooth space at R_c

ϕ_1 = pressure angle at R_c

ϕ_o = pressure angle at R_{oc}

δ = angle of tooth space at R_{oc}

T_c = arc space width of tooth at R_c, in.

R_c = nominal pitch radius of pinion-shaped cutter, in.

R_{oc} = outside radius of pinion-shaped cutter, in.

R_{bc} = radius of base circle of pinion-shaped cutter, in.

r_1 = momentary pitch radius of bevel pinion, in.

A = distance between origins of trochoids on cylinder, in.

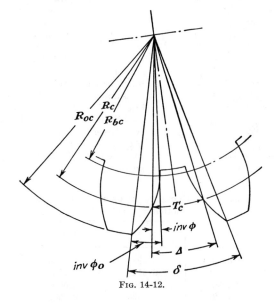

Fig. 14-12.

we have the following from the geometrical conditions shown in Fig. 14-12:

$$\text{arc } \Delta = T_c/R_c$$

$$\cos \phi_o = R_{bc}/R_{oc} \tag{14-26}$$

$$\text{arc } \delta = (T_c/R_c) + 2(\text{inv } \phi_o - \text{inv } \phi_1) \tag{14-27}$$

$$A = r_1\delta \tag{14-28}$$

Example of Approximation for Face-gear-tooth Form. As a definite example we shall use the following values:

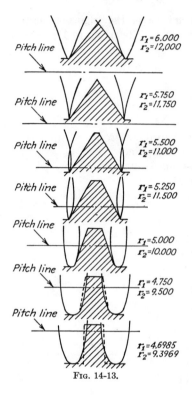

$r_1 = 6.000$
$r_2 = 12.000$

Pitch line

$r_1 = 5.750$
$r_2 = 11.750$

Pitch line

$r_1 = 5.500$
$r_2 = 11.000$

Pitch line

$r_1 = 5.250$
$r_2 = 11.500$

Pitch line

$r_1 = 5.000$
$r_2 = 10.000$

Pitch line

$r_1 = 4.750$
$r_2 = 9.500$

Pitch line

$r_1 = 4.6985$
$r_2 = 9.3969$

Fig. 14-13.

$N_1 = 20 \qquad N_2 = 40 \qquad R_1 = R_c = 5.000$
$R_2 = 10.000 \qquad T = T_c = 0.7854$
$\phi_1 = 20° \qquad c = 0.125$
$R_{r1} = 4.375 \qquad R_b = R_{bc} = 4.69846$
$R_{oc} = 5.625 \qquad \tan \gamma_p = {}^{20}\!/_{40} = 0.500$
$$\text{arc } \Delta = \frac{0.7854}{5.00} = 0.15708$$
$$\cos \phi_o = \frac{4.69846}{5.625} = 0.83528$$
$\phi_o = 33.355° \qquad \text{inv } \phi_o = 0.07610$
$\text{inv } \phi_1 = 0.01490$
$\text{arc } \delta = 0.15708 + 2(0.06120) = 0.27948$

With these values and the preceding equations, we obtain the values tabulated in Tables 14-5 and 14-6. These values are plotted in Fig. 14-13.

In this example, when the pressure angle is but a very little less than 20 deg, the conjugate face-gear tooth is undercut too much to be of any value. On the other side of the face gear, when the pressure angle is greater than 35 deg, the conjugate face-gear tooth becomes pointed. With increasing pressure angles, these teeth are reduced in height. As noted before, these conditions limit the effective face width of the face gear. With larger reduction ratios where the angle of the pitch cone of the pinion is smaller, the effective face width of the face gear becomes greater.

TABLE 14-5.   COORDINATES OF FACE-GEAR TEETH
(Plotted in Fig. 14-13)

r_1, in.	ϕ, deg	t, in.	a, in.	b, in.
4.6985	0.000	0.5980	0.1985	0.9265
4.7500	8.448	0.6148	0.2500	0.8750
5.0000	20.000	0.7854	0.5000	0.6250
5.2500	26.627	1.0526	0.7500	0.3750
5.5000	31.321	1.3785	1.0000	0.1250
5.7500	35.202	1.7797	1.2500	-0.1250
6.0000	38.456	2.2397	1.5000	-0.3750

Exact Analysis of Face-gear Teeth. We shall now determine the profile of the tooth of the face gear that meshes with an involute spur gear in a manner similar to that used for the analysis of the face gear that

TABLE 14-6. COORDINATES OF TROCHOIDAL FILLETS
(Plotted in Fig. 14-13)

x, in.	y, in.	x, in.	y, in.
$r_1 = 4.69846$ in. $\quad r_2 = 9.39692$ in. $A = 1.3131$ in.		$r_1 = 5.500$ in. $\quad r_2 = 11.000$ in. $A = 1.53714$ in.	
0.0000	0.0000	0.0000	0.0000
−0.0964	0.0308	−0.0123	0.0308
−0.1884	0.1229	−0.0198	0.1229
−0.2722	0.2753	−0.0183	0.2753
−0.3429	0.4863	−0.0001	0.4863
−0.3963	0.7536	0.0358	0.7536
−0.4264	1.0743	0.0976	1.0743
$r_1 = 4.750$ in. $\quad r_2 = 9.500$ in. $A = 1.3275$ in.		$r_1 = 5.750$ in. $\quad r_2 = 11.500$ in. $A = 1.6070$ in.	
0.0000	0.0000	0.0000	0.0000
−0.0908	0.0308	0.0140	0.0308
−0.1776	0.1229	0.0328	0.1229
−0.2558	0.2753	0.0614	0.2753
−0.3210	0.4863	0.1052	0.4863
−0.3682	0.7536	0.1694	0.7536
−0.3924	1.0743	0.2593	1.0743
$r_1 = 5.000$ in. $\quad r_2 = 10.000$ in. $A = 1.3974$ in.		$r_1 = 6.000$ in. $\quad r_2 = 12.000$ in. $A = 1.6769$ in.	
0.0000	0.0000	0.0000	0.0000
−0.0647	0.0308	0.0401	0.0308
−0.1250	0.1229	0.0854	0.1229
−0.1763	0.2753	0.1405	0.2753
−0.2140	0.4863	0.2113	0.4863
−0.2330	0.7536	0.3026	0.7536
−0.2539	1.0743	0.4201	1.0743
$r_1 = 5.250$ in. $\quad r_2 = 10.500$ in. $A = 1.4673$ in.			
0.0000	0.0000		
−0.0385	0.0308		
−0.0723	0.1229		
−0.0969	0.2753		
−0.1072	0.4863		
−0.0981	0.7536		
−0.0649	1.0743		

meshes with a lantern pinion.   Referring to Fig. 14-14, let
R_1 = nominal pitch radius on involute spur gear, in.
R_{b1} = radius of base circle of involute spur gear, in.
ϕ_1 = pressure angle of spur gear at R_1
r_1 = radius to contact point on spur gear, in.
ϕ_2 = pressure angle of spur gear at r_1
ϕ = momentary pressure angle of drive
α = initial position of origin of involute on spur gear
ϵ_1 = angle of rotation of spur gear
γ_p = pitch-cone angle of bevel pinion
R_2 = nominal pitch radius of face gear, in.
r_2 = radius to contact point on face gear, in.
F = position of pitch point from nominal pinion pitch circle, in.
G = height to pitch point and to contact point, in.
ϵ_2 = angle of rotation of face gear
θ_2 = vectorial angle to face-gear-tooth form
N_1 = number of teeth in involute gear
N_2 = number of teeth in face gear
T = arc space thickness of spur gear at R_1, in.

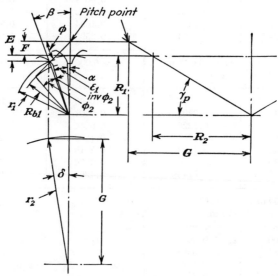

Fig. 14-14.

In order to establish the contact points between the involute spur gear and the face gear, we shall use a series of selected values of E that will cover the working depth of the tooth. With each value of E, we shall use a selected series of values of r_1 and the corresponding values of ϕ_2. Referring again to Fig. 14-14, we have the following from the conditions shown there:

$$\alpha = (T/2R_1) - \text{inv } \phi_1 \tag{14-29}$$
$$\cos \beta = (R_1 - E)/r_1 \tag{14-30}$$
$$\cos \phi_2 = R_{b1}/r_1 \tag{14-31}$$
$$\epsilon_1 = \beta - \alpha - \text{inv } \phi_2 \tag{14-32}$$
$$\phi = \beta + \phi_2 \tag{14-33}$$
$$F = (R_{b1}/\cos \phi) - R_1 \tag{14-34}$$
$$G = R_2 + F \cot \gamma_p \tag{14-35}$$
$$\tan \delta = r_1 \sin \beta/G \tag{14-36}$$
$$r_2 = G/\cos \delta \tag{14-37}$$
$$\theta_2 = \delta - \epsilon_2 \tag{14-38}$$

Example of Fellows Face Gear. As a definite example we shall use the same values as those used for the approximation. From these and the preceding equations, we

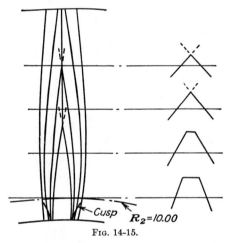

Fɪɢ. 14-15.

obtain the values tabulated in Table 14-7. These values are plotted in Fig. 14-15. In effect, they give a contour map of the teeth of the face gear.

There are cusps on two of the contour lines, which appear to start at a radius of about 9.500 in. on the face gear. Below this radius is the region of excessive undercut. For this particular example, the smallest effective radius of the face gear is something larger than 9.500 in.; the largest radius that retains nearly the full depth of tooth is about 12.00 in. Hence the maximum effective face width for this drive would be about 2.00 in.

Sections of the tooth form of the face gear are also shown in Fig. 14-15. These are obtained graphically from the contour lines. They do not appear to vary much from the approximation to the form, which is shown in Fig. 14-13.

TABLE 14-7. COORDINATES OF TEETH OF FACE GEAR
(Plotted in Fig. 14-15)

r_1, in.	ϕ, deg	ϵ_1, deg	r_2, in.	θ_2, deg
		$E = 0.500$		
4.750	27.116	14.960	10.6662	0.715
5.000	45.842	21.342	13.6638	−1.493
5.250	57.630	25.258	17.7589	−3.870
		$E = 0.250$		
5.500	1.049	−37.452	9.7989	2.291
5.250	1.418	−30.952	9.6622	2.095
5.000	1.805	−22.694	9.5304	1.918
4.750	8.448	− 3.708	9.5000	1.854
5.000	38.195	13.694	12.0538	0.592
5.250	51.836	19.465	15.3711	−1.368
5.500	61.593	23.092	19.9463	−3.557
		$E = 0.000$		
5.500	6.701	−31.810	9.7350	2.292
5.250	8.874	−23.497	9.6445	2.195
5.000	20.000	− 4.500	10.0000	2.250
5.250	44.380	12.009	13.2424	0.852
5.500	55.941	17.429	16.9345	−0.939
		$E = -0.250$		
5.500	13.974	−24.537	8.8214	2.657
5.250	26.627	− 5.744	10.5000	2.872
5.500	48.668	10.156	14.3229	1.497

CHAPTER 15

THE OCTOID FORM ON BEVEL GEARS

Bevel gears with interchangeable teeth similar to those of spur gears may be made. In such a case both gears of the pair must be conjugate to a crown rack of symmetrical form. Theoretically, as wide a variety of tooth forms may be used for the crown rack of bevel gears as can be used for the basic rack of spur gears. Practically, the choice is limited to those forms which can be generated readily. Bevel gears are generally made in pairs so that interchangeability to the degree sometimes necessary

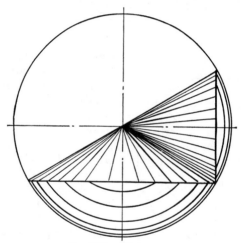

FIG. 15-1. Pitch cones.

for spur gears is seldom of major importance here. The most essential factors are the simplicity of the generating tools and of generating machines, with as great a reduction in variety as possible.

An exact analysis of the conjugate gear-tooth action of such bevel gears should be made on the surface of a sphere. The pitch cones contained in a sphere are shown in Fig. 15-1. The bevel gear equivalent to the basic rack of the spur gear is the crown rack whose pitch plane is the plane of the hemisphere as shown in Fig. 15-2.

The Octoid Form. The simplest and most commonly used crown-rack form for bevel gears is the one that generates bevel-gear teeth of

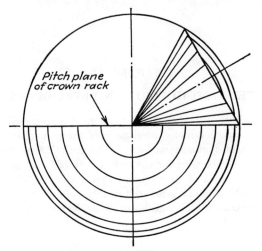

FIG. 15-2.

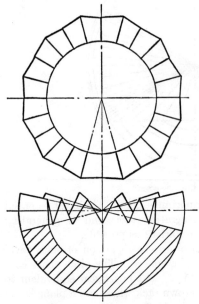

FIG. 15-3.   Crown rack.

octoid form.   The form is so called because its full path of contact is of the form of the figure eight.   Only a small part of the full path of contact is actually used for bevel-gear tooth action, and this portion is practically a straight line.   The crown rack of this system is composed of plane sides whose straight-line elements all converge to the cone center as indicated in Fig. 15-3.

We shall determine the projection of the path of contact of this crown-rack form on the surface of a sphere.   Referring to Fig. 15-4, let

A = radius of sphere, in.

ϕ = pressure angle of crown rack

ϵ = angle of rotation of bevel gear on crown rack

x = abscissa of projection of path of contact, in.

y = ordinate of projection of path of contact, in.

Other symbols are shown in Fig. 15-4.

From the geometrical conditions shown in Fig. 15-4, we have the following:

$$E = \sqrt{(A^2 - y^2)}$$
$$F = \sqrt{E^2 - y^2 \tan^2 \phi} = \sqrt{A^2 - y^2(1 + \tan^2 \phi)}$$
$$F = \frac{\sqrt{A^2 \cos^2 \phi - y^2}}{\cos \phi}$$
$$\tan \epsilon = \frac{y \tan \phi + (y/\tan \phi)}{F} = \frac{y(\tan^2 \phi + 1)}{F \tan \phi} = \frac{y}{F \sin \phi \cos \phi}$$

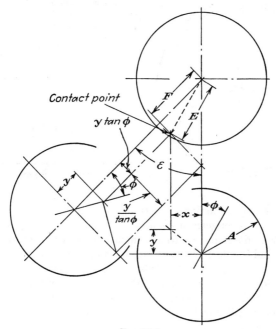

Contact point

$y \tan \phi$

$\frac{y}{\tan \phi}$

FIG. 15-4.

Substituting the value of F into this last equation, we obtain

$$\tan \epsilon = \frac{y}{\sin \phi \sqrt{A^2 \cos^2 \phi - y^2}}$$

Solving this last equation for y, we obtain

$$y = \frac{A \sin \epsilon \sin \phi \cos \phi}{\sqrt{\cos^2 \epsilon + \sin^2 \epsilon \sin^2 \phi}} \tag{15-1}$$

$$x = \frac{y \cos \epsilon}{\tan \phi} \tag{15-2}$$

When we solve these last two equations for a series of values of ϵ ranging from 0 to 90 deg, we obtain values for one quadrant of the projection of the path of contact of the octoid form on the surface of a sphere. The other three quadrants are of similar form.

Example of Octoid Path of Contact. As a definite example we shall use the following values:

$$A = 3.000 \qquad \phi = 14.50°$$

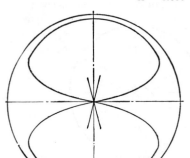

FIG. 15-5.

The values of the coordinates of the projection of this path of contact for one quadrant are as follows:

ϵ, deg	x, in.	y, in.
0	0.0000	0.0000
15	0.7171	0.1920
30	1.3739	0.4103
45	1.9046	0.6967
60	2.2063	1.1412
75	1.9579	1.9566
90	0.0000	2.9044

These tabulated values are plotted in Fig. 15-5.

Tregold's Approximation. Fortunately we do not have to resort to the solution of bevel-gear problems on the surface of a sphere because we have Tregold's approximation, which reduces the problem to one of spur gears, and is sufficiently accurate for all practical purposes for tooth numbers greater than about 8. This approximation is shown in Fig. 15-6. It uses the equivalent spur gears of the back cones of the bevel gears.

Let R_p = pitch radius of bevel pinion at large end, in.

R_g = pitch radius of bevel gear at large end, in.

γ_p = pitch-cone angle of bevel pinion

γ_g = pitch-cone angle of bevel gear

Σ = shaft angle

N_p = number of teeth in bevel pinion

N_g = number of teeth in bevel gear

R_{vp} = back-cone distance and pitch radius of equivalent spur pinion in Tregold's approximation, in.

R_{vg} = back-cone distance and pitch radius of equivalent spur gear in Tregold's approximation, in.

N_{vp} = number of teeth in equivalent spur pinion

N_{vg} = number of teeth in equivalent spur gear

A = cone distance and radius of sphere, in.

Referring to Fig. 15-6, we have the following:

$$\Sigma = \gamma_p + \gamma_g \tag{14-1}$$

$$\tan \gamma_p = \frac{\sin \Sigma}{(N_g/N_p) + \cos \Sigma} \tag{14-2}$$

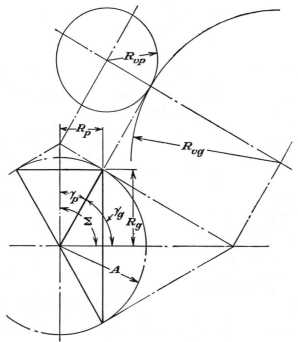

FIG. 15-6. Tregold's approximation.

When $\Sigma = 90°$, Eq. (14-2) reduces to

$$\tan \gamma_p = \frac{N_p}{N_g} \tag{14-3}$$

$$\tan \gamma_g = \frac{N_g}{N_p} \tag{14-4}$$

When P is the diametral pitch at large end of bevel gears,

$$R_p = \frac{N_p}{2P} \tag{15-3}$$

$$R_g = \frac{N_g}{2P} \tag{15-4}$$

$$A = \frac{R_p}{\sin \gamma_p} = \frac{R_g}{\sin \gamma_g} \tag{15-5}$$

$$R_{vp} = \frac{R_p}{\cos \gamma_p} = A \tan \gamma_p \qquad (15\text{-}6)$$

$$R_{vg} = \frac{R_g}{\cos \gamma_g} = A \tan \gamma_g \qquad (15\text{-}7)$$

$$N_{vp} = \frac{N_p}{\cos \gamma_p} \qquad (15\text{-}8)$$

$$N_{vg} = \frac{N_g}{\cos \gamma_g} \qquad (15\text{-}9)$$

The gear-tooth design and proportions for the bevel gears are determined in the same manner as those for spur gears. When these proportions have been established, they must then be translated into the equivalent angles on the bevel gears.

Bevel-gear-tooth Design. Bevel gears are commonly made of the same diametral pitches as those used for spur gears. These are based upon the size of the bevel gear at the large end. The tooth heights are measured along an element of the back cone. Referring to Fig. 15-7, let

R = pitch radius at large end, in.

a = addendum of tooth, in.

b = dedendum of tooth, in.

h_t = whole depth of tooth, in.

γ = pitch-cone angle

α = addendum angle

δ = dedendum angle

t = arc tooth thickness on pitch cone at R, in.

β = tooth angle (*i.e.*, angle for setting slide of cutters on bevel-gear generating machine. This is the angle between the bottom corner of the generating cutter and the center line of the tooth, measured from the cone center).

γ_o = face angle

γ_r = root angle

ϕ_c = pressure angle of generating cutter

We have the following from the geometrical conditions shown in Fig. 15-7:

$$\tan \alpha = \frac{a}{A} \qquad (15\text{-}10)$$

$$\tan \delta = \frac{b}{A} \qquad (15\text{-}11)$$

$$\gamma_o = \gamma + \alpha \qquad (15\text{-}12)$$

$$\gamma_r = \gamma - \delta \qquad (15\text{-}13)$$

$$\text{arc } \beta = \frac{(t/2) + b \tan \phi_c}{A} \qquad (15\text{-}14)$$

The subscripts p and g would be added to the several symbols when needed to distinguish the bevel-pinion values from those of the bevel gear.   The tooth proportions are taken directly from those of the equivalent spur gears in Tregold's approximation.

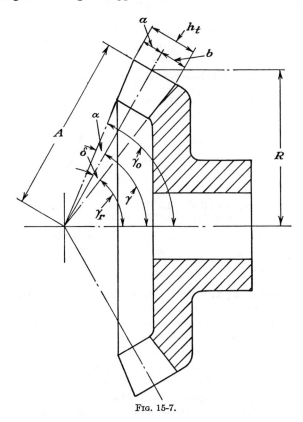

Fɪɢ. 15-7.

Example of Bevel-gear Design.   As a definite example we shall use a pair of 1-DP bevel gears of 20 and 40 teeth with a 20-deg pressure angle.   We shall use the tooth proportions of the spur-gear full-depth form, and the shaft angle will be 90 deg.   This gives the following values:

$$N_p = 20 \quad N_g = 40 \quad R_p = 10.000 \quad R_g = 20.000 \quad \phi_c = 20°$$
$$a = 1.000 \quad b = 1.157 \quad h_t = 2.157 \quad t = 1.5708 \quad \tan \phi_c = 0.36397$$
$$\tan \gamma_p = {}^{40}\!/_{20} = 0.5000$$
$$\gamma_p = 26.565° \quad \sin \gamma_p = 0.44721 \quad \cos \gamma_p = 0.89442$$
$$\gamma_g = 63.435° \quad \sin \gamma_g = 0.89442 \quad \cos \gamma_g = 0.44721$$

$$A = \frac{10}{0.44721} = 22.36068$$

$$R_{vp} = \frac{10}{0.89442} = 11.18034$$

$$N_{vp} = \frac{20}{0.89442} = 22.36068$$

$$R_{vg} = \frac{20}{0.44721} = 44.72136$$

$$N_{vg} = \frac{40}{0.44721} = 89.44272$$

If these bevel gears were to have tooth forms of special proportions, they would be designed as a pair of involute spur gears of 22.36068 teeth and 89.44272 teeth with pitch radii of 11.18034 in. and 44.72136 in., respectively. This example, however, uses standard or known tooth proportions; hence we proceed as follows:

$$\tan \alpha = \frac{1.00}{22.36068} = 0.04472 \qquad \alpha = 2.561°$$

$$\tan \delta = \frac{1.157}{22.36068} = 0.05017 \qquad \delta = 2.872°$$

$$\gamma_{op} = 26.565° + 2.561° = 29.126°$$
$$\gamma_{rp} = 26.565° - 2.872° = 23.693°$$
$$\gamma_{og} = 63.435° + 2.561° = 65.996°$$
$$\gamma_{rg} = 63.435° - 2.872° = 60.563°$$
$$\text{arc } \beta = \frac{0.7854 + (1.157 \times 0.36397)}{22.36068} = 0.05395$$

$$\beta = 3.091°$$

In this example, the value of the tooth angle is the same for both members of the pair.

Special Tooth Design for Bevel Gears. As with spur gears, a straight-toothed bevel gear may be generated with a given pressure angle of tool on one generating cone, and be operated with another bevel gear at a different pressure angle on a different pitch cone. Here, instead of changing the center distance of generation and of operation as with spur gears, the angles of the cones of generation and operation are changed. The tooth proportions of the gears are determined by Tregold's approximation exactly as for spur gears. After the equivalent-spur-gear proportions are determined, these values are then translated into the corresponding angular values for bevel gears. This translation is done for each gear separately. Thus let

A = cone distance, in.

R = pitch radius at large end, in.

γ = pitch-cone angle

γ' = generating cone angle

N = number of teeth in bevel gear

R_v = pitch radius of equivalent spur gear, in.

N_v = number of teeth in equivalent spur gear

a = addendum, in.
b = dedendum, in.
α = addendum angle
δ = dedendum angle
t = arc tooth thickness at R, in.
R'_v = generating radius of equivalent spur gear, in.
t' = arc tooth thickness at R'_v, in.
β = tooth angle
ϕ = operating pressure angle
ϕ_c = pressure angle of cutter
R_b = radius of base circle of equivalent spur gear, in.

Given the values of the bevel gear and the pressure angle of the cutter, we proceed as follows:

$$R_v = \frac{R}{\cos \gamma} \tag{15-6}$$

$$R_b = \frac{R_v}{\cos \phi}$$

$$R'_v = \frac{R_b}{\cos \phi_c} = \frac{R_v \cos \phi}{\cos \phi_c}$$

When ω is the difference between pitch and generating cone angles,

$$\omega = \gamma - \gamma'$$

$$\tan \omega = \frac{(R_v - R'_v)}{A} = R_v \left[1 - \frac{\cos \phi / (\cos \phi_c)}{A} \right]$$

But

$$\frac{R_v}{A} = \tan \gamma$$

Whence

$$\tan \omega = \left(1 - \frac{\cos \phi}{\cos \phi_c} \right) \tan \gamma \tag{15-15}$$

$$\gamma' = \gamma - \omega \tag{15-16}$$

We have the following from involute spur gears:

$$t' = 2R'_v \left(\frac{t}{2R_v} + \text{inv } \phi - \text{inv } \phi_c \right)$$

Whence

$$t' = \frac{2R_v \cos \phi}{\cos \phi_c} \left(\frac{t}{2R_v} + \text{inv } \phi - \text{inv } \phi_c \right) \tag{15-17}$$

When y is the distance from generating radius to root radius in inches,

$$y = b - (R_v - R'_v) = b - R_v\left(1 - \frac{\cos \phi}{\cos \phi_c}\right) \tag{15-18}$$

$$\text{arc } \beta = \frac{[(t'/2) + y \tan \phi_c]}{A} \tag{15-19}$$

$$\tan \alpha = \frac{a}{A} \tag{15-20}$$

$$\tan \delta = \frac{b}{A} \tag{15-21}$$

$$\gamma_o = \gamma + \alpha \tag{15-12}$$

$$\gamma_r = \gamma - \delta \tag{15-13}$$

Example of Machine Setting for Special Bevel Gear. As a definite example we shall determine the machine settings for generating the 20-deg bevel gears used for the preceding example with a 14½-deg generating tool. For the 20-tooth pinion, we have the following values:

$$R_v = 11.18034 \quad \phi = 20° \quad \cos \phi = 0.93969 \quad A = 22.36068$$
$$\phi_c = 14.500° \quad \cos \phi_c = 0.96815 \quad \tan \phi_c = 0.25862 \quad \tan \gamma = 0.5000$$
$$t = 1.5708 \quad a = 1.000 \quad b = 1.157 \quad \text{inv } \phi = 0.014904$$

$$\text{inv } \phi_c = 0.005545 \quad \frac{\cos \phi}{\cos \phi_c} = 0.97060$$

$$\tan \omega = 0.02940 \times 0.500 = 0.01470 \quad \omega = 0.842°$$
$$\gamma' = 26.565° - 0.842° = 25.723°$$

$$t' = \frac{22.36068 \times 0.93969}{0.96815}\left(\frac{1.5708}{22.36068} + 0.014904 - 0.005545\right) = 1.72774$$

$$y = 1.157 - (11.18034 \times 0.02940) = 0.82830$$

$$\text{arc } \beta = \frac{0.86387 + (0.82830 \times 0.25862)}{22.36068} = 0.04821$$

$$\beta = 2.762°$$

From the previous example we have

$$\alpha = 2.561° \quad \delta = 2.872° \quad \gamma_o = 29.126° \quad \gamma_r = 23.693°$$

For the 40-tooth bevel gear we have the following values:

$$R_v = 44.72136 \quad \phi = 20° \quad \phi_c = 14.500° \quad A = 22.36068$$
$$\gamma = 63.435° \quad t = 1.5708 \quad a = 1.000 \quad b = 1.157$$
$$\tan \omega = 0.02940 \times 2.000 = 0.05880 \quad \omega = 3.364°$$
$$\gamma' = 63.435° - 3.364° = 60.071°$$
$$t' = 2.33702$$
$$y = 1.157 - (44.72136 \times 0.02940) = -0.15781$$

In this example the generating cone is below or inside the root cone, hence this value of y is minus.

$$\text{arc } \beta = 0.05043 \quad \beta = 2.889°$$

From the preceding example we have

$$\alpha = 2.561° \quad \delta = 2.872° \quad \gamma_o = 65.996° \quad \gamma_r = 60.563°$$

TABLE 15-1. OPERATING PRESSURE ANGLES FOR 14½-DEG CROWN-RACK SYSTEM

$N_{vp} + N_{vg}$	ϕ, deg	cos ϕ/cos 14.5°	inv ϕ − inv 14.5°
24	26.10	0.92757	0.0288192
25	25.75	0.93033	0.0273756
26	25.40	0.93306	0.0259764
27	25.05	0.93574	0.0246207
28	24.70	0.93840	0.0233074
29	24.35	0.94102	0.0220358
30	24.00	0.94360	0.0208048
31	23.75	0.94543	0.0199500
32	23.50	0.94723	0.0191152
33	23.25	0.94902	0.0183000
34	23.00	0.95079	0.0175042
35	22.75	0.95254	0.0167276
36	22.50	0.95428	0.0159696
37	22.25	0.95599	0.0152302
38	22.00	0.95769	0.0145090
39	21.75	0.95937	0.0138056
40	21.50	0.96103	0.0131199
41	21.30	0.96234	0.0125837
42	21.10	0.96365	0.0120586
43	20.90	0.96494	0.0115442
44	20.70	0.96622	0.0110405
45	20.50	0.96749	0.0105473
46	20.30	0.96875	0.0100645
47	20.10	0.96999	0.0095920
48	19.90	0.97122	0.0091296
49	19.70	0.97244	0.0086771
50	19.50	0.97365	0.0082345
51	19.35	0.97455	0.0079080
52	19.20	0.97544	0.0075888
53	19.05	0.97633	0.0072740
54	18.90	0.97721	0.0069645
55	18.75	0.97808	0.0066602
56	18.60	0.97895	0.0063611
57	18.45	0.97981	0.0060672
58	18.30	0.98066	0.0057783
59	18.15	0.98151	0.0054944
60	18.00	0.98235	0.0052156
61	17.85	0.98318	0.0049417
62	17.70	0.98400	0.0046726
63	17.55	0.98482	0.0044084

TABLE 15-1. OPERATING PRESSURE ANGLES FOR 14½-DEG CROWN-RACK SYSTEM.
(Continued)

$N_{vp} + N_{vg}$	ϕ, deg	$\cos \phi / \cos 14.5°$	inv ϕ − inv 14.5°
64	17.40	0.98563	0.0041489
65	17.25	0.98644	0.0038941
66	17.10	0.98724	0.0036440
67	16.95	0.98803	0.0033986
68	16.80	0.98881	0.0031576
69	16.65	0.98959	0.0029212
70	16.50	0.99036	0.0026893
71	16.40	0.99087	0.0025372
72	16.30	0.99138	0.0023870
73	16.20	0.99189	0.0022387
74	16.10	0.99239	0.0020923
75	16.00	0.99289	0.0019479
76	15.90	0.99338	0.0018053
77	15.80	0.99387	0.0016646
78	15.70	0.99436	0.0015258
79	15.60	0.99485	0.0013888
80	15.50	0.99533	0.0012537
81	15.45	0.99557	0.0011868
82	15.40	0.99581	0.0011204
83	15.35	0.99605	0.0010544
84	15.30	0.99629	0.0009888
85	15.25	0.99653	0.0009237
86	15.20	0.99677	0.0008591
87	15.15	0.99700	0.0007949
88	15.10	0.99724	0.0007312
89	15.05	0.99747	0.0006678
90	15.00	0.99770	0.0006050
91	14.95	0.99793	0.0005425
92	14.90	0.99816	0.0004805
93	14.85	0.99839	0.0004190
94	14.80	0.99862	0.0003578
95	14.75	0.99885	0.0002971
96	14.70	0.99908	0.0002368
97	14.65	0.99931	0.0001770
98	14.60	0.99954	0.0001176
99	14.55	0.99977	0.0000586
100 and over	14.50	1.00000	0.0000000

14½-deg Crown-rack System. A series of pairs of bevel gears, all generated with the 14½-deg generating cutters, similar to the 14½-deg variable-center-distance system for spur gears can be developed readily. The Gleason Company has published a series of bevel-gear-tooth forms, using cutters of different pressure angles for different pairs, with proportions adjusted for effective action and contact. These are good designs, but the same results can be obtained with a single standard pressure angle of cutter.

The first step toward such an end will be the selection of suitable pressure angles for the various combinations of tooth numbers, based upon the equivalent spur gears of Tregold's approximation. These pressure angles will approximate those which are used for the spur gears. This has been done, and the values are tabulated in Table 15-1.

The tooth proportions of the 14½-deg crown rack will be as follows (these are the 1-DP values):

Inches

Addendum of 1-DP crown rack............................. 1.000
Dedendum of 1-DP crown rack............................. 1.188
Whole depth of tooth.. 2.188
Clearance.. 0.188

The next step is to determine the generating radii of the equivalent spur gears and their tooth proportions. This is done by the use of the spur-gear equations in Chap. 5, transposing them when necessary to suit our particular needs here.

The final step is to translate these spur-gear values into the angular values of the bevel gears.

Equations for Bevel-gear-tooth Design. Referring to Fig. 15-8, let

γ_p = pitch-cone angle of bevel pinion
γ_g = pitch-cone angle of bevel gear
A = cone distance, in.
R_{vp} = pitch radius of equivalent spur pinion, in.
R_{vg} = pitch radius of equivalent spur gear, in.
C_2 = center distance of Tregold's approximation, in.
C_1 = center distance for pressure angle of ϕ_c, in.
ϕ = operating pressure angle
ϕ_c = pressure angle of generating tool
a_p = addendum of bevel pinion, in.
b_p = dedendum of bevel gear, in.
a_g = addendum of bevel gear, in.
b_g = dedendum of bevel gear, in.
h_t = nominal whole depth of tooth, in.
a_c = nominal addendum of cutter, in.

c_c = nominal clearance on cutter, in.
h_{tc} = nominal whole depth of cutter, in.
R_{rp} = root radius of spur pinion, in.
R_{rg} = root radius of spur gear, in.
ω_p = difference between pitch and generating angle of pinion
ω_g = difference between pitch and generating angle of gear
γ'_p = generating angle of bevel pinion
γ'_g = generating angle of bevel gear
R'_{vp} = generating radius of spur pinion, in.
R'_{vg} = generating radius of spur gear, in.

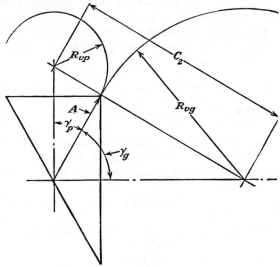

FIG. 15-8.

We have from Chap. 5

$$R_{rp} + R_{rg} = C_1 - 2a_c + \frac{C_1(\text{inv } \phi - \text{inv } \phi_c)}{\tan \phi_c} \qquad (5\text{-}48)$$

$$C_2 = C_1 \frac{\cos \phi_c}{\cos \phi} \qquad (5\text{-}35)$$

Transposing this last equation, we obtain

$$C_1 = C_2 \frac{\cos \phi}{\cos \phi_c}$$

From the conditions shown in Fig. 15-8, we have

$$C_2 = A(\tan \gamma_p + \tan \gamma_g) \qquad (15\text{-}22)$$

Substituting this value into the preceding equation, we obtain

$$C_1 = A \left(\frac{\cos \phi}{\cos \phi_c} \right) (\tan \gamma_p + \tan \gamma_g) \tag{15-23}$$

We already have

$$\tan \omega_p = \left(1 - \frac{\cos \phi}{\cos \phi_c} \right) \tan \gamma_p \tag{15-15}$$

$$\tan \omega_g = \left(1 - \frac{\cos \phi}{\cos \phi_c} \right) \tan \gamma_g \tag{15-15}$$

We shall now turn our attention to the tooth proportions. From Chap. 5 we have

$$h_t = \left(\frac{h_{tc}}{h_{tc} + c_c} \right) [C_2 - (R_{rp} + R_{rg})] \tag{5-51}$$

$$b_p = \frac{C_2 - (R_{rp} + R_{rg})}{1 + \sqrt{R_{vg}/R_{vp}}} \tag{5-49}$$

But

$$\frac{R_{vg}}{R_{vp}} = \frac{A \tan \gamma_g}{A \tan \gamma_p} = \frac{\tan \gamma_g}{\tan \gamma_p}$$

Whence

$$b_p = \frac{C_2 - (R_{rp} + R_{rg})}{1 + \sqrt{\tan \gamma_g/\tan \gamma_p}} \tag{15-24}$$

When $\Sigma = 90°$; then

$$\tan \gamma_p = \frac{1}{\tan \gamma_g} \quad \text{and} \quad \frac{\tan \gamma_g}{\tan \gamma_p} = \tan^2 \gamma_g$$

Thus when $\Sigma = 90°$,

$$b_p = \frac{C_2 - (R_{rp} + R_{rg})}{1 + \tan \gamma_g} \tag{15-25}$$

$$a_p = h_t - b_p \tag{15-26}$$

$$b_g = C_2 - (R_{rp} + R_{rg}) - b_p \tag{15-27}$$

$$a_g = h_t - b_g \tag{15-28}$$

To translate these values of the equivalent spur gears into the angle of the bevel gear and pinion, using the same symbols as before, we have

$$\tan \alpha_p = \frac{a_p}{A} \quad \tan \alpha_g = \frac{a_g}{A} \tag{15-10}$$

$$\tan \delta_p = \frac{b_p}{A} \quad \tan \delta_g = \frac{b_g}{A} \tag{15-11}$$

$$\gamma_{op} = \gamma_p + \alpha_p \quad \gamma_{og} = \gamma_g + \alpha_g \tag{15-12}$$

$$\gamma_{rp} = \gamma_p - \delta_p \quad \gamma_{rg} = \gamma_g - \delta_g \tag{15-13}$$

$$\text{arc } \beta_p = \frac{(\pi/2P) + a_c \tan \phi_c)}{A} \tag{15-14}$$

$$\text{arc } \beta_g = \frac{(\pi/2P) + a_c \tan \phi_c)}{A} \tag{15-14}$$

The foregoing are general equations and may be used for generating tools of any pressure angle and any tooth proportions.

SPECIFIC EQUATIONS FOR 1-DP, 14½-DEG CROWN-RACK SYSTEM. When the axes of the gears are at 90 deg and the following fixed proportions are used for the 1-DP crown rack of the system, the several equations reduce to the following:

$$h_{tc} = 2.188 \qquad a_c = 1.188 \qquad c_c = 0.188$$

$$A = \sqrt{R_p{}^2 + R_g{}^2} = \tfrac{1}{2} \sqrt{N_p{}^2 + N_g{}^2} \tag{15-29}$$

$$\tan \gamma_p = N_p/N_g \tag{14-3}$$

$$\tan \gamma_g = N_g/N_p \tag{14-4}$$

$$C_2 = A(\tan \gamma_p + \tan \gamma_g) \tag{15-22}$$

$$C_1 = C_2(\cos \phi/0.96815) \tag{15-30}$$

$$R_{rp} + R_{rg} = C_1 - 2.376 + 3.86671C_1 (\text{inv } \phi - 0.0055448) \tag{15-31}$$

$$h_t = 0.92087 [C_2 - (R_{rp} + R_{rg})] \tag{15-32}$$

$$b_p = [C_2 - (R_{rp} + R_{rg})]/(1 + \tan \gamma_g) \tag{15-25}$$

$$a_p = h_t - b_p \tag{15-26}$$

$$b_g = C_2 - (R_{rp} + R_{rg}) - b_p \tag{15-27}$$

$$a_g = h_t - b_g \tag{15-28}$$

$$\tan \omega_p = [1 - (\cos \phi/0.96815)] \tan \gamma_p \tag{15-15}$$

$$\tan \omega_g = [1 - (\cos \phi/0.96815)] \tan \gamma_g \tag{15-15}$$

$$\gamma'_p = \gamma_p - \omega_p \qquad \gamma'_g = \gamma_g - \omega_g \tag{15-16}$$

$$\tan \alpha_p = a_p/A \qquad \tan \alpha_g = a_g/A \tag{15-10}$$

$$\tan \delta_p = b_p/A \qquad \tan \delta_g = b_g/A \tag{15-11}$$

$$\gamma_{op} = \gamma_p + \alpha_p \qquad \gamma_{og} = \gamma_g + \alpha_g \tag{15-12}$$

$$\gamma_{rp} = \gamma_p - \delta_p \qquad \gamma_{rg} = \gamma_g - \delta_g \tag{15-13}$$

$$\beta_p = \beta_g \text{ (in degrees)} = 107.60378/A \tag{15-33}$$

Bevel-gear Generating-machine Setting. The cone angle or the generating angle to which the machine is set is not the actual pitch-cone angle (γ) of operation, but is the generating angle (γ').

$$\text{Block angle} = \gamma' - \gamma_r \tag{15-34}$$

The tooth angle is the value of β as calculated.

Example of Bevel-gear Design. As a definite example we shall use a 14-tooth bevel pinion and a 21-tooth bevel gear, 1 DP, with shafts at right angles to each other. From the foregoing equations, we obtain

$$\tan \gamma_p = {}^{14}\!/_{21} = 0.66667 \qquad \gamma_p = 33.690°$$
$$\tan \gamma_g = {}^{21}\!/_{14} = 1.50000 \qquad \gamma_g = 56.310°$$
$$A = \sqrt{(7)^2 + (10.5)^2} = 12.61943$$
$$C_2 = 12.61943 \times 2.16667 = 27.34214$$

The virtual number of teeth is equal to $2C_2 = 54.68428$.

We shall use the values from Table 15-1 for the sum of 55 teeth. These are as follows:

$$\phi = 18.75° \qquad \frac{\cos \phi}{\cos 14.50°} = 0.97808 \qquad \text{inv } \phi - \text{inv } 14.5° = 0.0066602$$

$$C_1 = 27.34214 \times 0.97808 = 26.74280$$

$$R_{rp} + R_{rg} = 26.74280 - 2.376 + 3.86671 (26.7426 \times 0.0066602) = 25.05551$$

$$\tan \omega_p = 0.02192 \times 0.66667 = 0.01461 \qquad \omega_p = 0.837°$$
$$\tan \omega_g = 0.02192 \times 1.5000 = 0.03288 \qquad \omega_g = 1.883°$$
$$h_t = 0.92087 (27.34212 - 25.05551) = 2.10569$$

$$b_p = \frac{2.28663}{(1 + 1.500)} = 0.91465 \qquad a_p = 2.10569 - 0.91465 = 1.19104$$

$$b_g = 2.28663 - 0.91465 = 1.37198 \qquad a_g = 2.10569 - 1.37198 = 0.73371$$

$$\beta_p = \beta_g = \frac{107.60378}{12.61943} = 8.526°$$

$$\tan \alpha_p = \frac{1.19104}{12.61943} = 0.09468 \qquad \alpha_p = 5.392°$$

$$\tan \delta_p = \frac{0.91465}{12.61943} = 0.07248 \qquad \delta_p = 4.146°$$

$$\tan \alpha_g = \frac{0.73371}{12.61943} = 0.05814 \qquad \alpha_g = 3.327°$$

$$\tan \delta_g = \frac{1.37198}{12.61943} = 0.10872 \qquad \delta_g = 6.205°$$

$$\gamma_{op} = 33.690° + 5.392° = 39.082°$$
$$\gamma_{rp} = 33.690° - 4.146° = 29.544°$$
$$\gamma'_p = 33.690° - 0.837° = 32.853°$$
$$\gamma_{og} = 56.310° + 3.327° + 59.637°$$
$$\gamma_{rg} = 56.310° - 6.205° = 50.105°$$
$$\gamma'_g = 56.310° - 1.883° = 54.427°$$

For the machine settings we have the following:

Generating cone angle for pinion = 32.853°
Block angle for pinion = 32.853° − 29.544° = 3.309°
Generating cone angle for gear = 54.427°
Block angle for gear = 54.427° − 50.105° = 4.322°
Tooth angle for gear and pinion = 8.526°

The 1-DP values for any pair of bevel gears with intersecting axes at right angles to each other can be determined in the same manner.

CHAPTER 16

SPIRAL BEVEL GEARS

The relation between a spiral bevel gear and a straight-toothed bevel gear is substantially the same as that between a helical gear and a spur gear with straight teeth. The pitch surfaces and the nature of the conjugate gear-tooth action is the same for the two types of bevel gears. With the spiral bevel gear as with the helical gear, the twisted tooth form tends to give a smoother engagement of teeth as they enter into contact.

FELLOWS SPIRAL FACE GEAR

The Fellows spiral-face-gear drive is the same as that with a spur gear except that a helical involute gear is substituted for the involute spur gear as the bevel-pinion member. In this case the face gear is generated by a helical involute pinion-shaped cutter of substantially the same size as the mating helical gear. With the substitution of a helical gear for the spur gear, Fig. 14-9 shows this type of drive also.

As before, the pitch surfaces of the drive are cones. The momentary pressure angle of the drive changes with a change in the distance from the center of the face gear. As with similar spur-gear drives, the active face width of the face gear cannot extend very much below the nominal pitch circle because the trochoidal fillet of the tip of the cutter tooth will cut away the conjugate gear-tooth profile of the face gear almost up to this nominal pitch circle.

On the other side, the active face width of the face gear is limited because of the pointed teeth and the reduced depth of tooth in the face gear, which result from the increasing pressure angle as the distance from the center of the face gear is increased.

The trochoidal fillets on the face gear at any specified distance from the center of the face gear are identical to those on the equivalent face gear that meshes with an involute spur pinion. The only difference is that whereas the center line of the tooth of the face gear that meshes with a spur pinion lies on a radial line of the face gear, in this case the center line of the face-gear tooth lies on a uniform rise or Archimedean spiral.

Exact Analysis of Spiral-face-gear Teeth. The analysis of this face gear is identical to that of one that meshes with a spur gear except for the introduction of the spiral form of the center line of the face-gear teeth.

The direction of this spiral will depend upon the direction of the helix of the helical gear.

Referring again to Fig. 14-14, let

R_1 = nominal pitch radius of involute helical gear, in.

R_{b1} = radius of base cylinder of helical gear, in.

ϕ_1 = pressure angle in plane of rotation of helical gear at R_1

r_1 = radius to contact point on helical gear, in.

ϕ_2 = momentary pressure angle at r_1

ϕ = momentary pressure angle of drive

α = initial position of origin of involute at distance R_2 from center of face gear

ϵ_1 = angle of rotation of helical gear

γ_p = pitch-cone angle of pinion

R_2 = nominal pitch radius of face gear, in.

r_2 = radius to contact point on face gear, in.

F = position of pitch point from nominal pinion pitch circle, in.

G = height to pitch point, in.

ϵ_2 = angle of rotation of face gear

δ = angle to contact point on face gear

θ_2 = vectorial angle to face-gear-tooth form

N_1 = number of teeth in involute helical gear

N_2 = number of teeth in face gear

T = arc space thickness of helical gear in its plane of rotation at R_1, in.

ψ_1 = helix angle of helical gear at R_1

ψ_b = helix angle of helical gear at base radius R_b

M = height to contact point, in.

L = lead of helical gear, in.

As noted before, the analysis is the same as for the spur-gear drive except for the twisting of the teeth, which results in an angular position of the normal to the point of contact that passes through the pitch point. The pitch point will be in the same position, but the actual point of contact will be shifted up or down because of the helix. We shall use the values R_1 and R_2 as the starting point, with the teeth at this point in the same position as before. Referring again to Fig. 14-14, we shall retain the following equations from the previous analysis:

$$\alpha = (T/2R_1) - \text{inv } \phi_1 \tag{14-29}$$
$$\cos \beta = (R_1 - E)/r_1 \tag{14-30}$$
$$\cos \phi_2 = R_b/r_1 \tag{14-31}$$
$$\phi = \beta + \phi_2 \tag{14-33}$$
$$F = (R_{b1}/\cos \phi) - R_1 \tag{14-34}$$
$$G = R_2 + F \cot \gamma_p \tag{14-35}$$

The normal to the point of contact, instead of being perpendicular to the axis of the pinion, is at an angle to it as indicated in Fig. 16-1. This figure shows the conditions for a right-handed helical gear. For a left-handed helical gear, the direction of the angle would be reversed. The position of the pitch point is the same as for the equivalent spur gear, but

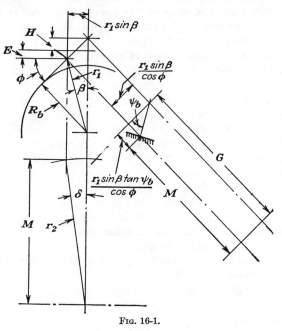

Fig. 16-1.

the contact point is moved up or down because of the angular position of the normal.

We have from the analysis of helical gears

$$\tan \psi_b = \tan \psi_1 \cos \phi_1 \qquad (8\text{-}6)$$

We have the following from the conditions shown in Fig. 16-1: For right-handed helical gears

$$M = G - (r_1 \sin \beta \tan \psi_b / \cos \phi) \qquad (16\text{-}1)$$

For left-handed helical gears

$$M = G + (r_1 \sin \beta \tan \psi_b / \cos \phi) \qquad (16\text{-}2)$$

For both cases

$$\tan \delta = r_1 \sin \beta / M \qquad (16\text{-}3)$$

$$r_2 = M / \cos \delta \qquad (16\text{-}4)$$

When y is the height of contact point above R_2 in inches,

$$y = M - R_2 \qquad (16\text{-}5)$$

For right-handed gear

$$\epsilon_1 = \beta - \alpha - \text{inv } \phi_2 - (2\pi y/L) \qquad (16\text{-}6)$$

For left-handed gear

$$\epsilon_1 = \beta - \alpha - \text{inv } \phi_2 + (2\pi y/L) \qquad (16\text{-}7)$$

$$\theta_2 = \delta - \epsilon_2 \qquad (16\text{-}8)$$

TABLE 16-1. COORDINATES OF SPIRAL-FACE-GEAR TEETH
(Plotted in Fig. 16-2)

r_1, in.	ϕ, deg	ϵ_1, deg	r_2, in.	θ_2, deg
		$E = 0.500$		
4.750	27.116	15.564	9.9920	0.970
5.000	45.842	8.492	12.8278	5.536
5.250	57.630	−1.678	15.7708	10.712
		$E = 0.250$		
5.500	1.049	−39.907	10.8642	5.170
5.250	1.418	−32.373	10.5322	3.929
5.000	1.804	−22.814	10.1455	2.555
4.750	8.448	−1.276	9.500	0.638
5.000	38.195	8.030	11.2730	3.946
5.250	51.836	1.152	13.9446	8.651
5.500	61.593	−13.046	17.6475	15.561
		$E = 0.000$		
5.500	6.701	−34.512	10.6316	4.810
5.250	8.874	−24.260	10.2824	3.174
5.000	20.000	−4.500	10.0000	2.250
5.250	44.380	1.042	12.3584	6.921
5.500	55.941	−7.610	15.3192	12.407
		$E = -0.250$		
5.500	13.974	−26.276	10.4866	4.141
5.250	26.627	−8.181	10.5000	4.091
5.500	48.668	−5.598	13.3394	9.860

Example of Fellows Spiral-face-gear Drive. As a definite example we shall use the same values as were used for the involute spur gear and face gear in Chap. 14. The helical gear will have a right-handed lead of 74.00 in. This gives the following values:

$$N_1 = 20 \qquad N_2 = 40 \qquad R_1 = 5.000 \qquad R_2 = 10.000 \qquad \phi_1 = 20°$$
$$R_{b1} = 4.69846 \qquad \cot \gamma_p = {}^{40}\!/_{20} = 2.000 \qquad \text{inv } \phi_1 = 0.01490$$
$$T = 0.7854 \qquad L = 74.00 \qquad \psi_1 = 23.000°$$
$$\alpha = \frac{0.7854}{10} - 0.01490 = 0.06364 \text{ radian} = 3.646°$$
$$\tan \psi_b = \tan 23° \cos 20° = 0.42447 \times 0.93969 = 0.39887$$

Using these values and the foregoing equations, we obtain the values tabulated in Table 16-1. These values are also plotted in Fig. 16-2.

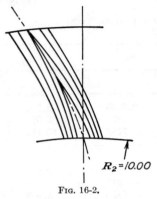

$R_2 = 10.00$

Fig. 16-2.

The opposite side of the face-gear teeth will be symmetrical to the calculated side in respect to a uniform-rise spiral that passes through the center of the tooth at R_2. The rise of this spiral will be in direct proportion to the lead of the helical gear and the numbers of teeth in the pair. In this example the lead of the helical gear is 74.00 in.; there are twice as many teeth in the face gear as there are in the helical pinion; hence the rise of this spiral per revolution of the face gear will be 2×74, which is equal to 148.00 in. This spiral is shown as a dot-and-dash line in Fig. 16-2.

With a larger reduction ratio, the effective face width of the crown gear would be increased. There are cusps in the form of these face-gear teeth similar to those on the face gear that meshes with a spur pinion. This indicates the presence of undercut tooth profiles. The effective face of this face gear stops just below the 10.00-in. radius of the nominal pitch circle.

OCTOID SPIRAL BEVEL GEAR

The study of the tooth forms of these spiral bevel gears is largely a study of the crown rack of the system. The spiral-bevel-gear teeth are conjugate to this crown rack; one spiral bevel gear is conjugate to the upper side of this crown rack, while the other or mating gear is conjugate to the lower side of this crown rack. In actual practice, when the gears are generated, two crown racks are used, identical in form except for the direction of the spiral. The spirals are of opposite hand for mating gears. The tooth form of the spiral bevel gear at any specific section is determined by the same methods as are used for bevel gears with straight teeth.

Theoretically, the ideal curve for the teeth of the crown rack is a logarithmic spiral because this curve gives a constant spiral angle at all diameters. However, with spiral bevel gears as with straight bevel gears, general practice and design is limited by the processes that are available to produce the product.

The most commonly used spiral bevel gear is the Gleason spiral bevel gear. This is produced by a circular cutter so that the curve of the crown-rack-tooth spiral is the arc of the circle described by the cutting edges of the tool, which are mounted on the periphery of the cutter body. The curve of such a crown rack is shown in Fig. 16-3.

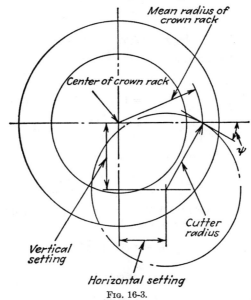

Fɪɢ. 16-3.

Unless otherwise specified, the spiral angle is measured at the mean radius of the crown-rack face. Referring to Fig. 16-3, when

ψ = spiral angle
R_m = mean radius of crown rack, in.
R_c = radius of cutter, in.
V = vertical setting of cutter, in.
H = horizontal setting of cutter, in.

we have the following from the conditions shown in Fig. 16-3:

$$V = R_c \cos \psi \qquad (16\text{-}9)$$
$$H = R_m - R_c \sin \psi \qquad (16\text{-}10)$$
$$\tan \psi = (R_m - H)/V \qquad (16\text{-}11)$$

Logarithmic Spiral. We shall examine the values of the logarithmic spiral to see how closely the arc of the cutter approximates it. Thus let

r = any radius of spiral, in.
θ = vectorial angle
m = cotangent of spiral angle
r_0 = starting radius[1] of spiral, in.

[1] This radius will be the radius to the middle of the crown rack, R_m.

ψ = angle of tangent to radius vector
e = base of natural logarithms
We have for the equation of the logarithmic spiral

$$r = r_0 e^{m\theta} \tag{16-12}$$

Whence

$$\log r = \log r_0 + m\theta \log e$$

But $\log e = 1.000$, whence

$$\log r = \log r_0 + m\theta$$

$$\frac{dr}{r\theta} = m\,d\theta \qquad \frac{r\,d\theta}{dr} = \tan\psi = \frac{1}{m} \qquad \frac{dr}{d\theta} = mr$$

$$\frac{d^2r}{d\theta\,dr} = m \qquad \frac{d^2r}{d\theta^2} = \frac{d^2r}{d\theta\,dr}\frac{dr}{d\theta} = rm^2$$

When R_c is the radius of curvature of spiral in inches,

$$R_c = \frac{[r^2 + (dr/d\theta)^2]^{3/2}}{r^2 - r(d^2r/d\theta^2) + 2(dr/d\theta)^2} \tag{1-34}$$

Substituting the values from the preceding equations into Eq. (1-34), combining, and simplifying, we obtain

$$R_c = r\sqrt{1 + m^2} \tag{16-13}$$

But

$$m^2 = \cot^2\psi$$

Whence

$$R_c = r/\sin\psi \tag{16-14}$$

If the radius of the cutter were made equal to the radius of curvature of the logarithmic spiral at the middle of the face of the crown rack, then the horizontal setting of the cutter would always be equal to zero, and the radius of the cutter would always be greater than the radius of the crown rack. This would require such a large cutter that some smaller radius must be used for practical reasons. The result of using smaller cutters is a greater change in the spiral angle across the face of the crown rack than would be present if the larger cutter could be used.

Example of Spiral. As a definite example we shall use the following values, and determine the change in spiral angle across the face of the crown rack:

Outside radius of crown rack............................... 8 in.
Inside radius of crown rack................................. 6 in.
Radius of cutter... 6 in.
Angle of spiral at mid-section............................. 30 deg

$$R_m = r_0 = 7.000 \qquad \sin 30° = 0.50000 \qquad \cos 30° = 0.86603$$

$$\text{Radius of curvature of logarithmic spiral} = \frac{7.00}{0.500} = 14.00$$

Referring to Fig. 16-4, when

ψ_1 = spiral angle at r

r = any radius, in.

and all other symbols are the same as before or as shown in the diagram, then

$$\tan \delta = \frac{H}{V} \qquad (16\text{-}15)$$

$$\tan \gamma = \frac{V}{H} \qquad (16\text{-}16)$$

$$\cos (\delta + \alpha) = \frac{V^2 + H^2 + R_c{}^2 - r^2}{2R_c \sqrt{V^2 + H^2}}$$

$$(16\text{-}17)$$

$$\cos (\gamma + \beta) = \frac{V^2 + H^2 + r^2 - R_c{}^2}{2r \sqrt{V^2 + H^2}}$$

$$(16\text{-}18)$$

$$\psi_1 = \alpha + \beta \qquad (16\text{-}19)$$

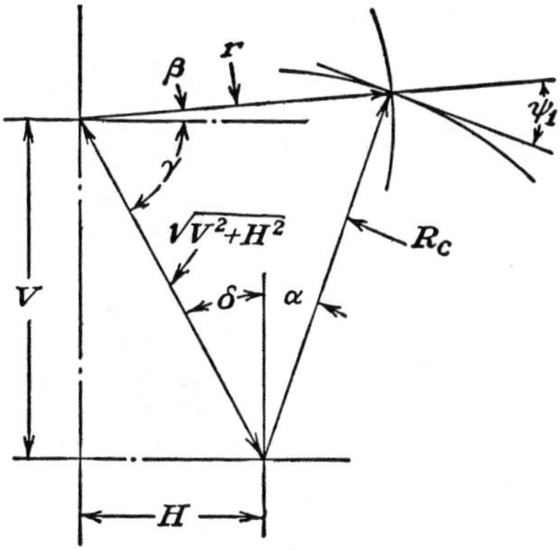

Fig. 16-4.

In this example, we have

$$V = 6 \times 0.86603 = 5.19618 \qquad\qquad V^2 = 27.000$$
$$H = 7.000 - (6 \times 0.50) = 4.000 \qquad H^2 = 16.000$$
$$\tan \delta = \frac{4.000}{5.19618} = 0.76980 \qquad\qquad \delta = 37.589°$$
$$\tan \gamma = \frac{5.19618}{4.000} = 1.29904 \qquad\qquad \gamma = 52.411°$$

When $r = 6.000$

$$\cos (\delta + \alpha) = \frac{43 + 36 - 36}{12 \sqrt{43}} = 0.54645 \qquad \delta + \alpha = 56.876°$$
$$\cos (\gamma + \beta) = \frac{43 + 36 - 36}{12 \sqrt{43}} = 0.54645 \qquad \gamma + \beta = 56.876°$$
$$\alpha = 56.876° - 37.589° = 19.287°$$
$$\beta = 56.876° - 52.411° = 4.465°$$
$$\psi_1 = 19.287° + 4.465° = 23.752°$$

When $r = 8.000$

$$\cos (\delta + \alpha) = \frac{43 + 36 - 64}{12 \sqrt{43}} = 0.19062 \qquad \delta + \alpha = 79.011°$$
$$\cos (\gamma + \beta) = \frac{43 + 64 - 36}{16 \sqrt{43}} = 0.67671 \qquad \gamma + \beta = 47.313°$$
$$\alpha = 79.011° - 37.589° = 41.422°$$
$$\beta = 47.313° - 52.411° = -5.098°$$
$$\psi_1 = 41.422° - 5.098° = 36.324°$$

If the cutter radius were the same as the radius of curvature of the logarithmic spiral at the mid-section, we would have

$$R_c = 14.000 \qquad\qquad\qquad R_c{}^2 = 196.000$$
$$V = 14.00 \times 0.86603 = 12.12442 \qquad V^2 = 147.000$$
$$H = 7.000 - (14 \times 0.500) = 0.000 \qquad H^2 = 0.000$$
$$\tan \delta = 0 \qquad \delta = 0 \qquad \tan \gamma = \infty \qquad \gamma = 90°$$

When $r = 6.000$

$$\cos(\delta + \alpha) = \frac{147 + 196 - 36}{28 \times 12.12442} = 0.90431 \qquad \delta + \alpha = 25.270°$$

$$\cos(\gamma + \beta) = \frac{147 + 36 - 196}{12 \times 12.12442} = -0.08935 \qquad \gamma + \beta = 95.126°$$

$$\alpha = 25.270° - 0 = 25.270° \qquad \beta = 95.126° - 90° = 5.126°$$
$$\psi_1 = 25.270° + 5.126° = 30.396°$$

When $r = 8.000$

$$\cos(\delta + \alpha) = \frac{147 + 196 - 64}{28 \times 12.12442} = 0.82184 \qquad \delta + \alpha = 34.730°$$

$$\cos(\gamma + \beta) = \frac{147 + 64 - 196}{16 \times 12.12442} = 0.07737 \qquad \gamma + \beta = 85.562°$$

$$\alpha = 34.730° - 0 = 34.730° \qquad \beta = 85.562° - 90° = -4.438°$$
$$\psi_1 = 34.730° - 4.438° = 30.292°$$

With the smaller cutter, the spiral angle varies from about $-6°$ at the inside radius of the crown gear to about $+6°$ at the outside radius. With the larger radius of cutter, the variation in the angle of the spiral is less than one-half of one degree.

Curvature and Contact. Thus far, in discussing the spiral or curvature of the crown-rack tooth, we have been referring to the conditions at the middle of the tip of the cutting tool. The actual radius of cutting changes slightly along the angular profile of the cutting edge. Thus there will be a difference in the actual radius of the crown-gear tooth between the two sides of the tooth space as indicated in Fig. 16-5. At most, this difference will be equal to the width of the tooth space. In most cases, different cutters are used to finish the two sides of

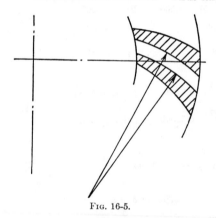

Fig. 16-5.

the teeth, and this difference is kept to a minimum.

The radius of curvature of the concave side of the tooth should be slightly larger than that of the convex side. Theoretically this would limit the contact to point contact. The relative difference in radius is so small, however, that the slight elastic deformation of the tooth surfaces under load will develop an appreciable length of contact. The actual contact area in service will show an area substantially as shown in Fig. 16-6.

In most cases, the design of bevel gears and their mountings is such that the pinion is overhung. This slight difference in curvature permits some slight bending or deflection without setting up contact at the edges of the gear faces.

That portion of the bevel gear tooth at the small end of the gear is called the *toe*, while the portion near the large end of the gear is called the *heel*. On heavily loaded spiral bevel gears, the gears are often so adjusted that under light loads, the contact is near the toe. Then when heavier loads are applied, the contact shifts towards the heel. Such a practice takes the greatest advantage of this characteristic of these spiral gears.

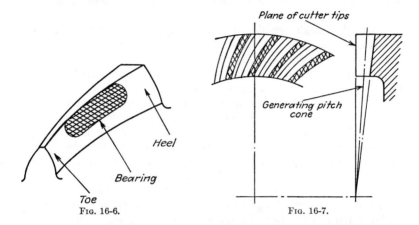

FIG. 16-6.   FIG. 16-7.

Form of Crown Rack. Theoretically, the pitch surface of the crown rack should be a plane. As actually made, however, the tips of the cutter lie in a plane so that the actual generating pitch surface of the Gleason spiral-bevel-gear system is a cone, as indicated in Fig. 16-7. This pitch-cone angle is very close to 90 deg, so that except for very small numbers of teeth in the pinion, the mathematical error introduced is generally much smaller than the normal and inevitable errors of actual production.

Gleason System of Spiral Bevel Gears. The proportions of the gear blanks for spiral bevel gears are determined in the same manner as those for bevel gears with straight teeth. The proportions of 1-DP crown-rack teeth at the large end are as follows:

	Inches
Addendum	0.850
Dedendum	1.038
Clearance	0.188
Working depth	1.700
Whole depth	1.888

Pressure angles of 14½ deg and 17 deg are used on these gears. The pressure angle employed depends upon the number of teeth in the pinion and on the gear ratio. The 14½-deg pressure angle is used on all combinations where the pinion has 14 or more teeth; for 13-tooth pinions when the gear has 20 or more teeth; for 12-tooth pinions when the gear has 20 or more teeth; for 11-tooth pinions where the gear has 25 or more teeth; and for 10-tooth pinions when the gear has 25 or more teeth. The 17-deg pressure angle is used for the smaller pinions that mesh with the smaller gears. A pinion of 10 teeth is the smallest standard gear of the system.

The spiral angles employed generally are between 30 and 35 deg. The hand of the spiral is denoted by its direction when the gear or pinion is viewed from its small end. A right-handed spiral curves off in a clockwise direction; a left-handed spiral curves off in a counterclockwise direction. The hand of spiral is reversed on mating gears.

The hand of spiral does not affect the quietness or smoothness of operation, but it does affect the intensities of the thrust loads. As regards the influence of the spiral alone, a right-handed spiral pinion driving in a clockwise direction tends to move the pinion toward the cone center. When this pinion drives in a counterclockwise direction, the spiral tends to move the pinion away from the cone center. A left-handed spiral driving pinion will act in the reverse direction in both cases.

Tooth Proportions. The Gleason Works have developed and published tables of tooth proportions for these spiral bevel gears. These proportions were determined by adjusting them until the amount of sliding during the approach action was about the same or slightly less than the amount of sliding during the recess action. When the addendum of the gear is reduced, the addendum of the mating pinion is increased a corresponding amount. The working depth and the whole depth of the teeth remain constant. The pressure angles are selected to give as low a pressure angle as possible without the presence of excessive undercutting of the pinion tooth.

Spiral Bevel Gears with Straight, Offset Teeth. Instead of using an arc of a circle for the approximation to the logarithmic spiral, we can use the straight line tangent to the spiral at the mid-section of the face of the crown gear as an approximation, as indicated in Fig. 16-8. This straight line will be tangent to a circle concentric with the crown rack. Referring to Fig. 16-8, when

R = radius to mid-section of face of crown rack, in.

ψ = spiral angle at R

R_b = radius of concentric circle to which straight line is tangent, in.

$$R_b = R \sin \psi \qquad (16\text{-}20)$$

Example of Straight, Offset Teeth. As an example we shall use the same values as were used for the example with curved teeth. Whence we have

$$R = 7.000 \qquad \psi = 30° \qquad \sin \psi = 0.500$$
$$R_b = 7.000 \times 0.500 = 3.500$$

Spiral Angle of Straight, Offset Teeth. We shall now determine the spiral angle at the outside and inside edges of the crown-gear face.
When r = radius to any point, in.
ψ_1 = spiral angle at r
then

$$\sin \psi_1 = R_b/r \qquad (16\text{-}21)$$

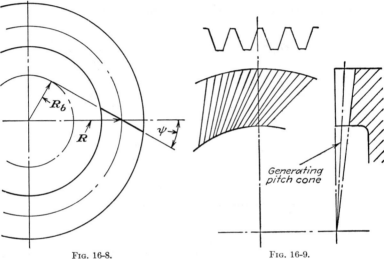

FIG. 16-8.   FIG. 16-9.

Example of Spiral Angle. Using the same values as before, we have

Outside radius of crown rack = 8.000 in.
Inside radius of crown rack = 6.000 in.

When $r = 6.000$

$$\sin \psi_1 = \frac{3.500}{6.000} = 0.58333 \qquad \psi_1 = 35.685°$$

When $r = 8.000$

$$\sin \psi_1 = \frac{3.500}{8.000} = 0.43750 \qquad \psi_1 = 25.945°$$

In this example, the straight-line form departs less from the logarithmic spiral than did the 6.000-in. radius used in the preceding example.

The form of the crown rack of this system is the same, except for the straight-line approximation instead of the arc of a circle, as the crown rack with curved teeth. This is shown in Fig. 16-9.

The tooth proportions of this system are identical to those of the system with curved teeth. In general, this type of tooth is used on large bevel gears where otherwise an extremely large circular cutter would be required.

FORMATE GEARS

Another type of spiral bevel gear, very similar to the foregoing, is also used. This type has been called the *formate gear*. In order to simplify the production of the gear, it is made with teeth of straight-sided profile. The approximation to the spiral may be an arc of a circle or a straight tangent line. This gear is produced by a simple forming process.

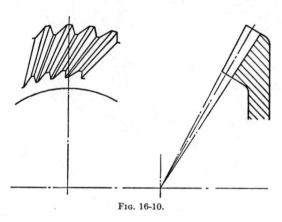

Fig. 16-10.

The pinion is generated with the same type of cutting tool as before. In its generation, however, it is made to roll, in effect, on the pitch cone of the mating gear instead of around the pitch plane of the crown rack. Thus the mating gear itself is substituted for the crown rack. We employ this same practice in the generation of the face gears to mesh with lantern pinions or with involute gears. The tooth forms are conjugate, and the meshing and operating conditions here are the same as for the equivalent type of spiral-bevel-gear drive. A formed gear with straight, offset teeth, which also represents the generating form for the pinion, is shown in Fig. 16-10.

The tooth proportions for these formate gears are the same as those for the spiral bevel gears that have already been described. The forms of the gear and pinion blanks, pitch-cone values, face and root angles, etc., for all these spiral bevel gears are determined in the same manner as the similar values for straight-toothed bevel gears.

SPIRAL BEVEL GEARS WITH FORMED TEETH

Some large bevel gears that are produced on template-forming machines are also made with spiral teeth. The same machine that is used for forming bevel gears with straight teeth is used for these spiral bevel gears by providing means to rock or oscillate the work blank as the cutting tool travels across the face of the blank. The rotation of the work is directly proportional to the travel of the tool along the face of the gear blank; hence the form of the spiral developed is a uniform rise or Archimedean spiral. The same templates that control the tooth form are used for both straight and spiral teeth.

CHAPTER 17

SKEW BEVEL OR HYPOID GEARS

A pair of nonparallel shafts that do not intersect may be driven by a pair of skew bevel or hypoid gears. Such gears have much in common with bevel gears and also begin to approach some of the conditions of spiral gears or worm gears. The pitch surfaces of skew bevel gears are hyperboloids of revolution. The action between these pitch surfaces is not true rolling but is a combination of rolling and sliding. We shall therefore start with a study of these hyperboloids of revolution.

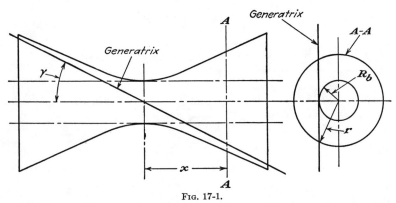

Fig. 17-1.

Hyperboloid of Revolution. A hyperboloid of revolution is the surface developed by a straight-line generatrix, tangent to a base cylinder and at an angle to the axis of the base cylinder, as it is revolved about the axis of the base cylinder. Such a surface is shown in Fig. 17-1. Referring to this figure, let

R_b = radius of base cylinder, in.

γ = angle of generatrix with axis of base cylinder

x = axial distance from point of tangency of generatrix with base cylinder, in.

r = radius of hyperboloid at r, in.

We have the following from the geometrical conditions shown in Fig. 17-1:

$$r = \sqrt{R_b{}^2 + x^2 \tan^2 \gamma} \qquad (17\text{-}1)$$

When two hyperboloids are tangent to each other, the locus of the points of tangency between them is along their straight-line generatrices. In other words, the line of tangency is where the two generatrices coincide. This line of tangency is the locus of all the pitch points of the conjugate gear-tooth profiles. For a given angle of axes and ratio of rotation, the angles of these generatrices are determined in the same manner as the angles of the pitch cones when the two axes intersect.

The diameters of the base cylinders of the pitch hyperboloids are, in effect, the pitch diameters of two spiral gears with helix angles equal to the angles of the generatrices. The sliding along the common generatrix (locus of pitch points) is similar to the sliding on the pitch surfaces of such spiral gears. The sum of the base cylinder radii is equal to the shortest distance between the axes of the two hyperboloids. This is the distance by which the axes are offset from each other, and it will be called the *center distance*. Thus when

N_p = number of teeth in hypoid pinion
N_g = number of teeth in hypoid gear
Σ = shaft angle
γ_p = angle of generatrix of pinion hyperboloid
γ_g = angle of generatrix of gear hyperboloid
C = center distance, in.
R_{bp} = radius of base cylinder of pinion hyperboloid, in.
R_{bg} = radius of base cylinder of gear hyperboloid, in.
r_p = radius of pinion hyperboloid, in.
r_g = radius of gear hyperboloid, in.

$$\Sigma = \gamma_p + \gamma_g \tag{14-1}$$

$$\tan \gamma_p = \frac{\sin \Sigma}{(N_g/N_p) + \cos \Sigma} \tag{14-2}$$

We shall restrict this analysis to the conditions that exist when the shaft angles are equal to 90 deg. Whence, when $\Sigma = 90°$,

$$\tan \gamma_p = \frac{N_p}{N_g} \qquad \tan \gamma_g = \frac{N_g}{N_p}$$

To determine the sizes of the base cylinders of the pitch hyperboloids, we shall treat them as spiral gears. Thus we have the following for spiral gears of 1 normal DP:

γ_p = helix angle of spiral pinion
γ_g = helix angle of spiral gear
R_{bp} = pitch radius of spiral pinion, in.
R_{bg} = pitch radius of spiral gear, in.

$$R_{bp} = \frac{N_p}{2 \cos \gamma_p} \qquad R_{bg} = \frac{N_g}{2 \cos \gamma_g} = \frac{N_g}{2 \sin \gamma_p}$$

$$C = R_{bp} + R_{bg} = \frac{N_p}{2 \cos \gamma_p} + \frac{N_g}{2 \sin \gamma_p}$$

$$\frac{N_p}{N_g} = \tan \gamma_p \quad \text{and} \quad N_g = \frac{N_p}{\tan \gamma_p}$$

$$C = \frac{N_p}{2}\left(\frac{1}{\cos \gamma_p} + \frac{1}{\sin \gamma_p \tan \gamma_p}\right) = \frac{N_p}{2 \cos \gamma_p} \frac{1}{\sin^2 \gamma_p}$$

Whence

$$C = \frac{R_{bp}}{\sin^2 \gamma_p}$$

and

$$R_{bp} = C \sin^2 \gamma_p \tag{17-2}$$

In similar manner we obtain

$$R_{bg} = C \cos^2 \gamma_p \tag{17-3}$$

$$\sin^2 \gamma_p = \frac{N_p{}^2}{N_p{}^2 + N_g{}^2} \qquad \cos^2 \gamma_p = \frac{N_g{}^2}{N_p{}^2 + N_g{}^2}$$

We have Eq. (17-1) to determine the radius of the hyperboloid at any point.

Example of Pitch Hyperboloids. As a definite example we shall use the following values:

$$N_p = 20 \qquad N_g = 40 \qquad \Sigma = 90° \qquad C = 3.000$$
$$\tan \gamma_p = {}^{20}\!/_{40} = 0.500 \qquad \gamma_p = 25.565° \qquad \sin^2 \gamma_p = 0.200$$
$$\tan \gamma_g = {}^{40}\!/_{20} \quad 2.000 \qquad \gamma_g = 63.435° \qquad \cos^2 \gamma_p = 0.800$$
$$R_{bp} = 3 \times 0.200 = 0.600 \qquad R_{bg} = 3 \times 0.800 = 2.400$$

From these values and a series of values of x in Eq. (17-1), we obtain the coordinates tabulated in Table 17-1. These two pitch surfaces are also plotted in Fig. 17-2. The projection of the locus of pitch points, which is also the projection of the common generatrix of the two hyperboloids, is also shown in Fig. 17-2.

Sliding on Hyperboloids. It will be noted from a comparison of the values shown in Table 17-1 that the radii of the contacting circles of the hyperboloids are not directly proportional to the numbers of teeth. These circles (axial sections of the hyperboloids) touch each other at the pitch point but do not rotate in the same plane or in planes at right angles to each other. The normal circular pitch of the two gears, however, must be identical, *i.e.*, the pitch points of successive teeth on both gears must coincide. Therefore sliding must be present between the pitch surfaces at the pitch point.

We shall consider the conditions at a pitch point at the gorge section. Here the two base cylinders of the hyperboloids touch each other, but

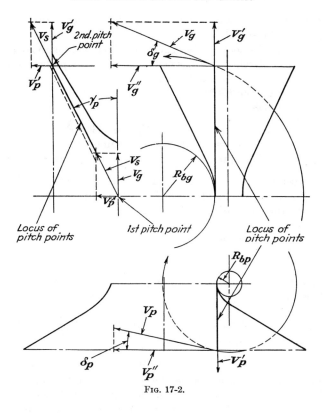

Fig. 17-2.

TABLE 17-1. COORDINATES OF HYPERBOLOIDS
(Plotted in Fig. 17-2)

Pinion hyperboloid		Gear hyperboloid	
x, in.	r_P, in.	x, in.	r_2, in.
0.0000	0.6000	0.0000	2.4000
1.0000	0.7810	0.5000	2.6000
2.0000	1.1662	1.0000	3.1241
3.0000	1.6155	1.5000	3.8419
4.0000	2.0881	2.0000	4.6648
5.0000	2.5710	2.5000	5.5462
6.0000	3.0594	3.0000	6.4622

they rotate on axes at right angles to each other. Hence the action here is all sliding. We shall let

V_p = peripheral velocity of base cylinder of pinion hyperboloid, ft/min
V_g = peripheral velocity of base cylinder of gear hyperboloid, ft/min
V_s = sliding velocity at pitch point, ft/min
n_p = rpm of pinion
n_g = rpm of gear

$$V_p = 0.5236 n_p R_{bp} \qquad V_g = 0.5236 n_g R_{bg}$$
$$n_g = (N_p/N_g)n_p = n_p \tan \gamma_p$$
$$V_p/V_g = n_p R_{bp}/n_p R_{bg} \tan \gamma_p = C \sin^2 \gamma_p/C \cos^2 \gamma_p \tan \gamma_p = \tan \gamma_p$$

These vectors are shown in Fig. 17-2. The vectors for sliding are shown in the direction of sliding on the pinion. Hence the direction of the sliding at this pitch point is along the common generatrix of the two hyperboloids.

From the foregoing we have

$$V_g = V_p/\tan \gamma_p$$

As these two components of the sliding velocity are at right angles to each other, we have

$$V_s = \sqrt{V_p{}^2 + V_g{}^2} = V_p \sqrt{1 + (1/\tan \gamma_p)^2} = V_p/\sin \gamma_p \quad (17\text{-}4)$$

This equation may also be written

$$V_s = 0.5236 n_p R_{bp}/\sin \gamma_p$$

This sliding velocity is the same as that between two spiral gears.

We shall now consider the conditions at a second pitch point that is at a distance x along the axis of the pinion. We shall use the same symbols as before.

$$V_p = 0.5236 n_p R_p \qquad V_g = 0.5236 n_g R_g$$

R_p = radius of pinion hyperboloid, in.
R_g = radius of gear hyperboloid, in.
V'_p, V'_g = sliding component of velocities
V''_p, V''_g = rolling component of velocities
δ_p, δ_g = angle of direction of velocities

$$
\begin{aligned}
V'_p &= V_p \sin \delta_p & V'_g &= V_g \sin \delta_g \\
V''_p &= V_p \cos \delta_p & V''_g &= V_g \cos \delta_g \\
\sin \delta_p &= R_{bp}/R_p & \sin \delta_g &= R_{bg}/R_g \\
\cos \delta_p &= x \tan \gamma_p/R_p & \cos \delta_g &= x/R_g
\end{aligned}
$$

Whence we obtain

$$V'_p = 0.5236 n_p R_{bp}$$

This is the same value as the sliding at the first pitch point.

$$V'_g = 0.5236 n_g R_{bg}$$

This is the same value as the sliding at the first pitch point. Hence this sliding is the same in amount and direction at all pitch points.

$$V''_p = 0.5236 n_p x \tan \gamma_p$$
$$V''_g = 0.5236 n_g x \quad \text{but} \quad n_g = n_p \tan \gamma_p$$

Whence

$$V''_g = 0.5236 n_p x \tan \gamma_p$$

This last value is the same as that for the pinion, hence the rolling velocities in the direction of rotation of the crown member of the pair is the same for both members.

Pitch Surface of Crown Member of Hypoid Gears. The pitch surface of the crown rack of bevel gears is a circular disk or plane that is tangent to the two pitch cones with its center at the cone center of the two pitch cones. The pitch surface of the crown member of hypoid gears, however, is a warped surface that is tangent to the two pitch hyperboloids. Its axis will be at the point where the two base cylinders of the tangent hyperboloids touch each other. This axis will lie in a plane that is parallel to the axes of the two hyperboloids.

We can use either member of the pair to determine the nature of this warped surface. We shall use the pinion as indicated in Fig. 17-3. We shall determine the lead angle

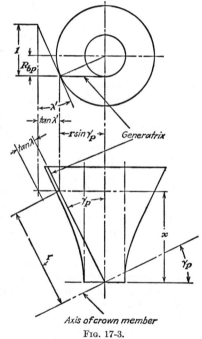

Fig. 17-3.

of this surface at any radius r of the warped surface. This angle will be the angle of the tangent to the hyperboloid at the given position of the generatrix. Thus let

R_{bp} = radius of base cylinder of pinion hyperboloid, in.

γ_p = angle of generatrix of pinion hyperboloid

λ' = angle of trace of tangent plane with plane of rotation of pinion

λ = lead angle of warped pitch surface of crown member of hypoid gears

r = any radius of warped pitch surface of crown member, in.

L = lead of pitch surface of crown member of hypoid gears, in.

Referring to Fig. 17-3, we have the following:

$$\tan \lambda' = R_{bp}/r \sin \gamma_p$$
$$\tan \lambda = \tan \lambda' \cos \gamma_p = R_{bp}/r \tan \gamma_p \qquad (17\text{-}5)$$

But

$$R_{bp} = C \sin^2 \gamma_p$$

whence

$$\tan \lambda = C \sin^2 \gamma_p/r \tan \gamma_p = C \sin \gamma_p \cos \gamma_p/r \qquad (17\text{-}6)$$

But

$$\tan \lambda = L/2\pi r$$

whence

$$L = 2\pi r \tan \lambda$$
$$L = 2\pi C \sin \gamma_p \cos \gamma_p \qquad (17\text{-}7)$$

Therefore this lead is the same for all diameters, and the pitch surface of the crown member of the hypoid gear is a screw helicoid.

Circular Pitch of Hypoid Gears.   It should be apparent that hypoid gears have much in common with spiral gears.   Each member of the pair will have its own circular pitch in its plane of rotation.   As the diameters of the pitch hyperboloids are not directly proportional to the numbers of teeth, these circular pitches will be different from each other. Their normal circular pitches, however, must be identical for any given pitch point, and must also be the same as the normal circular pitch of the crown member for the same pitch point.

We shall start with a pitch point at the axis of the crown member. Here the normal circular pitch does not reduce to zero, but will be equal to the lead of the crown member divided by the number of teeth in one complete turn of the crown member.   Thus let

N_p = number of teeth in hypoid pinion

N_g = number of teeth in hypoid gear

N_r = number of teeth in one complete revolution of crown member

p_n = normal circular pitch, in.

C = center distance, in.

r = any radius of helicoidal crown member

When $r = 0$

$$p_n = \frac{L}{N_r} = \frac{2\pi C \sin \gamma_p \cos \gamma_p}{N_r} \qquad (17\text{-}8)$$

$$N_r = \frac{2\pi C \sin \gamma_p \cos \gamma_p}{p_n} \qquad (17\text{-}9)$$

When p_p = circular pitch of pinion in its plane of rotation, in.

p_g = circular pitch of gear in its plane of rotation, in.

$$p_p = \frac{2\pi R_{bp}}{N_p} = \frac{2\pi C \sin^2 \gamma_p}{N_p} \qquad p_g = \frac{2\pi R_{bg}}{N_g} = \frac{2\pi C \cos^2 \gamma_p}{N_g}$$

But $N_g = N_p/\tan \gamma_p$, whence

$$p_g = \frac{2\pi C \sin \gamma_p \cos \gamma_p}{N_p}$$

$$p_n = p_p \cos \gamma_p = p_g \sin \gamma_p$$

For the pinion

$$p_n = \frac{2\pi C \sin^2 \gamma_p \cos \gamma_p}{N_p}$$

For the gear

$$p_n = \frac{2\pi C \sin^2 \gamma_p \cos \gamma_p}{N_p}$$

These last two values are identical. Substituting this value of the normal pitch into Eq. (17-9), we obtain

$$N_r = \frac{2\pi C N_p \sin \gamma_p \cos \gamma_p}{2\pi C \sin^2 \gamma_p \cos \gamma_p} = \frac{N_p}{\sin \gamma_p}$$

But

$$\sin \gamma_p = \frac{N_p}{\sqrt{N_p^2 + N_g^2}}$$

whence

$$N_r = \sqrt{N_p^2 + N_g^2} \qquad (17\text{-}10)$$

This value is the same as for the number of teeth in the crown rack of bevel gears.

When $r = x/\cos \gamma_p$. We shall now determine the normal circular pitch in the plane of rotation of the crown member for the crown member, the pinion, and the gear at any radius r of the crown member. The geometry for this is identical to that for the rolling velocities of the several members. Thus we have

For the crown member

$$p_{nr} = \frac{2\pi r}{N_r} = \frac{2\pi x}{\cos \gamma_p \sqrt{N_p^2 + N_g^2}} = \frac{2\pi x}{N_g} \qquad (17\text{-}11)$$

For the pinion

$$p_{np} = \frac{2\pi R_p \cos \delta_p}{N_p} = \frac{2\pi x \tan \gamma_p R_p}{N_p R_p} = \frac{2\pi x}{N_g} \qquad (17\text{-}11)$$

For the gear

$$p_{ng} = \frac{2\pi R_g \cos \delta_g}{N_g} = \frac{2\pi x R_g}{R_g N_g} = \frac{2\pi x}{N_g} \qquad (17\text{-}11)$$

All these values of the normal pitch in the plane of rotation of the crown member are identical. They are also the same as the circular pitch of the bevel gears in the plane of rotation of the crown rack.

Summary of Hypoid-gear Tooth Action. It should be apparent that the nature of the action between hypoid-gear teeth is, in effect, the combination of the rolling of bevel-gear tooth action and the sliding of spiral-gear or worm-gear tooth action. The amount of the sliding is dependent upon the amount of offset of the axes. The pitch surface of the crown member is a screw helicoid. When the offset is reduced to zero, the pitch surface of the crown member becomes a plane, and the resulting gears become bevel gears.

Because of the lead of the pitch surface of the crown member, if the pressure angle is the same on both sides of the tooth of the generating member, the effective pressure angle of operation in the plane of rotation of the crown member becomes greater by the amount of the lead angle on one side of the tooth, and less by the same amount on the other side of the tooth. This lead angle changes with the diameter of the crown member. Hence for substantially similar pressure angles of operation on both sides of the tooth, the nominal pressure angles of the two sides of the tooth of the generating member must be different.

Theoretically an interchangeable tooth form for the helicoid crown member of hypoid gears could be developed. The form must be held symmetrical in relation to the locus of pitch points on the helicoid surface of the crown member; the depth of the tooth must also vary with the diameter of the crown member, and reduce to zero at its axis. As a result, the development of the full form of such a crown member by means of a simple cutting tool controlled by the many different motions required to maintain its constantly changing position and alignment would involve a very complex mechanism.

The chances are that the more practical solution is to make one member of the pair of such a form that it can be readily reproduced by some simple motions of a suitable cutting tool, and then generate the mating member to be conjugate to this selected form. The arbitrary member selected may be the pinion member or the gear member. The available choice is quite extensive. The further discussion of hypoid gears will be devoted to such arbitrary combinations.

LANTERN PINION–HYPOID FACE GEAR

As the first example of a hypoid-gear drive, we shall use a lantern pinion and a hypoid face gear because this is the simplest one to analyze. The hypoid face gear could be generated in the same manner as indicated in Chap. 14 except that the axes of the lantern pinion and the face gear do not intersect each other.   Referring to Fig. 17-4, when

C = center distance, in.

N_p = number of teeth or pins in lantern pinion

N_g = number of teeth in hypoid face gear

R_p = radius to center of pins in lantern pinion, also nominal pitch radius of lantern pinion, in.

R_g = nominal pitch radius of hypoid face gear, in.

A = radius of pins of lantern pinion, in.

ϕ = momentary pressure angle of operation, plane of rotation of lantern pinion

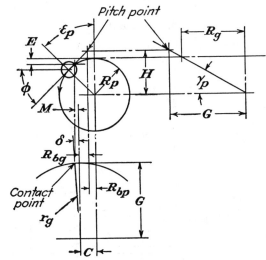

Fig. 17-4.

H = distance of pitch point from center line of lantern pinion, in.

G = height to pitch point and to contact point, in.

E = distance of contact point from tangent line to pin circle of lantern pinion, in.

r_g = radius to contact point on hypoid face gear, in.

δ = angle to contact point on hypoid face gear

ϵ_p = angle of rotation of lantern pinion

ϵ_g = angle of rotation of hypoid face gear

γ_p = angle of generatrix of hyperboloid pitch surface of pinion

R_{bp} = radius of base cylinder of pinion hyperboloid, in.

R_{bg} = radius of base cylinder of face gear hyperboloid, in.

M = distance of contact point from center line of face gear, in.

θ_g = vectorial angle to hypoid-face-gear tooth form

We have to start

$$\tan \gamma_p = N_p/N_g \qquad (14\text{-}3)$$
$$R_{bp} = C \sin^2 \gamma_p \qquad (17\text{-}2)$$
$$R_{bg} = C \cos^2 \gamma_p \qquad (17\text{-}3)$$

The method of analysis here is very similar to that of the lantern-pinion drive analyzed in Chap. 14. In this drive, however, the two sides of the teeth of the face gear are not symmetrical. We must therefore determine the forms of the two sides separately. Figure 17-4 shows the contact on the left-hand side of the face-gear tooth, or the side of the tooth that is towards the center of the face gear.

From the geometrical conditions shown in Fig. 17-4, which is similar in many respects to Fig. 14-3, we have the following:

$$\sin \phi = [R_p(1 - \cos \epsilon_p) - E]/A \qquad (17\text{-}12)$$
$$H = R_p \cos \epsilon_p + (R_p \sin \epsilon_p - R_{bp}) \tan \phi \qquad (17\text{-}13)$$
$$G = H/\tan \gamma_p \qquad (17\text{-}14)$$
$$M = R_p \sin \epsilon_p - A \cos \phi - C \qquad (17\text{-}15)$$
$$\tan \delta = M/G \qquad (17\text{-}16)$$
$$r_g = G/\cos \delta \qquad (17\text{-}17)$$
$$R_g = R_p/\tan \gamma_p \qquad (17\text{-}18)$$
$$\theta_g = \delta - \epsilon_g \qquad (17\text{-}19)$$

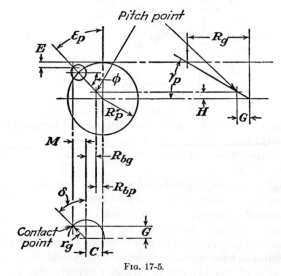

Fig. 17-5.

Figure 17-5 shows the conditions on the second side of the face-gear tooth. From this we have the following:

$$\sin \phi = [R_p(1 - \cos \epsilon_p) - E]/A \qquad (17\text{-}12)$$
$$H = R_p \cos \epsilon_p - (R_p \sin \epsilon_p - R_{bp}) \tan \phi \qquad (17\text{-}20)$$
$$G = H/\tan \gamma_p \qquad (17\text{-}14)$$
$$M = R_p \sin \epsilon_p + A \cos \phi - C \qquad (17\text{-}21)$$
$$\tan \delta = M/G \qquad (17\text{-}16)$$
$$r_g = G/\cos \delta \qquad (17\text{-}17)$$
$$\theta_g = \delta - \epsilon_g \qquad (17\text{-}19)$$

Example of Lantern-pinion–hypoid-face-gear Drive. As a definite example we shall use the same lantern pinion as before with a center distance of 1.500 in. This gives the following values:

$$N_p = 18 \qquad N_g = 36 \qquad R_p = 5.000 \qquad R_g = 10.000 \qquad A = 0.400$$
$$C = 1.500 \qquad R_{bp} = 0.300 \qquad R_{bg} = 1.200 \qquad \tan \gamma_p = {}^{18}\!\!/_{36} = 0.500$$

Using a series of values of E, ϵ_p, and ϵ_g, we obtain the values that are tabulated in Tables 17-2 and 17-3. These values are also plotted in Fig. 17-6. The coordinates for the second side of the tooth have been moved one tooth interval, or 10 deg, so as to give the contour of the tooth rather than that of the space.

As before, this figure gives a contour map of the tooth of the hypoid face gear. From this layout, sections of the gear tooth have been determined graphically.

All the contours for sections where the value of E is plus are not continuous curves but are formed of two intersecting curves. Contact does not exist on

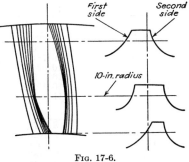

FIG. 17-6.

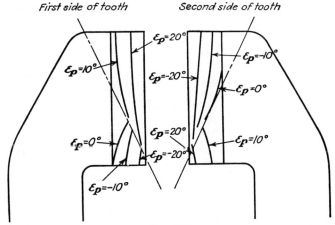

FIG. 17-7.

TABLE 17-2.   COORDINATES OF HYPOID FACE GEAR FOR LANTERN PINION—FIRST
SIDE OF TEETH
(Plotted in Fig. 17-6)

ϵ_p, deg	r_g, in.	θ_g, deg	G, in.
		$E = -0.300$	
−10	4.2365	−31.213	3.4181
− 5	8.2907	−12.292	8.0160
0	9.4731	−10.328	9.3197
5	10.3959	− 9.379	10.3210
10	12.9986	− 8.397	12.9757
		$E = -0.200$	
−15	4.2365	−50.402	1.8474
−10	8.0710	−14.212	7.6215
− 5	9.2812	−11.660	8.9991
0	9.8286	−10.828	9.6536
5	10.2356	−10.355	10.1396
10	10.9701	− 9.828	10.9311
15	14.5361	− 8.908	14.5317
		$E = -0.100$	
−15	7.4099	−17.136	6.7354
−10	9.1204	−12.387	8.7037
− 5	9.7817	−11.206	9.5031
0	10.0243	−10.852	9.8451
5	10.1501	−10.690	10.0466
10	10.4520	−10.451	10.4048
15	11.4937	− 9.996	11.4827
		$E = 0.000$	
−15	8.7477	−13.648	8.1586
−10	9.7928	−11.360	9.3963
− 5	10.1638	−10.787	9.8917
0	10.1789	−10.758	10.0000
5	10.0817	−10.848	9.9748
10	10.1188	−10.818	10.0679
15	10.6104	−10.568	10.5952
		$E = 0.100$	
−20	7.9779	−16.467	7.1417
−15	9.6325	−11.826	9.0897
−12	10.1543	−10.827	9.7195
12	9.8534	−11.009	9.8157
15	10.0324	−10.922	10.0145
20	11.0428	−10.702	11.0421
25	17.6493	−11.015	17.6433
		$E = 0.200$	
−25	6.1089	−27.291	4.6940
−20	9.0843	−13.326	8.3418
−17	9.9828	−11.177	9.3999
17	9.6806	−11.091	9.6706
20	10.1386	−10.999	10.1371
25	12.3499	−11.031	12.3458
		$E = 0.300$	
−25	7.8963	−17.732	6.8223
−20	10.0519	−11.047	9.3813
10	9.4097	−11.156	9.4078
25	10.7496	−11.166	10.7467

TABLE 17-3. COORDINATES OF HYPOID FACE GEAR FOR LANTERN PINION—SECOND
SIDE OF TEETH
(Plotted in Fig. 17-6)

ϵ_p, deg	r_g, in.	θ_g, deg	G, in.
		$E = -0.300$	
−10	16.4300	−2.799	16.2781
−5	12.0364	−5.883	11.9078
0	10.7593	−6.948	10.6803
5	9.6433	−7.755	9.6028
10	6.7391	−9.250	6.7205
		$E = -0.200$	
−15	17.6710	−1.101	17.4722
−10	12.2519	−4.758	12.0747
− 5	11.0414	−5.838	10.9247
0	10.4105	−6.362	10.3464
5	9.8114	−6.764	9.7842
10	8.7710	−7.248	8.7651
15	4.7872	−8.156	4.7869
		$E = -0.100$	
−15	12.8290	−3.734	12.5832
−10	11.1742	−5.343	10.9925
−5	10.5359	−5.982	10.4207
0	10.2157	−6.253	10.1549
5	9.9008	−6.452	9.8772
10	9.2955	−6.692	9.2914
15	7.8363	−6.850	7.8359
		$E = 0.000$	
−15	11.4219	−4.794	11.1600
−10	10.4875	−5.850	10.2999
−5	10.1490	−6.206	10.0321
0	10.0604	−6.277	10.0000
5	9.9711	−6.322	9.9490
10	9.6314	−6.438	9.6283
15	8.7248	−6.476	8.7234
		$E = 0.100$	
−20	12.0864	−3.710	11.7421
−15	10.5068	−5.706	10.2289
−12	10.0733	−6.264	9.8435
12	9.7475	−6.356	9.7473
15	9.3060	−6.348	9.3041
20	7.7716	−5.900	7.7517
25	0.9079	+45.369	0.4829
		$E = 0.200$	
−25	13.8340	−1.343	13.4322
−20	10.8266	−5.116	10.4520
−17	10.0580	−6.259	9.7261
17	9.4623	−6.311	9.4554
20	8.6775	−6.055	8.6569
25	5.8515	−3.557	5.7804
		$E = 0.300$	
−25	11.7619	−3.542	11.3039
−20	9.8230	−6.623	9.4125
20	9.4058	−6.281	9.3860
25	7.4437	−4.967	7.3795

these sections at the central position of the pin but starts at some distance on either side of this central position.

The projections of the contact lines between the two members at various angular positions can be determined as before in Chap. 14. For any given angular position, the value of E is one coordinate, and the value of G is the other. These contact lines are shown in Fig. 17-7 for the two sides of the tooth of the hypoid face gear.

The tooth forms and the contact conditions are different on the two sides of the teeth. According to the way this analysis has been made, if the lantern pinion is the driving member and the first side of the face-gear tooth is the loaded side, then the direction of rotation is in the opposite direction to that indicated in Fig. 17-4. Under these conditions, the plus values of ϵ_p are approach action, and the minus values are recess action. Under these circumstances, most of the surface of the face-gear tooth acts during the approach action.

If the lantern pinion is the driving member and the second side of the face-gear tooth is the loaded side, then the direction of rotation is reversed. It will then be in the same direction as indicated in Fig. 17-4. Here also, most of the surface of the face-gear tooth acts during the approach action. If the smoother conditions of recess action are needed, then this design of drive would be better for a step-up drive where the hypoid face gear is the driving member. Of the two sides of the face-gear tooth, the first side is the more favorable one. The conditions on both sides of the teeth would also be more favorable with a larger reduction ratio. With a smaller angle of the generatrix of the pinion hyperboloid, the face width could be increased, and more recess action would be present.

FELLOWS HYPOID-FACE-GEAR DRIVE

Another type of hypoid drive is the one where a spur pinion or helical involute pinion meshes with a face gear where the axes of the pair do not intersect. The hypoid face gear is generated with a pinion-shaped cutter of substantially the same size and form as the mating pinion. The relative positions of the hypoid face gear and the pinion-shaped cutter are the same as those of the face gear and the mating pinion of the hypoid drive, and their relative motions are the same.

The analysis of the contact conditions here is very similar to the analysis of the same type of gear drive with intersecting axes as given in Chap. 14. The pitch surfaces of these gears are hyperboloids of revolution. We will start with the analysis of the conditions when the pinion member is a spur gear.

Involute Spur Pinion as Hypoid Pinion. Referring to Figs. 17-8 and 14-14, let

R_1 = nominal pitch radius of involute spur pinion, in.

R_{b1} = radius of base cylinder of involute spur pinion, in.

ϕ_1 = pressure angle of spur pinion at R_1

r_1 = radius to contact point on spur pinion, in.

ϕ_2 = pressure angle of spur pinion at r_1

ϕ = momentary pressure angle of drive in plane of rotation of spur pinion

α = initial position of origin of involute on spur pinion

ϵ_1 = angle of rotation of spur pinion

γ_p = angle of generatrix of pinion hyperboloid

R_{bp} = radius of base cylinder of pinion hyperboloid, in.

C = center distance, in.

R_2 = nominal pitch radius of hypoid face gear, in.

r_2 = radius to contact point on hypoid face gear, in.

H = position of pitch point from pinion axis, in.

G = height to pitch point and to contact point, in.

M = distance of contact point from face-gear center line, in.

ϵ_2 = angle of rotation of hypoid face gear

δ = angle to contact point on hypoid face gear

θ_2 = vectorial angle to hypoid-face-gear tooth form

N_1 = number of teeth in involute pinion

N_2 = number of teeth in hypoid face gear

T = arc space thickness of spur pinion at R_1, in.

E = distance of contact point from R_1, in.

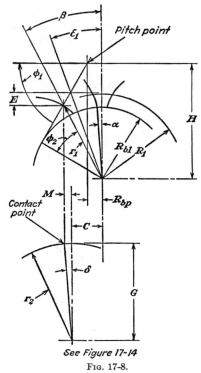

See Figure 17-14

Fig. 17-8.

The conditions here differ from those shown in Fig. 14-14 only in the displacement of the angular element that contains the pitch points and the displacement of the axis of the face gear. Thus we have from the geometrical conditions shown in Fig. 17-8 the following:

$$\alpha = (T/2R_1) - \text{inv } \phi_1 \tag{14-29}$$

$$\cos \beta = (R_1 - E)/r_1 \tag{14-30}$$

$$\cos \phi_2 = R_{b1}/r_1 \tag{14-31}$$

$$\epsilon_1 = \beta - \alpha - \text{inv } \phi_2 \tag{14-32}$$

$$\phi = \beta + \phi_2 \tag{14-33}$$

$$H = (R_{b1}/\cos \phi) - R_{bp} \tan \phi \tag{17-22}$$

$$G = H \cot \gamma_p \tag{17-23}$$

$$M = r_1 \sin \beta - C \qquad (17\text{-}24)$$
$$\tan \delta = M/G \qquad (17\text{-}25)$$
$$r_2 = G/\cos \delta \qquad (17\text{-}26)$$
$$\theta_2 = \delta - \epsilon_2 \qquad (17\text{-}27)$$

The foregoing equations apply to one side of the teeth of the hypoid face gear as shown in Fig. 17-8. The conditions on the two sides of the teeth are different. For the second side of the teeth of the hypoid face

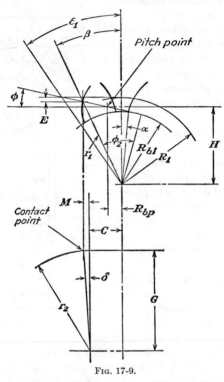

FIG. 17-9.

gear, we have the conditions shown in Fig. 17-9. From this we have the following:

$$\alpha = \text{inv } \phi_1 - (T/2R_1) \qquad (17\text{-}28)$$
$$\cos \beta = (R_1 - E)/r_1 \qquad (14\text{-}30)$$
$$\cos \phi_2 = R_{b1}/r_1 \qquad (14\text{-}31)$$
$$\epsilon_1 = \beta - \alpha + \text{inv } \phi_2 \qquad (17\text{-}29)$$
$$\phi = \phi_2 - \beta \qquad (17\text{-}30)$$
$$H = (R_{b1}/\cos \phi) + R_{bp} \tan \phi \qquad (17\text{-}31)$$

$$G = H \cot \gamma_p \qquad (17\text{-}23)$$
$$M = r_1 \sin \beta - C \qquad (17\text{-}24)$$
$$\tan \delta = M/G \qquad (17\text{-}25)$$
$$v_2 = G/\cos \delta \qquad (17\text{-}26)$$
$$\theta_2 = \delta - \epsilon_2 \qquad (17\text{-}27)$$

Example of the Fellows Spur-pinion–hypoid-gear Drive. As a definite example we shall use the following values:

$$N_1 = 20 \qquad N_2 = 60 \qquad R_1 = 5.000 \qquad R_2 = 15.000 \qquad \phi_1 = 20°$$
$$R_{b1} = 4.69846 \qquad \cot \gamma_p = {}^{60}\!\!/_{20} = 3.000 \qquad T = 0.7854$$
$$\text{inv } \phi_1 = 0.01490 \qquad C = 3.000 \qquad R_{bp} = 0.300$$
$$\alpha = \frac{0.7854}{10.00} - 0.01490 = 0.06364 \text{ radian} = 3.646°$$

Using these values together with a series of values of E and r_1 in the foregoing equations, we obtain the values for the two sides of the teeth of the hypoid face gear,

TABLE 17-4. COORDINATES OF HYPOID-FACE-GEAR TEETH—FIRST SIDE OF TEETH
(Plotted in Fig. 17-10)

r_1, in.	ϕ, deg	ϵ_1, deg	r_2, in.	θ_2, deg
		$E = 0.500$		
4.750	27.167	14.961	15.5419	−10.465
5.000	45.842	21.312	19.3239	−9.538
5.250	57.502	25.218	24.8241	−9.089
		$E = 0.250$		
5.500	1.050	−37.588	15.2185	−9.762
5.250	1.290	−30.994	15.0207	−10.070
5.000	1.805	−22.725	14.7947	−10.382
4.750	8.495	−3.711	14.4326	−10.760
5.000	38.195	13.665	17.2870	−9.329
5.250	51.708	19.424	21.6204	−8.500
5.500	61.596	22.958	27.9681	−8.118
		$E = 0.000$		
5.500	6.703	−31.935	15.0482	−9.942
5.250	8.859	−22.425	14.8526	−10.530
5.000	20.000	−4.530	14.9760	−10.046
5.250	44.139	11.855	18.9301	−8.221
5.500	55.943	17.305	23.8485	−7.471
		$E = -0.250$		
5.500	13.982	−24.656	15.0354	−9.755
5.250	26.499	−5.785	15.5926	−9.165
5.500	48.664	10.026	20.3636	−7.172

which are tabulated in Tables 17-4 and 17-5. These values are also plotted in Fig. 17-10.

The conditions here are very similar to those with the spur pinion meshing with a face gear. Thus there are cusps, which are not shown in the figure, on the contour of the teeth of the hypoid face gear in some cases. Below these cusps is the region of excessive undercutting. The conditions of undercut here are nearly the same as those shown in Chap. 14. They become more severe, however, as the center distance for the hypoid drive is increased. For this hypoid face gear, the smallest effective radius is about 15.00 in.

Sections of the teeth are also shown in Fig. 17-10. As with other face gears of this type, the teeth become pointed with increasing diameters so that the effective face

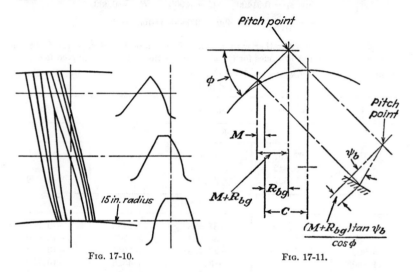

FIG. 17-10.   FIG. 17-11.

width of these gears is limited. The larger the ratios, or the smaller the angle of the generatrix of the pinion hyperboloid, the greater these effective face widths become.

Helical Involute Pinion as Hypoid Pinion.   This hypoid drive is the same as before with the substitution of a helical involute pinion for the involute spur pinion. The conditions here are very similar to those for a Fellows spiral face gear as analyzed in Chap. 16, with the shifting of the locus of pitch points to the generatrix of the hyperboloid pitch surface.

The direction of the spiral on the hypoid face gear must be from the pinion axis toward the face-gear axis. This is necessary because of the nature of the sliding on the pitch hyperboloids. Thus when the axes are offset as shown in Fig. 17-9, a right-handed helical pinion must be used.

The analysis of this spiral hypoid face gear is the same as for the preceding one with the introduction of the spiral form that develops from

TABLE 17-5. COORDINATES OF HYPOID-FACE-GEAR TEETH—SECOND SIDE OF TEETH
(Plotted in Fig. 17-10)

r_1, in.	ϕ, deg	ϵ_1, deg	r_2, in.	θ_2, deg
		$E = 0.500$		
4.750	27.167	−14.961	16.9213	−10.523
5.000	45.842	−21.312	21.7848	− 6.649
5.250	57.502	−25.218	28.2304	− 3.251
		$E = 0.250$		
5.500	1.050	37.588	14.1161	−13.451
5.250	1.290	30.994	14.1257	−13.428
5.000	1.805	22.725	14.2038	−13.388
4.750	8.495	3.711	14.6950	−13.016
5.000	38.195	−13.665	19.1931	− 9.193
5.250	51.708	−19.424	24.4539	− 5.888
5.500	61.596	−22.958	31.4993	= 2.907
		$E = 0.000$		
5.500	6.703	31.935	14.3157	−13.483
5.250	8.859	22.425	14.4746	−13.061
5.000	20.000	4.530	15.6184	−12.584
5.250	44.139	−11.855	21.0511	− 8.644
5.500	55.943	−17.305	27.0237	− 5.523
		$E = -0.250$		
5.500	13.982	24.656	14.8125	−13.488
5.250	26.499	5.785	16.4742	−12.420
5.500	48.664	−10.026	22.8409	− 8.378

the helical teeth of the pinion. Referring again to Fig. 17-8 and also to
Fig. 17-11, for the first side of the teeth, let

R_1 = nominal pitch radius of involute helical pinion, in.

R_{b1} = radius of base cylinder of helical pinion involute, in.

ϕ_1 = pressure angle, plane of rotation, of helical pinion at R_1

r_1 = radius to contact point on helical pinion, in.

ϕ_2 = pressure angle, plane of rotation, of helical pinion at r_1

ϕ = momentary pressure angle of drive, plane of rotation of pinion

α = initial position of origin of involute at distance R_2 from center
of face gear

ϵ_1 = angle of rotation of helical pinion

γ_p = angle of generatrix of pinion hyperboloid

R_{bp} = radius of base cylinder of pinion hyperboloid, in.

C = center distance, in.

R_2 = nominal pitch radius of face gear, in.

r_g = radius to contact point on hypoid face gear, in.

H = position of pitch point from pinion axis, in.

G = height to pitch point, in.

M = distance of contact point from face-gear center line, in.

P = height to contact point on face gear, in.

ϵ_2 = angle of rotation of hypoid face gear

δ = angle to contact point on face gear

θ_g = vectorial angle to hypoid-face-gear tooth form

N_1 = number of teeth in helical pinion

N_2 = number of teeth in hypoid face gear

T = arc space thickness of helical pinion, plane of rotation, at R_1, in.

E = distance of contact point from R_1, in.

ψ_1 = helix angle of helical pinion at R_1

ψ_b = helix angle of helical pinion at R_{b1}

As noted before, this analysis is the same as that for the spur pinion except for the twisting of the teeth, which gives an angular direction to the normal at the point of contact that passes through the pitch point. The pitch point will be in the same position as before, but the actual point of contact will be shifted up or down because of the helix. We will use the values R_1 and R_2 as the starting positions, with the teeth at this point in the same position as before. Referring again to Fig. 17-6, we retain the following equations from the previous analysis:

$$\alpha = (T/2R_1) - \text{inv } \phi_1 \tag{14-29}$$

$$\cos \beta = (R_1 - E)/r_1 \tag{14-30}$$

$$\cos \phi_2 = R_{b1}/r_1 \tag{14-31}$$

$$\phi = \beta + \phi_2 \tag{14-33}$$

$$H = (R_{b1}/\cos \phi) - R_{bp} \tan \phi \tag{17-22}$$

$$G = H \cot \gamma_p \tag{17-23}$$

$$M = r_1 \sin \beta - C \tag{17-24}$$

The normal from the pitch point to the point of contact, instead of being perpendicular to the axis of the pinion, is at an angle to it as indicated in Fig. 17-11. This figure shows the conditions for a right-handed helical pinion. The position of the pitch point is the same as for the equivalent spur pinion, but the contact point is moved up or down because of the angular position of the normal.

We have from the analysis of helical gears

$$\tan \psi_b = \tan \psi_1 \cos \phi_1 \tag{8-6}$$

We have the following from the conditions shown in Fig. 17-11:

For right-handed helical pinions

$$P = G - [(M + R_{bg}) \tan \psi_b / \cos \phi] \tag{17-32}$$

For left-handed helical pinions

$$P = G + [(M + R_{bg}) \tan \psi_b / \cos \phi] \tag{17-33}$$

$$\tan \delta = M/P \tag{17-34}$$
$$r_g = P/\cos \delta \tag{17-35}$$

Let y be the height of the contact point above R_2 in inches. Then

$$y = P - R_2 \tag{17-36}$$
$$\epsilon_1 = \beta - \alpha - \text{inv } \phi_2 - (2\pi y / L) \tag{17-37}$$

For a right-handed helical pinion

$$\epsilon_1 = \beta - \alpha - \text{inv } \phi_2 + (2\pi y / L) \tag{17-38}$$

For a left-handed helical pinion

$$\theta_g = \delta - \epsilon_2 \tag{17-39}$$

These are the equations for the first side of the teeth of the hypoid face gear.

For the second side of the teeth of the hypoid face gear, we have, referring to Fig. 17-9, the following:

$$\alpha = \text{inv } \phi_1 - (T/2R_1) \tag{17-28}$$
$$\cos \beta = (R_1 - E)/r_1 \tag{14-30}$$
$$\cos \phi_2 = R_{b1}/r_1 \tag{14-31}$$
$$\phi = \phi_2 - \beta \tag{17-30}$$
$$H = (R_{b1}/\cos \phi) + R_{bp} \tan \phi \tag{17-31}$$
$$G = H \cot \gamma_p \tag{17-23}$$
$$M = r_1 \sin \beta - C \tag{17-24}$$

Because of the angular position of the normal from the point of contact to the pitch point, we have the conditions indicated in Fig. 17-12. Whence

$$P = G - [(M + R_{bg}) \tan \psi_b / \cos \phi] \tag{17-32}$$

For right-handed helical pinions

$$\tan \delta = M/P \tag{17-34}$$
$$r_2 = P/\cos \delta \tag{17-35}$$
$$y = P - R_2 \tag{17-36}$$
$$\epsilon_1 = \beta - \alpha + \text{inv } \phi_2 - (2\pi y / L) \tag{17-40}$$

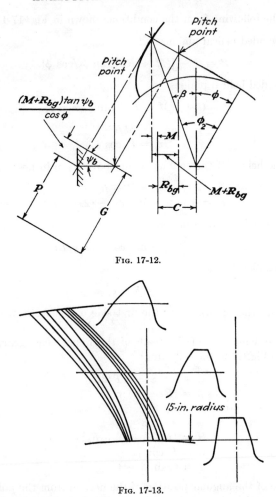

Fig. 17-12.

Fig. 17-13.

For a right-handed helical gear

$$\epsilon_1 = \beta - \alpha + \text{inv } \phi_2 + (2\pi y/L) \tag{17-41}$$

For a left-handed helical gear

$$\theta_g = \delta - \epsilon_2 \tag{17-39}$$

Example of Helical Pinion as Hypoid Pinion. As a definite example we shall use the following example:

$$N_1 = 20 \qquad N_2 = 60 \qquad R_1 = 5.000 \qquad R_2 = 15.000 \qquad \phi_1 = 20°$$
$$R_{b1} = 4.69846 \qquad \cot \gamma_p = {}^{60}\!/_{20} = 3.000 \qquad T = 0.7854 \qquad C = 3.000$$
$$R_{bp} = 0.300 \qquad R_{by} = 2.700 \qquad L = 74.000 \qquad \psi_1 = 23.000°$$

Using these values and the foregoing equations with a selected series of values for r_1 and E, we obtain the values for the two sides of the teeth of the hypoid face gear, which are tabulated in Tables 17-6 and 17-7. These values are also plotted in Fig. 17-13.

For this hypoid drive, the smallest effective radius of the face gear is about 15 in. The maximum effective radius is about 18-in., as shown. Sections of the teeth are shown as before. The first side of the teeth appears to be the more effective side for driving. The pressure angles here are somewhat greater than those on the second side of the teeth.

TABLE 17-6. COORDINATES OF HYPOID-FACE-GEAR TEETH—FIRST SIDE OF TEETH
(Plotted in Fig. 17-13)

r_1, in.	ϕ, deg	ϵ_1, deg	r_g, in.	θ_g, deg
		$E = 0.500$		
4.750	27.167	15.772	14.8970	−10.553
5.000	45.842	5.597	18.2487	− 4.443
5.250	57.502	−16.060	23.4866	4.635
		$E = 0.250$		
5.500	1.050	−39.083	16.3596	− 7.635
5.250	1.290	−31.437	15.9710	− 8.630
5.000	1.805	−21.835	15.5031	− 9.832
4.750	8.495	− 0.005	14.5508	−11.896
5.000	38.195	5.165	16.8089	− 6.632
5.250	51.708	− 6.655	20.3749	0.069
5.500	61.596	−30.039	25.8964	9.510
		$E = 0.000$		
5.500	6.703	−32.881	16.0893	− 8.240
5.250	8.859	−21.884	15.5805	− 9.842
5.000	20.000	− 3.556	15.1007	−10.274
5.250	44.139	1.347	17.2177	− 5.143
5.500	55.943	−18.791	22.4307	4.453
		$E = -0.250$		
5.500	13.982	−25.138	15.7958	− 8.702
5.250	26.499	− 7.902	15.7239	− 8.365
5.500	48.664	−19.060	21.0228	2.643

TABLE 17-7. COORDINATES OF HYPOID-FACE-GEAR TEETH—SECOND SIDE OF TEETH
(Plotted in Fig. 17-13)

r_1, in.	ϕ, deg	ϵ_1, deg	r_g, in.	θ_g, deg
		$E = 0.500$		
4.750	27.167	−25.291	17.7110	− 6.372
5.000	45.842	−58.189	23.1664	6.477
5.250	57.502	−97.602	30.4182	21.726
		$E = 0.250$		
5.500	1.050	−46.597	13.1296	−16.561
5.250	1.290	−39.043	13.3674	−16.290
5.000	1.805	−29.403	13.7030	−15.827
4.750	8.495	− 6.109	14.8141	−13.719
5.000	38.195	−35.985	20.1121	− 1.113
5.250	51.708	−69.601	26.0512	11.605
5.500	61.596	−113.066	34.0356	27.924
		$E = 0.000$		
5.500	6.703	39.240	13.5170	−16.086
5.250	8.859	27.851	13.9559	−15.079
5.000	20.000	− 2.317	15.7433	−11.767
5.250	44.139	−43.942	22.0781	2.646
5.500	55.943	−82.236	28.8364	16.839
		$E = -0.250$		
5.500	13.982	25.873	14.2642	−14.096
5.250	26.499	− 0.698	16.6057	−10.175
5.500	48.664	−51.555	23.9892	6.034

FORMATE HYPOID DRIVE

As another example of a hypoid drive we shall use a bevel gear with plane tooth surfaces and determine the tooth forms of the mating hypoid pinion. Such a bevel gear is shown in Fig. 17-14. The first step towards an analysis of this drive is to establish the values for the gear. Thus when

γ_g = angle of generatrix of gear hyperboloid

R_{bg} = radius of base cylinder of gear hyperboloid, in.

ϕ_1 = pressure angle of gear measured along hypoid element

ϕ_0 = pressure angle of gear measured from plane of rotation

ϕ_2 = angle of trace of gear measured from plane of rotation

we have

$$\tan \phi_3 = \tan \phi_1/\sin \gamma_g \qquad (17\text{-}42)$$
$$\tan \phi_2 = \tan \phi_1 \cos \gamma_g \qquad (17\text{-}43)$$

For the analysis of the hypoid pinion, we shall select a plane perpendicular to the axis of the pinion and determine the location of the contact points on this plane. The projection of the normal to the point of contact on any plane will be perpendicular to the trace of the plane gear-tooth surface of the bevel gear on that plane.

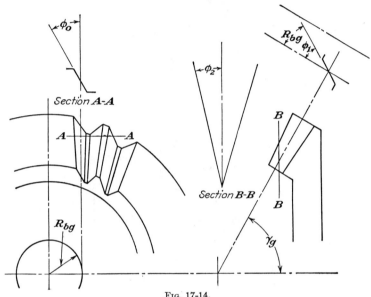

Fɪɢ. 17-14.

Referring to Fig. 17-15 for the contact on the first side of the thread, we have the following symbols:

N_1 = number of teeth in hypoid pinion

N_2 = number of teeth in bevel gear

R_1 = nominal pitch radius of hypoid pinion, in.

R_2 = nominal pitch radius of bevel gear, in.

γ_p = angle of generatrix of pinion hyperboloid

γ_g = angle of generatrix of gear hyperboloid

R_{bp} = radius of base cylinder of pinion hyperboloid, in.

R_{bg} = radius of base cylinder of gear hyperboloid, in.

ϕ = momentary pressure angle of drive, plane of rotation of pinion

ϵ_1 = angle of rotation of hypoid pinion

ϵ_2 = angle of rotation of bevel gear
A = projected length of element of gear hyperboloid to intersecting plane, in.
P = height to intersecting plane, in.
C = center distance, in.
G = height to pitch point, in.

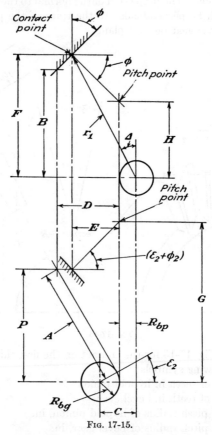

Fig. 17-15.

D = distance from locus of pitch points to intersection of gear-hyperboloid element with intersecting plane, in.
B = distance from pinion center line to intersection of gear-hyperboloid element with intersecting plane, in.
E = distance from locus of pitch points to contact point, in.
F = distance from pinion center line to contact point, in.

r_1 = radius on hypoid pinion to contact point, in.

Δ = angle from center line of pinion to contact point

H = distance from pinion center line to pitch point, in.

α = initial position of bevel-gear tooth element

Figure 17-15 shows any position of the bevel gear and the trace of the plane tooth surface of the bevel gear on the intersecting plane, which is at a height of P. The contact point is at some position E and F, as indicated. We know that the normal to the trace of the gear tooth on a plane perpendicular to the axis of the gear from some distance E must pass through the pitch point at a height of G from the axis of the bevel gear. Also we know that the normal to the trace of the bevel-gear tooth on the intersecting plane at the position E and F must pass through the pitch point at a distance of H from the axis of the hypoid pinion. We shall therefore set up equations for these conditions and solve for the unknown values of E and F.

We have the following from the geometrical conditions shown in Fig. 17-15:

$$A = \frac{P - R_{bg} \sin \epsilon_2}{\cos \epsilon_2} \tag{17-44}$$

$$B = A \tan \gamma_p$$
$$D = R_{bg}(1 - \cos \epsilon_2) + A \sin \epsilon_2 \tag{17-45}$$

The trace of the gear-tooth surface on a plane perpendicular to the axis of the gear is at an angle of $\epsilon_2 + \phi_2$, whence

$$E \tan (\epsilon_2 + \phi_2) + P = G$$

The trace of the gear-tooth surface on the intersecting plane is at an angle of ϕ, which is called the *momentary pressure angle* of the drive.

$$\tan \phi = \frac{\tan \phi_0}{\cos \epsilon_2} \tag{17-46}$$

$$B + \frac{D - E}{\tan \phi} - E \tan \phi = H$$

$$G = H \cot \gamma_p$$

Whence

$$\left(B + \frac{D - E}{\tan \phi} - E \tan \phi\right) \cot \gamma_p = G$$

Equating the two equations for G, solving for E, and simplifying, we obtain the following:

$$E = \frac{A \tan \phi - P \tan \phi + D \cot \gamma_p}{\tan (\epsilon_2 + \phi_2) \tan \phi + (\cot \gamma_p/\cos^2 \phi)} \tag{17-47}$$

$$F = A \tan \gamma_p + \frac{D - E}{\tan \phi} \tag{17-48}$$

$$\tan \Delta = \frac{E + R_{bn}}{F} \tag{17-49}$$

$$r_1 = \frac{F}{\cos \Delta} \tag{17-50}$$

We shall let

$$\alpha = \frac{90°}{N_2} \tag{17-51}$$

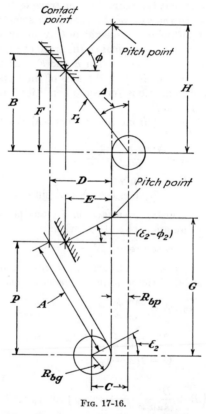

FIG. 17-16.

This holds true when the gear-tooth thickness is equal to one-half the circular pitch at the pitch element.

$$\theta_1 = \Delta - \epsilon_1 + \frac{N_2}{N_1} \alpha \tag{17-52}$$

The foregoing gives the equations for the first side of the teeth. The conditions for the second side of the teeth are shown in Fig. 17-16. The equations are derived in the same manner as before. Whence we obtain

$$A = \frac{P - R_{bg} \sin \epsilon_2}{\cos \epsilon_2} \tag{17-44}$$

$$D = R_{bg}(1 - \cos \epsilon_2) + A \sin \epsilon_2 \tag{17-45}$$

$$E = \frac{A \tan \phi - P \tan \phi - D \cot \gamma_p}{\tan (\epsilon_2 - \phi_2) \tan \phi - (\cot \gamma_p/\cos^2 \phi)} \tag{17-53}$$

$$F = A \tan \gamma_p - \frac{D - E}{\tan \phi} \tag{17-54}$$

$$\tan \Delta = \frac{E + R_{bp}}{F} \tag{17-49}$$

$$r_1 = \frac{F}{\cos \Delta} \tag{17-50}$$

$$\alpha = \frac{90°}{N_2} \tag{17-51}$$

$$\theta_1 = \Delta - \epsilon_1 - \frac{N_2}{N_1} \alpha \tag{17-55}$$

The foregoing gives the equations for the second side of the teeth of the hypoid pinion.

Example of Hypoid Pinion for Formate Gear. As a definite example we shall use the following values:

$N_1 = 20$ $N_2 = 60$ $R_1 = 5.000$
$R_2 = 15.000$ $C = 4.000$
$R_{bp} = 0.400$ $R_{bg} = 3.600$

For first side of teeth

$$\phi_1 = 20°$$

For second side of teeth

$$\phi_1 = 25°$$
$$\tan \gamma_g = {}^{60}\!/\!_{20} = 3.00 = \cot \gamma_p$$
$$\alpha = \frac{90°}{60} = 1.500°$$

Using a series of values of ϵ_2 for three positions of the intersecting plane and the foregoing values and equations, we obtain the values for the coordinates of the teeth of the hypoid pinion,

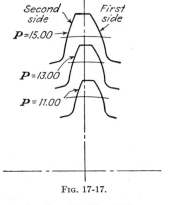

Fig. 17-17.

which are tabulated in Table 17-8. These values are also plotted in Fig. 17-17.

As with all other types of hypoid drives, the conditions are better on one side of the teeth than they are on the second side. In this example, the pressure angle of the second side of the teeth was increased in order to compensate, partially at least, for this difference.

Many other forms of bevel gears can be used for the mating members of hypoid pinions. The elements of the teeth may be in a straight or curved line, and this line may be at an angle to the projection of the locus of pitch points. The form must be such that it can be represented by the travel of the cutting tool during the generating process.

TABLE 17–8.   COORDINATE OF HYPOID PINION TEETH
(Plotted in Fig. 17–17)

ϵ_2, deg	First side of teeth		Second side of teeth	
	r_1, in.	θ_1, deg	r_1, in.	θ_1, deg
$P = 15.000$				
6.000	5.8430	3.310	4.6624	−1.337
4.000	5.5324	5.708	4.7296	−1.091
2.000	5.2551	7.635	4.8446	−0.645
0.000	5.0160	9.074	5.0160	0.074
−2.000	4.8197	10.030	5.2171	1.045
−4.000	4.6717	10.544	5.5087	2.667
−6.000	4.5769	10.692	5.8235	4.610
$P = 13.000$				
6.000	5.0739	3.814	4.0599	−0.564
4.000	4.8041	6.279	4.1097	−0.309
2.000	4.5620	8.272	4.2074	0.092
0.000	4.3517	9.774	4.3517	0.774
−2.000	4.1776	10.780	4.5411	1.804
−4.000	4.0445	11.335	4.7733	3.246
−6.000	3.9567	11.500	5.0451	5.124
$P = 11.000$				
6.000	4.3054	4.499	3.4580	0.499
4.000	4.0764	7.053	3.5377	0.514
2.000	3.8706	9.136	3.5713	1.108
0.000	3.6884	10.726	3.6884	1.726
−2.000	3.5366	11.811	3.8447	2.680
−4.000	3.4183	12.417	4.0386	4.034
−6.000	3.3376	12.607	4.2674	5.820

The Gleason hypoid drive is one where the bevel-gear member is a spiral bevel gear. This drive is described in a paper by Arthur L. Stewart and Ernest Wildhaber.[1]

[1] *Trans. SAE*, 1926.

CHAPTER 18

GEAR TEETH IN ACTION

The load-carrying capacity of any gear drive may be limited by any one or more of three factors. These are as follows:

1. Excessive heat of operation
2. Breaking of gear teeth
3. Excessive wear of gear-tooth surfaces

In addition to these, excessive noise in operation may make a gear drive unsuitable for use even though none of the three foregoing factors is involved.

Satisfactory gears must transmit power smoothly, with a minimum of vibration and noise, and must also have a reasonable length of useful life. In order to accomplish these ends, several essential requirements must be met. These requirements are of varying degrees of importance, depending largely upon the nature of the service the gears are to render. Some of them are requirements of the gears themselves; others have to do with their mounting and with the care and attention given to the drive in service.

Noise of Gears. It would be well to state at the outset that no metal gears in operation are absolutely noiseless. Quietness is a relative term. The most accurate gears when running under load at any appreciable speed will develop a certain amount of sound.

The noise of gears is of many kinds. It ranges from the unobtrusive hum of the better gears to rumbles, clashes, and squeals of varying pitches and intensities of the poorer grade. The exact causes of all conditions of noise are not, as yet, fully known. One fact, however, is certain: Excessive noise is evidence of improper conditions somewhere in the mechanism. In many respects, it is not a question of why gears are noisy but rather why they are ever quiet.

Noise is relative rather than absolute. It may be defined as an unpleasant or objectionable sound. This is certainly a proper definition of the noise of gears. There is only one sure method of reducing the amount of sound produced by gears. This method is to improve the accuracy and smoothness of profile of the gears themselves as well as to make their conditions of operation as favorable as possible. More effective designs of gear-tooth forms may help somewhat, but this improved design must be coupled with high-class workmanship in order to secure the full benefit of the improvement.

The search has been made for many years, and still continues, for some form of modification of gear-tooth profiles that will make unnecessary the need of extreme accuracy. This search has been fruitless in the past and probably will always be so. This does not mean that all modifications of gear-tooth profiles are always undesirable. Certain slight modifications can be made to advantage at times to minimize the effects of other small errors. The attempt should always be made to keep such modifications to a minimum. In most cases, this modification should be in the nature of a tolerance, *i.e.*, the direction of permissible errors on the tooth profiles should be in the direction that avoids edge contact at the beginning of mesh.

The form and material of the gear blank itself has a marked influence on the relative quietness of the gear at times. A blank that has the general characteristics of a bell will pick up and sustain a sound of its own vibration frequency that is created either by the engagement of the gear teeth or by some other outside agency. The characteristics of the material of which the blank is made are important factors in this respect.

The design of the gear housing is still another factor that may influence the noise of gears in operation. Some gear housings are effective resonators, while others are not. There is much to be done in this field of acoustics before we can design with certainty a gear housing that will absorb or muffle much of the sound of operation of the gears.

Insufficient backlash is sometimes the cause of excessive noise, heat, and wear. No gear drive can transmit loads continuously without backlash. Some misguided mechanics and others insist that the only good job is one with tight fits. The need of operating clearances holds true with gears as well as with almost every other type of operating mechanical element. Gears heat in operation. If sufficient backlash has not been provided to take care of the differential thermal expansions, the teeth will bind, with disastrous results.

Inadequate lubrication may also be a source of excessive heat, noise, and wear. At low speeds, the major purpose of the oil is lubrication. At high speeds, the major purpose of the oil is to act as a coolant and carry away the frictional heat of operation. Here the lubrication may be a secondary factor. Again, at high speeds, too much oil at the mesh point of the gears may be a source of excessive heat as the oil is expressed suddenly from between the teeth. In other words, the amount of oil alone is not the only consideration; careful attention may need to be given to the method of its application. Adequate lubrication, however, is required at all speeds.

Loads on Gear Teeth. The average load on gear teeth is, of course, the transmitted load. Both experience and actual tests, however, have made evident the condition that the actual working loads on gear teeth are greater than the average transmitted load. Some of this increase in load is the result of variations in the output, or a load variation set up by the conditions of service. Some drives are subjected to shock loads, while others have only to meet reasonably uniform loads. In addition, some increase in load is caused by the speed of engagement of the teeth and by the inaccuracies of the gear teeth themselves.

Dynamic Loads on Gear Teeth. The dynamic load is the maximum momentary load imposed on the gear teeth by the conditions of service, including the influence of errors in the gear teeth themselves. The nature and extent of these dynamic loads on gear teeth or on any other operating elements of a mechanism have long been open questions. In the absence of any evidence to the contrary, these dynamic loads have generally been considered as being directly proportional to the applied or average loads. Thus certain velocity factors have been established by running various mechanical elements, such as ball bearings and cast-iron gears, at varying velocities and imposing a sufficient load to cause failure. These applied loads which cause failure are then divided by the static load that would cause a similar failure, and the quotients so obtained have been used as velocity factors.

For example, if a pair of cast-iron gears running at a certain pitch-line velocity should fail under an applied load of 1,000 lb, and these same gears would fail under a static load of 3,000 lb, the velocity factor for this pitch-line velocity would be taken as equal to one-third. It is then assumed that under these same velocity conditions, the dynamic load would be equal to three times the applied load. Hence with an applied load of 250 lb, the dynamic load is assumed to be equal to 750 lb.

In 1879, John H. Cooper made an investigation of the strength of gear teeth and found that there were then in use about 48 well-established rules for horsepower and working strength, differing from each other in extreme cases by about 500 per cent. In 1886, Prof. William Harkness found from an examination of the literature of the subject, dating back to 1796, that according to the constants and formulas used by different authors, there were differences of 15 to 1 in the power that could be transmitted by a given pair of gears.

On October 15, 1892, Wilfred Lewis presented a paper before the Engineers' Club of Philadelphia entitled Investigation of the Strength of Gear Teeth. He appears to have been the first to use the form of the gear tooth as one of the factors in a formula for the strength of gear teeth. The Lewis formula, given below, soon became widely used.

LEWIS FORMULA.   When
W = transmitted load, lb
  s = safe working stress in material, psi
  p = circular pitch, in.
  F = face width of gears, in.
  y = tooth-form factor

$$W = spFy \qquad (18\text{-}1)$$

He then gave the following list of safe working stresses for cast iron and steel, which was determined from an English rule published in 1868: credited to E. R. Walker, Newcastle under Lyme.

Speed of teeth, ft/min	Safe working stress, psi	
	Cast iron	Steel
100 or less	8,000	20,000
200	6,000	15,000
300	4,800	12,000
600	4,000	10,000
900	3,000	7,500
1,200	2,400	6,000
1,800	2,000	5,000
2,000	1,700	4,300

Mr. Lewis continued his discussion with the following statement:

What fiber stress is allowable under different circumstances and conditions cannot be definitely settled at present, nor is it probable that any conclusions will be acceptable to engineers unless based on carefully made experiments.   Certain factors are given as applicable to certain speeds, and in the absence of any later or better light upon the subject, these have been constructed in a table to embody in convenient form the values recommended.   It cannot be doubted that slow speeds admit of higher working stresses than high speeds, but it may be questioned whether teeth running at 100 feet a minute are twice as strong as at 600 feet a minute or four times as strong as the same teeth at 1800 feet a minute.

BARTH EQUATION.   The foregoing values were later put into the form of an equation by Carl G. Barth.   This equation is as follows:
When s = safe working stress, psi
  s_1 = safe static stress, psi
  V = pitch-line velocity, ft/min

$$s = 600s_1/(600 + V) \qquad (18\text{-}2)$$

The foregoing values and the Barth equation were based on tests and experience with cast-iron gears with cast tooth forms prior to 1868.

With the introduction of cut and generated gears of a higher degree of accuracy, the Barth equation is often modified as follows:

$$s = 1{,}200s_1/(1{,}200 + V) \qquad (18\text{-}3)$$

HIGH-SPEED HERRINGBONE GEARS. With the introduction of higher speeds for prime movers, such as steam turbines and electric motors, and the use of reduction gear drives for connecting them with the driven mechanisms, such as generators, centrifugal pumps, fans, etc., it has been found by experience that after the gears reach a pitch-line velocity of the order of 5,000 ft/min, their load-carrying ability is practically constant for any higher speeds. This condition has led to the use of equations for such gears based on the limiting load per inch of tooth face. One such equation commonly used in the United States is the following:

When W = transmitted tooth load, lb

F = width of gear face, in.

k = constant depending upon materials and load conditions

D = pitch diameter of smaller gear, in.

$$W = FkD \qquad (18\text{-}4)$$

The following values of k are often employed in this equation for steel gears:

For single reduction gears, steady load, continuous service

$$k = 62.5$$

For single reduction gears, steady load, full load reached only occasionally

$$k = 100$$

Another similar equation that is widely used in England is the following:

$$W = Fk\sqrt{D} \qquad (18\text{-}5)$$

In this equation, for steel gears, the following values of k are often used:

For single reduction gears, steady load, continuous service

$$k = 175$$

For single reduction gears, steady load, full load reached only occasionally

$$k = 250$$

No allowance is made in these equations for the effect of pitch-line velocity. They are based on experience with high-speed gears in service.

When gears are made accurate enough to operate satisfactorily at about 5,000 ft/min pitch-line velocity, there seems to be but little difference, except at critical speeds, in either their quietness or load-carrying ability between that speed and up to over 10,000 ft/min.

MARX TESTS. It was not until about 1911 that any extensive and systematic tests were undertaken to obtain more reliable information on this subject of working loads; at that time Prof. Guido H. Marx at Stanford University made an extensive series of tests by running cast-iron gears to destruction. These tests were continued in 1915 with the assistance of Prof. Lawrence C. Cutter and included tests with pitch-line velocities running up to 2,000 ft/min. The results of these tests were reported in papers read before the ASME in 1912 and 1915. Some influence of the contact ratio was apparent in the results of these tests. Velocity factors that showed some variation from the commonly used values were established from these tests.

Ralph E. Flanders, in the discussion of the 1912 report, makes the following comments:

In regard to the dynamic qualities of the materials, is it safe to use the velocity coefficients given for all materials? Does not the strength of a gear running at high speed depend more upon the dynamic qualities of the metal than on the static strength? Would the coefficients derived from cast iron be correct when used for mild steel or when used for special heat-treated alloy steels, such as are used in automobile practice? It is also important to know how much the accuracy of the cutting affects the strength of the gears at high speed. The chances are that a high premium is put on accuracy from the standpoint of strength. If this is so, it should be definitely known, though it may not be practicable to include this factor in a formula.

In 1924, Franklin and Smith undertook a series of tests, under the supervision of Professor Marx, on cast-iron gears made to varying degrees of accuracy. The tests were made on the same apparatus as that used by Marx, and the results were presented in a paper read before the ASME in 1924. These tests showed that the accuracy of the gears had a marked influence on their strength at speed.

INCREMENT LOAD. The validity of the use of a velocity factor has been questioned from time to time. In the January, 1908, issue of *Machinery*, Ralph E. Flanders discusses the probable nature of the dynamic loads on gear teeth. He points out the possible effects of errors and the influence of the masses, and then states: "After some reflection, the writer has come to the conclusion that a variation in the strength of perfectly formed gearing, due to a variation in the velocity, can be due to but one thing—impact caused by the imperfect meshing of otherwise perfectly shaped teeth, deformed by the load they are transmitting."

The thought has been advanced that the actual dynamic load on gear teeth is the combination of two loads: first, the transmitted or useful load; and second, an additional or increment load caused by imperfect tooth profiles, unbalance, fluctuating applications of load, etc. Tooth action is made up of accelerations and deceleratoins caused by the deformation of the teeth under load, tooth-form errors, spacing errors, etc. At low speeds, these errors have but a relatively slight effect, but at high speeds they may develop increment loads many times greater than the applied load.

In an article published in *Zeitschrift des Vereines deutscher Ingenieure* in 1899, Oscar Lasche discussed the probable affect of errors and the large increment loads that might result from them at high pitch-line velocities. He gave certain calculated values for these increment loads based upon the assumption of rigid materials, but stated: "All such figures, however, depend upon assumptions which influence the results to a large extent and do not permit the determination of the results accurately." Considering rigid bodies, he states that the increment load caused by errors would be proportional to the square of the pitch-line velocity. He goes on to state, however:

The more elastic the teeth are, the greater the error that can be permitted. The differences in the velocities caused by the errors can be partially absorbed by the teeth themselves without disturbing the velocities of the rotating masses so much, and without causing such high increment loads. The duration of the changes in velocity is also spread over a longer period of time because of the springy action of the teeth, and consequently the acceleration of the masses is reduced, and the increment load is cut down.

In a paper read before the (British) Institution of Mechanical Engineers in May, 1916, Daniel Adamson discussed the probable value of the increment loads along similar lines to those followed by Oscar Lasche.

As a result of correspondence between Daniel Adamson, Wilfred Lewis, and Charles H. Logue, Lewis proposed in a paper read before the ASME in December, 1923, the design and construction of a testing machine that would enable these increment loads to be measured. He pointed out the difficulty of reconciling the results of actual breaking tests with the analytical work of Oscar Lasche, the one seeming to contradict the other. He stated:

In the writer's opinion, breaking tests are misleading and should be discouraged for the simple reason that when a gear is broken under any conditions as to load and speed, it is done for, and it becomes impossible to say what that gear would have shown under some other conditions. So also in regard to forming and spacing, it is a matter of vital importance that the minute errors, so detrimental to smooth running, be noted in terms of the velocity ratios which they produce, and as near as may be under actual working conditions.

But if it is true, and no doubt it is, that errors in forming and spacing are responsible for the loss of working strength at speed and that, as pointed out by Lasche, the variations in velocity ratio caused thereby give increment loads proportional to the speed squared for rigid forms in continuous rolling contact, a complete understanding of the effect of speed upon the strength of gear teeth requires a correlation of errors to the increment load or to the permissible speed, as indicated by Mr. Adamson in his paper.

ASME SPECIAL RESEARCH COMMITTEE ON THE STRENGTH OF GEAR TEETH. As a result of the efforts of Wilfred Lewis, the ASME Special Research Committee on the Strength of Gear Teeth was organized to complete the design of the testing machine, solicit funds for its construction, and to conduct tests. Preliminary studies and discussions considered the probable results of the tests and possible methods of analysis. A preliminary study made by Ernest Wildhaber deserves particular mention. He worked out an analysis on the basis of rigid bodies, following Oscar Lasche's method; then a similar one introducing the influence of elasticity with a constant effective mass; and third, the influence of elasticity with infinite mass, a condition that might be approached at very high speeds. He summed up these analyses as follows:

Summing up: We have now three conceptions of the effects of errors. According to the first, the increment loads will be proportional to the square of the velocity. According to the second conception, the increment loads will be directly proportional to the velocity; and according to the third, there is a limit increment load which is independent of the velocity. Furthermore the increment loads in the first case are directly proportional to the masses; in the second case, they would be proportional to the square root of the masses; and in the third case, the limit increment load would be independent of the masses. All of which shows we are badly in need of experimental data. And only after the gear tester has spoken is it possible to conceive a final, simplified, and practical theory of the effect of errors on the strength of gear teeth.

The Lewis gear-testing machine was built, and tests were run over a period of several years at the Massachusetts Institute of Technology. A description of the testing machine and the results of these tests were published in an ASME Research Publication, 1931, entitled Dynamic Loads on Gear Teeth. An analysis of the dynamic loads on spur gears based on this report is given here in Chap. 20.

Frictional Heat of Operation. The power loss at the tooth mesh of any gear drive is converted into heat, and this frictional heat must be carried away in some manner. Some of it is dissipated by direct radiation, and some of it is carried away by the lubricant. With a closed gear case, all the heat that is dissipated is radiated from the exposed surface of the gear case. The rate of this heat dissipation depends upon the exposed area of the gear case, the condition of the exposed surface, and the difference in temperature between the surface of the gear case and the surrounding air. When the rate of heat dissipation is equal to the rate at which the heat is created, we reach a condition of balanced temperature.

The entire exposed surface of the gear case will not all be at the same temperature. The exposed surface of the oil sump will have an appreciably higher temperature than the remainder of the exposed surface of the case. Thus any calculated value for the balanced temperature of the gear case will be an integrated average value. However, as the exposed area of the oil sump is generally small in comparison to the total exposed area, the calculated value of the temperature should be approximately that of the gear case.

The rate at which the heat will be radiated from the surface of the gear case is a very uncertain factor as it depends, in addition to other factors, upon the character and condition of the radiating surface and the conditions of air circulation around the gear case. It should also be appreciated that the temperature of the gears themselves, and that of other surfaces inside the gear case, will be much higher than the temperature of the outside of the gear case.

Under the more favorable conditions of radiating surface and air circulation about the case, we may assume a rate of 2.7 Btu per hour per square foot of exposed surface of the gear case for the rate at which this frictional heat is dissipated for a difference of 1°F between the temperature of the gear case and that of the surrounding air. This is equivalent to about 35 ft-lb per minute per square foot of exposed surface per 1°F difference in temperature. Thus when

W_f = power loss, ft-lb/min

T_d = difference in temperature between outside of gear case and room temperature, °F

A = area of exposed surface of gear case, sq ft

$$T_d = W_f/35A \qquad (18\text{-}6)$$

Under less favorable conditions, when the gear case is located in a corner or in a pit where there is little free circulation of air, the rate at which the heat will be dissipated will be materially less than the foregoing. Thus with the gear case in an unfavorable position, the rate of dissipation of heat may be reduced to or below about 1.8 Btu/(hr)(sq ft exposed surface) for each 1°F difference in temperature. This is equivalent to about 24 ft-lb/(min)(sq ft exposed area)(1°F temperature difference).   Under such conditions

$$T_d = W_f/24A \qquad (18\text{-}7)$$

Example of Heat Dissipation.   As a definite example we shall use a spiral-bevel-gear drive and assume that it transmits 25 hp with a power loss of 4,125 ft-lb/min, and that the area of the exposed surface of the gear case is equal to 2 sq ft.   Whence we have

$$A = 2.000 \qquad W_f = 4{,}125^*$$
$$T_d = \frac{4{,}125}{35 \times 2} = 59°F$$

When the amount of frictional heat is too great to be dissipated by the radiation from the exposed surface of the gear case, we must introduce a circulating-oil cooling system or its equivalent.   Such an equivalent might be a water-cooled jacket around the case.   In general, however, the circulating-oil system would be more effective, as this will take the heat directly from the gears and carry it away.   Here we direct a greater stream of oil against the gear blanks than is required for lubrication. Some of the heat may be dissipated by radiation from the exposed surface of the gear case, but we will assume for the present that all the heat must be carried away by the oil.   We will determine the amount of oil required to carry away the heat with a specified rise in temperature of the oil. Thus when

W_f = power loss, ft-lb/min

T_r = temperature rise permitted in the oil, °F

C = specific heat of oil

x = oil to be circulated, lb/min

Q = heat generated, Btu/min

$$Q = xT_rC \qquad (18\text{-}8)$$
$$1 \text{ Btu} = 777.5 \text{ ft-lb}$$
$$777.5Q = W_f$$
$$C = 0.40 \quad \text{(average for lubricating oils)}$$

Substituting these values into Eq. (18-8) and solving for x, we obtain

$$x = 0.00321(W_f/T_r) \qquad (18\text{-}9)$$
$$1 \text{ gal of oil (231 cu in.) weighs about 7.63 lb}$$
$$1 \text{ cu in. of oil weighs about 0.03292 lb}$$

Whence

$$\text{Pounds of oil per minute} = 0.00321(W_f/T_r) \qquad (18\text{-}9)$$
$$\text{Cubic inches of oil per minute} = 0.0975(W_f/T_r) \qquad (18\text{-}10)$$

---

* The friction losses of the bearings should also be added to those at the tooth mesh in order to determine the probable average temperature of the gear case.

Circulating-oil Cooling and Radiation. When part of the frictional heat is dissipated by radiation from the exposed surface of the gear case, then only the remainder will be carried away by the circulating oil. The temperature of the outside surface of the case should be taken as something less than the temperature of the oil. A few simple temperature measurements on any particular case would give the proper value. A fair guess might be that the temperature rise of the outside surface of the case will be about 75 per cent of the temperature rise of the circulating oil. Under these conditions, we must first determine the relative amounts of heat carried away by the two methods. For this we must combine the two sets of equations.

When W'_f = amount of power loss dissipated by case, ft-lb/min

$\quad\;\; W''_f$ = amount of power loss carried away by oil, ft-lb/min

$\quad\;\; A$ = area of exposed surface of gear case, sq ft

$\quad\;\; T_d$ = temperature rise of outside surface of gear case, °F

$\quad\;\; T_r$ = temperature rise of circulating oil, °F

$\quad\;\; W_f$ = total power loss, ft-lb/min

$\quad\;\; x$ = oil to be circulated, lb

we already have the following:

$$T_d = W'_f/35A \qquad (18\text{-}6)$$

Whence

$$W'_f = 35A T_d \qquad (18\text{-}11)$$
$$W''_f = W_f - W'_f$$
$$x = 0.00321(W''_f/T_r) \qquad (18\text{-}9)$$

Whence we have the following:

$$\text{Pounds of oil per minute} = 0.00321(W_f - W'_f)/T_r \quad (18\text{-}13)$$
$$\text{Cubic inches of oil per minute} = 0.0975(W_f - W'_f)/T_r \quad (18\text{-}14)$$

Examples of Circulating-oil Cooling. *First Example.* For the first example we shall assume that all the frictional heat is to be carried away by the circulating oil. We will use the following values:

$$\text{Inlet oil} = 100°F \qquad \text{Outlet oil} = 200°F \qquad W_f = 30,000$$

Whence

$$T_r = 100°F$$
$$\text{Cubic inches of oil per minute} = 0.0875 \frac{30,000}{100} = 29.95 \text{ cu in./min}$$

Second Example. As a second example we shall assume that the exposed area of the gear case in the preceding example is equal to 5 sq ft, that the average temperature of the outside of the gear case is 175°F, the room temperature is 75°F, and that all other factors are the same as before. This gives the following values:

$$T_d = 100° \qquad T_r = 100° \qquad A = 5.00 \qquad W_f = 30,000$$

From Eq. (18-11) we have

$$W'_f = 35 \times 5 \times 100 = 17,500$$

From Eq. (18-12) we have

$$W''_f = 30,000 - 17,500 = 12,500$$

From Eq. (18-14) we obtain

$$\text{Cubic inches of oil per minute} = 0.0975 \, \frac{12,500}{100} = 12.19 \text{ cu in/min}$$

CHAPTER 19

EFFICIENCIES OF GEARS

Analyses of the efficiencies of gears have been made by Reuleaux[1] and Weisbach.[2] Leutwiler[3] gives an approximation based on Weisbach's analysis that meets all ordinary requirements. These equations are general in nature and must be adapted or arranged to solve any specific problem of gear contact. Substantially the same results are obtained by an unpublished analysis made by Prof. William Howard Clapp of the efficiency of involute gears. The following analysis for involute gears is that of Professor Clapp:

It is customary to express the efficiencies of many power-transmitting elements in terms of a coefficient of friction. Familiar examples are power screws, worm-gear drives, belts, friction clutches, and the various types of bearings. This method has the merit of bringing out relationships that would not otherwise be obvious. The effects of changes in design, whether in proportions of parts or in materials, where frictional contact occurs, are indicated by the equations and are sufficient warrant for their use. In addition, the heat of operation and the question of the possible need for cooling often makes the use of such an equation imperative.

In any analysis some simplifying assumptions must be made. The first trial solution may be oversimplified in order to gain a better understanding of the problem and to reduce the number of variables. Then other conditions may be introduced in order to approach closer to the actual conditions. We will start, therefore, with the following assumptions, none of which may be absolutely true in practice:

ASSUMPTIONS.

1. Perfectly shaped and equally spaced involute teeth.

2. A constant normal pressure at all times between the teeth in engagement.

3. When two or more pairs of teeth carry the load simultaneously, the normal pressure is shared equally between them.

[1] REULEAUX, "The Constructor," 4th ed., p. 134 (Suplee, 1893). Phila.

[2] WEISBACH and HERMAN, "The Mechanics of Machinery," translated by J. V. Klein, Vol. III, p. 347 (Wiley, New York, 1894).

[3] LEUTWILER, "Elements of Machine Design," p. 319 (McGraw-Hill, New York, 1917).

Analysis of Friction Loss. Referring to Fig. 19-1, in which one gear driving clockwise drives another, and where the subscript 1 is used on

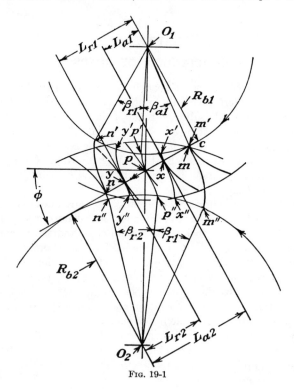

Fig. 19-1

the symbols for the driver and the subscript 2 is used on those for the follower, we will let

R_{b1} = radius of base circle of driver, in.

R_{b2} = radius of base circle of follower, in.

ϕ = pressure angle of gears

mn = the path of contact, with m the point of first contact and n the point of last contact

x = any point of contact along the path of approach

y = any point of contact along the path of recess

p = pitch point

m', n', p', x', y' = corresponding points of the origin of the involute curve on the base circle of the driver, so that the distance mn is equal to the arc distance $m'n'$, etc.

m'', n'', p'', x'', y'' = corresponding points of the origin of the involute curve on the base circle of the follower

$p'O_1$ = reference radius from which the angles of action of the driver are measured

$p''O_2$ = reference radius for angle of action of follower

β_{a1} = angle of approach for driver (angle $p'O_1m'$)

β_{r1} = angle of recess for driver (angle $p'O_1n'$)

β_1 = any angle of action such as $p'O_1x'$ or $p'O_1y'$

β_{a2}, β_{r2}, β_2 = corresponding angles for follower

f' = coefficient of friction, considered constant

f_a = average coefficient of friction of approach

f_r = average coefficient of friction of recess

W_n = total normal pressure acting between the teeth, lb

L_{a1} = friction lever arm for driver during approach, in.

L_{r1} = friction lever arm for driver during recess, in.

L_{a2}, L_{r2} = corresponding lever arms for follower, in.

T_{a1}, T_{r1} = torque exerted by driver during approach and recess, respectively, in.-lb

T_{a2}, T_{r2} = torque exerted on follower during approach and recess, respectively, in.-lb

W_{a1}, W_{r1} = work output of driver during approach and recess, respectively

W_{a2}, W_{r2} = work input of follower during approach and recess

W_f = friction loss of both gears

ω_1 = angular velocity of driver, radians/min

N_1, N_2 = number of teeth on driver and follower, respectively

From assumption 3, it can be shown that the total friction loss and the power input to the follower during the engagement of one pair of mating teeth are the same as when one pair of mating teeth carry the entire load throughout their period of engagement.

Work Output of Driver during the Engagement of One Pair of Teeth. Referring again to Fig. 19-1, during approach, considering any position of contact as at x, the normal force W_n opposes the rotation of the driver, while the frictional force fW_n assists rotation. The torque exerted by the driver at any approach position is as follows:

$$T_{a1} = W_nR_{b1} - fW_nL_{a1}$$
$$L_{a1} = cx = cp - px = cp - p'x' = R_{b1} \tan \phi - R_{b1}\beta_{x1}$$
$$T_{a1} = W_nR_{b1}[1 - f(\tan \phi - \beta_{x1})]$$

The work output of the driver during approach is as follows:

$$W_{a1} = \int T_{a1}\, d\beta_1 = W_n R_{b1} \int_{-\beta_{al}}^{0} [1 - f(\tan\phi - \beta_1)]\, d\beta_1$$

During recess, the direction of the sliding between the teeth is reversed, so that

$$T_{r1} = W_n R_{b1} + f W_n L_{r1}$$
$$W_{r1} = \int T_{r1}\, d\beta_1 = W_n R_{b1} \int_{0}^{\beta_{r1}} [1 + f(\tan\phi + \beta_1)]\, d\beta_1$$

Work Input to Follower during Contact of One Pair of Teeth. During approach, the normal force and the frictional force oppose each other.

$$T_{a2} = W_n R_{b2} - f W_n L_{a2}$$
$$L_{a2} = R_{b2} \tan\phi + R_{b2}\beta_{x2}$$
$$W_{a2} = W_n R_{b2} \int_{-\beta_{a2}}^{0} [1 - f(\tan\phi + \beta_2)]\, d\beta_2$$

During recess, both the normal force and the frictional force assist the rotation of the follower.

$$T_{r2} = W_n R_{b2} + f W_n L_{r2}$$
$$L_{r2} = R_{b2} \tan\phi - R_{b2}\beta_{x2}$$
$$W_{r2} = W_n R_{b2} \int_{0}^{-\beta_{r2}} [1 + f(\tan\phi - \beta_2)]\, d\beta_2$$

Case I. f Considered as Constant. Let us assume first that the coefficient of friction remains constant throughout engagement. Integrating the equations for work, we obtain

$$W_{a1} = W_n R_{b1}\left[\beta_{a1} - f'(\tan\phi)\,\beta_{a1} + \frac{f'}{2}\,\beta_{a1}{}^2\right]$$

$$W_{r1} = W_n R_{b1}\left[\beta_{r1} + f'(\tan\phi)\,\beta_{r1} + \frac{f'}{2}\,\beta_{r1}{}^2\right]$$

$$W_{a2} = W_n R_{b2}\left[\beta_{a2} - f'(\tan\phi)\,\beta_{a2} - \frac{f'}{2}\,\beta_{a2}{}^2\right]$$

$$W_{a2} = W_n R_{b1}\left[\beta_{a1} - f'(\tan\phi)\,\beta_{a1} - \frac{f' R_{b1}}{2R_{b2}}\,\beta_{a1}{}^2\right]$$

$$W_{r2} = W_n R_{b2}\left[\beta_{r2} + f'(\tan\phi)\,\beta_{r2} - \frac{f'}{2}\,\beta_{r2}{}^2\right]$$

$$W_{r2} = W_n R_{b1}\left[\beta_{r1} + f'(\tan\phi)\,\beta_{r1} - \frac{f' R_{b1}}{2R_{b2}}\,\beta_{r1}{}^2\right]$$

The efficiency of the gears is equal to

$$\frac{W_{a2} + W_{r2}}{W_{a1} + W_{r1}} = \frac{(\beta_{a1} + \beta_{r1}) - f' \tan \phi \,(\beta_{a1} - \beta_{r1}) - (f'/2m)(\beta_{a1}{}^2 + \beta_{r1}{}^2)}{(\beta_{a1} + \beta_{r1}) - f' \tan \phi \,(\beta_{a1} - \beta_{r1}) + (f'/2)(\beta_{a2}{}^2 + \beta_{r1}{}^2)}$$

The friction loss per minute is equal to

$$W_f = \frac{W_n \omega_1 R_{b1}}{W_{a1} + W_{r1}} \left[\frac{f'}{2} \,(\beta_{a1}{}^2 + \beta_{r1}{}^2) \left(1 + \frac{1}{m}\right) \right] \qquad (19\text{-}1)$$

The efficiency can be written more simply and almost exactly by considering the work input to be equal to

$$W_{a1} + W_{r1} = W_n \omega_1 R_{b1}$$

whence

$$\text{Efficiency} = 1 - \left[\frac{1 + (1/m)}{\beta_{a1} + \beta_{r1}} \right] \frac{f'}{2} \,(\beta_{a1}{}^2 + \beta_{r1}{}^2) \qquad (19\text{-}2)$$

Case II. f Considered as Variable. Actually, the coefficient of friction is not constant but varies with different loads, speeds, lubricants, and gear materials, as well as with different types of finish and probably with many other factors.

Actual tests on gears indicate that the general form of the curves representing the average coefficients of friction plotted against sliding or pitch-line velocities is very much the same as on similar graphs representing the performance of plain bearings. Here, at low speeds, the values of the coefficient of friction are high, reducing rapidly to a minimum with increasing speed, and then rising again slowly with further increases in speed.

We have already seen that the nature of the sliding between involute gear teeth consists of sliding in one direction during approach, reducing to zero at the pitch point where the direction of sliding changes, and increasing again as the contact progresses through the recess action.

Since the direction of sliding changes at the pitch point, we may conclude that the coefficient of friction will never lie wholly within the field of perfect film lubrication during the period of engagement of a pair of mating teeth.

We may set up these efficiency equations in a variety of ways. The chances are, however, that the most we can determine by experiment is to establish some average values for approach action and average values for recess action. These conditions could be studied by testing gears that have all approach action in one case and all recess action in another case, with all other factors remaining constant.

From the observation of the behavior of gears on the Lewis gear-testing machine at very low speeds up to about 5 ft/min pitch-line velocity, the friction of approach appeared to be about double that of recess on hobbed, milled, and shaped gears of cast iron, soft steel, bronze, and aluminum. On hardened and ground steel gears, however, there appeared to be no appreciable difference between the friction of approach and that of recess. There did appear to be, however, a momentary jump in the friction load to about 150 per cent of the average value when the contact passed through the pitch point. Whether or not these conditions would hold true when the gears are operating at higher speeds is an open question. Yet regardless of this, it would appear that some average values of the two coefficients of friction, approach and recess, would give us a reasonably close measure of the truth.

We shall therefore introduce different average values for the coefficients of friction of approach and recess. Thus we obtain

$$W_{a1} = W_n R_{b1} \left[\beta_{a1} - f_a \, (\tan \phi) \, \beta_{a1} + \frac{f_a}{2} \, \beta_{a1}{}^2 \right]$$

$$W_{r1} = W_n R_{b1} \left[\beta_{r1} + f_r \, (\tan \phi) \, \beta_{r1} + \frac{f_r}{2} \, \beta_{r1}{}^2 \right]$$

$$W_{a2} = W_n R_{b1} \left[\beta_{a1} - f_a \, (\tan \phi) \, \beta_{a1} - \frac{f_r}{2m} \, \beta_{a1}{}^2 \right]$$

$$W_{r2} = W_n R_{b1} \left[\beta_{r1} + f_r \, (\tan \phi) \, \beta_{r1} - \frac{f_r}{2m} \, \beta_{r1}{}^2 \right]$$

The friction loss per minute is equal to

$$W_f = \frac{W_n \omega_1 R_{b1}}{\beta_{a1} + \beta_{r1}} \left[\left(1 + \frac{1}{m} \right) \left(\frac{f_a}{2} \, \beta_{a1}{}^2 + \frac{f_r}{2} \, \beta_{r1}{}^2 \right) \right] \qquad (19\text{-}3)$$

Considering the work input to be equal to

$$W_{a1} + W_{r1} = W_n \omega_1 R_{b1}$$

then the efficiency will be equal to

$$\text{Efficiency} = 1 - \left[\frac{1 + (1/m)}{\beta_{a1} + \beta_{r1}} \right] \left(\frac{f_a}{2} \, \beta_{a1}{}^2 + \frac{f_r}{2} \, \beta_{r1}{}^2 \right) \qquad (19\text{-}4)$$

Summary of Efficiency Equations. To bring the foregoing material into a more condensed form for ready use, we make the following summary (we shall also repeat the equations for the sliding velocity and for the arcs of approach and recess):

When N_1, N_2 = number of teeth on driver and follower, respectively

m = gear ratio

β_a, β_r = arc of approach and recess on driver, respectively

f = average coefficient of friction

f_a = average coefficient of friction of approach action

f_r = average coefficient of friction of recess action

$$m = \frac{N_2}{N_1} \tag{19-5}$$

When the coefficient of friction is assumed as constant

$$\text{Efficiency} = 1 - \left[\frac{1 + (1/m)}{\beta_a + \beta_r} \right] \frac{f}{2} (\beta_a{}^2 + \beta_r{}^2) \tag{19-6}$$

When the average coefficients of friction of approach and recess are different

$$\text{Efficiency} = 1 - \left[\frac{1 + (1/m)}{\beta_a + \beta_r} \right] \left(\frac{f_a}{2} \beta_a{}^2 + \frac{f_r}{2} \beta_r{}^2 \right) \tag{19-7}$$

Arc of Approach and Recess

R_1, R_2 = pitch radius of driver and follower, respectively, in.

R_{o1}, R_{o2} = outside radius of driver and follower, respectively, in.

R_{b1}, R_{b2} = base-circle radius of driver and follower, respectively, in.

C = center distance, in.

ϕ = pressure angle

$$\beta_a = \frac{\sqrt{R_{o2}{}^2 - R_{b2}{}^2} - R_2 \sin \phi}{R_{b1}} \tag{4-17}$$

$$\beta_r = \frac{\sqrt{R_{o1}{}^2 - R_{b1}{}^2} - R_1 \sin \phi}{R_{b1}} \tag{4-18}$$

Sliding Velocity. When

V = pitch-line velocity of gears, ft/min

V_s = sliding velocity, ft/min

n = rpm of driver

r = any radius of driver, in.

and all other symbols are the same as before

$$V = 2\pi R_1 n/12 = 0.5236 R_1 n \tag{19-8}$$

$$V_s = V[(1/R_1) + (1/R_2)] (\sqrt{r^2 - R_{b1}{}^2} - R_1 \sin \phi) \tag{4-14}$$

As noted before, the sliding velocity varies from zero when the contact is at the pitch point to a maximum in either direction at the beginning and ending of mesh. This variation in sliding velocity is uniform along the path of contact. We shall use the average sliding velocity, which will be one-half of the maximum value, to select the coefficient of friction. Thus when

V'_s = average sliding velocity, ft/min

V_{sa} = average approach sliding velocity, ft/min

V_{sr} = average recess sliding velocity, ft/min

$$V_{sr} = (V/2) [(1/R_1) + (1/R_2)] (\sqrt{R_{o1}^2 - R_{b1}^2} - R_1 \sin \phi)$$

This is equal to

$$V_{sr} = (V/2) [(1/R_1) + (1/R_2)] R_{b1}\beta_r$$

But

$$R_{b1} = R_1 \cos \phi$$

Introducing this value into the foregoing equation, we obtain

$$V_{sr} = (V/2) [1 + (N_1/N_2)] \beta_r \cos \phi \tag{19-9}$$

In a similar manner we obtain

$$V_{sa} = (V/2) [1 + (N_1/N_2)] \beta_a \cos \phi \tag{19-10}$$

In most cases, when we need some measure of the efficiency in order to determine the amount of heat to be dissipated, the use of an average sliding velocity for the entire tooth engagement and an average coefficient of friction will be adequate. It is primarily when we wish to compare the relative merits of different tooth designs that we need to consider the approach and recess separately. Thus, for the value of the average sliding velocity, we have

$$V'_s = V \cos \phi [1 + (N_1/N_2)] [(\beta_a + \beta_r)/4] \tag{19-11}$$

Coefficients of Friction on Spur Gears. Several tests of the power losses with spur gears are reported on pages 51 to 59, inclusive, of the ASME Research Publication, 1931, entitled Dynamic Loads on Gear Teeth. Applying the foregoing analysis to the results of these tests, we can obtain some tentative values for the coefficients of friction for use until more extensive and more reliable values can be obtained.

Referring to this report, on the run N, both pairs of gears in the testing machine were the same. The master gears used on this run were also used on all other test runs. Hence we will divide the power losses here by 2 on the assumption that the losses were the same on both similar pairs of gears.

Starting with run N-24 where there was a tooth load of 1,092 lb (from table on page 59 of report) and the pitch-line velocity was 25 ft/sec or 1,500 ft/min, we have for these 3-DP, 14½-deg-form, milled cast-iron gears the following values:

		Pounds
Power loss, total	....	19.25*
Power loss, zero load	....	5.25†
Difference	....	14.00†
Master-gear loss	....	7.00†
Test-gear loss	....	7.00†

* At pitch radius.
† At pitch line.

These gears have the following values:

$$\beta_a = 0.3691 \text{ radian} \qquad \beta_r = 0.3045 \text{ radian} \qquad N_1 = 18 \qquad N_2 = 48$$
$$\text{Output load} = 1{,}092 \text{ lb}$$
$$\text{Input load} = 1{,}099 \text{ lb}$$
$$\text{Efficiency} = \frac{1{,}092}{1{,}099} = 0.9936$$

Transposing Eq. (19-6) to solve for f, we have

$$f = \frac{2(1 - 0.9936)(\beta_a + \beta_r)}{(\beta_a{}^2 + \beta_r{}^2)(1 + 0.375)} = 0.0272$$

In like manner, the values for the coefficients of friction for various speeds on runs R and S have been determined. These values are tabu-

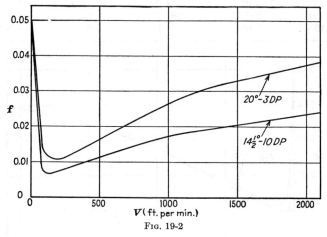

Fig. 19-2

lated in Table 19-1. The gears in run R were 20-deg, 3-DP gears of soft steel and cast iron. The gears in run S were 14½-deg, 10-DP gears of soft steel and cast iron.

These values of f are plotted in Fig. 19-2 against the values of the pitch-line velocities. Here there is a distinct difference between the two graphs; that for the finer pitch is definitely below that for the coarser pitch. In other words, the finer pitch gears are more efficient.

TABLE 19-1. VALUES OF f FOR RUNS R AND S
(Plotted in Figs. 19-2 and 19-3)

V, ft/min	Run R		Run S	
	f	V'_s, ft/min	f	V'_s, ft/min
0	0.0517	0.00	0.0496	0.00
150	0.0112	27.75	0.0066	12.00
300	0.0122	55.50	0.0088	24.00
450	0.0155	83.25	0.0110	36.00
600	0.0188	111.00	0.0127	48.00
750	0.0216	138.75	0.0143	60.00
900	0.0244	166.50	0.0160	72.00
1,050	0.0273	194.25	0.0173	84.00
1,200	0.0296	222.00	0.0185	96.00
1,350	0.0315	249.74	0.0196	108.00
1,500	0.0329	277.50	0.0206	120.00
1,650	0.0343	305.25	0.0216	132.00
1,800	0.0357	333.00	0.0225	144.00
1,950	0.0371	360.75	0.0234	156.00
2,100	0.0385	388.50	0.0242	168.00

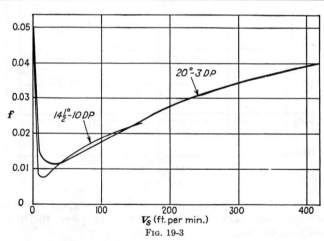

FIG. 19-3

These values of f are plotted in Fig. 19-3 against the average sliding velocities. Here the two graphs intersect each other. Averaging the several test values, we obtain the values of f that are plotted in Fig. 19-4 against the average sliding velocities.   These values will be used for soft gears of conventional design where the arcs of approach and recess are substantially equal.

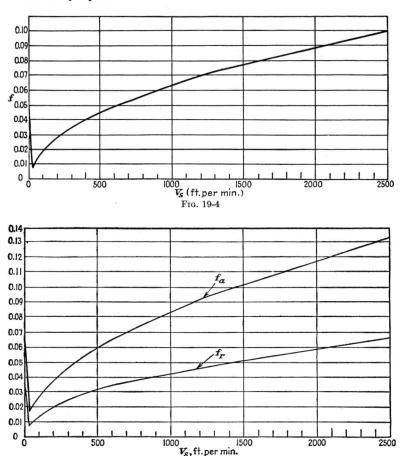

Fig. 19-4

Fig. 19-5

Values of f_a and f_r for soft gears are plotted in Fig. 19-5.   These will be used when the design gives a preponderance of recess or approach action.   In these cases, the average velocity of sliding will be different on the approach from that on the recess.   These values are based on the differences noted in very low speed tests.   It is hoped that further tests can be made to obtain more reliable values.   The value of f_r will be used for both approach and recess on hardened-steel gears.

These values are given very closely by the following empirical equation:

When f = average coefficient of friction

f_a = average coefficient of friction of approach

f_r = average coefficient of friction of recess

V_s = sliding velocity, ft/min

e = base of natural logarithms

$$f = \frac{0.050}{e^{0.125V_s}} + 0.002 \sqrt{V_s} \qquad (19\text{-}12)$$

TABLE 19-2. COEFFICIENTS OF FRICTION

V_s, ft/min	f	f_a	f_r
0	0.0500	0.0667	0.0333
10	0.0207	0.0276	0.0138
20	0.0130	0.0174	0.0087
30	0.0121	0.0162	0.0081
40	0.0130	0.0174	0.0087
50	0.0142	0.0190	0.0095
60	0.0155	0.0206	0.0103
70	0.0167	0.0222	0.0111
80	0.0179	0.0238	0.0119
90	0.0190	0.0254	0.0127
100	0.0200	0.0267	0.0133
150	0.0245	0.0326	0.0163
200	0.0283	0.0378	0.0189
250	0.0316	0.0422	0.0211
300	0.0346	0.0462	0.0231
400	0.0400	0.0534	0.0267
500	0.0447	0.0596	0.0298
600	0.0490	0.0654	0.0327
700	0.0529	0.0706	0.0353
800	0.0566	0.0754	0.0377
900	0.0600	0.0800	0.0400
1,000	0.0632	0.0842	0.0421
1,500	0.0775	0.1034	0.0517
2,000	0.0894	0.1192	0.0596
2,500	0.1000	0.1333	0.0667

$$f_a = \frac{4f}{3} \qquad (19\text{-}13)$$

$$f_r = \frac{2f}{3} \qquad (19\text{-}14)$$

Values determined by the use of these equations are tabulated in Table 19-2.

Examples of Spur-gear Efficiency. *First Example.* As a first example we shall compare the efficiencies of a pair of 4-DP, 14½-deg gears with a similar pair of 20-deg gears. All dimensions will be the same except the pressure angles. We shall use the following values:

$$N_1 = 24 \qquad N_2 = 120 \qquad R_1 = 3.000 \qquad R_2 = 15.000 \qquad V = 1,500$$
$$R_{o1} = 3.250 \qquad R_{o2} = 15.250 \qquad C = 18.000 \qquad m = 5.000$$

For the 14½-deg gears we have

$$R_{b1} = 2.90445 \qquad R_{b2} = 14.52225 \qquad \beta_a = 0.3095 \qquad \beta_r = 0.2400$$

For the 20-deg gears we have

$$R_{b1} = 2.81907 \qquad R_{b2} = 14.09535 \qquad \beta_a = 0.2449 \qquad \beta_r = 0.2073$$

For the 14½-deg gears we obtain

$$V_{sa} = 750(1.20)(0.3095 \times 0.96815) = 270$$

whence

$$f_a = 0.044$$
$$V_{sr} = 750(1.20)(0.2400 \times 0.96815) = 209$$

whence

$$f_r = 0.020$$

Introducing these values into Eqs. 19-7, we obtain

$$\text{Efficiency} = 0.99415 \qquad \text{Power loss} = 0.585\%$$

For the 20-deg gears we obtain

$$V_{sa} = 750(1.20)(0.2449 \times 0.93969) = 197$$

whence

$$f_a = 0.038$$
$$V_{sr} = 750(1.20)(0.2073 \times 0.93969) = 175$$

whence

$$f_r = 0.018$$
$$\text{Efficiency} = 0.99595 \qquad \text{Power loss} = 0.405\%$$

Under these conditions, the 20-deg gears show a slightly higher efficiency than do the 14½-deg gears. Their power loss is 0.405 per cent as compared with 0.585 per cent on the 14½-deg gears. Their sliding velocities are lower and their arcs of approach and recess are less than those of the 14½-deg gears.

Second Example. As a second example we shall use the same 20-deg gears as before, but will use them as a step-up drive instead of as a reduction drive. We will then compare their efficiency with the same gears used as a reduction drive. In this case we can simply exchange the sliding velocities of approach and recess to determine the respective coefficients of friction. Thus we have

$$V_{sa} = 175 \qquad f_a = 0.036 \qquad V_{sr} = 197 \qquad f_r = 0.019$$

In this example, as the larger gear is the driver, we have

$$R_{b1} = 14.09535 \qquad R_{b2} = 2.81907 \qquad \beta_a = 0.0415 \qquad \beta_r = 0.0490 \qquad m = 0.20$$

Introducing these values into Eq. (19-7), we obtain

$$\text{Efficiency} = 0.99643 \qquad \text{Power loss} = 0.357\%$$

In this example, the efficiency of the step-up drive is a little higher than that of the same gears used for a reduction drive, primarily because of its greater recess action. When the amount of approach and recess are equal, there will be no difference. When the recess action of the driver on a reduction drive is greater than the approach action, then the corresponding step-up drive will be less efficient than the reduction drive.

Third Example. As a third example we shall compare the efficiency of an 8-DP, 20-deg reduction drive of the same diameters as before with the 4-DP, 20-deg drive used in the first example. We already know from the results of tests that the finer pitch gears will have a smaller power loss because of their lower sliding velocities. We will attempt to obtain some measure of this difference. We have the following values to start:

$$N_1 = 48 \qquad N_2 = 240 \qquad R_1 = 3.000 \qquad R_2 = 15.000 \qquad V = 1,500 \qquad m = 5.00$$
$$R_{o1} = 3.125 \qquad R_{o2} = 15.125 \qquad C = 18.000 \qquad R_{b1} = 2.81907 \qquad R_{b2} = 14.09535$$
$$\beta_a = 0.1258 \qquad \beta_r = 0.1140$$

Introducing these values into Eq. (19-10), we obtain

$$V_{sa} = 750(1.20)(0.1258 \times 0.93969) = 106$$

whence

$$f_a = 0.028$$
$$V_{sr} = 750(1.20)(0.1140 \times 0.93969) = 96$$

whence

$$f_r = 0.0130$$

Introducing these values into Eq. (19-7), we obtain

$$\text{Efficiency} = 0.99847 \qquad \text{Power loss} = 0.153\%$$

In this example, the power loss is equal to 0.153 per cent as compared to a power loss of 0.405 per cent for the 4-DP gears. Thus by reducing the circular pitch to one-half of the original value, the power loss has been reduced to about 38 per cent of the original amount.

Fourth Example. For the fourth example we shall determine the power loss and efficiency of a pair of 4-DP, 20-deg, hardened and ground steel gears of the same size as before. Here all the values will be the same as before except the coefficients of friction. For these we have

$$f_a = f_r = 0.018$$

Introducing these values into Eq. (19-6), we obtain

$$\text{Efficiency} = 0.99754 \qquad \text{Power loss} = 0.246\%$$

In this case, the power loss amounts to 0.246 per cent as compared with a power loss of 0.406 per cent for the soft gears.

As a matter of interest, the calculated power loss, using this analysis and the specified values for the coefficients of friction, on a pair of hardened and ground steel gears of special tooth design transmitting over 3,000 hp at a pitch-line velocity of over 6,000 ft/min was about 0.60 per cent. After these gears were made, they were run on a dynamometer test, and the measured power loss was about 0.50 per cent.

EFFICIENCY OF INTERNAL GEARS WITH STRAIGHT TEETH

If an analysis is made for the efficiency of an internal-gear drive like the one made for external spur gears, it will be found that the normal force and the frictional torque either assist or oppose gear rotation in exactly the same way as was shown to occur in the case of external gears. There is, however, one difference in the two cases. This is in respect to the distance to the lever arms of the frictional forces. For external gears, the lever arms of the driver are greater on the approach than on the recess. For internal gears, this condition is reversed. Thus let

m = gear ratio
f = average coefficient of friction
f_a = average coefficient of friction of approach
f_r = average coefficient of friction of recess
β_a, β_r = arc of approach and recess of driver, respectively

When a single average coefficient of friction is used, then

$$\text{Efficiency} = 1 - \left[\frac{1 - (1/m)}{\beta_a + \beta_r} \right] \frac{f}{2} [\beta_a{}^2 + \beta_r{}^2] \qquad (19\text{-}15)$$

When different coefficients of friction are used for approach and recess, then

$$\text{Efficiency} = 1 - \left[\frac{1 - (1/m)}{\beta_a + \beta_r} \right] \left[\frac{f_a}{2} \beta_a{}^2 + \frac{f_r}{2} \beta_r{}^2 \right] \qquad (19\text{-}16)$$

Arc of Approach and Recess

R_1, R_2 = pitch radius of driver and internal follower, respectively, in.
R_{o1} = outside radius of driver, in.
R_i = inside radius of internal follower, in.
R_{b1}, R_{b2} = radius of base circles, in.
C = center distance, in.
ϕ = pressure angle

$$\beta_a = (R_2 \sin \phi - \sqrt{R_i{}^2 - R_{b2}{}^2})/R_{b1} \qquad (6\text{-}13)$$
$$\beta_r = (\sqrt{R_{o1}{}^2 - R_{b1}{}^2} - R_1 \sin \phi)/R_{b1} \qquad (6\text{-}14)$$

Sliding Velocity

V = pitch-line velocity, ft/min
V_s = sliding velocity, ft/min
n = number of rpm of driver
r_1 = any radius of driver, in.

$$V = 0.5236 R_1 n \qquad (19\text{-}8)$$
$$V_s = V[(1/R_1) - (1/R_2)] \, (\sqrt{r_1{}^2 - R_{b1}{}^2} - R_1 \sin \phi) \qquad (6\text{-}18)$$

As before, the average sliding velocity will be equal to one-half the maximum sliding velocity of approach or recess.

When V'_s = average sliding velocity, ft/min

V_{sa} = average sliding velocity of approach, ft/min

V_{sr} = average sliding velocity of recess, ft/min

$$V'_s = V \cos \phi \, [1 - (N_1/N_2)] \, [(\beta_a + \beta_r)/4] \qquad (19\text{-}17)$$
$$V_{sa} = (V/2) \, [1 - (N_1/N_2)] \, \beta_a \cos \phi \qquad (19\text{-}18)$$
$$V_{sr} = (V/2) \, [1 - (N_1/N_2)] \, \beta_r \cos \phi \qquad (19\text{-}19)$$

Coefficient of Friction for Internal Gears. In the absence of any test data on the performance of internal-gear drives, we shall use the same values of the coefficient of friction here as for external spur gears with straight teeth.

Examples of Efficiency of Internal-gear Drive. *First Example.* As a first example we shall use the same values as were used for the 4-DP, 20-deg spur gears. This gives the following:

$N_1 = 24$ $N_2 = 120$ $R_1 = 3.000$ $R_2 = 15.000$ $V = 1,500$ $m = 5.00$
$R_{o1} = 3.250$ $R_i = 14.750$ $C = 12.000$ $R_{b1} = 2.81907$ $R_{b2} = 14.09535$

Introducing these values into Eqs. (6-13) and (6-14), we obtain

$$\beta_a = 0.2783 \qquad \beta_r = 0.2073$$

Introducing these values into Eqs. (19-18) and (19-19), we obtain

$$V_{sa} = 157 \qquad f_a = 0.034 \qquad V_{sr} = 117 \qquad f_r = 0.015$$

Introducing these values into Eq. (19-16), we obtain

$$\text{Efficiency} = 0.99740 \qquad \text{Power loss} = 0.260\%$$

In this example, the inside radius of the internal gear is held to conventional proportions. This results, among other things, in an increased arc of approach. Even so, the power loss here is equal to 0.260 per cent as compared with a loss of 0.405 per cent for the equivalent spur-gear drive.

Second Example. As a second example we shall use the same size of gears as before, but will proportion the teeth as recommended in Chapter 6 on internal gears. This gives the two following changed values:

$$R_{o1} = 3.4125 \qquad R_i = 14.850$$

Introducing these changed values into the various equations, we obtain

$$\beta_a = 0.1619 \qquad \beta_r = 0.3181 \qquad V_{sa} = 91 \qquad f_a = 0.025 \qquad V_{sr} = 179$$
$$f_r = 0.018$$

Whence

$$\text{Efficiency} = 0.99794 \qquad \text{Power loss} = 0.206\%$$

This change in tooth proportions brings the value of the power loss here to about one-half that of the equivalent spur-gear drive.

EFFICIENCY OF HELICAL GEARS

Equation (19-1) applies also to helical gears except that the influence of the helix angle on the normal tooth load must be considered. Thus when

W = tangential applied load, lb

W'_n = normal force acting on helical tooth profile, lb

W_f = friction loss per minute

ϕ = pressure angle in plane of rotation

ϕ_n = normal pressure angle

ψ = helix angle at pitch radius

$$W'_n = \frac{W}{\cos \phi_n \cos \psi} \tag{19-20}$$

$$\tan \phi_n = \tan \phi \cos \psi \tag{19-21}$$

Substituting the value of W'_n for W_n in Eq. (19-1) for the friction loss per minute, we obtain

$$W_f = \frac{W'_n \omega_1 R_{b1}}{\beta_{a1} + \beta_{r1}} \left[\frac{f'}{2} (\beta_{a1}{}^2 + \beta_{r1}{}^2) \left(1 + \frac{1}{m} \right) \right]$$

Considering as before the work input as equal to $W \omega_1 R_1$

$$\text{Efficiency} = 1 - \left(\frac{\cos \phi}{\cos \phi_n \cos \psi} \right) \left[\frac{1 + (1/m)}{\beta_{a1} + \beta_{r1}} \right] \frac{f'}{2} (\beta_{a1}{}^2 + \beta_{r1}{}^2) \tag{19-22}$$

Summary. To bring the foregoing into the same form as the similar material for spur gears, we have the following: The equations for the arcs of approach and recess are the same as for spur gears. The values of the pressure angle in the plane of rotation must be used in these equations. When N_1, N_2 = number of teeth in driver and follower, respectively

m = gear ratio

β_a, β_r = arc of approach and recess of the driver, respectively

f = average coefficient of friction

f_a = average coefficient of friction of approach

f_r = average coefficient of friction of recess

ψ = helix angle at pitch line

ϕ = pressure angle in plane of rotation

ϕ_n = normal pressure angle

$$m = \frac{N_2}{N_1} \tag{19-5}$$

When the coefficient of friction is assumed as constant

$$\text{Efficiency} = 1 - \left(\frac{\cos \phi}{\cos \phi_n \cos \psi} \right) \left[\frac{1 + (1/m)}{\beta_a + \beta_r} \right] \frac{f}{2} (\beta_a{}^2 + \beta_r{}^2) \tag{19-23}$$

When the coefficients of friction of approach and recess are different

$$\text{Efficiency} = 1 - \left(\frac{\cos \phi}{\cos \phi_n \cos \psi}\right)\left[\frac{1 + (1/m)}{\beta_a + \beta_r}\right]\left(\frac{f_a}{2}\beta_a{}^2 + \frac{f_r}{2}\beta_r{}^2\right) \quad (19\text{-}24)$$

Coefficients of Friction on Helical Gears. In the absence of experimental data on the power losses on helical gears, we shall use the same values for the coefficients of friction here as are used for spur gears.

Example of Efficiency of Helical Gears. As a definite example we shall use a pair of helical gears, 4 DP, 20 deg in the plane of rotation, with 24 and 120 teeth, and a helix angle of 30 deg. The values will be the same as those for the similar pair of spur gears. We will also use the same speed of operation. If we multiply the power loss of the spur gears by the value of the bracket in Eq. (19-24), it will give the value of the power loss for these helical gears. Whence we obtain

Power loss $= 0.405 \times 1.1377 = 0.461\%$ Efficiency $= 0.99539$

EFFICIENCY OF SPIRAL GEARS

The contact between a pair of helical gears that operate together on nonparallel axes is point contact. The action between the teeth is primarily sliding, a type of action that exists between the teeth of all gears operating on nonparallel, nonintersecting axes. The sliding on the tooth profiles because of their different lengths is also present, but this sliding here is so little in comparison to the amount of peripheral sliding that it will be ignored. Its influence could be introduced if necessary, but this would result in a long and involved equation, and would be a refinement not justified by the present state of our knowledge about the coefficients of friction and other factors.

We have already derived equations for the sliding velocity between the two mating basic racks of a spiral-gear system. We shall use this as the average sliding velocity between the teeth of the gears. The work output will be taken as equal to the product of the tangential applied load and the pitch-line velocity of the driver in its plane of rotation. The friction loss will be the product of the normal load on the teeth, the coefficient of friction, and the average sliding velocity on the gear teeth. The work input will be taken as the sum of the work output and the friction loss.

Sliding Velocity of Spiral Gears

R_1 = radius of pitch cylinder of driver, in.

R_2 = radius of pitch cylinder of follower, in.

n = number of rpm of driver

V = pitch-line velocity of driver in its plane of rotation, ft/min

V_s = sliding velocity between basic racks, ft/min

ψ_1 = helix angle of driver at R_1

ψ_2 = helix angle of follower at R_2

Σ = shaft angle

$$V = 0.5236 R_1 n \tag{4-12}$$

$$V_s = V \sin \Sigma / \cos \psi_2 \tag{9-21}$$

When $\Sigma = 90°$, then

$$V_s = V / \cos \psi_2 \tag{9-22}$$

Efficiency of Spiral Gears

f = average coefficient of friction

W = tangential applied load on driver, lb

W_n = normal tooth load, lb

W_f = frictional work, ft-lb/min

ϕ = pressure angle, plane of rotation of driver

ϕ_n = pressure angle of basic racks and normal pressure angle of spiral gears

For the work output we have

$$\text{Work output} = WV$$

For the friction loss we have

$$W_f = W_n f V_s$$

But

$$W_n = \frac{W}{\cos \phi_n \cos \psi_1} \tag{19-25}$$

and

$$V_s = \frac{V \sin \Sigma}{\cos \psi_2} \tag{9-21}$$

whence

$$W_f = \frac{WVf \sin \Sigma}{\cos \phi_n \cos \psi_1 \cos \psi_2}$$

$$\text{Work input} = WV \left(1 + \frac{f \sin \Sigma}{\cos \phi_n \cos \psi_1 \cos \psi_2} \right)$$

$$\text{Efficiency} = \frac{\cos \phi_n \cos \psi_1 \cos \psi_2}{\cos \phi_n \cos \psi_1 \cos \psi_2 + f \sin \Sigma} \tag{19-26}$$

When $\Sigma = 90°$, then $\cos \psi_2 = \sin \psi_1$, and

$$\text{Efficiency} = \frac{\cos \phi_n \sin 2\psi_1}{\cos \phi_n \sin 2\psi_1 + 2f} \tag{19-27}$$

Coefficient of Friction for Spiral Gears. In the absence of definite test data on the efficiency of spiral-gear drives, we shall consider that the conditions here are similar enough to those on worm-gear drives to use test data from that source. Test data on the performance of different worm-gear drives show a considerable variation. The minimum coefficients of friction appear to be at sliding velocities between 300 and about 500 ft/min. Under the lighter loads, the minimum value is at the lower velocity. Under the heavier loads, the minimum value is at the higher sliding velocity.

With the theoretical point contact of spiral gears, the unit load on the gear-tooth surface is high, even under small applied loads. The experimental values for the coefficients of friction of worm gears under the heavier loads will be used for spiral-gear drives. These values are given very closely by the following empirical equation:

When V_s = sliding velocity, ft/min

e = base of natural logarithms

f = average coefficient of friction

$$f = \frac{0.200}{e^{0.17\sqrt{V_s}}} + 0.0013 \sqrt{V_s} \qquad (19\text{-}28)$$

Values of the coefficients of friction obtained by the use of this equation are tabulated in Table 19-3. These values will also be used for worm-gear drives.

Examples of Efficiency of Spiral-gear Drives. *First Example.* As a definite example we shall use a pair of spiral gears, 16 and 64 teeth, with a normal rack form of 10 DP, 14½ deg, with a 90-deg shaft angle and a helix angle of 45 deg for both gears.

TABLE 19-3. COEFFICIENTS OF FRICTION FOR SPIRAL GEARS AND FOR WORM GEARS

V_s, ft/min	f	V_s, ft/min	f
0	0.2000	750	0.0375
10	0.1209	1,000	0.0420
20	0.0993	1,250	0.0465
30	0.0859	1,500	0.0506
40	0.0764	1,750	0.0545
50	0.0693	2,000	0.0582
60	0.0637	2,500	0.0650
70	0.0591	3,000	0.0712
80	0.0553	4,000	0.0822
90	0.0522	5,000	0.0919
100	0.0495	6,000	0.1007
150	0.0408	7,000	0.1088
200	0.0365	8,000	0.1163
300	0.0330	9,000	0.1233
400	0.0327	10,000	0.1300
500	0.0358		

The driver will operate at a speed of 2,000 rpm. This gives the following values:

$$\Sigma = 90° \qquad \psi_1 = \psi_2 = 45° \qquad R_1 = 1.1313 \qquad R_2 = 4.5254 \qquad n = 2,000$$
$$V = 0.5236 \times 1.1313 \times 2,000 = 1,185$$
$$V_s = \frac{1185}{0.70711} = 1,675$$

whence

$$f = 0.0533$$
$$\text{Efficiency} = \frac{0.96815}{0.96815 + 0.1066} = 0.900 \qquad \text{Power loss} = 10.0\%$$

Second Example. As a second example we shall use the same values as before except that

$$\psi_1 = 60° \qquad \psi_2 = 30°$$

This gives the following values for the pitch radii:

$$R_1 = 1.600 \qquad R_2 = 3.650$$
$$V = 0.5236 \times 1.60 \times 2,000 = 1,675 \qquad V_s = \frac{1675}{0.86602} = 1,934 \qquad f = 0.0572$$
$$\text{Efficiency} = \frac{0.96815 \times 0.86602}{0.96815 \times 0.86602 + 0.1144} = 0.879 \qquad \text{Power loss} = 12.1\%$$

A comparison of this power loss with that of the previous example gives some indication of the value of selecting, whenever possible, helix angles equal to one-half the shaft angle.

Third Example. For the third example we shall use the same values as before but with a shaft angle of 60 deg. We shall use helix angles of 30 deg for both gears. This gives the following values:

$$\Sigma = 60° \qquad \psi_1 = \psi_2 = 30° \qquad R_1 = 0.9238 \qquad R_2 = 3.6950 \qquad n = 2,000$$
$$V = 0.5236 \times 0.9238 \times 2,000 = 967 \qquad V_s = \frac{967 \times 0.86602}{0.86602} = 967$$

Whence, from Table 19-3,

$$f = 0.0414$$
$$\text{Efficiency} = \frac{0.96815 \times 0.86602 \times 0.86602}{0.72611 + (0.0414 \times 0.86602)} = 0.953 \qquad \text{Power loss} = 4.7\%$$

In this drive, we are part of the way back toward a parallel-shaft drive, and so the sliding velocity is reduced materially. As a result, the power loss is reduced.

EFFICIENCY OF WORM-GEAR DRIVES

The contact conditions on a worm-gear drive depend upon many factors. Among them are the thread angle of the worm, its lead angle, and the position of the pitch plane of the worm. Here, as with all other types of screw gearing, the action is primarily sliding. The actual sliding velocities will be different, to a greater or lesser extent, on different drives because of differences in the nature of the contact. If we wish to be as precise as possible, we must make a complete and detailed contact analysis of every drive to determine the average sliding velocity.

One major use of any efficiency equation for worm-gear drives is to obtain some reasonable estimate of the amount of frictional heat that must be dissipated. We shall therefore consider a worm-gear drive as a development from a spiral-gear drive where one member of the pair has been made to envelop the other partially so as to obtain line contact instead of point contact. We will, therefore, use the same equations for the average sliding velocities here as are used for spiral-gear drives.

In the case of worms, we use the lead angle instead of the helix angle. Also, on the worm gear there is no uniform axial lead to the teeth, and hence this gear has neither a lead angle nor a helix angle. We shall therefore rearrange the equations for sliding to use the worm values.

Average Sliding Velocity for Worm-gear Drives

R_1 = radius to pitch plane of worm, in.

λ = lead angle of worm at R_1

Σ = shaft angle

ϕ_n = normal thread angle (one-half included angle of thread)

ϕ_x = axial thread angle

V = peripheral velocity of worm at R_1, ft/min

V_s = average sliding velocity, ft/min

n = number of rpm of worm

$$V = 0.5236R_1n \qquad (4\text{-}12)$$

Rearranging Eq. (9-21) to use the worm values, we obtain

$$V_s = V \sin \Sigma/\sin (\Sigma - \lambda) \qquad (19\text{-}29)$$

When $\Sigma = 90°$,

$$V_s = V/\cos \lambda \qquad (19\text{-}30)$$

Efficiency of Worm-gear Drives. We shall rearrange the efficiency equations for spiral gears for use on worm-gear drives. The values of the coefficients of friction in Table 19-3 were determined from experimental data from worm-gear tests, and these values will be used here. We will let f be the average coefficient of friction.

$$\tan \phi_n = \tan \phi_x \cos \lambda \qquad (19\text{-}31)$$

$$\text{Efficiency} = \frac{\cos \phi_n \sin \lambda \sin (\Sigma - \lambda)}{\cos \phi_n \sin \lambda \sin (\Sigma - \lambda) + f \sin \Sigma} \qquad (19\text{-}32)$$

When $\Sigma = 90°$, then $\sin (\Sigma - \lambda) = \cos \lambda$.

$$\text{Efficiency} = \frac{\cos \phi_n \sin 2\lambda}{\cos \phi_n \sin 2\lambda + 2f} \qquad (19\text{-}33)$$

Examples of Efficiency of Worm-gear Drive. *First Example.* As a definite example we shall use the following: A single-thread worm, 90-deg shaft angle, 30-deg normal thread angle, ½-in. lead, 3.00-in. nominal pitch diameter, running at 1,000 rpm. This gives the following values:

$$R_1 = 1.500 \qquad \phi_n = 30° \qquad n = 1,000 \qquad \lambda = 3.037° \qquad \sin 2\lambda = 0.10581$$
$$V = 0.5236 \times 1.500 \times 1,000 = 785$$
$$V_s = \frac{785}{0.99859} = 786$$

From Table 19-3, $f = 0.0381$.

$$\text{Efficiency} = \frac{0.86602 \times 0.10581}{0.09163 + 0.0762} = 0.546 \qquad \text{Power loss} = 45.4\%$$

Second Example. As a second example we shall use a 3-start worm of the same size as before, with all other values the same. This gives the following values:

$$R_1 = 1.500 \qquad n = 1,000 \qquad \lambda = 9.043° \qquad \cos \lambda = 0.98757 \qquad \sin 2\lambda = 0.31044$$
$$V = 785 \qquad V_s = \frac{785}{0.98757} = 795 \qquad f = 0.0383$$
$$\text{Efficiency} = \frac{0.86602 \times 0.31044}{0.26885 + 0.0766} = 0.778 \qquad \text{Power loss} = 22.2\%$$

Third Example. As a third example we shall use a 6-start worm of the same size and at the same speed as before. This gives the following:

$$R_1 = 1.500 \qquad \lambda = 17.657° \qquad \cos \lambda = 0.95289 \qquad \sin 2\lambda = 0.57806$$
$$V = 785 \qquad V_s = \frac{785}{0.95289} = 824 \qquad f = 0.0389$$
$$\text{Efficiency} = \frac{0.86602 \times 0.57806}{0.50061 + 0.0778} = 0.865 \qquad \text{Power loss} = 14.5\%$$

Fourth Example. As a fourth example we shall use a 12-start worm of the same size and at the same speed as before. This gives the following:

$$R_1 = 1.500 \qquad \lambda = 32.482° \qquad \cos \lambda = 0.84356 \qquad \sin 2\lambda = 0.90604$$
$$V = 785 \qquad V_s = \frac{785}{0.84356} = 930 \qquad f = 0.0408$$
$$\text{Efficiency} = \frac{0.86602 \times 0.90604}{0.78465 + 0.0816} = 0.905 \qquad \text{Power loss} = 9.5\%$$

Fifth Example. As a fifth example we shall use an 18-start worm of the same size as before. This gives the following values:

$$R_1 = 1.500 \qquad \lambda = 43.679° \qquad \cos \lambda = 0.72322 \qquad \sin 2\lambda = 0.99894$$
$$V = 785 \qquad V_s = \frac{785}{0.72322} = 1,085 \qquad f = 0.0435$$
$$\text{Efficiency} = \frac{0.86602 \times 0.99894}{0.86510 + 0.0870} = 0.909 \qquad \text{Power loss} = 9.1\%$$

It will be noted that very little is gained in efficiency with increasing lead angles after the value has reached 30 deg or over.

EFFICIENCIES OF BEVEL GEARS

The tooth action of bevel gears with straight teeth is very similar to that of spur gears. The beveled faces of the gears introduce axial thrusts, which are absent on spur gears, but these thrusts are components of the normal tooth loads, and the resulting axial thrusts are carried by the bearings. This increases the bearing friction losses over those on spur gears of equal size and loads, but the efficiency equations we are concerned with here deal only with the friction losses at the tooth mesh.

We shall therefore use the spur-gear efficiency equations for these bevel gears. The arcs of approach and recess will be determined from the equivalent spur gears of Tregold's approximation. We shall therefore start with a summary of Tregold's approximation and the calculation of the arcs of approach and recess.

Tregold's Approximation

N_p, N_g = number of teeth in bevel pinion and gear, respectively

γ_p, γ_g = pitch angle of bevel pinion and gear, respectively

Σ = shaft angle

R_p, R_g = pitch radius of bevel pinion and gear at large end, in.

R_{vp}, R_{vg} = pitch radius of equivalent spur pinion and gear, in.

$$\Sigma = \gamma_p + \gamma_g \tag{14-1}$$
$$R_{vp} = R_p/\cos \gamma_p \tag{15-6}$$
$$R_{vg} = R_g/\cos \gamma_g \tag{15-7}$$

We shall use the mean diameters of the faces of the bevel gears to determine the average pitch-line and sliding velocities. We shall use the diameters at the large ends to determine the arcs of approach and recess because they are the simplest to use and these values in radians are the same for all parts of the face.

Arcs of Approach and Recess

R_{vp}, R_{vg} = pitch radius of equivalent spur pinion and gear, in.

a_p, a_g = addendum of pinion and gear, respectively, in.

R_{op}, R_{og} = outside radius of equivalent spur pinion and gear, in.

R_{bp}, R_{bg} = base radius of equivalent spur pinion and gear, in.

ϕ = pressure angle

β_a = arc of approach of pinion

β_r = arc of recess of pinion

$$R_{op} = R_{vp} + a_p \tag{19-34}$$
$$R_{og} = R_{vg} + a_g \tag{19-35}$$
$$R_{bp} = R_{vp} \cos \phi \tag{19-36}$$
$$R_{bg} = R_{vg} \cos \phi \tag{19-37}$$

Substituting these symbols into Eq. (4-17) and (4-18), we obtain

$$\beta_a = (\sqrt{R_{og}^2 - R_{bg}^2} - R_{vg} \sin \phi)/R_{bp} \tag{19-38}$$
$$\beta_r = (\sqrt{R_{op}^2 - R_{bp}^2} - R_{vp} \sin \phi)/R_{bp} \tag{19-39}$$

Sliding Velocity

V = pitch-line velocity at R_m, ft/min
V'_s = average sliding velocity, ft/min
V_{sa} = average sliding velocity of approach, ft/min
V_{sr} = average sliding velocity of recess, ft/min
R_m = mean pitch radius of bevel pinion, in.
n = rpm of bevel pinion driver

$$V = 0.5236R_m n \tag{19-40}$$

$$V'_s = V \cos \phi \, [1 + (R_{vp}/R_{vg})] \, [(\beta_a + \beta_r)/4] \tag{19-41}$$

$$V_{sa} = \frac{V}{2}\left(1 + \frac{R_{vp}}{R_{vg}}\right) \beta_a \cos \phi \tag{19-42}$$

$$V_{sr} = \frac{V}{2}\left(1 + \frac{R_{vp}}{R_{vg}}\right) \beta_r \cos \phi \tag{19-43}$$

Coefficients of Friction for Bevel Gears. We shall use the same coefficients of friction here as are used for spur gears. As with spur gears, the values of f will be used on soft gears when the arcs of approach and recess are substantially the same, and also in those cases where some approximation for the amount of the power loss is needed to determine the frictional heat of operation. Values of f_a and f_r will be used to compare the probable or relative performance of different tooth designs used for soft gears. The values of f_r will be used for both approach and recess on hardened-steel bevel gears.

Efficiency of Bevel Gears. When m is the equivalent-spur-gear ratio

$$m = \frac{R_{vg}}{R_{vp}} \tag{19-44}$$

When a constant coefficient of friction is used, then

$$\text{Efficiency} = 1 - \left[\frac{1 + (1/m)}{\beta_a + \beta_r}\right]\left(\frac{f}{2}\right)(\beta_a{}^2 + \beta_r{}^2) \tag{19-45}$$

When coefficients of friction of approach and recess are used, then

$$\text{Efficiency} = 1 - \left[\frac{1 + (1/m)}{\beta_a + \beta_r}\right]\left(\frac{f_a}{2}\beta_a{}^2 + \frac{f_r}{2}\beta_r{}^2\right) \tag{19-46}$$

Examples of Efficiency of Bevel Gears. *First Example.* For a definite example we shall use a pair of 16-tooth, 4-DP bevel gears operating at a speed of 1,000 rpm. The face widths of the gears will be taken as 0.750 in. The axes of the gears are at 90 deg. The arcs of approach and recess will be calculated from the 1-DP values, as follows:

$$R_p = R_g = 8.000 \qquad \gamma_p = \gamma_g = 45° \qquad a_p = a_g = 0.9338$$

From Tregold's approximation we obtain

$$R_{vp} = R_{vg} = 11.3135 \qquad \phi = 20.50° \qquad R_{op} = R_{og} = 12.2473$$
$$R_{bp} = R_{bg} = 10.5970 \qquad \cos \phi = 0.93667 \qquad \sin \phi = 0.35021$$

Whence

$$\beta_a = \beta_r = 0.2055$$

For the 4-DP values we have

$$R_m = 2.000 - 0.2620 = 1.738 \qquad n = 1,000$$
$$V = 0.5236 \times 1.738 \times 1,000 = 910 \qquad m = \frac{2.8284}{2.8284} = 1.000$$
$$V_{sa} = V_{sr} = (^{910}\!\!/\!_2)(1 + 1)(0.2055 \times 0.93667) = 175 \qquad f = 0.0264$$

As the arcs of approach and recess are equal, and these are soft gears, we would use the value of f in any case. Substituting these values into Eq. (19-45), we obtain

$$\text{Efficiency} = 0.9946 \qquad \text{Power loss} = 0.54\%$$

Second Example.   As a second example we shall use a 16-tooth bevel pinion and a 32-tooth bevel gear. The other values will be the same as before. We shall use the following 1-DP values:

$$R_p = 8.000 \qquad R_g = 16.000 \qquad \gamma_p = 26.565° \qquad \gamma_g = 63.435°$$
$$a_p = 1.3948 \qquad a_g = 0.6035$$

From Tregold's approximation we obtain

$$R_{vp} = 8.9442 \qquad R_{vg} = 35.7773 \qquad \phi = 15.05° \qquad R_{op} = 10.3390$$
$$R_{og} = 36.3806 \qquad \sin \phi = 0.25966 \qquad \cos \phi = 0.96570 \qquad R_{bp} = 8.6374$$
$$R_{bg} = 34.5501 \qquad \beta_a = 0.2442 \qquad \beta_r = 0.3983$$

For the 4-DP values we have

$$R_m = 2.000 - 0.1677 = 1.8323 \qquad n = 1,000$$
$$V = 0.5236 \times 1.8323 \times 1,000 = 959 \qquad m = \frac{35.7773}{8.9442} = 4.000$$
$$V_{sa} = (^{959}\!\!/\!_2)(1.25)(0.2442 \times 0.96570) = 141 \qquad f_a = 0.0316$$
$$V_{sr} = (^{959}\!\!/\!_2)(1.25)(0.3983 \times 0.96570) = 230 \qquad f_r = 0.0202$$

Substituting these values into Eq. (19-46), we obtain

$$\text{Efficiency} = 0.9957 \qquad \text{Power loss} = 0.41\%$$

In this case, the larger reduction drive shows a slightly smaller power loss than the pair of miter gears. This is largely because of the greater amount of recess action and a lesser amount of approach action on the last pair of bevel gears.

EFFICIENCY OF SPIRAL BEVEL GEARS

The tooth action of spiral bevel gears has much in common with that of helical gears. We shall therefore determine the equivalent helical gear by the use of Tregold's approximation, and use the efficiency equations for helical gears to obtain a measure of the efficiencies of spiral bevel gears. For the value of the helix angle, we shall use that of the spiral angle of the crown rack of the spiral-bevel-gear system.

As the contact on the commonly used spiral bevel gear does not extend entirely across the tooth face, and as the actual action is a combination of conjugate gear-tooth action and rolling action along the spiral, we shall not attempt to separate the friction of approach and recess, but shall use a single average value for the coefficient of friction. We shall, however, determine the arcs of approach and recess at the middle of the gear face in order to obtain some measure of the average amount and velocity of the sliding action. These conditions change across the face of the spiral bevel gear because of the changing spiral angle, but the value at the center of the face of the gear should give a reasonable average value.

The form of the equivalent helical gear and the arcs of approach and recess are determined in exactly the same manner as for bevel gears with straight teeth. The value of the pressure angle in the plane of rotation of the equivalent helical gear is used for this purpose.

When ψ = spiral angle at middle of tooth face

ϕ = pressure angle in plane of rotation of equivalent helical gear

ϕ_n = normal pressure angle at middle of face

and all other symbols are the same as those for bevel gears

$$\tan \phi = \frac{\tan \phi_n}{\cos \psi} \tag{19-47}$$

$$m = \frac{R_{vg}}{R_{vp}} \tag{19-44}$$

$$V = 0.5236 R_m n \tag{19-40}$$

$$V'_s = V \cos \phi \left(1 + \frac{R_{vp}}{R_{vg}}\right)\left(\frac{\beta_a + \beta_r}{4}\right) \tag{19-41}$$

$$\text{Efficiency} = 1 - \left(\frac{\cos \phi}{\cos \phi_n \cos \psi}\right)\left[\frac{1 + (1/m)}{(\beta_a + \beta_r) \cos \psi}\right]\left(\frac{f}{2}\right)(\beta_a{}^2 + \beta_a{}^2) \tag{19-48}$$

The values of f in Table 19-2 will be used for spiral bevel gears made of soft materials, while the value of f_r will be used for spiral bevel gears made of hardened steel.

Example of Efficiency of Spiral Bevel Gear. As a definite example we shall use a 16-tooth spiral bevel pinion and a 32-tooth gear, 4 DP, with the driving pinion running at a speed of 1,000 rpm, and shall compare their performance with the similar pair of bevel gears with straight teeth. These gears will be of soft steel for the purpose of comparison. To determine the arcs of approach and recess, we have the following 1-DP values:

$$R_p = 8.000 \qquad R_g = 16.000 \qquad \gamma_p = 26.565° \qquad \gamma_g = 63.435°$$
$$a_p = 1.150 \qquad a_g = 0.550 \qquad \psi = 30° \qquad \phi_n = 14.50° \qquad R_{vp} = 8.9442$$
$$R_{vg} = 35.7773 \qquad \tan \phi = \frac{0.25862}{0.86602} = 0.29863 \qquad \phi = 16.627°$$
$$R_{bp} = 8.5702 \qquad R_{bg} = 34.2814 \qquad R_{op} = 10.0942 \qquad R_{og} = 36.3273$$

Whence

$$\beta_a = 0.1992 \qquad \beta_r = 0.3237$$

For the 4-DP values, we have as before

$$R_m = 1.8323 \qquad V = 959 \qquad V'_s = (959 \times 0.95819)(1.250)\left(\frac{0.5229}{4}\right) = 150$$

Whence from Table 19-2 we obtain

$$f = 0.0245$$

Substituting these values into Eq. (19-48), we obtain

$$\text{Efficiency} = 0.9950 \qquad \text{Power loss} = 0.50\%$$

The reduced tooth heights of this form reduce the arcs of approach and recess and the sliding velocity, as compared with the bevel gears with straight teeth. As a result, the power loss here is practically the same as that on the straight-toothed bevel gears of the same size. With hardened-steel gears, the power loss would be about two-thirds of this amount, or equal to 0.33 per cent.

EFFICIENCY OF HYPOID GEARS

The tooth action of hypoid gears is complex. It has some of the characteristics of spiral-bevel-gear tooth action and some of the characteristics of worm-gear tooth action. The worm-gear tooth action develops from the sliding of the hyperbolic pitch surfaces on each other. As the distance between the axes of the gears increases, this sliding also increases, and the tooth action approaches closer to that of a worm-gear drive. As this distance becomes less, the sliding is reduced, and the action approaches closer to that of spiral bevel gears. When this distance becomes zero, then we have a pair of spiral bevel gears.

Any general analysis of the tooth action of hypoid gears made for the purpose of determining actual sliding velocities and the resulting efficiencies would probably be an approximation at best, or else too complex for general use. Until we have further information on measured power losses here, any such analysis or approximation will be open to question. However, in order to complete this general analysis of the efficiencies of different types of gears, we shall venture on a general approximation for this purpose.

To keep this analysis relatively simple, we shall make the following assumptions:

1. We shall assume that the conjugate gear-tooth action here is substantially the same as that on equivalent spiral bevel gears, and shall therefore use the spiral-bevel-gear analysis for this part of the power loss.

2. We shall assume that the sliding of the pitch surfaces in the planes of rotation of the two members is the controlling factor of the worm-gear action, and that the results of this sliding are commensurate with that on

worm gears. We shall therefore use many of the factors from the worm-gear analysis here.

3. We shall assume that the total power loss is equal to the sum of the two foregoing elements of power loss. We shall call them the *spiral-bevel-gear loss* and the *worm-gear loss*.

Spiral-bevel-gear Power Loss

ψ = spiral angle of gear at middle of tooth face

R_m = mean pitch radius of hypoid pinion, in.

ϕ = pressure angle in plane of rotation of equivalent helical gear

ϕ_n = normal pressure angle at middle of gear face

R_{vp}, R_{vg} = pitch radius of equivalent helical pinion and gear, in.

m = gear ratio of equivalent helical gears

β_a, β_r = arc of approach and recess of equivalent helical gears

f = average coefficient of friction (Table 19-2)

V = pitch-line velocity, ft/min

V'_s = average sliding velocity, ft/min

n = rpm of driving pinion

R_{hp}, R_{hg} = radius of base cylinder of pinion and gear hypoid, in.

C = center distance

R_p, R_g = pitch radius of hypoid pinion and gear at large ends, in.

R'_p, R'_g = pitch radius of equivalent bevel pinion and gear, in.

N_p, N_g = number of teeth in hypoid pinion and gear, respectively

γ_p, γ_g = angle of generatrix of pinion and gear hypoids

$$R_p = \sqrt{(R'_p)^2 + R_{hp}{}^2} \tag{19-49}$$

$$R_g = \sqrt{(R'_g)^2 + R_{hg}{}^2} \tag{19-50}$$

$$R_{vp} = \frac{R'_p}{\cos \gamma_p} \tag{19-51}$$

$$R_{vg} = \frac{R'_g}{\cos \gamma_g} \tag{19-52}$$

$$\tan \phi = \frac{\tan \phi_n}{\cos \psi} \tag{19-47}$$

$$m = \frac{R_{vg}}{R_{vp}} \tag{19-44}$$

$$V = 0.5236 R_m n \tag{19-40}$$

$$V'_s = V \cos \phi \left(1 + \frac{R_{vp}}{R_{vg}}\right)\left(\frac{\beta_a + \beta_r}{4}\right) \tag{19-41}$$

We will let

$$A = \frac{\text{power loss}}{\text{transmitted power}}$$

$$A = \frac{\cos \phi}{\cos \phi_n \cos \psi}\left[\frac{1 + (1/m)}{(\beta_a + \beta_r)\cos \psi}\right]\left(\frac{f}{2}\right)(\beta_a{}^2 + \beta_r{}^2) \tag{19-53}$$

Worm-gear Power Loss. We might develop some extended equations to determine the sliding velocities along the teeth of hypoid gears, the sliding action that is the result of the sliding between the two pitch hyperboloids of these gears. It is doubtful at this time, however, if we could do more than make a series of approximations. Therefore in order to have some simple approximation that will enable us to estimate the conditions of frictional heat, we shall treat this action as though it were worm-gear action, and use as a measure of the sliding velocity the component of the slding velocity between the pitch hyperboloids in their planes of rotation. In other words, this sliding velocity will be considered the same as the peripheral velocity of the radius to the pitch plane of the worm. Thus when

R_{bp} = radius of base cylinder of pinion hypoid, in.

γ_p = angle of generatrix of pinion

ϕ_n = normal pressure angle at middle of face

V = peripheral velocity of pinion at R_{bp}, ft/min

V_s = average sliding velocity, ft/min

n = number of rpm of driving pinion

f = average coefficient of friction (Table 19-3)

$$V = 0.5236 R_{bp} n \qquad (19\text{-}54)$$

$$V_s = \frac{0.5236 R_{bp} n}{\sin \gamma_p} \qquad (19\text{-}55)$$

We will let

$$B = \frac{\text{power loss}}{\text{transmitted power}}$$

$$B = \frac{2f}{\cos \phi_n \sin 2\gamma_p + 2f} \qquad (19\text{-}56)$$

Total Power Loss on Hypoid Gears

$$\frac{\text{Total power loss}}{\text{Transmitted power}} = A + B \qquad (19\text{-}57)$$

$$\text{Efficiency} = \frac{1}{1 + A + B} \qquad (19\text{-}58)$$

Example of Efficiency of Hypoid Drive. As a definite example we shall use a hypoid drive of the same size, speed, and ratio as was used in the example for spiral bevel gears. Here the spiral-bevel-gear loss will be the same as before. This gives the following values:

$N_p = 16 \quad N_g = 32 \quad \gamma_p = 26.565° \quad \gamma_g = 63.435° \quad \psi = 30°$

$\phi_n = 14.50° \quad R'_p = 2.000 \quad R'_g = 4.000 \quad C = 1.500 \quad R_{bp} = 0.300$

$R_{bg} = 1.200 \quad n = 1,000 \quad R_p = 2.0223 \quad R_g = 4.1761$

From the example on spiral bevel gears, we have

$$A = 0.0050$$

For the worm-gear power loss we have

$$V_s = 0.5236 \times 0.300 \times \frac{1,000}{0.44721} = 586 \qquad f = 0.0416$$

$$B = \frac{0.0832}{0.96815 \times 0.800 + 0.0832} = 0.0970$$

$$A + B = 0.0050 + 0.0970 = 0.1020$$

$$\text{Efficiency} = \frac{1}{1.102} = 0.9074 \qquad \text{Power loss} = 9.26\%$$

If this approximation is reasonably accurate, the power loss on these hypoid gears is almost twenty times the amount of the power loss on the equivalent spiral-bevel-gear drive. Even so, the actual amount of these losses is small. The most important matter is the amount of the frictional heat that must be dissipated.

CHAPTER 20

ANALYSIS OF DYNAMIC LOADS ON SPUR-GEAR TEETH

The following chapter is an analysis of the tests made on the Lewis gear-testing machine by the ASME Special Research Committee on the Strength of Gear Teeth. This committee was organized under the chairmanship of Wilfred Lewis and was directed by him until his death in 1929. A description of the testing machine and details of the tests has been published in the ASME Research Publication, 1931, entitled Dynamic Loads on Gear Teeth. Some further development of this material is included in the following chapter.

Errors on gear-tooth profiles, caused by elastic deformation under load or by inaccuracies of production, or both, act to change the relative velocities of the mating members. This varying velocity of the rotating members results in a varying load cycle on the teeth of the gears; the amount of this load variation depends largely upon the extent of the effective masses of the revolving gears, the extent of the effective errors, and the speed of the gears. If the gears were made of rigid materials, the acceleration loads would vary as the square of the velocity. With elastic materials, however, the deformation of the teeth will also increase with an increase in load and tend to reduce the amount of change in velocity, and this will reduce the intensity of the high momentary acceleration load.

It appears from a study of the charts made on the testing machine that give a measure of the accuracy of the gears, that the effective error seems to act primarily as the load is being transferred from one pair of mating teeth to the next pair. The errors may be of any type, yet their influence seems to be greatest during the transfer of the load from tooth to tooth.

When a positive error, or high spot, is present and comes into mesh, it acts to slow down the driving gear and to speed up the follower. The relative change in velocity of either member will depend upon the amount of the effective masses acting at the pitch line of each gear; the change in velocity of the member with the greater effective mass will be the smaller of the two. If the masses are equal, the change in velocity will be divided equally between the two gears.

At the instant that the second pair of mating tooth profiles have taken over the full load, and the accelerating action of the error has ceased to

426

act, the masses are moving at different velocities such that the bearing surfaces of the mating teeth tend to separate. This relative motion away from each other is opposed by the power input and torsional deflection on one side and the applied load on the other side. Eventually the two teeth come together again with an impact, the intensity of which is the maximum momentary load on the gear teeth, or the dynamic load. In other words, the change in momentum set up by the action of the effective error is absorbed by elastic impact, and this impact load is always the maximum load value of the cycle. Thus we have two load surges at every tooth engagement: the acceleration load, which is set up by the first phase of the tooth engagement, followed by the impact or dynamic load, which is the reaction to the acceleration load.

Acceleration Load. If the materials were rigid, the acceleration load would vary as the square of the pitch-line velocity. As the materials are elastic, when the load required to deform the teeth the amount of the effective error is less than that required to accelerate the effective masses, the teeth will be deformed, and the acceleration of the masses will be reduced accordingly. At infinite speeds, the momentum of the masses would be infinite, and no change in the velocity of the masses would be possible. Hence, under these conditions, the teeth would be deformed the full amount of the effective error, and the maximum load would be that required to deform the teeth that amount. This gives a limiting or asymptotic value of the acceleration load. Thus when

f_a = force acting at acceleration, lb

V = pitch-line velocity, ft/min

C_1 = value representing the reactions of rigid bodies

C_2 = asymptotic load or force required to deform the teeth the full amount of the effective error, lb

$$f_1 = C_1 V^2 \qquad f_2 = C_2$$

Then

$$1/f_a = (1/f_1) + (1/f_2) \qquad (20\text{-}1)$$

Solving for f_a, we obtain

$$f_a = f_1 f_2/(f_1 + f_2) \qquad (20\text{-}2)$$

Reactions of Rigid Bodies. We shall direct our attention first to the equation of the parabola that represents the reactions of rigid bodies. In order to obtain numerical values for this equation, we must have some measure of the effective error, the distance in which it acts, and the effective masses of the mating gears. The effective error is in the nature of a high spot or foreign body that tends to separate the mating tooth surfaces. It is measured by the displacement of the relative positions of the gears at the pitch line. Thus when

E = effective error in action of the gears, ft

m = effective mass at pitch line of gears (polar moment of inertia divided by the square of the pitch radius)

v = pitch-line velocity, ft/sec

a = acceleration, ft /sec^2

Then

$$f_1 = ma$$

This accelerating force acts for a very short time while the load is being transferred from one pair of mating teeth to the next pair. Thus when

S = distance a point on pitch line travels while the accelerating force acts, ft

D_1 = distance a point on pitch line would travel in the same time if no error were present, ft

t = time in which accelerating force acts, sec

v_0 = initial pitch-line velocity, ft/sec

$$S = v_0 t + (at^2/2)$$

whence

$$a = 2(S - v_0 t)/t^2$$

But

$$S = D_1 + E \qquad \text{and} \qquad t = D_1/v_0$$

Substituting these values into the equation for the acceleration, we obtain

$$a = 2Ev_0{}^2/D_1{}^2$$

Whence

$$f_1 = ma = 2mEv_0{}^2/D_1{}^2 \tag{20-3}$$

We shall transform Eq. (20-3) into the following units, which are the ones commonly used in gear calculations:

V = pitch-line velocity, ft/min = $60v_0$

e = effective error, in. = $12E$

D = distance in which error acts = $12D_1$

Substituting these values into Eq. (20-3), we obtain

$$f_1 = meV^2/150D^2 \tag{20-4}$$

The results of the tests on the Lewis machine indicate that as far as work done is involved, it is the amount of the error rather than its exact nature that is the determining factor in the magnitude of the loads required to keep the teeth of the test gears from separating far enough to break the electrical circuit that is passed through the meshing teeth.

The observed conditions would indicate that the value of e/D^2 is practically a constant for a given size and tooth form of gear. The following graphical analysis gives us a value for this expression that agrees closely with the results of the tests.

In order to obtain some expression for the relationship e/D^2, we shall consider the two gears separately and as each gear meshing with a common basic rack. We shall also consider the error to be divided between them. We shall then determine the value for each gear meshing with this basic rack and add the two expressions for the final answer.

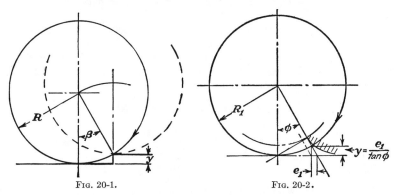

Fig. 20-1.   Fig. 20-2.

The first problem is to establish some reasonable relationship between the two factors. As a preliminary move, we shall consider the conditions of a hoop or circle rolling along a straight line until it meets an obstruction, as indicated in Fig. 20-1.

Referring to Fig. 20-1 where

y = height of obstruction, in.

R = radius of circle, in.

β = angle of rotation to rise on obstruction

D = circumferential rotation of point on circle through angle β, in.

$$\cos \beta = (R - y)/R$$

whence

$$y = R(1 - \cos \beta) \qquad D = R\beta$$

Whence

$$y/D^2 = (1 - \cos \beta)/R\beta^2$$

In this preliminary example, the obstruction or error is perpendicular to the direction of rolling of the circle. To meet the conditions on a gear, it must be in the same direction. Such a condition for the pinion is indicated in Fig. 20-2. We shall assume that the form of the pinion tooth is full, or extends beyond the theoretical form, so that it moves the basic rack the additional distance e_1 when it is rotated through the angle ϕ. In this case

R_1 = pitch radius of pinion, in.
R_2 = pitch radius of gear, in.
ϕ = pressure angle
e_1 = error on pinion profile, in.
e_2 = error on gear profile, in.

$$y = e_1/\tan \phi \qquad e_1 = y \tan \phi \qquad D = R_1\phi$$

Whence

$$e_1/D^2 = \tan \phi(1 - \cos \phi)/R_1\phi^2$$

We shall assume that the relationship between the error on the gear and the distance through which it acts is the same as the similar relationship on the pinion. Whence we have

$$e_2/D^2 = \tan \phi(1 - \cos \phi)/R_2\phi^2$$

But

$$e = e_1 + e_2$$

Whence

$$e/D^2 = [(1/R_1) + (1/R_2)] [\tan \phi(1 - \cos \phi)/\phi^2] \qquad (20\text{-}5)$$

Substituting this value into Eq. (20-4), we obtain

$$f_1 = [\tan \phi(1 - \cos \phi)/150\phi^2] [(1/R_1) + (1/R_2)]mV^2 \qquad (20\text{-}6)$$

When ϕ is equal to $14\frac{1}{2}$ deg, this equation becomes

$$f_1 = 0.00086 [(1/R_1) + (1/R_2)]mV^2 \qquad (20\text{-}7)$$

When ϕ is equal to 20 deg, this equation becomes

$$f_1 = 0.00120 [(1/R_1) + (1/R_2)]mV^2 \qquad (20\text{-}8)$$

These values satisfy the test results and will be used for the further analysis of the dynamic loads.

Asymptotic Load. We must now determine the value of f_2, or the value of the load required to deform the teeth the amount of the effective error. To do this, we must determine the amount of deformation of the gear teeth under load. Formulas for the calculation of the elastic deformation of gear teeth are given in a paper by S. Timoshenko and R. V. Baud.[1] These equations are as follows:

[1] The Strength of Gear Teeth, *Mech. Eng.*, November, 1926.

Compressive Deformation. Starting with the compressive deformation, when

b = width of strip of contact between two cylinders under load, in.

P = load per inch of face, lb

E = modulus of elasticity of material

r_1 = radius of first cylinder, in.

r_2 = radius of second cylinder, in.

$$b = 3.04 \sqrt{\frac{P}{E} \frac{r_1 r_2}{r_1 + r_2}} \tag{20-9}$$

When d_1 = compressive deformation, in.

m = Poisson's ratio for the material

$$d_1 = \frac{2(1 - m^2)}{E} \frac{P}{\pi} \left(\frac{2}{3} + \log \frac{4r_1}{b} + \log \frac{4r_2}{b} \right) \tag{20-10}[1]$$

The radius of curvature on an involute gear-tooth profile is changing constantly as the diameter changes. However, when a pair of involute gears are meshed, the sum of the radii of curvature $(r_1 + r_2)$ on the mating profiles is constant and is equal to the center distance multiplied by the sine of the pressure angle. Equations (20-9) and (20-10) may be combined and simplified for such involute gears as follows:

$$\log (4r_1/b) + \log (4r_2/b) = \log \frac{16r_1 r_2}{b^2}$$

$$b^2 = 10.336 \left(\frac{P}{E} \frac{r_1 r_2}{r_1 + r_2} \right)$$

$$d_1 = \frac{2(1 - m^2)}{E} \frac{P}{\pi} \left[\frac{2}{3} + \log \frac{1.548E (r_1 + r_2)}{P} \right] \tag{20-11}$$

It will be seen from Eq. (20-11) that the amount of compression depends upon the sum of the radii of curvature of the surfaces in contact. As this sum is a constant on a pair of mating involute gear-tooth profiles, the deformation caused by the compression will be constant over the entire profile as long as the load and pressure angle are constant.

According to Eq. (20-11), the amount of compression increases with increasing values for the radii of curvature, all other factors remaining the same. For the compressive deformation of the curved profiles of gear teeth, this is contrary to all expectations. The foregoing condition is due to the derivation of Eq. (20-10) through integration from one center of curvature to the other. In most cases on gear teeth, these centers of curvature are outside the tooth form, particularly as the radius

---

[1] All logarithms are to base e.

of curvature becomes larger. It is believed that the approximate formula given later as Eq. (20-13), which satisfies measured conditions of deformation under known test loads, gives a closer measure of the truth than does Eq. (20-11). The approximate equation (20-13) will be used in the further analysis.

Bending Deflection. We shall now consider the bending deflections of the gear teeth.

When d_2 = deflection caused by bending and shear of the gear tooth, in.

L = length of tooth to sharp point, in.

a = distance from sharp point to point of application of load, in.

h_0 = thickness of tooth at base, in.

h = thickness of tooth at point of application of load, in.

$$d_2 = \frac{12PL^3}{Eh_0{}^3} \left[\left(\frac{3}{2} - \frac{a}{2L} \right) \left(\frac{a}{L} - 1 \right) + \log \frac{L}{a} \right] + \frac{4P(L - a)(1 + m)}{(h + h_0)E}$$

(20-12)

Equation (20-12) is that for a cantilever beam of variable depth. The first term on the right-hand side represents the deflection due to the bending moment, and the second term represents the deflection due to the shearing force.

When we use the foregoing equations, the total deformation of a pair of loaded gear teeth will be equal to the sum of the compressive deformation of (20-11) and the bending deflection of both members from (20-12).

Approximate Equation for Compression and Bending. Equations (20-11) and (20-12) are used to determine the relative deflections on a pair of gear teeth as the load is applied at different points over their profiles. For working values, when the contact is at the middle of the gear teeth, the following approximation, which is based on experimental measured values of different tooth forms and different materials, will be used. The value gives the combined bending and compressive deformation of the mating pair of gear teeth.

When z = elasticity form factor of gear teeth

y = Lewis tooth-form factor

E = modulus of elasticity of material

d = total elastic deformation of the pair of mating tooth profiles, 1-in. face width, at middle of profile, in.

F = face width of gears, in.

W = applied tangential load, lb

$$d = (W/F) \, [(E_1 z_1 + E_2 z_2)/E_1 z_1 E_2 z_2]$$ (20-13)

$$z = y/(0.242 + 7.25y)$$ (20-14)

Limiting Acceleration Load. The limiting acceleration load, as noted before, is the load that is required to deform the teeth by the amount of the effective error. With perfectly formed teeth, the effective error would be the amount of deformation under the applied load. When errors are present, the effective error is the combination of the original error and the deformation under the applied load. Thus when

e = measured error on pair of mating teeth, in.

d = total deformation of mating teeth under applied load, in.

f_2 = limiting acceleration load, lb

W = applied tangential load, lb

$$f_2 = W[(e/d) + 1] \qquad (20\text{-}15)$$

Effective Mass. The effective mass acting at the pitch line of any pair of gears attached to shafts carrying other rotating masses is variable. The amount of this variation will depend largely upon the speed of the gears, the extent of the effective error in the gear teeth, and the elasticity of the shaft or coupling between the gears and the other rotating masses. This variation in effective mass is caused by the elasticity of the connecting member. If the shaft were rigid and all the other rotating masses were rigidly connected to it, the effective mass would be constant, and all variations in velocity caused by the imperfect meshing of the gear teeth would be imparted to all the connected rotating bodies. The momentum of these bodies would then set up greater acceleration loads than those which are created by the masses of the gear blanks alone, and the greater the masses of these connected bodies, the greater this additional load would be.

However, the shafts or other connections are not rigid but elastic, so that when it takes less force to twist the shaft than to accelerate the connected masses, the shaft will twist, and the acceleration of the connected masses will be correspondingly reduced.

We shall now attempt to determine the influence of the elasticity of the shaft or coupling on the amount of the acceleration load of the connected masses that will be felt at the pitch line of the gears. Thus when

f_1 = force required to accelerate connected masses, acting at radius equal to pitch radius of gear, lb

f_2 = force, acting at radius equal to pitch radius of gear, required to twist shaft by amount of displacement, lb

f_r = resultant force required to accomplish a combined acceleration of the connected masses and twisting of shaft, lb

$$f_r = (f_1 f_2)/(f_1 + f_2) \qquad (20\text{-}16)$$

Here we have a condition similar to that for the acceleration load on gear teeth. We shall now direct our attention to the value of f_1, or the force required to accelerate the connected masses.

e'' = additional distance at pitch radius to be moved, in.

D = distance along pitch circle in which acceleration takes place, in.

m_a = mass effect of connected masses at pitch radius = I_0/R^2

V = pitch-line velocity of gears, ft/min

From Eq. (20-4) we have

$$f_1 = m_a e'' V^2 / 150 D^2 \qquad (20\text{-}17)$$

The previous analysis brings out the condition that the relationship for any gear of e''/D^2 appears to be a value independent of the extent of the error. We shall therefore use the same relationship here. Whence

$$e''/D^2 = [(1/R_1) + (1/R_2)] [\tan \phi \, (1 - \cos \phi)/\phi^2]$$

We have for $14\frac{1}{2}$-deg gears, from (20-7),

$$H = 0.00086[(1/R_1) + (1/R_2)]$$

We have for 20-deg gears, from (20-8),

$$H = 0.00120[(1/R_1) + (1/R_2)]$$

Then

$$f_1 = H m_a V^2 \qquad (20\text{-}18)$$

We shall now direct our attention to the force required to twist the shaft or coupling an amount equal to e'' measured at the pitch radius of the gear. Thus when

Z = elasticity factor of shaft or coupling

P = load applied at pitch radius to twist connecting member, lb

T = torsional deflection at pitch radius under load P, in.

$$Z = P/T \qquad (20\text{-}19)$$

For flexible couplings and other complex forms, the value of Z may be determined experimentally. For solid cylindrical shafts, however, this factor may be calculated as follows:

When R = radius where load is applied (pitch radius), in.

L = length of shaft, in.

d = diameter of shaft, in.

F = torsional modulus of elasticity

$$T = (P/F)(32R^2L/\pi d^4) \qquad (20\text{-}20)$$

Whence

$$Z = \pi d^4 F / 32 R^2 L \qquad (20\text{-}21)$$

Then

$$f_2 = e''Z \tag{20-22}$$

The ratio of the total effective mass of the connected bodies to the amount that will be felt at the tooth mesh of the gears will then be equal to f_r/f_1. Thus when

m_a = total effective mass of connected bodies = I_0/R^2

m_b = mass effect of connected bodies at the tooth mesh

$$m_b/m_a = f_r/f_1$$

Substituting the value of f_r from Eq. (20-16), we have

$$m_b/m_a = f_2/(f_1 + f_2) \tag{20-23}$$

We must now direct our attention to the distance that one gear moves in relation to its mating gear because of the effective error. This distance will depend upon the extent of the error and the relative masses of the mating gears at their pitch lines. Thus when

m_1 = effective mass acting at pitch line of first gear

m_2 = effective mass acting at pitch line of second gear

e = effective error in action, in.

e'' = amount of movement at pitch line of first gear, in.

then

$$e'' = m_2 e/(m_1 + m_2) \tag{20-24}$$

The value of m_1 is the sum of the effective mass of the first gear blank itself, which is a constant, and the mass effect m_b of the connected masses. This last value will be a variable. In addition, there will be some part of the mass of the rotating shaft and other parts, but these additional mass effects will be so slight in most cases that they will be ignored in this analysis. When m_p is the effective mass of the pinion blank at the pitch line

$$m_1 = m_p + m_b \tag{20-25}$$

Substituting this value into Eq. (20-24), we obtain

$$e'' = m_2 e/(m_p + m_b + m_2) \tag{20-26}$$

In this equation, we have two unknown values, m_b and e''. Substituting the value of f_1 from Eq. (20-18) and the value of f_2 from Eq. (20-22) into Eq. (20-23), and solving for e'', we obtain

$$e'' = H m_a m_b V^2/Z(m_a - m_b) \tag{20-27}$$

Equation (20-27) has the same two unknown values as Eq. (20-26). By equating these two equations for e'' and solving for m_b, we obtain

$$m_b{}^2(H m_a V^2) + m_b[H m_a V^2(m_p + m_2) + m_2 e Z] - m_a m_2 e Z = 0 \tag{20-28}$$

The solution of Eq. (20-28) for m_b would give a long and extended equation. We shall therefore let

$$A = H m_a V^2$$
$$B = A(m_p + m_2) + m_2 e Z$$
$$C = m_a m_2 e Z$$

Then

$$m_b = (\sqrt{B^2 + 4AC} - B)/2A \qquad (20\text{-}29)$$

When appreciable masses are connected to both gears of a pair, a similar analysis must be made for each gear. The effective mass of one gear blank can be used to find the mass effect m_b of the connected masses of the other gear. Then the calculated value of the effective mass of the first gear will be used to determine the mass effect of the connected masses of the second gear. Any attempt to solve for both values at the same time leads to indeterminate equations. Hence, in effect, we solve this problem by trial.

We shall now consider the effective mass influence acting at the pitch line of the meshing gears. Thus when

m = effective mass influence at pitch line of gears

m_1 = effective mass acting at pitch line of pinion

m_2 = effective mass acting at pitch line of gear

$$m = (m_1 m_2)/(m_1 + m_2) \qquad (20\text{-}30)$$

This equation gives the value of m that is used in Eq. (20-6).

Separation of Tooth Surfaces. At the instant that the engaging pair of tooth profiles have come fully into mesh and the acceleration load has ceased to act, the masses are revolving at slightly different velocities. As noted before, the movement of the driving member has been slowed down, while that of the follower has been speeded up. In effect, they are moving away from each other because of this difference in velocity, which has been imparted to them by the accelerating force. This relative movement apart is resisted by the input torque and torsional deflection of various members and the applied load or work that the mechanism is performing. The conditions here are roughly represented by Fig. 20-3. This represents a rigid slide with a cam surface moving in the direction shown by the arrow. The mass m is held against it by the force W. The rise or cam on the slide with a height e represents the amount of the error in action. When the cam surface on this slide reaches the contact finger of the mass, an acceleration load is set up, and the mass is set in motion in a vertical direction. While the mass is moving up the cam surface, the accelerating force f_a is acting through the vertical distance e. The additional work done by this cam is equal to $f_a e$. The distance k that the mass will move away from the slide depends upon the intensity of the applied load W. This force will act through the distance k to absorb the additional work of acceleration. Hence for rigid bodies

$$f_a e = Wk$$

Hence

$$k = (f_a/W)e \qquad (20\text{-}31)$$

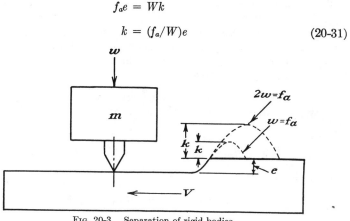

FIG. 20-3. Separation of rigid bodies.

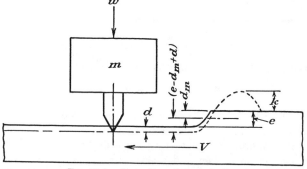

FIG. 20-4. Separation of elastic bodies.

Thus when W is equal to f_a, these rigid bodies will separate a distance equal to e. When W is double the value of f_a, the distance k will be one-half the distance e, etc. This separation is indicated by the dotted line in Fig. 20-3.

With elastic bodies, however, the accelerating force, in addition to imparting a certain difference in velocity between the masses, will also cause additional deformation. Under these conditions, assuming that all the deformation is on the slide, we would have the conditions indicated in Fig. 20-4. In this case, the work that is stored in the deformed material will also act on the separation of the two bodies, in addition to the separating effect of the change in momentum imparted by the action of the accelerating force, while the applied load W will act as before to hold them together. Thus when

W = applied load, lb
f_a = accelerating force, lb
d = deformation set up by the applied load, in.
d_m = maximum deformation set up by the total load, in.
e = error or rise on undeformed profile, in.
k = separation of surfaces, in.

$$\text{Work done by accelerating force alone} = f_a(e - d_m + d)$$
$$\text{Work stored in elastically deformed profile} = \tfrac{1}{2}\,(f_a + W)d_m$$

Whence

$$W(k + d_m) = f_a(e - d_m + d) + (\tfrac{1}{2})(f_a + W)d_m$$

Solving this equation for k, we obtain

$$k = (f_a/W)(e - d_m + d) + (\tfrac{1}{2})[(f_a/W) + 1]d_m - d_m$$

But

$$d_m = [(f_a/W) + 1]d$$

Substituting and simplifying, we obtain

$$k = (f_a/W)e - (d/2)[(f_a/W)^2 + 1] \tag{20-32}$$

Under static conditions, $f_a = 0$, whence

$$k = -d/2$$

When the value of k is minus, it indicates deformation instead of actual separation, and its value represents the distance the centers of mass of the compressed contacting surfaces have moved from their relative positions in contact under zero load.

Impact Loads. The following analysis of elastic impact was made by Carl G. Barth, a member of the ASME Special Research Committee on the Strength of Gear Teeth.

After the gear teeth have separated because of the relative changes in velocity imparted to the gears by the action of the accelerating load, and have reached the point of maximum separation, they will come together again with an impact. At the instant of maximum separation, both gears will be traveling at the same velocity.

Figure 20-5 is a representation of three successive instants of the impact action. Thus when

k = amount of separation, in.
W = applied load, lb
m_1 = effective mass of driver
m_2 = effective mass of follower
t = time required for m_1 to overtake m_2, min
s_1 = space traveled by m_1, ft
s_2 = space traveled by m_2, ft
V_c = common velocity of gears, ft/min
V_1 = velocity of m_1 when it overtakes m_2, ft/min
V_2 = velocity of m_2 when overtaken by m_1, ft/min

$$V_1 = V_c + at = V_c + \frac{Wt}{m_1} \qquad V_2 = V_c - at = V_c - \frac{Wt}{m_2}$$

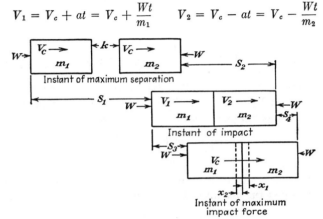

FIG. 20-5

whence

$$V_1 - V_2 = Wt \left(\frac{1}{m_1} + \frac{1}{m_2} \right) \tag{20-33}$$

$$s_1 = (V_c + V_1)\frac{t}{2} \qquad s_2 = (V_c + V_2)\frac{t}{2}$$

$$s_1 - s_2 = k = \tfrac{1}{2}(V_1 - V_2)t \qquad t = 2k/(V_1 - V_2)$$

Substituting this value of t into Eq. (20-33), we obtain

$$(V_1 - V_2)^2 = 2kW \left(\frac{1}{m_1} + \frac{1}{m_2} \right) \tag{20-34}$$

Considering now the intensity of the impact force and referring to the last two phases of the impact shown in Fig. 20-5, when
V_c = common velocity of the two gears at the instant of maximum impact, ft/min

W_d = maximum intensity of the impact load, lb
x_1 = corresponding maximum deformation of m_1, in.
x_2 = corresponding maximum deformation of m_2, in.
z_1 = elasticity form factor of m_1
z_2 = elasticity form factor of m_2
E_1 = modulus of elasticity of m_1
E_2 = modulus of elasticity of m_2
F = face width of gears, in.

then we have

$$\frac{W_d}{F} = z_1 E_1 x_1 = z_2 E_2 x_2$$

$$x_1 = \frac{W_d}{z_1 E_1 F} \qquad z_2 = \frac{W_d}{z_2 E_2 F}$$

$$x_1 + x_2 = \frac{W_d}{F}\left(\frac{1}{z_1 E_1} + \frac{1}{z_2 E_2}\right) \tag{20-35}$$

The total energy, which is all kinetic energy at the instant of impact, is equal to $\frac{1}{2}(m_1 V_1^2 + m_2 V_2^2)$. The total kinetic energy at the instant of maximum impact force is equal to $\frac{1}{2}(m_1 + m_2)V_c^2$. But

$$V_c = \frac{m_1 V_1 + m_2 V_2}{m_1 + m_2}$$

Substituting this value of V_c into the preceding equation, then the total kinetic energy at the instant of the maximum impact force is equal to

$$\frac{m_1^2 V_1^2 + 2m_1 m_2 V_1 V_2 + m_2^2 V_2^2}{2(m_1 + m_2)}$$

Ignoring internal friction, the loss of kinetic energy during impact is equal to the amount of potential energy stored in the deformed elastic bodies. If no outside force were acting during impact, this loss of kinetic energy would be equal to $(W_d/2)(x_1 + x_2)$. When a constant force W acts on the bodies during impact, some account must be taken of the work done by this force through the relative distance these bodies move while impact is taking place. The potential energy stored in the deformed bodies would then be equal to

$$\left(\frac{W_d}{2} - W\right)(x_1 + x_2)$$

whence we have

$$(W_d - 2W)(x_1 + x_2) = (m_1 V_1^2 + m_2 V_2^2) - \frac{(m_1 V_1 + m_2 V_2)^2}{m_1 + m_2}$$

Combining and simplifying, we obtain

$$(W_d - 2W)(x_1 + x_2) = \frac{(V_1 - V_2)^2 m_1 m_2}{m_1 + m_2}$$

Substituting the value of $x_1 + x_2$ from Eq. (20-35) into the foregoing, combining, and solving for W_d, we obtain

$$W_d = W + \sqrt{W^2 + \left(\frac{z_1 E_1 z_2 E_2}{z_1 E_1 + z_2 E_2}\right)\left(\frac{m_1 m_2}{m_1 + m_2}\right)(V_1 - V_2)^2 F} \quad (20\text{-}36)$$

From Eq. (20-13) we have

$$d = \frac{W}{F}\left(\frac{z_1 E_1 + z_2 E_2}{z_1 E_1 z_2 E_2}\right) \quad (20\text{-}13)$$

Whence

$$\left(\frac{z_1 E_1 z_2 E_2}{z_1 E_1 + z_2 E_2}\right) = \frac{W}{Fd}$$

Substituting the foregoing value into Eq. (20-36), we obtain

$$W_d = W + \sqrt{W^2 + \frac{W}{d}\left(\frac{m_1 m_2}{m_1 + m_2}\right)(V_1 - V_2)^2} \quad (20\text{-}37)$$

But

$$(V_1 - V_2)^2 = 2kW\left(\frac{m_1 + m_2}{m_1 m_2}\right) \quad (20\text{-}34)$$

Substituting this value into Eq. (20-37), we obtain

$$W_d = W + \sqrt{W^2 + \frac{2W^2 k}{d}} = W\left(1 + \sqrt{1 + \frac{2k}{d}}\right) \quad (20\text{-}38)$$

For a suddenly applied load, the value of k would be equal to zero. Substituting this value into Eq. (20-38), we have

$$W_d = 2W$$

For a static load, the value of k would be equal to $-d/2$. Substituting this value into Eq. (20-38), we have

$$W_d = W$$

Equation (20-38) thus appears to be a general equation for all conditions of load, impact, suddenly applied, and static. For conditions of variable loads, greater than static ones but less than a suddenly applied one, the use of the minus value for k, which depends upon the amount of elastic deformation (or preload) at the instant of reversal of load, should give us a measure of the maximum intensity of the loading.

This equation is of the same form as that derived by Mansfield Merriman about 60 years ago for impact loads on beams. For this purpose, Merriman derived the following:

$$T = S + S \sqrt{1 + \frac{2h}{f}}$$

where T = maximum flexural unit stress produced by the impact
S = unit stress that is caused by the static load P
f = deflection caused by the static load P
h = height above beam from which weight P falls

Merriman discusses the possible use of a time factor to modify the value of the radical in this equation in order to have it apply to loads of less severity than suddenly applied loads. It would seem, however, that a minus value for k in Eq. (20-38) when the bodies are elastically deformed at the instant of reversal or change of loading would be a more practical solution.

Summary of Analysis of Dynamic Loads on Spur-gear Teeth. The foregoing analysis of dynamic loads on spur-gear teeth includes several troublesome factors, which make necessary the use of assumptions and approximations. For one thing, the amount of deformation of the tooth profile is variable, depending upon the position on the profile where the load is applied, and also upon whether one pair of mating teeth or two pairs are carrying the load. The foregoing analysis assumes that but a single pair of teeth are carrying the load at the critical phase of the load transfer.

Again, it has been assumed that the acceleration load has a constant value. This assumption is probably never exactly true, although the influence of the elasticity of the materials will tend to make the acceleration load approach this condition. Furthermore, the time factor has been eliminated from the equations so that they represent work done during acceleration rather than the actual intensities of the acceleration loads. In other words, the calculated acceleration load represents the mean effective pressure on the gear teeth during acceleration. Actually the maximum acceleration load may closely approach the severity of the impact or dynamic load at times, but it can never exceed it.

The assumptions and approximations used have been noted in the foregoing analysis. We shall assemble here the specific equations needed for the solution of definite problems. The analysis of the dynamic loads on spur-gear teeth requires the determination of the following:

1. Effective mass acting at pitch line of gears
2. Acceleration load

3. Amount of separation of profiles

4. Impact or dynamic load

We would reduce the amount of computing if we could determine the intensity of the impact directly from the acceleration load and its components. The following equation gives a very close approach to the value of the dynamic load, and will be used:

$$W_d = W + \sqrt{f_a(2f_2 - f_a)} \qquad (20\text{-}39)$$

Effective Mass. For the determination of the effective mass, we have

m_a = full effective mass of connected bodies at $R_1 = I_a/R_1{}^2$

m_b = mass effect of m_a at pitch line of pinion

m_p = effective mass of pinion blank at $R_1 = I_p/R_1{}^2$

m_1 = effective mass acting at pitch line of pinion

m_2 = effective mass acting at pitch line of gear

m = effective mass influence at pitch line of gears

V = pitch-line velocity of gears, ft/min

Z = elasticity factor of connecting member

e = measured error in action of gears, in.

W = tangential applied load, lb

F = face width of gears, in.

d = deformation of teeth at pitch line under applied load W, in.

R_1 = pitch radius of pinion, in.

R_2 = pitch radius of gear, in.

z_1, z_2 = elasticity form factors of gear teeth

E_1, E_2 = modulus of elasticity of materials

y = Lewis tooth-form factor

$$z = y/(0.242 + 7.25y) \qquad (20\text{-}14)$$
$$d = (W/F)\,[(1/E_1z_1) + (1/E_2z_2)] \qquad (20\text{-}13)$$
$$m_b = (\sqrt{B^2 + 4AC} - B)/2A \qquad (20\text{-}29)$$

where $A = Hm_aV^2$

$B = (m_p + m_2)A + em_2Z$

$C = em_am_2Z$

For 14½-deg gears

$$H = 0.00086[(1/R_1) + (1/R_2)]$$

For 20-deg gears

$$H = 0.00120[(1/R_1) + (1/R_2)]$$
$$Z = P/T \qquad (20\text{-}19)$$

where P = load applied to shaft or coupling at radius R_1, lb

T = torsional deflection at R_1, in.

$$m_1 = m_p + m_b \tag{20-25}$$
$$m = m_1 m_2 / (m_1 + m_2) \tag{20-30}$$

The values for the tooth deformation d for gear teeth of conventional design can be determined very closely by the following empirical equations:

For $14\frac{1}{2}$-deg gears

$$d = 9.345(W/F)\,[(1/E_1) + (1/E_2)] \tag{20-40}$$

For 20-deg full-depth form

$$d = 9.000(W/F)\,[(1/E_1) + (1/E_2)] \tag{20-41}$$

For 20-deg-stub tooth form

$$d = 8.70(W/F)\,[(1/E_1) + (1/E_2)] \tag{20-42}$$

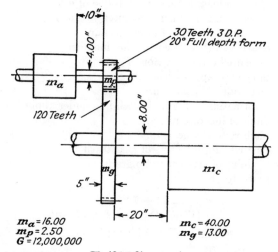

FIG. 20-6.

Example of Effective Mass. As a definite example of the determination of the value for the effective mass we shall use the example shown in Fig. 20-6. From the values given there we have

$$Z_1 = \frac{3.1416 \times 4^4 \times 12,000,000}{32 \times 5^2 \times 10} = 1,206,400$$

$$Z_2 = \frac{3.1416 \times 8^4 \times 12,000,000}{32 \times 20^2 \times 20} = 603,200$$

$$A_1 = 0.00120(\tfrac{1}{5} + \tfrac{1}{20})16 \times 1,000^2 = 4,800$$

$B_1 = [(2.50 + 13.00)4,800] + [0.001 \times 13 \times 1,206,400] = 90,083$

$C_1 = 0.001 \times 16 \times 13 \times 1,206,400 = 250,931$

$$m_b = \frac{\sqrt{B^2 + 4A_1C_1} - B_1}{2A_1} = \frac{113,722 - 90,083}{9,600} = 2.46$$

$$m_1 = 2.50 + 2.46 = 4.96$$

We shall now make similar calculations to determine the influence of the members connected to the gear shaft.

$A_2 = 0.00120 \times 0.250 \times 40 \times 1,000^2 = 12,000$

$B_2 = (13 + 4.96)12,000 + 0.001 \times 4.96 \times 603,200 = 218,512$

$C_2 = 0.001 \times 40 \times 4.96 \times 603,200 = 119,680$

$$m_d = \frac{\sqrt{B_2^2 + 4A_2C_2} - B_2}{2A_2} = \frac{231,284 - 218,512}{24,000} = 0.53$$

$$m_2 = m_g + m_d = 13.00 + 0.53 = 13.53$$

$$m = \frac{4.96 \times 13.53}{4.96 + 13.53} = 3.63$$

In order to show the variations in the value of the effective mass with changes in velocity, values of m_b, m_1, m_d, m_2, and m have been calculated for different pitch-line velocities. These values are tabulated in Table 20-1. Most of them are plotted in Fig. 20-7.

A study of Fig. 20-7 will give a good idea of how the mass influence of the connected masses reduces with increasing velocity. If the connections are less elastic, such as results from larger or shorter shafts between the connected masses and the gears, the influence of these connected masses will persist into higher pitch-line velocities.

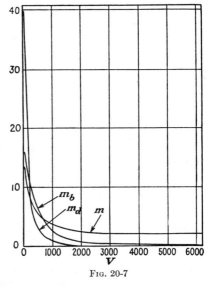

Fig. 20-7

Acceleration Loads. For the determination of the acceleration load we have the following:

When $f_1 =$ force required to accelerate the masses as rigid bodies, lb

$f_2 =$ force required to deform teeth amount of effective error, lb

$f_a =$ acceleration load on gear teeth, lb

$W =$ applied tangential load, lb

$m =$ effective mass acting at pitch line of gears

$e =$ measured error in action, in.

$d =$ deformation of gear teeth under load W, in.

TABLE 20-1.   EFFECT OF SPEED ON MASS FACTORS
(Plotted in Fig. 20-7)

V, ft/min	m_b	m_1	m_d	m_2	m
0	16.00	18.50	40.00	53.00	13.71
100	14.64	17.14	37.30	50.30	12.78
200	11.97	14.47	14.95	27.95	9.53
300	9.48	11.98	7.57	20.57	7.57
400	7.52	10.02	4.40	17.40	6.36
500	6.04	8.54	2.72	15.72	5.53
1,000	2.46	4.96	0.53	13.53	3.63
2,000	0.76	3.26	0.10	13.10	2.61
3,000	0.36	2.86	0.04	13.04	2.35
4,000	0.21	2.71	0.022	13.022	2.24
5,000	0.13	2.63	0.013	13.013	2.19
6,000	0.09	2.59	0.009	13.009	2.16
7,000	0.07	2.57	0.006	13.006	2.14

$$f_1 = HmV^2 \qquad (20\text{-}43)$$

For 14½-deg gears

$$H = 0.00086[(1/R_1) + (1/R_2)]$$

For 20-deg gears

$$H = 0.00120[(1/R_1) + (1/R_2)]$$
$$f_2 = W[(e/d) + 1] \qquad (20\text{-}15)$$
$$f_a = f_1 f_2/(f_1 + f_2) \qquad (20\text{-}2)$$

Example of Acceleration Load.   As a definite example we shall use the values shown in Fig. 20-6.   Whence we have

$$f_1 = 0.00120\ (\tfrac{1}{5} + \tfrac{1}{20})3.63 \times 1,000^2 = 1,089$$
$$d = 9.0 \times \frac{1,000}{5}\left(\frac{1}{15,000,000} + \frac{1}{15,000,000}\right) = 0.00024$$
$$f_2 = 1,000\left(\frac{0.001}{0.00024} + 1\right) = 5,167$$
$$f_a = \frac{1,089 \times 5,167}{1,089 + 5,167} = 899$$

In order to show the variation in the values of f_1, f_2, and f_a with changes in velocity, these values have been calculated for different pitch-line velocities.   They are tabulated in Table 20-2 and are plotted in Fig. 20-8.

Dynamic Load.   For the determination of the maximum intensity of the impact load, which is the dynamic load, we have the following: When W_d = dynamic load, lb

$$W_d = W + \sqrt{f_a(2f_2 - f_a)} \qquad (20\text{-}39)$$

Example of Dynamic Load. As a definite example we shall use the values shown in Fig. (20-6). Whence we have

$$W_d = 1,000 + \sqrt{899(2 \times 5,167 - 899)} = 3,912 \text{ lb}$$

Values of the dynamic load at different pitch-line velocities have been calculated and are tabulated in Table 20-2 together with the values for the acceleration loads. These values are plotted in Fig. 20-8.

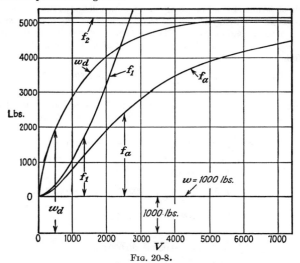

Fig. 20-8.

TABLE 20-2.   ACCELERATION LOADS
(Plotted in Fig. 20-8)
$f_2 = 5,167$ lb

V, ft/min	f_1, lb	f_a, lb	W_d, lb
0	0	0	1,000
100	38	38	1,625
200	114	111	2,065
300	204	196	2,410
400	305	288	2,701
500	415	384	2,955
1,000	1,089	899	3,912
2,000	3,132	1,950	5,043
3,000	6,345	2,848	5,617
4,000	10,752	3,490	5,887
5,000	16,425	3,930	6,017
6,000	23,328	4,230	6,081
7,000	31,458	4,438	6,115

Dynamic Loads on Small Gears. Small gears are generally mounted on such small shafts that the influence of the connected masses is negligible and can be ignored. When the pinion shaft is less than 2 in. in diameter, we can use the same analysis as before, but use only the values of the effective masses of the gear blanks themselves in Eq. (20-43).

Example of Dynamic Load on Small Gears. As a definite example we shall use a pair of 24-DP gears, having 24 and 72 teeth, ¼-in. face, running at 600 rpm, and transmitting a tooth load of 25 lb. These will be of 20-deg full-depth form of conventional design. Both gear blanks will be of steel and of plain disk form. This gives the following values:

$$N_1 = 24 \qquad N_2 = 72 \qquad R_1 = 0.500 \qquad R_2 = 1.500 \qquad F = 0.250$$
$$W = 25 \qquad V = 157 \text{ ft/min}$$

The weights of the gear blanks will be 0.055 lb for the pinion and 0.491 lb for the gear. The effective weight at the pitch line will be one-half the total weight; hence

$$m_1 = \frac{0.055}{64} = 0.00086 \qquad m_2 = \frac{0.491}{64} = 0.00767$$
$$m = \frac{0.00086 \times 0.00767}{0.00086 + 0.00767} = 0.00077$$
$$f_1 = 0.00120 \left(\frac{1}{0.5} + \frac{1}{1.5}\right) 0.00077 \times 157^2 = 0.061 \text{ lb}$$

We shall assume the maximum error in action to be 0.002 in.

$$d = 9.00 \frac{25}{0.25} \left(\frac{1}{30,000,000} + \frac{1}{30,000,000}\right) = 0.00006$$
$$f_2 = 25 \left(\frac{0.002}{0.00006} + 1\right) = 857 \text{ lb}$$
$$f_a = \frac{0.061 \times 857}{0.061 + 857} = 0.061 \text{ lb}$$
$$W_d = \sqrt{0.061(1,714 - 0.061)} + 25 = 35.22 \text{ lb}$$

DYNAMIC LOADS AND INFLUENCE OF FINE PITCH AND HIGH SPEED

Thus far we have considered the dynamic load conditions when the flow of power through the tooth mesh has been a direct reflection of the irregularities of the tooth mesh. Under some conditions, however, the momentum of the revolving parts will act to maintain a substantially constant velocity, and the gear teeth will be in intermittent contact only enough to restore the energy that is lost between successive impulses.

When the flow of power follows the full irregularities of the tooth mesh we have, as pointed out before, two peak loads for each tooth engagement: the acceleration load followed by the impact or dynamic load. The dynamic load absorbs in elastic impact the change of momentum set up by the action of the acceleration load. As the time interval between successive tooth engagements becomes very small because of high pitch-line speeds or because of fine pitches at lower pitch-line velocities, the gears may not have time enough to complete both load cycles. In effect, this would tend to reduce the effective error, and also act to reduce the intensity of the dynamic load.

The major problem here is to determine the amount of time required for the gears to complete the full double-load cycle for each tooth engagement. We know that the average load must be equal to the transmitted load. We know also that time is consumed in the separation of the teeth as well as for the return, or impact. The best that we can do at the present time is to·set up some reasonable hypothesis and to try it; then to check the results against definite applications in service. If the assumed conditions appear to be less severe than the actual ones, then the assumed time interval should be increased. If the assumed conditions appear to be more severe than the actual ones, then the assumed time interval should be reduced. In other words, the following assumptions represent a first trial and must be checked against actual working conditions before too much reliance is placed on the actual numerical values obtained.

We know that the total time interval for each tooth mesh is equal to the time between successive tooth contacts. We shall assume that the applied load w acts for the same length of time in overcoming the separating action as it does in returning the teeth into contact again. We would then have the condition that the greatest possible effective error would be the distance that the applied load could move the masses of the gear blanks in a time interval equal to one-half the time between successive tooth meshes. If this distance is greater than the error in action, we shall assume that the time is sufficient for the double-load cycle and that the conditions of the preceding analysis will prevail. If, on the other hand, this distance is less than the error in action we shall assume that the time is not sufficient for the double-load cycle and that the effective error is therefore reduced to the distance that the gear blanks can be moved by the applied load in the limited time available.

Under these conditions, it is apparent that for all gear drives, regardless of the extent of the actual error in action, there will be a speed at which the dynamic load is independent of the actual error in action. It may be that the dynamic load will reach a maximum value at some speed and then reduce with a further increase in speed. In such cases the gears must be strong enough to carry the loads through this maximum value without failure in order to be able to operate at the higher speeds. Thus when

n = rpm of driver

N_1 = number of teeth in driver

t = one-half the time between successive tooth meshes, sec

m = mass effect at pitch line of gears

W = applied tangential load, lb

S = distance along pitch line that gears can be moved by force W in time t, ft

e' = distance on pitch line that gears can be moved by force W in time t, in.

a = acceleration of gears, ft/sec^2

$$W = ma \qquad a = W/m \qquad S = at^2/2 = Wt^2/2m$$
$$t = 30/nN_1 \tag{20-44}$$
$$e' = 12Wt^2/2m = 6Wt^2/m \tag{20-45}$$

When e' is greater than the measured or assumed error in action, e, the value of e will be used in the dynamic load equations. When e' is less than e, then the value of e' will be used in the dynamic load equations.

If the gears are to run at the higher speeds except when starting and stopping, then stresses greater than the endurance limits of the materials could be permitted for the load at its maximum value. On the other hand, if the gears are to run at a varying range of speeds above and below this maximum or critical value, then the stresses must be within the endurance limits of the materials if the gears are to have a reasonable length of useful life.

In any event we should be able to determine this critical value directly. This can be done and the speed at which this critical load occurs can be determined directly by a rearrangement of Eq. (20-28) and (20-45). This solution will give the value of n when the value of e' is equal to e. Thus when

n_c = rpm of driver when e' is equal to e

and all other symbols are the same as before

$$n_c = (30/N_1) \sqrt{6W/em} \tag{23-46}$$

We shall disregard the influence of any connected masses here. Any error in so doing will make the value of e' larger than it would otherwise be. We shall then use the value of e' as the effective error in the dynamic load equation we may be using.

Example of Dynamic Load at High Speed. As a definite example we shall use the same gears as those in the preceding example. This gives the following values:

$$N_1 = 24 \qquad N_2 = 72 \qquad R_1 = 0.500 \qquad R_2 = 1.500 \qquad F = 0.25 \qquad e = 0.002$$
$$m = 0.00077 \qquad W = 25$$
$$f_1 = 0.0032 \times 0.00077 \times V^2$$
$$f_2 = 25 \left(\frac{e}{0.00006} + 1 \right)$$

We shall first determine the value of n_c.

$$n_c = \frac{30}{24} \sqrt{\frac{150}{0.002 \times 0.00077}} = 12,950$$

We will use a value of $n = 20,000$.

$$t = \frac{30}{20,000 \times 24} = 0.0000625$$

$$e' = \frac{150(0.0000625)^2}{0.00077} = 0.00076$$

$$V = 0.5236 \times 0.50 \times 20,000 = 5,236$$

$$f_1 = 0.0032 \times 0.00077(5,236)^2 = 67.54$$

$$f_2 = 25 \left(\frac{0.00076}{0.00006} + 1\right) = 342$$

$$f_a = \frac{67.54 \times 342}{67.54 + 342} = 56.40$$

$$W_d = \sqrt{56.40(682 - 56.40)} + 25 = 213\,\text{lb}$$

If we had used the full value of the error, we would have

$$W_d = \sqrt{56.40(1,714 - 56.40)} + 25 = 331\,\text{lb}$$

In this example, the limited time between successive tooth meshes has reduced the dynamic load by an amount equal to 118 lb.

At the critical velocity where $n_c = 12,950$, we would have

$$V = 3,390 \qquad f_1 = 28.32 \qquad f_2 = 857 \qquad f_a = 27.41$$

Whence

$$W_d = \sqrt{27.41(1,714 - 27.41)} + 25 = 233\,\text{lb}$$

Second Example of Dynamic Loads at High Speed. As a second example we shall use the gears shown in Fig. 20-6. We shall determine the critical value and values of W_d for velocities where the value of e' is less than the value of e. For this we have the following values:

$$N_1 = 30 \qquad R_1 = 5.00 \qquad R_2 = 20.000 \qquad F = 5.000 \qquad e = 0.001$$
$$m_p = 2.50 \qquad m_g = 13.00 \qquad W = 1,000$$

We will use

$$m = \frac{2.50 \times 13.00}{2.50 + 13.00} = 2.10$$

$$f_1 = 0.0003 \times 2.10 \times V^2$$

$$f_2 = 1,000 \left(\frac{e}{0.00024} + 1\right)$$

$$n_c = \frac{30}{30} \sqrt{\frac{6,000}{0.001 \times 2.10}} = 1,690$$

$$V_c = 0.5236 \times 5 \times 1,690 = 4,425$$

$$f_1 = 0.00063 \times 1,690^2 = 12,335$$

$$f_2 = 1,000 \left(\frac{0.001}{0.00024} + 1\right) = 5,167$$

$$f_a = \frac{12,335 \times 5,167}{12,335 + 5,167} = 3,642$$

$$W_d = \sqrt{3,642(10,334 - 3,642)} + 1,000 = 5,936$$

Values obtained for e' for various pitch-line velocities are tabulated in Table 20-3. Corresponding values of W_d are also tabulated together with the values of the dynamic loads for the full value of e. These values are also plotted in Fig. 20-9.

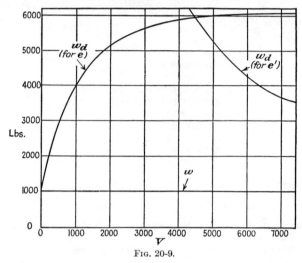

Fig. 20-9.

In this last example, at a speed of 1,690 rpm or a pitch-line velocity of 4,425 ft/min, the time is just sufficient to permit the complete double-load cycle, if the assumptions on which this analysis is based are reasonably correct. Beyond this speed, the effective error is reduced and the value of the dynamic load is also reduced with further increases in speed. For these higher speeds with the lower dynamic loads, however, the gears must be strong enough to carry themselves through the maximum load conditions.

TABLE 20-3. VALUES OF W_d FOR VALUES OF e AND e'
(Plotted in Fig. 20-9)

V, ft/min	n, rpm	e', in	W_d, lb	
			For e	For e'
1,000	382	0.019568	3,912	11,138
2,000	764	0.004892	5,043	10,556
3,000	1,146	0.002184	5,617	8,756
4,000	1,528	0.001223	5,887	6,973
4,425	1,690	0.001000	5,936	5,936
5,000	1,910	0.000786	6,017	5,283
6,000	2,292	0.000546	6,081	4,249
7,000	2,673	0.000400	6,115	3,659

CHAPTER 21

DYNAMIC LOADS ON GEAR TEETH

The analysis of the dynamic loads on spur-gear teeth is given in the preceding chapter. We shall now attempt to apply the results of this analysis to the dynamic loads on other types of gears, starting with internal-gear drives.

DYNAMIC LOADS ON INTERNAL-GEAR TEETH

The dynamic loads on internal-gear teeth are the same as those on spur gears except that the directions of curvature of the pitch circles of the internal gears follow those of the mating pinions, instead of moving away from them. Hence the value of the pitch radius of the internal gear is minus instead of plus as regards all conditions of relative curvatures.

Effective Mass. The determination of the effective mass for internal-gear drives is exactly the same as that for spur gears except that the value of H required for the solution of Eq. (20-29) is as follows:
For 20-deg gears

$$H = 0.00120[(1/R_1) - (1/R_2)] \tag{21-1}$$

where R_1, R_2 = pitch radius of spur pinion and internal gear, in.

Acceleration Loads. For the determination of the acceleration load on internal-gear teeth we have the following:
When f_1 = force required to accelerate mass as rigid body, lb
f_2 = force required to deform teeth amount of error, lb
f_a = acceleration load, lb
W = applied tangential load, lb
m = effective mass acting at pitch line of gears
e = measured error acting at pitch line of gears, in.
d = deformation of teeth under applied load W, in.
V = pitch-line velocity, ft/min

$$f_1 = HmV^2 \tag{20-43}$$

For 20-deg gears

$$H = 0.00120[(1/R_1) - (1/R_2)] \tag{21-1}$$
$$f_2 = W[(e/d) + 1] \tag{20-15}$$
$$d = (W/F)[(1/E_1z_1) + (1/E_2z_2)] \tag{20-13}$$

Where z_1, z_2 = elasticity form factors for gear teeth

E_1, E_2 = modulus of elasticity of materials

F = face width of gears, in.

For 20-deg gears of conventional design

$$d = 9.00(W/F)[(1/E_1) + (1/E_2)] \qquad (20\text{-}41)$$
$$f_a = f_1 f_2 / (f_1 + f_2) \qquad (20\text{-}2)$$

Example of Acceleration Load on Internal-gear Drive. As a definite example we shall use the same values as were used for the spur-gear drive. This gives the following values:

$R_1 = 5.000$ $R_2 = 20.000$ $W = 1,000$ $V = 1,000$ $e = 0.001$ $m = 3.63$

$$H = 0.00120(\tfrac{1}{5} - \tfrac{1}{20}) = 0.00018$$
$$f_1 = 0.00018 \times 3.63 \times 1,000^2 = 653$$
$$f_2 = 1,000 \left(\frac{0.001}{0.00024} + 1 \right) = 5,167$$
$$f_a = \frac{653 \times 5,167}{653 + 5,167} = 580 \text{ lb}$$

Dynamic Load on Internal-gear Drive. For the determination of the maximum intensity of the dynamic load we have the following: When W_d = dynamic tooth load, lb

$$W_d = W + \sqrt{f_a(2f_2 - f_a)} \qquad (20\text{-}39)$$

Example of Dynamic Load on Internal-gear Drive. Continuing the preceding example, we have

$$W_d = 1,000 + \sqrt{580(10,334 - 580)} = 3,378 \text{ lb}$$

The dynamic load on the equivalent pair of spur gears is equal to 3,912 lb.

DYNAMIC LOADS ON HELICAL-GEAR TEETH

Experience teaches us that with all other factors equal, a helical-gear drive will run more smoothly than a pair of spur gears with straight teeth. Here the tooth action is a combination of conjugate gear-tooth profile action and a camming action along the mating helices. As regards smoothness of running alone, continuous action can be transmitted by the contact along the helices regardless of the form of the tooth profiles. If these mating profiles are not conjugate, then the action will be between the two mating helices that extend the greatest distance beyond the theoretical conjugate gear-tooth profiles. Under such conditions, however, the contact approaches point contact, and the load capacity of the drive is limited by the amount of load this small surface area can carry without excessive wear. With correct tooth profiles, the contact between them is line contact, and then the whole active profiles of the teeth of both gears will come into contact. In other words, the major effect of profile errors on helical gears is to restrict the area of the actual tooth contact and thus reduce its ultimate load-carrying capacity.

On the other hand, when the gears are made of plastic materials, there may be some corrective influence during the running-in period if the initial loads applied are not too great. The high stresses set up by any restricted contact will cause local plastic flow and thus tend to increase the actual contact areas. This type of action is absent on spur gears, and plastic flow on such gears tends to increase the error in the tooth profile.

If the initial loads on helical gears made of the more plastic metals are too great, excessive plastic flow of the surface material will be the result. This may increase the error in action, thus increasing both the noise of operation and also the intensity of the dynamic loads.

We shall attempt to introduce into the equations for the dynamic loads the influence of the helical action on the intensities of these loads. One way in which this might be done is to use as the velocity of engagement the normal velocity of the basic-rack form as it theoretically engages the teeth of both gears simultaneously. In such a case, when

V = pitch-line velocity in plane of rotation, ft/min

V_n = pitch-line velocity of normal basic-rack form, ft/min

ψ = helix angle of gears at pitch line

$$V_n = V \cos \psi \qquad (21\text{-}2)$$

Another factor to be considered is the elastic deformation of the gear teeth under load. We have used an average deflection value for a given basic-rack system for the analysis of the dynamic loads on spur-gear teeth, and this system or its equivalent is the normal basic-rack form of the helical gears. The tangential load normal to the helix angle will be greater than the tangential load in the plane of rotation, but the developed length or face width of the helical gear will be greater in the same proportion. However, the deflection in the plane of rotation will be greater than the deflection in the direction normal to the helix. Thus when

d_n = normal deflection of tooth form under load W, in.

W = applied tangential load in plane of rotation, lb

d = deflection of tooth form at pitch line in plane of rotation, in.

$$d = d_n/\cos \psi$$

The foregoing equation might apply if the load were concentrated at the pitch line. But with helical gears of involute form, the contact line is at an angle to the trace of the pitch surface, and the resulting deflection will be greater than that given by the foregoing equation. This increase in deflection will be greater with an increase in the helix angle.

We will therefore use the following equation as a measure for the deflection on helical gears in their plane of rotation:

$$d = d_n/\cos^2 \psi \qquad (21\text{-}3)$$

As the errors in form or spacing act, there may be some shifting of the masses in an axial direction as well as a change in velocity in the plane of rotation. This is particularly true for a herringbone-gear drive where the pinion is usually left free to float in an axial direction and find its own position. It would appear, however, that the value of the mass influence would be affected but little by this action. We shall therefore introduce the influence of the helix angle, as expressed in Eqs. (21-2) and (21-3), into the equations for the dynamic loads on spur-gear teeth to obtain equations for use on helical gears.

Effective Mass. For the determination of the effective mass we have the following:

When m_a = full effective mass of connected bodies on pinion shaft at $R_1 = I_a/R_1{}^2$

$\quad m_b$ = mass effect of m_a on tooth profile of pinion

$\quad m_p$ = effective mass of pinion blank at $R_1 = I_p/R_1{}^2$

$\quad m_1$ = effective mass acting at pitch line of pinion

$\quad m_c$ = full effective mass of connected bodies on gear shaft at $R_2 = I_c/R_2{}^2$

$\quad m_d$ = mass effect of m_c on tooth profile of gear

$\quad m_g$ = effective mass of gear blank at $R_2 = I_g/R_2{}^2$

$\quad m_2$ = effective mass acting at pitch line of gear

$\quad m$ = effective mass influence at pitch line of gears

$\quad V$ = pitch-line velocity, ft/min

$\quad Z_1$ = elasticity factor of connecting member on pinion shaft

$\quad Z_2$ = elasticity factor of connecting member on gear shaft

$\quad e$ = measured error in action, in.

$\quad W$ = tangential applied load at pitch line, lb

$\quad F$ = active face width of gears, in.

$\quad d$ = deformation of teeth under applied load W, in.

$\quad R_1$ = pitch radius of pinion, in.

$\quad R_2$ = pitch radius of gear, in.

$\quad \psi$ = helix angle at pitch line

z_1, z_2 = elasticity form factors of gears

E_1, E_2 = modulus of elasticity of materials

$\quad y$ = Lewis tooth-form factor

$$z = \frac{y}{0.242 + 7.25y} \qquad (20\text{-}14)$$

$$d = \frac{(W/F)[(1/E_1 z_1) + (1/E_2 z_2)]}{\cos^2 \psi} \qquad (21\text{-}4)$$

$$m_b = \frac{\sqrt{B^2 + 4A_1 C_1} - B_1}{2A_1} \qquad (20\text{-}29)$$

where $A_1 = Hm_a V^2 \cos^2 \psi$
$\quad B_1 = (m_p + m_g)A_1 + em_g Z_1$
$\quad C_1 = em_a m_g Z_1$
For $14\frac{1}{2}$-deg gears

$$H = 0.00086 \left(\frac{1}{R_1} + \frac{1}{R_2} \right)$$

For 20-deg gears

$$H = 0.00120 \left(\frac{1}{R_1} + \frac{1}{R_2} \right)$$

$$Z = \frac{P}{T}$$

where P = load applied to shaft or coupling at pitch line of gear, lb
$\quad T$ = torsional deflection at pitch line of gear, in.

$$m_d = \frac{\sqrt{B_2^2 + 4A_2 C_2} - B_2}{2A_2} \qquad (21\text{-}5)$$

where $A_2 = Hm_c V^2 \cos^2 \psi$
$\quad B_2 = (m_g + m_1)A_2 + em_1 Z_2$
$\quad C_2 = em_c m_1 Z_2$

$$m_1 = m_p + m_b \qquad (20\text{-}25)$$
$$m_2 = m_g + m_d \qquad (21\text{-}6)$$
$$m = \frac{m_1 m_2}{m_1 + m_2} \qquad (20\text{-}30)$$

The value for the tooth deformation d can be determined very closely for gears of conventional design by the following equations:
$14\frac{1}{2}$-deg gears

$$d = \frac{9.345(W/F)\,[(1/E_1) + (1/E_2)]}{\cos^2 \psi} \qquad (21\text{-}7)$$

20-deg full depth

$$d = \frac{9.000(W/F)\,[(1/E_1) + (1/E_2)]}{\cos^2 \psi} \qquad (21\text{-}8)$$

20-deg stub tooth

$$d = \frac{8.700(W/F)\,[(1/E_1) + (1/E_2)]}{\cos^2 \psi} \qquad (21\text{-}9)$$

Example of Effective Mass for Helical Gears.   As a definite example we shall use the values given in Fig. 20-6 with a 30-deg helix angle.   We have from the spur-gear example:

$$Z_1 = 1,206,400 \qquad Z_2 = 603,200$$
$$A_1 = 0.00120(\tfrac{1}{5} + \tfrac{1}{20})16 \times 1,000^2 \times 0.86603^2 = 3,600$$
$$B_1 = [(2.50 + 13)3,600] + [0.001 \times 13 \times 1,206,400] = 71,483$$
$$C_1 = 0.001 \times 16 \times 13 \times 1,206,400 = 250,931$$
$$m_b = \frac{93,398 - 71,483}{7,200} = 3.04$$
$$A_2 = 0.00120 \times 0.25 \times 40 \times 1,000^2 \times 0.86603^2 = 9,000$$
$$B_2 = (5.54 + 13)9,000 + 0.001 \times 5.54 \times 603,200 = 166,860$$
$$C_2 = 0.001 \times 40 \times 5.54 \times 603,200 = 250,931$$
$$m_d = \frac{180,705 - 166,860}{18,000} = 0.77$$
$$m_1 = 2.50 + 3.04 = 5.54$$
$$m_2 = 13.00 + 0.77 = 13.77$$
$$m = \frac{5.54 \times 13.77}{5.54 + 13.77} = 3.95$$

Acceleration Load on Helical Gears.   For the determination of the acceleration load we have the following:

When f_1 = force required to accelerate the masses as rigid bodies, lb

f_2 = force required to deform teeth amount of error, lb

f_a = acceleration load on gears, lb

W = applied tangential load, lb

m = effective mass of gears

e = measured or assumed error in action, in.

d = deformation of gear teeth under load W, lb

$$f_1 = HmV^2 \cos^2 \psi \qquad (21\text{-}10)$$

For $14\tfrac{1}{2}$-deg gears

$$H = 0.00086[(1/R_1) + (1/R_2)]$$

For 20-deg gears:

$$H = 0.00120[(1/R_1) + (1/R_1)]$$
$$f_2 = W[(e/d) + 1] \qquad (20\text{-}15)$$
$$f_a = f_1 f_2/(f_1 + f_2) \qquad (20\text{-}2)$$

Example of Acceleration Load on Helical Gears.   As a definite example we shall use the values shown in Fig. 20-6 with a 30-deg helix angle.   Whence we have

$$f_1 = 0.00120 \times 0.25 \times 3.95 \times 1,000^2 \times 0.86603^2 = 889 \text{ lb}$$
$$f_2 = 1,000 \left(\frac{0.001}{0.00032} + 1 \right) = 4,125 \text{ lb}$$
$$f_a = \frac{889 \times 4,125}{889 + 4,125} = 731 \text{ lb}$$

Dynamic Loads on Helical-gear Teeth.   For the determination of the maximum intensity of the impact load, which is the dynamic load, on helical-gear teeth, we have the following:

When W_d = dynamic tooth load, lb

$$W_d = W + \sqrt{f_a(2f_2 - f_a)} \tag{20-39}$$

Example of Dynamic Load on Helical-gear Teeth. As a definite example we shall continue with the preceding problem. Whence we have

$$W_d = 1,000 + \sqrt{731(8,250 - 731)} = 3,334 \text{ lb}$$

Dynamic Loads on Fine Pitches and at High Speeds. When the time interval between successive tooth contacts is too small to permit the full double-load cycle to act, then we have the following:

When n = rpm of driver

$\quad N_1$ = number of teeth in driver

$\quad t$ = one-half the time between successive tooth contacts, sec

$\quad m$ = effective mass of gear blanks at pitch line of gears

$\quad e$ = measured or assumed error in action, in.

$\quad e'$ = distance on pitch line that gears can move in time t, in.

$$t = 30/nN_1 \tag{20-44}$$
$$e' = 6Wt^2/m \tag{20-45}$$

When e' is greater than e, the value of e will be used in the dynamic load equations. When e' is less than e, then the value of e' will be used as the effective error in the dynamic load equations.

When n_c = rpm of driver when e' is equal to e
and all other symbols are the same as before

$$n_c = (30/N_1)\sqrt{6W/em} \tag{20-46}$$

Example. Using the same values as before, we have

$$W = 1,000 \qquad m = 2.50 \qquad N_1 = 30 \qquad e = 0.001$$
$$n_c = \frac{30}{30}\sqrt{\frac{6,000}{0.001 \times 2.50}} = 1,548 \text{ rpm}$$

In this example, at speeds above about 1,550 rpm, the effective error will be less than the measured one.

DYNAMIC LOADS ON SPIRAL GEARS

The primary tooth action on a pair of spiral gears is a screwing or camming action. The theoretical contact is point contact. When the materials of the gears are soft and plastic, there is an appreciable amount of corrective action or running-in, so that the error in action is reduced with continued use provided that a destructive abrasive action is not set up. In general, it is probable that when the extent of the error in the direction of the action of the theoretical basic rack does not exceed about 0.002 in. on soft materials, the larger part of this error will be corrected in service. Under such conditions, the effective error in action may be primarily the elastic deformation of the tooth surfaces under load, and the dynamic load will then be substantially proportional to the applied load. In any event, we can safely assume that the eventual error in action after thorough running-in on the softer materials will be of the order of 0.0010 in., and the dynamic loads so determined will be greater, if anything, than the actual ones.

With spiral gears of hardened steel, on the other hand, although some corrective wear may occur, the larger part of the initial error will persist. In general, with carefully cut gears, and including probable distortions in hardening, the extent of this initial error will be of the order of 0.002 in. Under these conditions, the extent of the initial error will have a major influence on the intensity of the dynamic load.

Eventually it may prove necessary to set up two equations for the dynamic loads on spiral-gear teeth: one for use when both gears are made of hardened steel; and another for use when one or both of them are made of softer and plastic metal. We are in need of experimental data not only to complete the analysis of the dynamic loads on spiral-gear drives but also to determine the limiting wear-load conditions. In the absence of such information, however, we shall set up tentative equations for use until more reliable information is available.

Deformation of Spiral-gear-tooth Profiles The permissible tooth loads on spiral-gear drives are relatively small, and the greater part of the deformation will be the surface deformation at the point contact. We shall therefore ignore the small amount of the bending of the teeth under these small loads. As a reasonable approximation, we shall assume that the contact here is similar to that of a pair of crossed cylinders with their axes at right angles to each other. The actual tooth surfaces are curved in two directions relative to the contact point, and the determination of these actual radii of curvature would be a very complex operation. With definite experimental data to analyze, such calculations are justified. In the absence of such information, a simple approximation will be as useful and effective as a more complex analysis.

When d = tooth deformation, in.

  P = load on crossed cylinders, lb

  R_{c1} = smaller radius of curvature, in.

  R_{c2} = larger radius of curvature, in.

E_1, E_2 = modulus of elasticity of materials

m_1, m_2 = Poisson's ratio for materials

  A = value depending upon value of R_{c2}/R_{c1} (see Table 21-1)

  ψ = helix angle of gear

  ϕ = pressure angle of normal basic rack

R_1 = pitch radius of driver, in.
R_2 = pitch radius of follower, in.

$$R_c = \frac{R \sin \phi}{\cos^2 \psi} \tag{21-11}$$

TABLE 21-1

R_{c2}/R_{c1}	A
1.000	2.080
1.500	2.060
2.000	2.025
3.000	1.950
4.000	1.875
6.000	1.770
10.000	1.613

The approximate radius of curvature of both gears will be determined, and the larger value will be R_{c2} and the smaller value will be R_{c1}.

$$d = A \sqrt[3]{\frac{P^2}{\{[E_1/(1-m^2)] + [E^2/(1-m^2)]\}^2}\left(\frac{R_{c1}+R_{c2}}{2R_{c1}R_{c2}}\right)} \tag{21-12}$$

Let e = error in action on normal basic rack, in.

C = load required to deform teeth amount of error, lb

Substituting the symbol e for d and C for P in Eq. (21-12), and solving for C, we obtain

$$C = \left(\frac{e}{A}\right)^{3/2}\left(\frac{E_1}{1-m_1^2} + \frac{E_2}{1-m_2^2}\right)\sqrt{2\left(\frac{R_{c1}R_{c2}}{R_{c1}+R_{c2}}\right)} \tag{21-13}$$

Dynamic Loads on Spiral Gears. We shall determine the dynamic load on spiral gears as the maximum momentary load normal to the theoretical basic-rack form. We shall ignore the influence of the connected masses and use only the effective masses of the gear blanks themselves. Thus when

W = tangential applied load on driver, plane of rotation, lb
W_n = normal tooth load, lb
ψ_1 = helix angle of driver
ϕ = pressure angle of normal basic rack
V = pitch-line velocity of driver, plane of rotation, ft/min
V_r = pitch-line velocity of normal basic rack, ft/min
m_1 = effective mass of driver blank at pitch line
m_2 = effective mass of follower blank at pitch line
m = effective mass acting at pitch line of gears
R_1 = pitch radius of driver, in.
R_2 = pitch radius of follower, in.
e = error in action along normal basic rack, in.

C = load required to deform tooth amount of error, lb [value determined by use of Eq. (21-13)]

W_{dn} = normal dynamic tooth load, lb

$$W_n = W/\cos \psi_1 \cos \phi \qquad (21\text{-}14)$$
$$V_r = V \cos \psi_1 \qquad (21\text{-}15)$$
$$m = m_1 m_2/(m_1 + m_2) \qquad (20\text{-}30)$$
$$f_1 = H m V_r^2 \qquad (21\text{-}16)$$

For 14½-deg basic-rack form

$$H = 0.00086[(\cos^2 \psi_1/R_1) + (\cos^2 \psi_2/R_2)] \qquad (21\text{-}17)$$
$$f_2 = C + W_n \qquad (21\text{-}18)$$
$$f_a = f_1 f_2/(f_1 + f_2) \qquad (20\text{-}2)$$
$$W_{dn} = W_n + \sqrt{f_a(2f_2 - f_a)} \qquad (21\text{-}19)$$

Examples of Dynamic Load on Spiral Gears. *First Example.* As a definite example we shall use a pair of hardened-steel spiral gears of the following sizes:

$$N_1 = 12 \qquad N_2 = 48 \qquad P_n = 10 \qquad \psi_1 = 60° \qquad \psi_2 = 30° \qquad \phi = 14.50°$$
$$\text{Face width} = 1.000 \qquad W = 20 \text{ lb} \qquad n = 2,000 \qquad e = 0.002$$

Whence

$$R_1 = {}^6\!/_{10} \cos 60° = 1.200 \qquad R_2 = {}^{24}\!/_{10} \cos 30° = 2.7713$$
$$V_r = 0.2618 \times 2.40 \times 2,000 \times 0.500 = 628 \text{ ft/min}$$

When 0.265 lb is the weight per cubic inch of steel

$$m_1 = \frac{3.1416 \times 1.2^2 \times 1 \times 0.265}{2 \times 32.2} = 0.018$$
$$m_2 = \frac{3.1416 \times 2.7713^2 \times 1 \times 0.265}{2 \times 32.2} = 0.032$$
$$m = \frac{0.018 \times 0.032}{0.018 + 0.032} = 0.0115$$
$$W_n = \frac{20}{0.500 \times 0.96814} = 41.3 \text{ lb}$$
$$R_{c2} = \frac{1.20 \times 0.25038}{0.25} = 1.202$$
$$R_{c1} = \frac{2.7713 \times 0.25038}{0.75} = 0.925$$
$$\frac{R_{c2}}{R_{c1}} = \frac{1.202}{0.925} = 1.3$$

Whence

$$A = 2.074$$
$$C = \left(\frac{0.002}{2.047}\right)^{3\!/_2} \left(\frac{60,000,000}{0.900}\right) \sqrt{2\left(\frac{0.925 \times 1.202}{0.925 + 1.202}\right)} = 2,000 \text{ lb}$$
$$H = 0.00086 \left(\frac{0.250}{1.200} + \frac{0.750}{2.7713}\right) = 0.00032$$
$$f_1 = 0.00032 \times 0.0115 \times 628^2 = 1.5 \text{ lb}$$
$$f_2 = 2,000 + 41.3 = 2,041.3 \text{ lb}$$
$$f_a = \frac{1.5 \times 2,041.3}{1.5 + 2,041.3} = 1.5 \text{ lb}$$
$$W_{dn} = 41.3 + \sqrt{1.5(4,082.6 - 1.5)} = 119.5 \text{ lb}$$

This dynamic load of 119.5 lb would develop a maximum compressive stress of over 240,000 psi in the hardened-steel materials.

Second Example. As a second example we shall use the same values as before, but with cast iron as the material for both gears. This would give the following values:

$$V_r = 628 \qquad m = 0.0115 \qquad f_1 = 1.5 \qquad R_{c1} = 0.925 \qquad R_{c2} = 1.202$$
$$E = 15{,}000{,}000 \qquad A = 2.074 \qquad W_n = 41.3$$

These gears will receive corrective wear so that we shall assume an eventual error in action of 0.0005 in. along the normal basic rack; whence

$$C = \left(\frac{0.0005}{2.074}\right)^{3/2}\left(\frac{30{,}000{,}000}{0.900}\right)\sqrt{2\left(\frac{0.925 \times 1.202}{0.925 + 1.202}\right)} = 125 \text{ lb}$$
$$f_2 = 125 + 41.3 = 166.3$$
$$f_a = \frac{1.5 \times 166.3}{1.5 + 166.3} = 1.5 \text{ lb}$$
$$W_{dn} = 41.3 + \sqrt{1.5(332.6 - 1.5)} = 63.6 \text{ lb}$$

This dynamic load of 63.6 lb would develop a maximum compressive stress of about 63,000 psi in the cast-iron materials.

DYNAMIC LOADS ON WORM GEARS

The tooth action on a worm-gear drive is similar to that on a pair of spiral gears except that the contact is line contact instead of point contact. The exact nature of this line contact depends upon several factors, such as the thread angle, the lead angle of the worm, and the position of the pitch plane of the worm in reference to the thread depth. Another variable factor is the changing form of the worm-gear tooth across its face. All these variables make it a complex task to calculate the local deformations under load and also to determine with any degree of accuracy the influence of other conditions that affect the intensity of the dynamic load. We shall therefore use only approximations for this purpose.

In common with spiral-gear drives, when the extent of the error in action on a worm-gear drive does not exceed about 0.002 in., it is probable that a large part of this error will be corrected in service by the plastic flow of the material of the worm gear. We shall therefore use a value of 0.001 in. for the effective error on worm gears as a measure of the conditions after the initial running-in of the drive.

As with spiral gears, we shall ignore the influence of the connected masses and use only the effective masses of the worm and worm-gear blanks. Thus when

W = transmitted axial tooth load, lb

V_r = normal pitch-line velocity, ft/min

λ = lead angle of worm

L = lead of worm thread, in.

n = rpm. of worm

F = face width of worm gear, in.
R_1 = nominal pitch radius of worm, in.
R_2 = pitch radius of worm gear, in.
E_1, E_2 = modulus of elasticity of materials
m_1 = effective mass of worm blank at pitch radius
m_2 = effective mass of worm gear blank at pitch radius
m = effective mass influence at pitch line of drive
W_d = dynamic axial tooth load, lb
C = load per inch of face required to deform teeth by amount of error, lb
e = error in action, in.

$$m = \frac{m_1 m_2}{m_1 + m_2} \tag{20-30}$$

$$V_r = \frac{nL \cos \lambda}{12} \tag{21-20}$$

$$C = \frac{0.120e}{(1/E_1) + (1/E_2)} \tag{21-21}$$

$$f_1 = HmV_r^2 \tag{21-16}$$

For worms of 14½-deg pressure angle

$$H = 0.00086 \left(\frac{\sin^2 \lambda}{R_1} + \frac{\cos^2 \lambda}{R_2} \right) \tag{21-22}$$

For worms of 20-deg pressure angle

$$H = 0.00120 \left(\frac{\sin^2 \lambda}{R_1} + \frac{\cos^2 \lambda}{R_2} \right) \tag{21-23}$$

For worms of 25-deg pressure angle

$$H = 0.00153 \left(\frac{\sin^2 \lambda}{R_1} + \frac{\cos^2 \lambda}{R_2} \right) \tag{21-24}$$

For worms of 30-deg pressure angle

$$H = 0.00188 \left(\frac{\sin^2 \lambda}{R_1} + \frac{\cos^2 \lambda}{R_2} \right) \tag{21-25}$$

$$f_2 = FC + W \tag{21-26}$$

$$f_a = \frac{f_1 f_2}{f_1 + f_2} \tag{20-2}$$

$$W_d = W + \sqrt{f_a(2f_2 - f_a)} \tag{20-39}$$

Examples of Dynamic Load on Worm-gear Drive. *First Example.* As a definite example we shall use a hardened and ground steel worm and a bronze worm gear with the following values: 6-start worm and 48-tooth worm gear, 1-in. axial pitch.

$R_1 = 1.910$ $\lambda = 22.565°$ $\phi = 30°$ Length of worm $= 4.00$ $L = 6.00$
$R_2 = 7.6394$ $m_1 = 0.20$ $m_2 = 1.68$ $n = 2,000$ $W = 1,000$ $F = 2.125$

$$V_r = 2,000 \times 6 \times \frac{0.92344}{12} = 923 \text{ ft/min}$$

$$C = \frac{0.120 \times 0.001}{(1/30,000,000) + (1/15,000,000)} = 1,200 \text{ lb}$$

$$H = 0.00188 \left(\frac{0.14725}{1.910} + \frac{0.85275}{7.6394} \right) = 0.000355$$

$$m = \frac{0.20 \times 1.68}{0.20 + 1.68} = 0.179$$

$$f_1 = 0.000355 \times 0.179 \times 923^2 = 54 \text{ lb}$$

$$f_2 = 2,550 + 1,000 = 3,550 \text{ lb}$$

$$f_a = \frac{54 \times 3,550}{54 + 3,550} = 54 \text{ lb}$$

$$W_d = 1,000 + \sqrt{54(7,100 - 54)} = 1,617 \text{ lb}$$

Second Example. As a second example we shall use a cast-iron worm and bronze worm gear of the same size as before. Whence we have

$$W = 1,000 \qquad V_r = 923 \text{ ft/min}$$

$$C = \frac{0.120 \times 0.001}{(1/15,000,000) + (1/15,000,000)} = 900 \text{ lb}$$

$$H = 0.000355 \qquad m = 0.179$$

$$f_1 = 0.000355 \times 0.179 \times 923^2 = 54 \text{ lb}$$

$$f_2 = 1,912 + 1,000 = 2,912 \text{ lb}$$

$$f_a = \frac{54 \times 2,912}{54 + 2,912} = 54 \text{ lb}$$

$$W_d = 1,000 + \sqrt{54(5,824 - 54)} = 1,558 \text{ lb}$$

DYNAMIC LOADS ON BEVEL-GEAR TEETH

The tooth action on bevel gears with straight teeth is very similar to that on spur gears. We can use the same analysis here when it is adjusted to the equivalent spur gears of Tregold's approximation.

Effective Mass. For the determination of the effective mass we have Eq. (20-29) from the spur-gear analysis. The pitch radii of the gears will be the mean pitch radii, or the radii of the pitch cones at the middle of the tooth faces. For the value of H we have, when

R_{mp} = mean pitch radius of bevel pinion, in.

R_{mg} = mean pitch radius of bevel gear, in.

γ_p = pitch-cone angle of bevel pinion

γ_g = pitch-cone angle of bevel gear

For $14\frac{1}{2}$-deg gears

$$H = 0.00086[(\cos \gamma_p/R_{mp}) + (\cos \gamma_g/R_{mg})] \qquad (21\text{-}27)$$

For 20-deg gears

$$H = 0.00120[(\cos \gamma_p/R_{mp}) + (\cos \gamma_g/R_{mg})] \qquad (21\text{-}28)$$

Acceleration Load. For the determination of the acceleration load we have the following:

When f_1 = force required to accelerate the masses as rigid bodies, lb

$\quad f_2$ = force required to deform teeth by amount of effective error, lb

$\quad f_a$ = acceleration load, lb

$\quad W$ = applied tangential load, lb

$\quad m$ = effective mass of gears

$\quad m_1$ = effective mass acting at pitch line of bevel pinion

$\quad m_2$ = effective mass acting at pitch line of bevel gear

$\quad e$ = measured or assumed error in action, in.

$\quad C$ = load per inch of face required to deform teeth by amount of error, lb

$\quad F$ = face width of gears, in.

$$m = \frac{m_1 m_2}{m_1 + m_2} \qquad (20\text{-}30)$$

$$f_1 = HmV^2 \qquad (20\text{-}43)$$

Use Eqs. (21-27) and (21-28) for values of H.

$$f_2 = FC + W \qquad (21\text{-}26)$$

For 14½-deg gears

$$C = \frac{0.107e}{(1/E_1) + (1/E_2)} \qquad (21\text{-}29)$$

For 20-deg gears

$$C = \frac{0.111e}{(1/E_1) + (1/E_2)} \qquad (21\text{-}30)$$

$$f_a = \frac{f_1 f_2}{f_1 + f_2} \qquad (20\text{-}2)$$

Dynamic Load on Bevel-gear Teeth. For the determination of the maximum intensity of the impact load on bevel-gear teeth, which is the dynamic load, we have the following:

When W_d = dynamic tooth load, lb

$$W_d = W + \sqrt{f_a(2f_2 - f_a)} \qquad (20\text{-}39)$$

Example of Dynamic Load on Bevel-gear Teeth. As a definite example, we shall use a pair of 6-DP bevel gears of 24 and 48 teeth, 20-deg full-depth form, with a face width of 1 in., with a tooth load of 300 lb, and with the pinion running at 1,200 rpm. Both gears are of steel. We shall use the following values:

$N_p = 24$	$\gamma_p = 26.565°$	$\cos \gamma_p = 0.89442$	$R_p = 2.000$	$F = 1.000$
$N_g = 48$	$\gamma_g = 63.435°$	$\cos \gamma_g = 0.44721$	$R_g = 4.000$	$W = 300$
	$n = 1,200$	$e = 0.002$		

We will use $m = 0.070$.

$$R_{mp} = R_p - \frac{F}{2} \sin \gamma_p \tag{21-31}$$

$$R_{mg} = R_g - \frac{F}{2} \sin \gamma_g \tag{21-32}$$

$R_{mp} = 2.000 - (0.500 \times 0.44721) = 1.7764$

$R_{mg} = 4.000 - (0.500 \times 0.89442) = 3.5528$

$V = 0.5236 R_{mp} n \tag{21-33}$

$V = 0.5236 \times 1.7764 \times 1,200 = 1,116$ ft/min

$H = 0.00120 \left(\dfrac{0.89442}{1.7764} + \dfrac{0.44721}{3.5528} \right) = 0.00075$

$f_1 = 0.00075 \times 0.070 \times 1,116^2 = 65.4$ lb

$C = \dfrac{0.111 \times 0.002}{(1/30,000,000) + (1/30,000,000)} = 3,330$ lb

$f_2 = 3,330 + 300 = 3,630$ lb

$f_a = \dfrac{65.4 \times 3,630}{65.4 + 3,630} = 64.2$ lb

$W_d = 300 + \sqrt{64.2(7,260 - 64.2)} = 980$ lb

Dynamic Loads at High Speeds on Bevel Gears. As with spur gears, when the time interval between successive tooth contacts is too short because of fine pitch or high pitch-line velocities to permit the full double-load cycle, the extent of the effective error will be reduced, and the intensity of the dynamic load will also be reduced. Thus when

n = rpm of bevel pinion

N_p = number of teeth in bevel pinion

t = one-half the time between tooth contacts, sec

m = mass effect at pitch line of gears

W = applied tangential load, lb

e = measured or assumed error, in.

e' = effective error, in.

n_c = rpm of pinion when e' is equal to e

$$t = 30/nN_p \tag{20-44}$$

$$e' = 6Wt^2/m \tag{20-45}$$

$$n_c = (30/N_p) \sqrt{6W/em} \tag{20-46}$$

When e' is greater than e, the value of e will be used in the dynamic load equations. When e' is less than e, then the value of e' will be used in the dynamic load equations.

The maximum value of the dynamic load will be when the bevel pinion is running at a speed of n_c. If the gears are to operate most of the time at higher speeds, then stresses greater than the endurance limits of the materials could be permitted for the load at its maximum value. If the gears are to run at a varying range of speeds above and below this critical value, then the stresses must be within the endurance limits at this critical value if the gears are to have a reasonable length of useful life.

Example of High-speed Bevel-gear Drive. As a definite example we shall use the same values as before, but with a reduced error in action of 0.001 in., and a higher speed of operation. We shall use the following values:

$$N_p = 24 \quad \gamma_p = 26.565° \quad R_{mp} = 1.7764 \quad N_g = 48 \quad \gamma_g = 63.435°$$
$$R_{mg} = 3.5528 \quad F = 1.000 \quad W = 300 \quad n = 7{,}500 \quad e = 0.001 \quad m = 0.070$$
$$H = 0.00075 \quad C = 1{,}665$$

$$n_c = \frac{30}{24} \sqrt{\frac{1{,}800}{0.00007}} = 6{,}345 \text{ rpm}$$

We shall determine the critical dynamic loads at the speeds of 6,345 rpm and 7500 rpm.

When V_c = pitch-line velocity at critical load, ft/min

$$V_c = 0.5236 \times 1.7764 \times 6{,}345 = 5{,}900 \text{ ft/min}$$
$$f_1 = 0.00075 \times 0.070 \times 5{,}900^2 = 1{,}828 \text{ lb}$$
$$f_2 = 1{,}665 + 300 = 1{,}965 \text{ lb}$$
$$f_a = \frac{1{,}828 \times 1{,}965}{1{,}828 + 1{,}965} = 947 \text{ lb}$$
$$W_d = 300 + \sqrt{947(3{,}930 - 947)} = 1{,}980 \text{ lb}$$

When $n = 7{,}500$, then

$$V = 0.5236 \times 1.7764 \times 7{,}500 = 6{,}976 \text{ ft/min}$$
$$t = \frac{30}{7{,}500 \times 24} = 0.000167 \text{ sec}$$
$$e' = 6 \times 300 \times \frac{0.000167^2}{0.070} = 0.000714 \text{ in.}$$
$$C = \frac{0.111 \times 0.000714}{(1/30{,}000{,}000) + (1/30{,}000{,}000)} = 1{,}188$$
$$f_1 = 0.00075 \times 0.070 \times 6{,}976^2 = 2{,}555 \text{ lb}$$
$$f_2 = 1{,}188 + 300 = 1{,}488 \text{ lb}$$
$$f_a = \frac{2{,}555 \times 1{,}488}{2{,}555 + 1{,}488} = 940 \text{ lb}$$
$$W_d = 300 + \sqrt{940(2{,}976 - 940)} = 1{,}684 \text{ lb}$$

In this example, the operating dynamic load is about 300 lb less than the maximum dynamic load at the critical point.

DYNAMIC LOADS ON SPIRAL BEVEL GEARS

The tooth action of spiral bevel gears has much in common with the tooth action on helical gears. We shall therefore use the same analysis here, adjusted to the equivalent helical gears by the use of Tregold's approximation.

Effective Mass. For the determination of the effective mass we have Eqs. (20-29) and (21-5) from the helical-gear analysis. The pitch radii of the spiral bevel gears will be their mean pitch radii. For the value of H we have the following:

When R_{mp} = mean pitch radius of spiral bevel pinion, in.

R_{mg} = mean pitch radius of spiral bevel gear, in.

γ_p, γ_g = pitch-cone angles of spiral bevel pinion and gear

For 14½-deg gears

$$H = 0.00086[(\cos \gamma_p/R_{mp}) + (\cos \gamma_g/R_{mg})] \qquad (21\text{-}27)$$

For 20-deg gears

$$H = 0.00120[(\cos \gamma_p/R_{mp}) + (\cos \gamma_g/R_{mg})] \qquad (21\text{-}28)$$

Acceleration Load on Spiral Bevel Gears. For the determination of the acceleration load on spiral bevel gears, we have the following:

When f_1 = force required to accelerate masses as rigid body, lb

f_2 = force required to deform teeth by amount of error, lb

f_a = acceleration load, lb

m = effective mass acting at pitch line of gears

m_1 = effective mass acting at pitch line of pinion

m_2 = effective mass acting at pitch line of gear

e = measured or assumed error in action, in.

C = load per inch of face to deform teeth by amount of error, lb

F = face width of gears, in.

V = pitch-line velocity, ft/min

ψ = spiral angle at middle of tooth face

n = rpm of spiral bevel pinion

W = applied tangential load, lb

$$m = \frac{m_1 m_2}{m_1 + m_2} \qquad (20\text{-}30)$$

$$V = 0.5236 R_{mp} n \qquad (21\text{-}33)$$

$$f_1 = H m V^2 \cos^2 \psi \qquad (21\text{-}10)$$

We already have Eqs. (21-27) and (21-28) for the value of H.

$$f_2 = FC \cos^2 \psi + W \qquad (21\text{-}34)$$

For 14½-deg gears

$$C = \frac{0.107e}{(1/E_1) + (1/E_2)} \qquad (21\text{-}29)$$

For 20-deg gears

$$C = \frac{0.111e}{(1/E_1) + (1/E_2)} \qquad (21\text{-}30)$$

$$f_a = \frac{f_1 f_2}{f_1 + f_2} \qquad (20\text{-}2)$$

Dynamic Loads on Spiral Bevel Gears. For the determination of the dynamic loads on spiral bevel gears, we have the following:

When W_d = dynamic tooth load, lb

$$W_d = W + \sqrt{f_a(2f_2 - f_a)} \qquad (20\text{-}39)$$

Example of Dynamic Load on Spiral Bevel Gears. As a definite example we shall use the same values as those of the bevel-gear example with straight teeth, but with a spiral angle of 30 deg. This gives the following values:

$N_p = 24$ $\gamma_p = 26.565°$ $\cos \gamma_p = 0.89442$ $R_p = 2.000$ $F = 1.000$
$N_g = 48$ $\gamma_g = 63.435°$ $\cos \gamma_g = 0.44721$ $R_g = 4.000$ $W = 300$
$n = 1,200$ $e = 0.002$ $m = 0.070$ $\psi = 30°$ $\cos \psi = 0.86603$

$$R_{mp} = R_p - \frac{F}{2} \sin \gamma_p \tag{21-31}$$

$$R_{mg} = R_g - \frac{F}{2} \sin \gamma_g \tag{21-32}$$

$R_{mp} = 1.7764$ $R_{mg} = 3.5528$ $V = 1,116$ $H = 0.00075$ $C = 3,330$

$$f_1 = 0.00075 \times 0.070 \times 1,116^2 \times 0.7500 = 49.0 \text{ lb}$$
$$f_2 = 3,330 \times 0.7500 + 300 = 2.798 \text{ lb}$$
$$f_a = \frac{49 \times 2,798}{49 + 2,798} = 48 \text{ lb}$$
$$W_d = 300 + \sqrt{48(5,596 - 48)} = 816 \text{ lb}$$

Dynamic Loads on Spiral Bevel Gears at High Speeds.

In common with other types of gears, when the time interval between successive tooth contacts is too short to permit the full double-load cycle to act, either because of fine pitches or because of high pitch-line velocities, the extent of the effective error will be reduced, and the intensity of the dynamic tooth load will also be reduced. Thus when

n = rpm of spiral bevel pinion
N_p = number of teeth in spiral bevel pinion
t = one-half the time between successive tooth contacts, sec
m = mass effect at pitch line of gears
W = applied tangential load, lb
e = measured or assumed error, in.
e' = effective error, in.

$$t = 30/nN_p \tag{20-44}$$
$$e' = 6Wt^2/m \tag{20-45}$$
$$n_c = (30/N_p) \sqrt{6W/em} \tag{20-46}$$

When e' is greater than the measured error e, the value of e will be used in the dynamic load equations. When the value of e' is less than e, then the value of e' will be used in the dynamic load equations.

The maximum value of the dynamic load will occur when the spiral bevel pinion is running at a speed of n_c. If the gears are to operate most of the time at higher speeds, and pass through this critical value only occasionally when starting and stopping, then stresses greater than the endurance limits of the materials used for the gears could be permitted for the load at its maximum value. If the gears are to run at varying speeds, above and below this critical value, then the stresses here must be within the endurance limits of the materials if the gears are to have a reasonable length of useful life.

Example of Spiral Bevel Gears at High Speeds. As a definite example we shall use the same values as before, but with a reduced error in action of 0.001 in., and a higher speed of operation. We shall use the following values:

$$N_p = 24 \quad R_{mp} = 1.7764 \quad N_g = 48 \quad R_{mg} = 3.5528 \quad F = 1.000$$
$$W = 300 \quad n = 7,500 \quad e = 0.001 \quad m = 0.070$$
$$n_c = \frac{30}{24} \sqrt{\frac{1,800}{0.00007}} = 6,345 \text{ rpm}$$

We shall determine the critical dynamic load at the speed of 6,345 rpm and also the dynamic load at the speed of 7,500 rpm.

When V_c = pitch-line velocity at critical point, ft/min

$$V_c = 0.5236 \times 1.7764 \times 6,345 = 5,900 \text{ ft/min}$$
$$f_1 = 0.00075 \times 0.070 \times 5,900^2 \times 0.750 = 1,371 \text{ lb}$$
$$f_2 = 1,665 \times 0.750 + 300 = 1,549 \text{ lb}$$
$$f_a = \frac{1,371 \times 1,549}{1,371 + 1,549} = 727 \text{ lb}$$
$$W_d = 300 + \sqrt{727(3,098 - 727)} = 1,613 \text{ lb}$$

When $n = 7,500$, then

$$V = 0.5236 \times 1.7764 \times 7,500 = 6,973 \text{ ft/min}$$
$$t = \frac{30}{7,500 \times 24} = 0.000167 \text{ sec}$$
$$e' = 6 \times 300 \times \frac{0.000167^2}{0.070} = 0.000714$$
$$C = \frac{0.111 \times 0.000714}{(1/30,000,000) + (1/30,000,000)} = 1,188 \text{ lb}$$
$$f_1 = 0.00075 \times 0.070 \times 6,973^2 \times 0.750 = 1,916 \text{ lb}$$
$$f_2 = 1,916 \times 0.750 + 300 = 1,191 \text{ lb}$$
$$f_a = \frac{1,916 \times 1,191}{1,916 + 1,191} = 734 \text{ lb}$$
$$W_d = 300 + \sqrt{734(2,382 - 734)} = 1,400 \text{ lb}$$

DYNAMIC LOADS ON HYPOID GEARS

The tooth action on hypoid gears varies somewhat from that on spiral bevel gears because of the sliding action that develops as a result of the offset axes of the gears. We shall, however, treat them here as spiral bevel gears with an increased spiral angle because of the offset contact. Thus when

R_{mg} = mean pitch radius of hypoid gear, in.
R_{bg} = radius of base cylinder of gear hyperboloid, in.
γ_g = angle of generatrix of gear hyperboloid
ψ = spiral angle of hypoid gear at mean pitch radius
ψ_c = effective spiral angle of hypoid gear at mean pitch radius

$$\psi_c = \psi + \sin^{-1}(R_{bg} \sin \gamma_g / R_{mg}) \tag{21-35}$$

Effective Mass of Hypoid Gears. The effective mass of hypoid gears will be determined in the same manner as the masses for spiral bevel gears.

Acceleration Load on Hypoid Gears. We have the following for the determination of the acceleration load on hypoid gears.

When f_1 = force required to accelerate the masses as rigid bodies, lb

$\qquad f_2$ = force required to deform teeth by amount of effective error, lb

$\qquad f_a$ = acceleration load, lb

$\qquad m$ = effective mass of gears at point of mesh

$\qquad \cdot m_1$ = effective mass acting at pitch line of hypoid pinion

$\qquad m_2$ = effective mass acting at pitch line of hypoid gear

$\qquad e$ = measured or assumed error in action, in.

$\qquad C$ = load per inch of face to deform teeth by amount of error, lb

γ_p, γ_g = angle of generatrix of pinion and gear hyperboloids

$\qquad F$ = face width of gears, in.

$\qquad R_p$ = pitch radius of hypoid pinion, in.

$\qquad R_g$ = pitch radius of hypoid gear, in.

$\qquad R_{mp}$ = mean pitch radius of hypoid pinion, in.

$\qquad R_{mg}$ = mean pitch radius of hypoid gear, in.

$\qquad V$ = pitch-line velocity, ft/min

$\qquad \psi_c$ = effective spiral angle at middle of tooth face

$\qquad n$ = rpm of hypoid pinion

$\qquad W$ = applied tangential load, lb

$$R_{mp} = R_p - \frac{F}{2} \sin \gamma_p \qquad (21\text{-}31)$$

$$R_{mg} = R_g - \frac{F}{2} \sin \gamma_g \qquad (21\text{-}32)$$

$$m = \frac{m_1 m_2}{m_1 + m_2} \qquad (20\text{-}30)$$

$$V = 0.5236 R_{mp} n \qquad (21\text{-}33)$$

$$f_1 = Hm V^2 \cos^2 \psi_c \qquad (21\text{-}10)$$

For 14½-deg gears

$$H = 0.00086 \left(\frac{\cos \gamma_p}{R_{mp}} + \frac{\cos \gamma_g}{R_{mg}} \right) \qquad (21\text{-}27)$$

For 20-deg gears

$$H = 0.00120 \left(\frac{\cos \gamma_p}{R_{mp}} + \frac{\cos \gamma_g}{R_{mg}} \right) \qquad (21\text{-}28)$$

$$f_2 = FC \cos^2 \psi_c + W \qquad (21\text{-}34)$$

For 14½-deg gears

$$C = \frac{0.107e}{(1/E_1) + (1/E_2)} \qquad (21\text{-}29)$$

For 20-deg gears

$$C = \frac{0.111e}{(1/E_1) + (1/E_2)} \qquad (21\text{-}30)$$

$$f_a = \frac{f_1 f_2}{f_1 + f_2} \qquad (20\text{-}2)$$

Dynamic Load on Hypoid Gears. For the determination of the dynamic load on hypoid gears we have the following:

When W_d = dynamic tooth load, lb

$$W_d = W + \sqrt{f_a(2f_2 - f_a)} \qquad (20\text{-}39)$$

Example of Dynamic Load on Hypoid Gears. As a definite example we shall use the same values as those of the spiral-bevel-gear example with an offset or center distance of 1.500 in. This gives the following values:

$$N_p = 24 \qquad \gamma_p = 26.565° \qquad \cos \gamma_p = 0.89442 \qquad R_{bp} = 0.500$$
$$N_g = 48 \qquad \gamma_g = 63.435° \qquad \cos \gamma_g = 0.44721 \qquad R_{bg} = 1.000$$
$$F = 1.000 \qquad W = 300 \qquad n = 1{,}200 \qquad e = 0.002 \qquad m = 0.070 \qquad \psi = 30°$$
$$R_p = \sqrt{(0.500)^2 + (4.00)^2(0.500)^2} = 2.0615$$
$$R_g = \sqrt{(1.000)^2 + (2.00)^2(2.00)^2} = 4.1230$$
$$R_{mp} = 2.0615 - (0.50 \times 0.44721) = 1.8379$$
$$R_{mg} = 4.1230 - (0.50 \times 0.89442) = 3.6758$$
$$\psi_c = 30° + \sin^{-1}\left(\frac{1 \times 0.44721}{3.6758}\right) = 36.988°$$
$$\cos \psi_c = 0.79876 \qquad \cos^2 \psi_c = 0.63802$$
$$V = 0.5236 \times 1.8379 \times 1{,}200 = 1{,}155 \text{ ft/min}$$
$$H = 0.00075 \qquad C = 3{,}330$$
$$f_1 = 0.00075 \times 0.070 \times 1{,}155^2 \times 0.63802 = 44.7 \text{ lb}$$
$$f_2 = (3{,}330 \times 0.63802) + 300 = 2{,}425 \text{ lb}$$
$$f_a = \frac{44.7 \times 2{,}425}{44.7 + 2{,}425} = 43.9 \text{ lb}$$
$$W_d = 300 + \sqrt{43.9(4{,}850 - 43.9)} = 759 \text{ lb}$$

CHAPTER 22

BEAM STRENGTH OF GEAR TEETH

As noted earlier, the load-carrying ability of any gear drive may be limited by one or more of the following factors:

1. Heat of operation
2. Beam strength of teeth
3. Wear-load capacity of materials of gears

In other words, a satisfactory gear drive must have the ability to dissipate the frictional heat of operation, must have teeth sufficiently strong to carry the dynamic loads without breaking or shearing, and must be made of materials whose surface-endurance properties are adequate to carry the dynamic loads without excessive wear.

When we study the beam strength of the teeth, the gear teeth are considered as cantilever beams. The most severe conditions of loading would be when the full load is carried at the tips of the teeth. On the more accurate gears, the full load will not be carried there, because with a slight amount of elastic deformation, the load will be shared by a second pair of mating gear teeth. However, when the requirements of weight and size of gears are not critical, a condition that includes the great majority of gears used in machine design, we shall certainly be safe if we assume that the load may be carried on the tip of a single gear tooth.

BEAM STRENGTH OF SPUR-GEAR TEETH

Lewis Formula. Wilfred Lewis appears to have been the first to use the form of the gear tooth as one of the factors in a formula for the strength of gear teeth. This formula, which has become the one most widely used today, was presented in a paper read before the Engineers' Club of Philadelphia on Oct. 15, 1892. This formula is as follows:

When W_s = safe bending load on gear tooth, lb

s = safe working stress of material, psi

p = circular pitch of gear, in.

F = face width of gears, in.

y = tooth-form factor

$$W_s = spFy \qquad (18\text{-}1)$$

474

Tooth-form Factor. The tooth-form factor y is obtained by considering the gear tooth as a beam, fixed at one end and loaded at the other. These factors are usually determined graphically as follows: Referring to Fig. 22-1, when

    s = maximum fiber stress, psi
    b = thickness of beam, in.
    h = height of beam or tooth, in.

we have for such a beam

$$W_s = \frac{sFb^2}{6h} \qquad (22\text{-}1)$$

It can be shown by similar triangles in Fig. 22-1 that

$$\frac{x}{b/2} = \frac{b/2}{h}$$

whence

$$x = \frac{b^2}{4h}$$

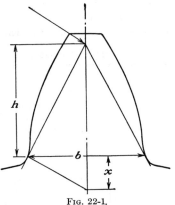

Fig. 22-1.

Substituting this value into Eq. (22-1), we have

$$W_s = sF\frac{2x}{3} = spF\frac{2x}{3p} = spFy$$

From whence

$$y = \frac{2x}{3p} \qquad (22\text{-}2)$$

Such tooth-form factors for several different gear-tooth systems are tabulated in Table 22-1. These values are for use when the load is assumed to be concentrated at the tips of the gear teeth.

Maximum Load at Middle of Tooth. A study of the dynamic loading of gear teeth indicates that the impact load acts shortly after the engagement of the succeeding pair of mating teeth, and appears to be imposed somewhere near the middle of the tooth depth. This impact load is the maximum momentary load, although under some conditions the maximum intensity of the acceleration load may be nearly equal to the intensity of the impact load. The acceleration load is imposed near the tip of the driven gear. However, at this point, when the elastic tooth deflection is equal to the error in action, this load will be shared by two pairs of teeth. In the tests on the Lewis gear-testing machine, the conditions of the tooth surfaces of the test gears indicated that the maximum punishment of the tooth surfaces occurred near the middle of the tooth depth. It shifted slightly with changes in speed, but never appeared near the tip of either of the two mating teeth. Hence when size and weight become more critical factors, we can use tooth-form factors for the more accurate gears that are based on the line of force passing through the middle of the tooth, as shown in Fig. 22-2. Values of y based on these conditions are tabulated in Table 22-2.

TABLE 22-1. VALUES OF TOOTH-FORM FACTOR LOAD AT TIP OF TOOTH

N	y					
	14½-deg form	14½-deg variable center distance	20-deg full-depth form	20-deg stub tooth form	Internal gears	
					Spur pinion	Internal gear
12	0.067	0.125	0.078	0.099	0.104	*
13	0.071	0.123	0.083	0.103	0.104	*
14	0.075	0.121	0.088	0.108	0.105	*
15	0.078	0.120	0.092	0.111	0.105	*
16	0.081	0.120	0.094	0.115	0.106	*
17	0.084	0.120	0.096	0.117	0.109	*
18	0.086	0.120	0.098	0.120	0.111	*
19	0.088	0.119	0.100	0.123	0.114	*
20	0.090	0.119	0.102	0.125	0.116	*
21	0.092	0.119	0.104	0.127	0.118	*
22	0.093	0.119	0.105	0.129	0.119	*
24	0.095	0.118	0.107	0.132	0.122	*
26	0.098	0.117	0.110	0.135	0.125	*
28	0.100	0.115	0.112	0.137	0.127	0.220
30	0.101	0.114	0.114	0.139	0.129	0.216
34	0.104	0.112	0.118	0.142	0.132	0.210
38	0.106	0.110	0.122	0.145	0.135	0.205
43	0.108	0.108	0.126	0.147	0.137	0.200
50	0.110	0.110	0.130	0.151	0.139	0.195
60	0.113	0.113	0.134	0.154	0.142	0.190
75	0.115	0.115	0.138	0.158	0.144	0.185
100	0.117	0.117	0.142	0.161	0.147	0.180
150	0.119	0.119	0.146	0.165	0.149	0.175
300	0.122	0.122	0.150	0.170	0.152	0.170
Rack	0.124	0.124	0.154	0.175		

* Internal gears with less than 28 teeth must be designed specially for the particular application, and their values of y must be determined for each one individually.

TABLE 22-2. VALUES OF y WHEN THE LOAD IS NEAR THE MIDDLE OF THE TOOTH

N	y				
	14½-deg form	20-deg full-depth form	20-deg stub tooth form	Internal gears	
				Spur pinion	Internal gear
12	0.113	0.132	0.158	0.207	*
13	0.120	0.141	0.164	0.208	*
14	0.127	0.149	0.172	0.209	*
15	0.132	0.156	0.177	0.210	*
16	0.137	0.160	0.184	0.211	*
17	0.142	0.163	0.187	0.215	*
18	0.146	0.166	0.192	0.218	*
19	0.150	0.170	0.196	0.222	*
20	0.153	0.173	0.200	0.225	*
21	0.156	0.176	0.203	0.228	*
22	0.158	0.178	0.206	0.230	*
24	0.162	0.182	0.211	0.233	*
26	0.166	0.187	0.216	0.236	*
28	0.170	0.190	0.219	0.239	0.400
30	0.172	0.193	0.222	0.242	0.395
34	0.176	0.200	0.227	0.246	0.387
38	0.180	0.207	0.232	0.250	0.380
43	0.183	0.214	0.235	0.253	0.372
50	0.187	0.221	0.241	0.256	0.364
60	0.192	0.227	0.246	0.260	0.356
75	0.195	0.234	0.252	0.264	0.348
100	0.198	0.241	0.257	0.268	0.340
150	0.202	0.248	0.264	0.272	0.332
300	0.207	0.255	0.272	0.276	0.325
Rack	0.210	0.262	0.280		

* Internal gears with less than 28 teeth must be designed specially for the particular application; their values of y must be determined for each one individually.

Double Tooth Contact. In very critical drives, special tooth proportions and tooth designs that give a contact ratio greater than two may be used. In such cases, we can make some reasonably close approximation to the way in which the load will be shared between the two pairs of mating teeth. The contact on the first pair of mating teeth will be near the tip of the tooth of the driving member when the contact on the second pair is near the middle of these teeth. With a definite amount of error, the tooth profiles of the highest pair must deform by the amount of this error before the second pair of teeth will begin to share this load.   If the

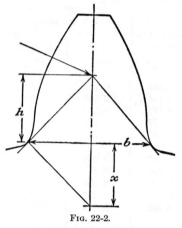

FIG. 22-2.

amount of deflection under a given load were the same at the tip and at the middle of the tooth, the remainder of the load would be shared equally. But the unit deflection at the tip is always greater than at the middle of the teeth.   Hence the contact at the tip will carry a proportionately lesser part of the remaining load than does the contact at the middle of the teeth.

The nature of the error may be such that the initial contact takes place at the tip or at the middle of the tooth.   Therefore we must assume that it may be at either place, determine the possible loads and stresses accordingly, and use the value that gives the greater bending stress. This will generally be when the initial contact is assumed at the tip of the tooth.

From the preceding chapter, we have

   C = load per inch of face to deform teeth by amount of error   lb
   d = deformation at middle of teeth under load W, in.
   W = applied tangential load, lb
   F = face width of gears, in.
   z_1 = deformation factor of pinion tooth
   z_2 = deformation factor of gear tooth
E_1, E_2 = modulus of elasticity of materials
   y = Lewis tooth-form factor
   e = measured or assumed error, in.

For the deformation at the middle of the mating gear tooth profiles we have

$$d = (W/F)[(1/E_1z_1) + (1/E_2z_2)] \qquad (20\text{-}13)$$
$$z = y/(0.242 + 7.25y) \qquad (20\text{-}14)$$
$$FC/W = e/d$$

whence

$$FC = eW/d \qquad (22\text{-}3)$$

In general, the deformation at the tip of the tooth will be about 150 per cent of the deformation at the middle.   Hence when one pair of teeth has been deformed by the amount of the error, the remainder of the load will be shared so that about 40 per cent of it will be applied at the tip of the first pair and 60 per cent at the middle of the second pair. Thus when

d_t = deformation at tip contact, in.

$$d_t = 1.50d \qquad (22\text{-}4)$$

Example of Double Tooth Contact. As a definite example we shall use the following assumed values:

$$
\begin{array}{llll}
W = 23{,}360 & W_d = 37{,}430 & e = 0.0005 & F = 2.750 \\
y_1 \text{ (at tip)} = 0.092 & y_1 \text{ (at middle)} = 0.231 & y_2 \text{ (at tip)} = 0.122 \\
y_2 \text{ (at middle)} = 0.210 & z_1 = 0.10121 & z_2 = 0.10830 & p = 0.5236 \\
E_1 = E_2 = 30{,}000{,}000 & d = 0.00541 & C = 785 & FC = 2{,}160
\end{array}
$$

We shall assume first that the high point is at the tip of the driving pinion. The load required to deform the teeth here by the amount of the error will be

$$\tfrac{2}{3}FC = \tfrac{2}{3} \times 2{,}160 = 1{,}440 \text{ lb}$$

Whence we have

	Pounds
Total tooth load..	37,430
Load to deform teeth to double contact.....................	1,440
Load to be shared.......................................	35,990
Carried at tip, 40% of shared load........................	14,396
Carried at middle, 60% of shared load.....................	21,594
Total load at tip........................ 1,440 + 14,396 =	15,836
Total load at middle.....................................	21,594

We shall now assume that the high point is at the middle of the tooth. The load required to deform the teeth here by the amount of the error will be $FC = 2{,}160$ lb. Whence we have

	Pounds
Total tooth load..	37,430
Load required to obtain double contact.....................	2,160
Load to be shared.......................................	35,270
Carried at tip, 40% of shared load........................	14,108
Carried at middle, 60% of shared load.....................	21,162
Total load at tip.......................................	14,108
Total load at middle.................... 2,160 + 21,162 =	23,322

From these values we have

$$
\begin{aligned}
\text{Maximum load at tip} &= 15{,}836 \text{ lb} \\
\text{Maximum load at middle} &= 23{,}322 \text{ lb}
\end{aligned}
$$

We shall now compute the bending stresses. Transposing the Lewis formula to solve for stress, we obtain

$$s = W/pFy \qquad (22\text{-}5)$$

For the load at the tip of the pinion tooth, we have

$$s = \frac{15,836}{0.5236 \times 2.75 \times 0.092} = 120,000 \text{ psi}$$

For the load at the middle of the pinion tooth, we have

$$s = \frac{23,322}{0.5236 \times 2.75 \times 0.231} = 70,150 \text{ psi}$$

In this example the smaller load at the tip of the pinion tooth develops the higher bending stress.

Stress Concentration at Fillet.   Whenever there is a rapid increase of section in a stressed body made of elastic material, there will be an increased local stress, or stress concentration, at the region of increase of section.   The intensity of this stress concentration depends largely upon the rate of change of section.   Thus the actual maximum local stresses at the root of a loaded gear tooth are larger than the average stresses as determined by any bending formula such as the Lewis equation.

Photoelastic studies of the stress concentrations at the roots of gear teeth have been made by several investigators.   Similar studies have been made upon the effects of the size of fillets on the stress concentrations in test bars.   These results have been compared with the results on similar metal bars subjected to fatigue tests.   Physical fatigue tests on soft steel indicate that this material yields under stress, and the material is stress- or work-hardened and strengthened at times to a greater or lesser degree, and thus it apparently shows a lower stress concentration than that obtained from photoelastic tests.   On hardened-steel samples, on the other hand, the results of the physical fatigue tests agree very closely with those of the photoelastic tests.   The influence of these stress concentrations is not apparent under simple tension or bending tests; it requires the repeated loading of the fatigue tests for the influence of these high local stresses to make themselves felt.

Because of the local yielding and work-hardening of the softer materials, and their apparently lower values of the stress concentration, gears made of such materials as cast iron, bronze, and soft steel would use a lower stress-concentration factor than those obtained from photoelastic tests.   For gears made of hardened steel, however, the full value of the photoelastic stress-concentration factors should be used.

In Bulletin 288, December, 1936, of the Engineering Experiment Station of the University of Illinois, written by Paul H. Black, are given the following stress-concentration factors for the fillets of gear teeth. These values were determined from photoelastic tests on from 12- to 24-tooth pinions, 14½-deg composite form, where

k_t = stress-concentration factor in tension
k_c = stress-concentration factor in compression

DP	k_t	k_c
4	1.47	1.61
5	1.47	1.61
6	1.42	1.57
7	1.35	1.50
8	1.345	1.500

Other investigators have noted that the stress concentrations at the fillet of the nonloaded or compression side of the gear tooth are larger than those on the loaded or tension side of the tooth. Some claim is made that there is no fatigue under compressive stresses, yet the great majority of broken gear teeth show the fracture extending from the fillet on the compression side into the gear blank and up and out on the loaded or tension side of the tooth above the fillet, as indicated in Fig. 22-3.

Stress Concentrations at Keyways. The foregoing photoelastic tests also included a study of the stress concentrations at the keyways of the bores.

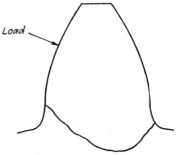

Fig. 22-3. Typical tooth fracture.

Black draws the following conclusions from the results of these photoelastic tests:

The best location of the keyway for solid spur gears on the basis of lowest stress is that in which the center line of the keyway is in line with the center line of a tooth space. The poorest location is that in which the center line of the keyway is in line with the center line of a tooth.

The maximum bore that should be used in a solid spur gear having Brown and Sharpe teeth for equal maximum tensile stresses in the gear, based on the photoelastic method, is given by the following equation, where

d = maximum allowable bore in gear, inches
D = pitch diameter of gear, inches
N = number of teeth on the gear

$$d = D(0.50 + 0.0344 \sqrt{N - 12}) \qquad (22\text{-}6)$$

This equation, which allows for a square key whose sides are one-quarter the diameter of the shaft, applies only when the center line of the keyway is in line with the center line of a tooth space. In addition, it strictly applies only for the numbers of teeth covered in this investigation, namely, 12 to 24. For gears having a larger number of teeth than 24, the equation may be used to obtain an approximate maximum bore by taking N equal to 24 in the equation.

Stress-concentration Factors for Gear-tooth Fillets. A later investigation at the University of Illinois was reported by Thomas J. Dolan.[1] Dolan gives as an equation that meets the results of his tests, the following: When r = radius of fillet, in.

t = thickness of tooth at root, in.

h = height of load position on tooth above weakest section, in.

For 14½-deg gears

$$k_t = 0.22 + (t/r)^{0.2}(t/h)^{0.4} \qquad (22\text{-}7)$$

For 20-deg gears

$$k_t = 0.18 + (t/r)^{0.2}(t/h)^{0.4} \qquad (22\text{-}8)$$

Dolan points out the fact that for purpose of design, any such values should also consider the behavior of the material, and should be carefully checked against laboratory tests of the materials and compared with actual service experience.

Fillet Stress Concentration. *First Example.* We shall apply Dolan's analysis to a definite example. We shall use the same gears that were used in the previous example. These are 20-deg gears of hardened steel. Judging from the results of other tests, the photoelastic stresses and the actual stresses in hardened steel appear to agree very closely.

The outline of this pinion tooth is the one shown in Fig. 22-3. We shall first determine the values of the stress-concentration factors for the load applied at the tip of the tooth and also for the load applied at the middle of the succeeding tooth. For this we have the following values:

$t = 0.337$ $r = 0.040$ h (at tip) = 0.400 h (at middle) = 0.200
$t/r = 8.425$ t/h (load at tip) = 0.8425 t/h (load at middle) = 1.685

With load at tip of tooth

$$k_t = 0.18 + (8.425)^{0.2}(0.8425)^{0.4} = 1.61$$

With load at middle of tooth

$$k_t = 0.18 + (8.425)^{0.2}(1.685)^{0.4} = 2.07$$

We shall now determine the intensities of the stress concentrations under the foregoing conditions. In the previous example, we obtained the following computed stresses: with load at tip of tooth, 120,000 psi; with load at middle of tooth, 70,140 psi. If the foregoing stress-concentration factors apply, we would have the following as a measure of the actual maximum local stresses:

With load at tip

$$120,000 \times 1.61 = 193,200$$

[1] DOLAN, THOMAS J., Influence of Certain Variables on the Stresses in Gear Teeth, *J. Applied Phys.*, Vol. 12, No. 8, pp. 384–391, August, 1941.

With load at middle

$$70,150 \times 2.07 = 145,200$$

If we increased the radius at the fillet to 0.60 in., we would have

$$t/r = 5.616 \text{ with all other factors the same.}$$

At tip

$$k_t = 0.18 + (5.616)^{0.2}(0.8425)^{0.4} = 1.49$$

At middle

$$k_t = 0.18 + (5.616)^{0.2}(1.685)^{0.4} = 1.92$$

Then the corresponding stress concentrations at the fillets would be

Stress with load at tip $= 178,800$ psi
Stress with load at middle $= 134,700$ psi

Second Example. As a second example we shall determine the stress-concentration factors for a 7-DP soft-steel gear of 20-deg full-depth tooth form. A layout of this profile will give the following values:

$$r = 0.033 \qquad t = 0.250$$

To tip

$$h = 0.285$$

To middle

$$h = 0.143$$

We shall determine the stress-concentration factors for the loading at these two positions.

$$t/r = 7.576$$

At tip

$$t/h = 0.877$$

At middle

$$t/h = 1.608$$

When the load is applied at the tip of the tooth

$$k_t = 0.18 + (7.576)^{0.2} (.877)^{0.4} = 1.69$$

When the load is applied at the middle of the tooth

$$k_t = 0.18 + (7.576)^{0.2} (1.608)^{0.4} = 1.89$$

With soft-steel gears, the plastic yielding of the material will reduce the intensity of the actual stress concentration, or its apparent intensity. This may be the result of increased physical properties because of the work-hardening. Values for this must be determined by experiment and by analyzing actual service data, wherever it may be obtained. It may be that the reduction of area of the material in a tensile test will give a measure of this factor. In the absence of more definite information, and judging from results on other types of machine elements made of soft steel, we shall assume that it will vary uniformly from about 40 per cent of the increase in the calculated or photoelastic value for steel of about 200 Brinell hardness to about 80 per cent for steel of about 400 Brinell hardness. We shall assume a hardness of 200 Brinell for this gear. Under these conditions we have

1. Increase by photoelastic test when load is applied at tip of tooth is equal to 0.60.

$$0.60 \times 0.40 = 0.24$$

We shall use $k_t = 1.24$.

2. Increase by photoelastic test when load is applied at middle of tooth is equal to 0.89.

$$0.89 \times 0.40 = 0.35$$

We shall use $k_t = 1.35$.

When more reliable test data is collected, it should be possible to establish some reasonably reliable relationship of this nature that would enable more accurate calculations or estimates to be made.

Working Stresses. One of the most difficult factors to select for the design of gears is that of a suitable value for the working stress for the material. One reason for this difficulty is the absence of precise information about the actual conditions of the transmitted load in service. For example, a thousand different lathes of a given size and make may be in service, yet no two of them will be subjected to the same working loads in their use and abuse throughout their useful life. To meet these conditions of uncertainty, we may assume the worst possible conditions of loading, and design accordingly. Again, we may asume an average condition of loading and then select a low working stress for the material to give us some margin with which to meet the extreme conditions.

In those cases where size and weight are not critical, we can so design as to be generally safe. Here, in many cases, we shall have much more load capacity than we actually need. If the volume of production is small, such a procedure is probably the best.

On the other hand, when the volume of production is large, and also where size and weight are critical, the only direct answer is to "try and see"; *i.e.*, to make experimental models and test them. In addition, the performance of the units in service must be watched and checked, because no laboratory test is ever a complete substitute for the actual conditions of service.

Values obtained from service on one type of mechanism can seldom be used safely on different types of mechanisms or widely different conditions of service unless their conditions of service and construction are substantially identical. Thus the manufacturers of any specialized type of product should establish from their own experience with the designs and materials they are actually using the values for the working stresses in their designs. Test data and the experience of others may give many valuable clues, but each manufacturer must himself determine his own limiting working-stress values.

In the case where size and weight are not critical, we can use the flectional-endurance limit of the material as the safe working stress in the equation for the beam strength of the gear tooth. When the gears are always loaded in the same direction, then the stress range will be from about zero to a maximum, and the endurance limit under such conditions is generally about 150 per cent of the endurance limit when the stress range is plus and minus a definite amount. In general, for soft steel, the endurance limit for reversed bending is about one-third of the ultimate strength in tension. With hardened steel, there does not appear to be a definite endurance limit. In such cases we must know the load-life characteristics of the materials and select a stress that will ensure the desired length of useful life.

Several texts are available on the subject of the fatigue of materials. The flectional-endurance limits of most of the materials commonly used in gear construction have been established and published. As design conditions become more critical, however, endurance tests should be made on the materials actually used, and these specific test values should be used as the basis for the selection of the working stress that is to be used in design.

When size and weight are critical, we must determine by experiment and experience some factors of use and abuse to guide us in our selection of the working stresses. For example, in a paper by J. O. Almen of the General Motors Corporation, presented before the American Gear Manufacturers' Association in 1941, is given the following schedule of the life requirements at maximum stress of certain mechanical elements as established by service experience:

	Cycles
Automobile rear-axle gears	100,000
Automobile transmission low gear	100,000
Automobile chassis springs	100,000
Automobile transmission second gear	300,000
Truck rear-axle gear	500,000
Bus rear-axle gear	1,000,000

Lewis Equation with Fillet Stress-concentration Factor. When a suitable working stress has been selected, the limiting beam load for the gear tooth can be computed from the Lewis equation and the fillet stress-concentration factor. This limiting beam load should be greater than the dynamic load to provide a margin of safety. Such a margin will be a measure of the additional load that can be carried. Some margin of safety is always desirable. With spur gears, a broken tooth means a failure of the gear drive. This margin of safety should be enough, at least, to cover the probable increase of error in action because of wear. The softer or the more plastic the materials of the gears may be, the greater the chances are that the error in action will increase with continued service.

The Lewis equation may be modified to include the fillet stress-concentration factor as follows:

When W_s = limiting beam load, lb

s = working stress of the material, psi

p = circular pitch, in.

F = face width of gears, in.

y = Lewis form factor

k_t = fillet stress-concentration factor, tension side

$$W_s = spFy/k_t \qquad (22\text{-}9)$$

Example of Limiting Beam Load on Spur Gear. As a definite example we shall use the following values: 24-tooth, 7-DP gear of 20-deg full-depth tooth form, made of soft steel, 200 Brinell hardness, with radius of fillet = 0.033 in. The values of k_t are from a previous example (page 483).

At tip

$$k_t = 1.24 \qquad y = 0.107$$

At middle

$$k_t = 1.35 \qquad y = 0.182$$
$$p = 0.4488 \qquad F = 2.000 \qquad s = 50{,}000 \text{ (for one-way loading)}$$

Whence we have for the load at the tip

$$W_s = \frac{50{,}000 \times 0.4488 \times 2.00 \times 0.107}{1.24} = 3{,}872 \text{ lb}$$

For the load at the middle of the tooth we have

$$W_s = \frac{50{,}000 \times 0.4488 \times 2.00 \times 0.182}{1.35} = 6{,}050 \text{ lb}$$

As the dynamic load is applied near the middle of the tooth, we shall use the last value as the limiting beam load for this gear. If this value is less than the dynamic load, we should either increase the circular pitch of the gears or increase the face width of the gears. If this load is very much greater than the dynamic load, we should decrease the circular pitch of the gears because the finer pitches are inherently smoother running and more efficient than the coarser pitches.

Load Distribution across the Face of the Gears.

If the gears were made of rigid materials, the tooth load would be distributed uniformly across the face of the gear. But the materials are elastic and deform under load. This deformation is of three major types: deformation of the gear teeth, torsional deflection of the gear blanks, and bending deflection of the gear blanks. Considering at first only the deformation of the gear teeth, including bending and compressive surface deformation, and the torsional deflection of the gear blank, we have the condition where the difference in the tooth deformation on opposite sides of the gear face will be equal to the torsional deflection of the gear blanks. When we introduce the effects of the bending of the gear blanks, this will tend to reduce the intensity of the loading at the middle of the gear blank and increase it correspondingly at the ends of the gear face.

All these variables are interdependent; they depend upon each other. The torsional deflection introduces a variable load distribution across the face of the gear. The nature of this load distribution influences the amount of torsional deflection. For example, if the load is distributed uniformly across the face of the gear, the total torsional deflection between the end where torque is applied and the free end would be equal to the torsional deflection that would result from the application of one-half the total load at the free end. Again, if the load varied uniformly across the face of the gear from a maximum at the driving end to zero at the free end, the torsional deflection at the free end would be equal to that developed by one-third of the total load applied at the free end of the gear blank.

A very similar relationship develops from the bending. If we ignore the torsional deflection for the moment, and assume a uniform distribution of the load across the face of the gear, the deflection at the middle of the face would be equal to that developed by the application of one-half the total load at the middle of the gear face, assuming bearings at each end of the face of the gear. This would reduce the tooth deformation at the center of the gear face with a corresponding reduction in the applied load there and an increase in the load intensity at the ends of the gear face. This in turn would influence the nature of the bending deflection.

Differential equations have been set up for this condition of loading Each drive requires a specific integration, because the position of the bearings in relation to the face of the gear affects the bending deflection. In general, these show a maximum intensity of loading at the driving end, a minimum intensity near the middle of the face, and some increase again toward the free end of the gear blank.

In most cases, we are interested primarily in the maximum intensity of the loading, which will always be at the driving end. In view of the fact that our knowledge of even the applied load conditions on most gear drives is only approximate, a simple approximation here for the load distribution across the face of the gears that will give a good approximation for the value of the maximum intensity of loading should be adequate for most uses. We shall therefore set up an approximation based on the following incomplete assumptions:

1. The torsional deflection of the gear blank will be so small that it can be ignored.

2. The bending deflection of both the gear and pinion blanks will also be ignored.

3. The torsional deflection of the pinion blank will be taken as equal to that developed by one-half the applied load at the free end of the pinion.

4. The difference in load intensity at the two ends of the pinion face will be measured by the difference in tooth deflection at opposite ends, which is equal to the torsional deflection of the pinion blank.

Torsional Deflection of Pinion Blank. For the torsional deflection of the pinion blank, we have the following:

Let R = pitch radius of pinion, in.

R_r = root radius of pinion, in.

W_d = dynamic tooth load, lb

T = torsional deflection at free end of pinion at pitch radius under load $W_d/2$, in.

F = face width of pinion, in.

E_s = shearing modulus of elasticity = 12,000,000 for steel = 6,000,000 for cast iron

When the pinion is solid

$$T = (W_d/2E_s)(2R^2F/\pi R_r^4) \tag{22-10}$$

When the pinion is hollow and

R_h = radius of bore, in.

$$T = (W_d/2E_s)[2R^2F/\pi(R_r^4 - R_h^4)] \tag{22-11}$$

Deformation of Gear Teeth. For the tooth deformation, we have the following:

When w' = load per inch of face at driving end of pinion, lb

w'' = load per inch of face at free end of pinion, lb

w = average load per inch of face, lb

W_d = total dynamic tooth load, lb

F = face width of pinion, in.

Δw = difference in unit load between that at driving end and that at free end, lb

E_1, E_2 = modulus of elasticity of materials

$$\Delta w = w' - w''$$

For 14½-deg gears

$$\Delta w = \frac{0.107T}{(1/E_1) + (1/E_2)} \tag{22-12}$$

For 20-deg full-depth tooth form

$$\Delta w = \frac{0.111T}{(1/E_1) + (1/E_2)} \tag{22-13}$$

For 20-deg-stub tooth form

$$\Delta w = \frac{0.115T}{(1/E_1) + (1/E_2)} \tag{22-14}$$

Load Distribution across Face.

$$w = W_d/F \qquad (22\text{-}15)$$
$$w = (w' + w'')/2$$
$$w' = 2w - w'' = w + (\Delta w/2) \qquad (22\text{-}16)$$
$$w'' = w' - \Delta w \qquad (22\text{-}17)$$

With this approximation, the value of w' appears to be slightly greater that that obtained from the use of the integration of the differential equation, while the value of w'' is appreciably smaller than the value obtained from a more exact analysis. However it is the value of w' that is the more critical and important one.

Examples of Load Distribution across Face of Gear. *First Example.* As a definite example we shall use the following: 48-tooth pinion of 8 DP, 20-deg full-depth form, 18-in. face width, solid steel pinion, with a dynamic load of 20,000 lb. Whence we have

$$R = 3.000 \qquad R_r = 2.850 \qquad W_d = 20,000 \qquad F = 18.000$$
$$T = \frac{20,000}{24,000,000} \frac{2 \times 9 \times 18}{3.1416 \times 65.975} = 0.0013$$
$$\Delta w = 0.111 \times \frac{0.0013}{2/30,000,000} = 2,165 \text{ lb}$$
$$w = \frac{20,000}{18} = 1,111 \text{ lb/(in. face)}$$
$$w' = 1,111 + \frac{2,165}{2} = 2,194 \text{ lb/(in. face)}$$
$$w'' = 2,194 - 2,165 = 29 \text{ lb/(in. face)}$$

In this example, the intensity of the load at the driving end is nearly double the average intensity of loading. The intensity of the loading at the free end is almost nothing. This face width is too wide for this diameter of pinion.

Second Example. As a second example we shall use the same pinion as before, but with a face width of 12 in. and a dynamic load of 13,333 lb. This gives the following values:

$$R = 3.000 \qquad R_r = 2.850 \qquad W_d = 13,333 \qquad F = 12.000$$
$$T = \frac{13,333}{2 \times 12,000,000} \frac{2 \times 9 \times 12}{3.1416 \times 65.975} = 0.00058$$
$$\Delta w = \frac{0.111 \times 0.00058}{2/30,000,000} = 966 \text{ lb}$$
$$w = \frac{13,333}{12} = 1,111 \text{ lb/(in. face)}$$
$$w' = 1,111 + {}^{96}\!6\!\!/_2 = 1,594 \text{ lb/(in. face)}$$
$$w'' = 1,594 - 966 = 628 \text{ lb/(in. face)}$$

Third Example. As a third example we shall use the following: 96-tooth pinion of 8 DP, 20-deg full-depth form, 24-in. face width, hollow steel pinion with 8-in.-diameter bore, with a dynamic load of 48,000 lb. Whence we have

$$R = 6.000 \qquad R_r = 5.850 \qquad R_h = 4.000 \qquad F = 24,000 \qquad W_d = 48,000$$

$$T = \frac{48,000}{24,000,000} \frac{2 \times 36 \times 24}{3.1416(1,171.265 - 256)} = 0.00092$$

$$\Delta w = \frac{0.111 \times 0.00092}{2/30,000,000} = 1,531 \text{ lb}$$

$$w = \frac{48,000}{24} = 2,000 \text{ lb/(in. face)}$$

$$w' = 2,000 + {}^{1531}\!/_2 = 2,765 \text{ lb/(in. face)}$$

$$w'' = 2,765 - 1,531 = 1,234 \text{ lb/(in. face)}$$

BEAM STRENGTH OF HELICAL-GEAR TEETH

The contact between helical-gear teeth is along a line that is at an angle to the trace of the pitch cylinder. The load along this line is not uniform, because of the difference in the amount of elastic deformation at the different heights on the teeth. However, the stress at the root of the teeth will be influenced more by the total load that is applied than by the local conditions of loading.

Because of this angular contact line, which travels across the face of the gear in an axial direction as the contact progresses, the ends of the teeth at either side of the gear face may be subjected to higher momentary loading at the beginning and ending of mesh. However, if there is an overlap on the face contact, elastic deformation here will bring a second pair of teeth into contact, and this second contact will be in a position away from the ends of the gear face.

The normal tangential load on helical gears is greater than the tangential load in the plane of rotation. But the length of the tooth along the helix is also greater than the face width, and here the influence of one factor will exactly counterbalance the other.

The beam strength of helical gears is usually calculated in the same manner as that for spur gears, using the normal circular pitch and the normal pressure angle and the virtual number of teeth in the helical gear. Thus when

W_s = limiting beam load for helical-gear teeth, lb

p = circular pitch in plane of rotation, in.

p_n = normal circular pitch, in.

F_a = active face width, in.

s = working stress of materials, psi

ψ = helix angle at pitch line

N = number of teeth in gear

N_v = virtual number of teeth in gear

y = Lewis tooth-form factor

k_t = stress-concentration factor, tension side

$$N_v = N/\cos^3 \psi \qquad\qquad (22\text{-}18)$$

The value of y will be determined from the virtual number of teeth and the normal pressure angle. These values are the same as those for spur gears which are given in Tables 22-1 and 22-2.

Although the conditions of local loading will make some difference in the value of the stress-concentration factor, in the absence of definite test data, we shall use the same values here as for spur gears. Whence we have

$$p_n = p \cos \psi \qquad (22\text{-}19)$$
$$W_s = sp_nF_ay/k_t \qquad (22\text{-}20)$$

In general, we can use values of y that are based upon the application of the dynamic load at the middle of the tooth form.

Working Stresses. We have the same problem here as to the selection of the safe or suitable working stresses as we have for spur gears. Where size and weight are not critical, we can use the flexional-endurance limits of the materials. Where these factors are critical, we must determine from experiment and experience suitable values for use.

The limiting beam strength of the gear tooth should always be greater by a suitable margin of safety than the dynamic load. The conditions of service are the determining factors for the extent of such margins of safety.

Example of Limiting Beam Load on Helical-gear Teeth. As a definite example we shall use the following values: 48-tooth pinion, 8 DP normal, 14½-deg normal tooth form, 30-deg helix angle, 10-in. active face width, steel of 200 Brinell hardness. This gives the following values:

$$N = 48 \qquad \psi = 30° \qquad F_a = 10.000 \qquad p_n = 0.3927$$
$$r = \text{radius of fillet} = 0.020 \qquad t = \text{thickness of tooth at base} = 0.220$$
$$h = \text{height to point of application of load} = 0.125$$
$$\frac{t}{r} = 11.00 \qquad \frac{t}{h} = 1.760$$
$$k_t = 0.22 + (11.00)^{0.2} (1.760)^{0.4} = 2.47$$

Because of the plastic yielding and work-hardening of this soft steel, the probable value of the stress-concentration factor may be about 40 per cent of this *increase* in stress.
Hence

$$1.47 \times 0.40 = 0.588$$

We will use $k_t = 1.59$.

$$N_n = \frac{48}{0.64952} = 73.9$$

whence

$$y = 0.195$$

We will use $s = 50,000$ (for one-way loading). whence we have

$$W_s = \frac{50,000 \times 0.3927 \times 10 \times 0.195}{1.59} = 24,080 \text{ lb}$$

Load Distribution across Face of Helical Gear. For single helical gears, the load distribution across the face of the gear will be very much the same as that for spur gears except that the form of the normal basic rack will be used to determine the normal tooth deformation, and this normal deformation must be converted into the deflection in the plane of rotation of the gears.

Torsional Deflection of Pinion—Single Helical Gears. For the torsional deflection of the pinion blank, we have the following:

Let R = pitch radius of pinion, in.

R_r = root radius of pinion, in.

W_d = dynamic tooth load, lb

T = torsional deflection at free end at R under load $W_d/2$, lb

F_a = active face width of pinion, in.

E_s = shearing modulus of elasticity

R_h = radius of bore in pinion, in.

When the pinion is solid, we have the following:

$$T = (W_d/2E_s)(2R^2F_a/\pi R_r^4) \qquad (22\text{-}10)$$

When the pinion is hollow, we have the following:

$$T = (W_d/2E_s)[2R^2F_a/\pi(R_r^4 - R_h^4)] \qquad (22\text{-}11)$$

Tooth Deformation of Single Helical Gears. For the tooth deformation of single helical gears, we have the following:

When w' = load per inch of face at driving end, lb

w'' = load per inch of face at free end, lb

w = average load per inch of face, lb

Δw = difference in unit loading between ends of pinion, lbs.

ψ = helix angle at pitch radius

E_1, E_2 = modulus of elasticity of materials

For 14½-deg normal gear-tooth form

$$\Delta w = \frac{0.107T \cos^2 \psi}{(1/E_1) + (1/E_2)} \qquad (22\text{-}21)$$

For 20-deg normal full-depth form

$$\Delta w = \frac{0.111T \cos^2 \psi}{(1/E_1) + (1/E_2)} \qquad (22\text{-}22)$$

For 20-deg normal stub tooth form

$$\Delta w = \frac{0.115T \cos^2 \psi}{(1/E_1) + (1/E_2)} \qquad (22\text{-}23)$$

Load Distribution across Face of Single Helical Gear. For the load distribution across the face of a single helical gear, we have the following:

$$w = W_d/F_a \tag{22-15}$$
$$w' = w + (\Delta w/2) \tag{22-16}$$
$$w'' = w' - \Delta w \tag{22-17}$$

Example of Load Distribution across Face of Single Helical Gear. As a definite example we shall use the values from the preceding example with a dynamic load of 20,000 lb. This gives the following values:

$$R = 3.464 \qquad R_r = 3.315 \qquad F_a = 10.000 \qquad W_d = 20,000$$
$$\psi = 30° \qquad \cos^2 \psi = 0.7500$$
$$T = \frac{20,000}{2 \times 12,000,000} \frac{2 \times 11.999 \times 10}{3.1416 \times 120.758} = 0.00052$$
$$w = \frac{20,000}{10} = 2,000 \text{ lb/(in. face)}$$
$$\Delta w = \frac{0.107 \times 0.00052 \times 0.75}{2/30,000,000} = 626 \text{ lb}$$
$$w' = 2,000 + {}^{626}\!\!/_2 = 2,323 \text{ lb/(in. face)}$$
$$w'' = 2,323 - 626 = 1,697 \text{ lb/(in. face)}$$

Load Distribution across Face of Herringbone Gears.

With herringbone gears, the teeth will tend to center themselves on the two helices of opposite hand, and so the effective torsional deflection will be reduced to one-half of that of the full active face width. Otherwise it will be the same as for single helical gears. Using the same symbols as before we have

For solid pinions
$$T = (W_d/2E_s)(R^2F_a/\pi R_r^4) \tag{22-24}$$

For hollow pinions
$$T = (W_d/2E_s)[R^2F_a/\pi(R_r^4 - R_h^4)] \tag{22-25}$$

Example of Load Distribution across Face of Herringbone Gear. Using the same example as before, but as a herringbone gear with a total face width of 10 in., we have

$$T = \frac{0.00052}{2} = 0.00026$$
$$w = 2,000 \text{ lb/(in. face)}$$
$$\Delta w = {}^{626}\!\!/_2 = 323 \text{ lb}$$
$$w' = 2,000 + {}^{323}\!\!/_2 = 2,162 \text{ lb/(in. face)}$$
$$w'' = 2,162 - 323 = 1,839 \text{ lb/(in. face)}$$

BEAM STRENGTH OF SPIRAL-GEAR TEETH

The contact between spiral-gear teeth is point contact, which moves from the bottom to the top of the tooth of the driver as the action progresses. Here the tooth load is concentrated at a point. The load capacity of spiral gears is small and is limited primarily by the ability of the material to resist excessive wear. It is seldom that the beam strength of spiral gears becomes a limiting factor. In any case of question, however, the beam strength of a spiral gear will be calculated in exactly the same manner as that for a single helical gear.

BEAM STRENGTH OF WORM-GEAR TEETH

Because of the many variations in the design and contact conditions on worm-gear drives, any general formula for the beam strength of the teeth of these gears can be but an approximation. The nature of the contact, for example, depends upon the lead angle and the thread angle of the worm and upon the position of the pitch plane of the worm in relation to the depth of the thread. The contact lines on the worm threads may vary from approximately concentric arcs from the worm center to lines that are almost radial to the worm center. Again, the form of the tooth of the worm gear across its face is constantly changing in size, in form, and in thickness. Thus at the edges of the gear face, the tooth forms are much thicker and shorter than they are at the central plane.

Furthermore, the beam strength of the teeth of worm-gear drives is a deciding factor only for slow-speed gears or when heavily loaded drives are used only intermittently. Here the limiting factor may be either the beam strength or the shear strength of the teeth. Thus we need some estimate of the bending strength and of the shear strength, and should always use the weaker of the two. The worm-thread form is always much stronger than that of the tooth of the worm gear, and so we shall consider only the probable strength of the tooth of the worm gear.

Limiting Load for Beam Strength of Tooth of Worm Gear. Although the number of teeth in the worm gear will have some influence on the limiting load for beam strength, we shall set up approximations and use a single tooth-form factor for each thread form of worm. In addition, these approximations will not be close enough to justify the use of any stress-concentration factors here, although we must not overlook the fact that we should always use as large a fillet as possible at the roots of the gear teeth to keep these stress concentrations to a minimum. These fillets are governed largely by the radius at the tip of the tooth of the hob.

On critical drives where a complete contact analysis is made, and the actual contours of the teeth of the worm gear are established, we can then determine values for the form factor and also for the fillet radius at the various sections. We could then check the stresses on the basis of the unit loading applied at the weakest sections, and obtain some measure of the probable maximum bending stress.

Thus when

W_s = limiting beam strength of tooth of worm gear, lb

p_x = axial pitch of worm, in.

p_n = normal pitch of worm, in.

λ = lead angle of worm

F = face width of worm gear, in.

s = safe working stress for material, psi

y = tooth-form factor

ϕ_n = normal thread angle of worm

$$p_n = p_x \cos \lambda \qquad (22\text{-}26)$$
$$W_s = s p_n F y \qquad (22\text{-}27)$$

We shall use the following values of y:

ϕ_n	y
14½	0.100
20	0.125
25	0.150
30	0.175

Working Stresses. In most cases, a worm-gear drive is not used where size and weight are critical factors. Here we can use the flectional-endurance limits of the materials for our working stresses. For a reversing drive, we should use the flectional-endurance limit for reversed bending. For a one-way drive, we can use the endurance limit for stresses from zero to a maximum value. These values will be about 50 per cent higher than those for reversed bending.

In the exceptional case where the size or weight of the worm-gear drive must be kept to a minimum, and some degree of limited life is acceptable, we must find by experiment the stress limits to which we must keep. In this case we are in the same position as before with spur and helical gears.

Materials for Worm Gears. The choice of materials for worm gears is more limited than for most other types of gears. The worm gear is never completely generated in the hobbing process to the degree necessary for satisfactory operation; the final generation or finishing will be by plastic flow and cold-working of the surface material of the enveloping member or worm gear during its running-in or during its initial service operation. Thus the helicoid member of the pair must be hard enough to hold its form while the enveloping member is being cold-worked to its final form. This limits the choice of materials for the enveloping member, which is usually the larger member of the pair, to the softer and more plastic materials. Cast iron is sometimes used for slow-speed worm-gear drives for the enveloping member, but it is not generally as satisfactory as most of the other materials because of its limited plasticity. Gear bronzes are the most widely used materials for the enveloping member. Some aluminum alloys are used for worm gears; they have excellent plastic qualities, but any failure of lubrication is generally disastrous. A cast-iron worm with an aluminum-alloy worm gear might be a good combination.

We shall therefore consider here the following: cast iron, gear bronze, and a tin-free antimony-copper alloy, which has excellent bearing characteristics but a low tensile strength. The physical properties of the different gear bronzes will vary. Bronze is one of the most temperamental of materials; a great deal depends upon the technique employed in its alloying, melting, and casting. In any case of question, specific tests should be made to determine its physical characteristics, particularly its fatigue characteristics. For general purposes, the following values may be used for the working stresses:

Material	Working stress, psi	
	One-way drive	Reversing drive
Cast iron......................	12,000	8,000
Gear bronze..................	24,000	16,000
Antimony bronze..............	15,000	10,000

Example of Limiting Beam Load for Worm Gear. As a definite example we shall use a hardened and ground steel worm and a phosphor-bronze worm gear with the following values: 6-start worm and 48-tooth worm gear, 1-in. axial pitch.

$$R_1 = 1.910 \qquad \lambda = 26.565° \qquad \phi_n = 30° \qquad L = 6.000 \qquad p_x = 1.000$$
$$p_n = 0.894 \qquad F = 2.250 \qquad y = 0.175 \qquad s = 24,000 \text{ (one-way drive)}$$
$$W_s = 24,000 \times 0.894 \times 2.250 \times 0.175 = 8,448 \text{ lb}$$

Shear Strength of Teeth of Worm Gear. At slow speeds, in particular, and when heavy shock loads must be carried, the limiting factor for the load on a worm-gear drive may be the shear strength of the worm-gear teeth. We shall therefore set up an approximation for this shear strength.

When F = face width of worm gear, in.

p_x = axial pitch of worm, in.

s_s = shear strength of materials, psi

A = approximate area of root section of tooth of worm gear, sq in.

we shall use the following approximations:

For 14½-deg worms

$$A = 0.60Fp_x \tag{22-28}$$

For 20-deg worms

$$A = 0.70Fp_x \tag{22-29}$$

For 25-deg worms

$$A = 0.75Fp_x \tag{22-30}$$

For 30-deg worms

$$A = 0.75Fp_x \tag{22-30}$$

When W'_s = limiting shear strength of tooth of worm gear, lb

$$W'_s = \tfrac{2}{3}As_s \tag{22-31}$$

We shall tentatively use the following values for the shear strength of the materials:

Material	s_s, *psi*
Cast iron	10,000
Gear bronze	10,000
Antimony bronze	6,000

Example of Shearing Strength of Tooth of Worm Gear. As a definite example we shall use the same worm-gear drive as before. This gives the following values:

$$\phi_n = 30° \qquad p_x = 1.000 \qquad F = 2.250 \qquad s_s = 10,000$$
$$A = 0.75 \times 2.250 \times 1.000 = 1.68 \text{ sq in.}$$
$$W'_s = \tfrac{2}{3} \times 1.68 \times 10,000 = 11,200 \text{ lb}$$

This value should be more, by a suitable margin of safety, than either the dynamic load or any momentary overload that may exist in starting.

The foregoing does not apply to a progressive shearing that may result from a fatigue crack that starts at the region where the pitch line crosses the root curve of the tooth of the worm gear, a place where the size of the root fillet is at its smallest value.

BEAM STRENGTH OF BEVEL-GEAR TEETH

If the materials of the bevel gears were rigid and the gears were rigidly mounted, the load would be distributed uniformly across the face of the gears. But both the materials and the mounting are elastic. Considering only the elastic deformation of the gear teeth, if the contact line extended is to pass through the cone center of the pair, the teeth must deflect more at the large ends of the gears than they do at the small ends. This would require a correspondingly greater unit load at the large end. We should consider now the elasticity of the mounting, because most bevel-gear drives have an overhung pinion. Elastic deformation here would tend to increase the unit load at the large end still more. Thus it should be evident that there is a variable load intensity across the face of the bevel-gear drive. This is one reason why it is desirable to use as narrow a face as possible on a bevel-gear drive. When the face width is adequate to carry the load, any increase here is a liability and not an asset. In no event should it be greater than one-third of the cone distance. One-quarter would be a better maximum value. It is good practice to make it as much less than this as is possible.

We can set up equations for the beam strength of bevel-gear teeth similar to those for spur gears. The original Lewis equation for bevel gears is as follows:

When W_s = limiting beam strength of bevel-gear teeth, lb

D = pitch diameter of bevel gear at large end, in.

d = pitch diameter of bevel gear at small end, in.

s = safe working stress for the material, psi

p = circular pitch at large end, in.

F = face width of bevel gears, in.

y = tooth-form factor, based on equivalent spur gear

$$W_s = spFy \frac{D^3 - d^3}{3D^2(D - d)} \tag{22-32}$$

or more simply

$$W_s = spFy \frac{d}{D} \tag{22-33}$$

which gives almost identical results when d is not less than two-thirds of D.

As the cone distance from any diameter is directly proportional to the diameter, we can use the following:

When A = cone distance from large end of gear, in.

γ = pitch-cone angle

R = pitch radius at large end of gear, in.

$$A = \frac{R}{\sin \gamma} \tag{22-34}$$

$$W_s = spFy \frac{A - F}{A} \tag{22-35}$$

The value of y, based on the virtual number of teeth, or the number of teeth in the equivalent spur gears of Tregold's approximation, will be the same as for spur gears. If the suggested system of $14\frac{1}{2}$-deg bevel gears are used, then the values of y for the $14\frac{1}{2}$-deg variable-center-distance system of spur gears would be used here. The Gleason Works have published such values for gears of their design. Thus when

N = number of teeth in bevel gear

N_v = virtual number of teeth in bevel gear

$$N_v = \frac{N}{\cos \gamma} \tag{22-36}$$

We also use the same stress-concentration factors for the bevel gears as are used on spur gears. Thus when

k_t = stress-concentration factor

$$W_s = spFy \frac{[(A - F)/A]}{k_t} \qquad (22\text{-}37)$$

Working Stresses. We must meet the same problems here about the working stresses and the margin of safety as we must meet on spur gears. Hence the discussion of these factors for spur gears applies directly to bevel gears also. Where size and weight are not critical factors, we can use the flectional endurance limits of the materials for our working stresses. Where these are critical factors, we must establish the working stresses by experiment and experience with the units in actual service.

Example of Limiting Beam Strength of Bevel Gears. As a definite example, we shall use a pair of 6-DP bevel gears of 24 and 48 teeth with a face width of 1 in. Both gears are of steel, 250 Brinell hardness. We shall use the following values:

$N_p = 24$ $N_g = 48$ $\gamma_p = 26.565°$ $\sin \gamma_p = 0.44721$ $\cos \gamma_p = 0.89414$
$F = 1.000$ $p = 0.5236$ $R_p = 2.000$ $s = 60,000$ (one-way drive)
Radius of fillet = 0.033   Height to point of loading = 0.340
Thickness of tooth at base = 0.315

$$k_t = 0.22 + \left(\frac{0.315}{0.033}\right)^{0.2} \left(\frac{0.315}{0.340}\right)^{0.4} = 1.54$$

As these materials are plastic, for the specified hardness we shall use a factor that gives only 50 per cent of the calculated *increase* in stress; whence

$$0.5 \times 0.54 = 0.27$$

We will use $k_t = 1.27$.

$$N_v = \frac{24}{0.89414} = 26.8$$

We shall use the value of y for the 14½-deg variable-center-distance system; whence

$$y = 0.117$$
$$A = \frac{2.000}{0.44721} = 4.472$$
$$W_s = 60,000 \times 0.5236 \times 1.00 \times \frac{0.117[(4\,472 - 1.00)/4.472]}{1.27} = 2,245 \text{ lb}$$

This value should be greater, by a suitable margin of safety, than the dynamic load.

BEAM STRENGTH OF SPIRAL-BEVEL-GEAR TEETH

The relationship between bevel gears with straight teeth and spiral bevel gears is essentially the same as that between spur gears with straight teeth and helical gears. We shall therefore set up equations for the beam strength of spiral bevel gears based on those for bevel gears with straight teeth, adjusted to the spiral angle. These same equations will also be used for hypoid gears where the gear member is substantially the same as that for a spiral-bevel-gear drive. Thus let

W_s = limiting beam strength of spiral-bevel-gear teeth, lb

p = circular pitch in plane of rotation at large end of gear, in.

p_n = normal circular pitch at large end of gear, in.

F = face width, in.

s = working stress of material, psi

ψ = spiral angle at middle of gear face

γ = pitch-cone angle of gear

N = number of teeth in gear

y = Lewis tooth-form factor

k_t = stress-concentration factor, tension side

A = cone distance from large end, in.

R = pitch radius at large end, in.

ϕ_n = normal pressure angle at middle of tooth face

The Gleason system of spiral bevel gears has its own tooth proportions, and values of y for these gears, based on the actual number of teeth in the gear, have been published by the Gleason Works. These published values are also used for the Gleason hypoid gears, although they are only approximate for the hypoid gears. In general, the values for the hypoid gears would be something larger.

The values for the stress concentration factor k_t will also be determined in the same manner as for spur gears.

$$p_n = p \cos \psi \tag{22-19}$$

$$W_s = \frac{s p_n F y \left[(A - F)/A \right]}{k_t} \tag{22-38}$$

The working stresses will be the same as those for spur, helical, and bevel gears. The limiting beam load should be greater than the dynamic load by a suitable margin of safety. The same margins of safety that may be found adequate for spur, helical, and bevel gears should also be suitable for the spiral bevel gears.

Example of Limiting Beam Strength of Spiral-bevel-gear Teeth. As a definite example we shall use a 6-DP spiral-bevel-gear drive of 24 and 48 teeth, 30-deg spiral angle, both gears of steel, 250 Brinell hardness. This gives the following values:

$$N_p = 24 \qquad \gamma_p = 26.565° \qquad \cos \gamma_p = 0.89442 \qquad R_p = 2.000$$
$$N_g = 48 \qquad \gamma_g = 63.435° \qquad \cos \gamma_g = 0.44721 \qquad R_g = 4.000$$
$$F = 1.000 \qquad \psi = 30° \qquad \cos \psi = 0.86603 \qquad \phi_n = 14.500°$$
$$p = 0.5236 \qquad s = 60{,}000 \text{ (one-way drive)} \qquad y = 0.124$$
$$\text{Radius of fillet} = 0.033 \qquad \text{Height to point of loading} = 0.285$$
$$\text{Thickness of tooth at base} = 0.285$$
$$k_t = 0.22 + \left(\frac{0.285}{0.033} \right)^{0.2} \left(\frac{0.285}{0.285} \right)^{0.4} = 1.76$$

As these materials are plastic, for the specified hardness we shall use a factor that gives only 50 per cent of the calculated *increase* in stress; whence

$$0.50 \times 0.76 = 0.38$$

We shall use $k_t = 1.38$.

$$A = \frac{2.000}{0.44721} = 4.472$$

$$p_n = 0.5236 \times 0.86603 = 0.4534$$

$$W_s = 60,000 \times 0.4534 \times 1.00 \times \frac{0.124 \,[(4.472 - 1.00)/4.472]}{1.38} = 1,900 \text{ lb}$$

This value should be greater, by a suitable margin of safety, than the dynamic load.

CHAPTER 23

SURFACE-ENDURANCE LIMITS OF MATERIALS

As pointed out before, the load-carrying ability of any gear dri e may be limited by excessive heat of operation, breakage of teeth, or excessive wear. We shall now consider the subject of wear on gear teeth. Our first need is some definition of the term *wear*. For the purposes of this discussion, we shall consider anything that alters the form, size, or surface smoothness of the gear-tooth profiles as wear. Thus we may have beneficial as well as destructive wear. The cold-working of the tooth surfaces of gears made of the softer and more plastic materials that results in smoother and harder tooth profiles without increasing the error in action is one type of beneficial wear. On plain bearings this condition is sometimes referred to as *running-in* or *wearing-in*. Destructive wear starts when the bearing surfaces begin to wear out, and the transition point is not always easy to establish.

Considering the character or the smoothness of the surface, there are six distinct types of wear that have been noted on gear-tooth profiles in service. We believe that we have identified the cause and effects of these types of wear, because we have set up conditions in the laboratory that we believe are responsible for these types of wear and have obtained results comparable with those observed in service. No claim is made, however, that we have observed and identified all types of wear. Many other conditions, or combinations of conditions, of wear have been observed in service that have not yet been classified or identified.

Types of Wear. Some of the identified types of wear appear to be caused by failure of the lubricant. Other types of wear appear to be substantially independent of the lubrication, although some of them may be accentuated by inadequate lubrication. We shall call these identified types of wear by the following terms:

1. Pitting
2. Abrasion
3. Scoring or cutting
4. Spalling
5. Galling or scuffing
6. Seizing

In addition, several other types have been noted, such as a flaking at the corners or edges of hardened-steel gears, and a burning effect that

502

appears on heavily loaded and high-speed gears. This last appears to be the result of the condition where the heat is created faster than the coolant can carry it away. Possibly the local temperature is so high that the lubricant does not remain in liquid form. For example, on hardened-steel gears when the product of the unit tooth load and the maximum sliding velocity, PV_s, exceeds a value of around 3,000,000, this condition appears. On soft-steel gears it may appear at a lower value. It appears to be some measure of the rate of creation of frictional heat.

Types of Wear Not Caused by Failure of the Lubrication. We shall consider first several of the foregoing types of wear that do not appear to be caused by failure of the lubrication, starting with pitting.

Pitting. There appear to be at least two general types of pitting: one a shallow surface failure, and the other a deeper destructive pitting. The surface pitting appears to be of several kinds. For one, there is an "incipient" or "superficial" pitting that appears to start as a crack on the surface, generally at right angles to the direction of rolling. When sliding is present, particularly on hardened steel under heavy loads, these surface cracks sometimes appear at an angle that reflects the influence of the sliding. These cracks are a possible result of the elastic wave ahead of the contact, which subjects the surface material to reversed bending. The resulting pits are shallow, seldom more than 0.005 in. deep, and the shapes of the craters appear to depend upon the structure of the material. Some are microscopic in size, only a few thousandths of an inch across; others are $\frac{1}{16}$ in. or more across, and of irregular shape. At all events, this incipient pitting does not appear to be the cause of any great concern. If the loads imposed are below the limiting loads for the materials, this incipient pitting appears early in the life of the gear, appears to progress to a certain extent, and then to cease.

Another type of pitting develops on some soft-steel gears. Cracks will start below the surface of the material at a depth of from 0.005 to 0.010 in., and eventually break to the surface. These may be of considerable extent. The depth is much less than the depth to the point of maximum shear. These cracks appear to be the result of the plastic flow of the surface material. The surface lamina tends to creep in the direction of rolling and sliding under repeated loads, a condition that, added to the shear stress set up by the compressive effect of the loading, results in cracks substantially parallel to the surface. Sometimes a double line of cracks develops, and with continued loading, the material between the two lines of cracks becomes shattered, apparently along the crystal boundaries of the materials. These conditions have become evident upon the microscopic examination of samples under test to determine their surface-endurance properties.

There are other types of surface pitting, some of which may destroy the usefulness of the gear. At times a severe surface pitting appears at the pitch-line area of the tooth profile, which may continue until the usefulness of the surface is destroyed. This appears to be more common on the softer materials. The whole subject of the surface disintegration of materials under rolling and sliding contact deserves much more attention than has yet been given to it.

Simultaneously with the foregoing phenomena, and probably with many more that have not yet been detected, shear stresses are repeatedly imposed upon the material, and the intensity of these shear stresses is different at different depths, the maximum shear stress being at some distance below the surface of the material. When the loads imposed develop stresses beyond the surface-endurance limits of the material, particles or flakes will be sheared out of the surface of the material. In tests to date, the thickness of these flakes has generally been equal to or greater than the depth to the point of maximum shear. With case-hardened steel, if the depth of case is not equal to about double the depth to the point of maximum shear, subsurface cracks develop along the line of case and core, and sections of the case are sheared out. In other words, the depth of case must be below the region of high shear stress.

The exact sequence of the different phenomena is a matter of question. Probably in some cases a crack may start under the surface in the region of high shear stress before the surface disintegration of the material has progressed very far. In other cases the reverse may be true. Again, both types of failure may progress together. Probably the physical characteristics of the materials influence these conditions.

Lubrication does not appear to be a controlling factor here. With sliding in particular, inadequate lubrication appears to accentuate the conditions and result in a lower surface-endurance limit. In these cases, the sliding appears to develop higher surface stresses in the direction of sliding so that the resultant combined stress is higher.

Abrasion. Abrasion is caused by the presence of foreign matter such as grit or metallic particles between the rubbing surfaces. On gear teeth it results in scratches or fine grooves spaced more or less at random and running in the direction of the sliding between the surfaces. Use of a heavier lubricant may reduce the effect of abrasion but will not eliminate it. If the abrasive is carried between the rubbing surfaces by the lubricant, the oil should be adequately filtered to prevent this condition. If the abrasive consists of small particles of the materials that have been released by the pitting, abrasion may be reduced or eliminated by reducing the unit loads sufficiently so that the particles are not released and so the abrasion will not take place. If the abrasive comes from some outside source, such as abrasive particles in the air of a cement mill or free abrasive released from grinding wheels in operation, effective guards should be provided to prevent the abrasive from reaching the contacting surfaces of the gears. Information as to the relative resistance of different materials to abrasion is of value, but as regards this type of wear in the majority of cases, the major problem is to design so as to avoid any likelihood of its occurrence.

Scoring or Cutting. Scoring or cutting takes place when, because of other types of wear, rough surface finish, misalignment of parts, or other imperfections, sharp corners or edges are present that cut through the oil film and score the mating surface. At times a scored surface may appear to be very similar to one that has suffered from abrasion. Generally, when scoring is present, some abrasion is present also, because the particles of metal gouged out will act as an abrasive. In general, scratches resulting from scoring will have a more regular pattern than those resulting from abrasion. Also the scratches are generally deeper. In the majority of cases, scoring is the result of poor workmanship, so that the most effective way to avoid it is by exercising more care in producing the mating surface.

One type of scoring on gear teeth takes place when, because of wear, error in tooth form or spacing, or poorly designed tooth profiles, coupled with deformation of the teeth under load, the entering tip of the tooth of the driven member makes premature contact with the flank of the tooth of the driving gear. This entering tip travels in a trochoidal path in relation to the driving gear, a path whose form almost coincides with the profile of the flank of the tooth of the driving gear, so that a very slight displacement because of error, wear, or deformation under load will permit premature contact.

When this condition exists, the flank of the tooth of the driving gear will be gouged away and will tend to be shaped to the concave form of the trochoidal path of the tip of the entering tooth, thus resulting in a worm tooth form of double curvature. On worm-gear drives, particularly when the thread angle is small in relation to the lead angle of the worm, a similar cutting often takes place, so that comparatively large flakes of the bronze of the worm gear are found at the bottom of the gear case.

Spalling. Spalling is a type of wear or surface failure that sometimes takes place on the more ductile materials. It occurs when the shear stresses set up by the movement of the elastic and plastic wave ahead of the contact area between the curved surfaces exceed the shear strength of the material. In many respects it is similar to some types of destructive pitting and results in the shearing out of flakes of material of appreciable size, but it does not appear to be a phenomenon that is caused by fatigue.

In some cases spalling occurs when heavily loaded gears made of soft materials are first operated, but often, continued operation results in the cold-working of the surface material, increasing its physical properties while reducing the amount of plastic deformation, so that the spalling ceases. In many cases, a suitable running-in period under increasing loads will cold-work the surface material so that this spalling will not take place.

With the softer and more plastic materials tested, a definite plastic flow of the surface material occurs, even though particles are not sheared out of the surface. Sometimes this plastic flow develops into a series of waves on the surface. With the introduction of sliding between the two surfaces in generally rolling contact, this corrugation effect is increased greatly. Inadequate lubrication will also accentuate this condition. Such an effect is often found on gear teeth in the form of a hollow at the pitch-line area of the driver and a ridge at the pitch-line area of the follower, at the place where the direction of the sliding reverses.

None of the four preceding types of wear is caused primarily by failure of lubrication. Some borderline cases may show up with inadequate lubrication that would not exist under more favorable conditions of lubrication, notably the plastic flow of the surface with the development of hollows and ridges. Any or all of them may exist under the best possible conditions of lubrication.

Types of Wear Caused by Failure of Lubrication. There are several types of wear that are directly caused by failure of lubrication. When adequate lubrication is present, none of them will exist. We shall now consider these types of wear, starting with galling.

Galling or Scuffing. Galling results from a momentary failure of the oil film, which sometimes causes high local temperatures and also a plastic flow of the surface of the material. When particles of material are dragged out of the surface by this action, some abrasion will also take place. Too heavy a lubricant on gear teeth often results in such galling because the time between successive contacts of the same teeth is not sufficient for the lubricant to flow back again over the contacting surfaces. In such cases the use of a lighter lubricant will often overcome this trouble.

Galling is difficult to prevent between sliding surfaces which have only a slight amount of motion between them or which have a frequently reversing direction of relative sliding because, under such conditions, a continuous film of oil cannot be formed and maintained. At the other extreme, galling often takes place when the loads are heavy and the rate of sliding is high because of the frictional heat and the necessity, on gears at least, of using relatively light lubricants under such operating conditions.   In such cases, the oil film is not strong enough to support the load, and as a consequence, momentary metallic contact between the sliding surfaces is of frequent occurrence.   This condition is often met at the tips of the teeth of gears running at high speed, and is often called a *wiredrawing effect*.   The high speed tends to throw the oil off from the tooth surfaces.   In such cases, a baffle may sometimes be arranged to block the space on the in-meshing side of the gears, and sufficient oil may be fed to minimize or to overcome this galling.

Seizing.   Seizing, in many respects, is an extreme case of galling. In this case, local temperatures are so high during the momentary failures of the oil film that particles of metal are actually welded or brazed onto the contacting surfaces.   On gear teeth, particles so welded to the tooth surfaces then act to score the mating tooth surfaces.   In the case of plain cylindrical bearings, such particles of material may actually weld the two members together in spots.   To minimize or overcome both galling and seizing, many special lubricants, commonly known as *EP* (*extreme-pressure*) *lubricants*, have been developed.   These may be divided into two major types: the one where some solid lubricant such as graphite or lead soap has been introduced, and the other where sulfur or chlorine or other element has been added to prevent welding.   Where sulfur or chlorine is added to the lubricant for this purpose, the element acts to contaminate the surface material and thus acts as an antiflux, preventing the particles from welding to the contacting surfaces.

Of the six foregoing types of wear, failure of lubrication is directly responsible for only the last two types, namely, galling and seizing.

"Gear-tooth Contact Wears Away Small Particles of Metal."   The statement is made in a standard specification for gears to the effect that every time a pair of gear teeth makes contact, small particles of material are worn away, thus resulting in a slow but relatively uniform wearing away of the surfaces.   Such a condition exists only in the case of abrasion, provided that the quantity and the quality of the abrasive is constant; or, in the case of scoring, provided that the cutting edges are not worn down or dulled.

In most cases, wear on gear teeth is of a periodic nature.   For example, if the tooth surfaces are rough, wear will be rapid at the start because of the scoring and the resulting abrasion.   After the rough surfaces or sharp edges have been worn down and the abrasive particles have worked free of the contacting surfaces, practically no further wear is evident for a greater or lesser period of time, depending upon the intensity of the tooth load.   Under heavy load, the next cycle of wear will start when more particles of material are released because of the surface fatigue of the material. Then further abrasion takes place, often followed by scoring or cutting, which is caused by the entering tip of the tooth of the driven gear, which gouges out the flank of the tooth of the driving gear. After this tip has cleared its path and the released particles of the material have worked free of the contacting surfaces, another period free from wear ensues until the same cycle is repeated again. Under suitable conditions of load and lubrication, gear teeth may run indefinitely without any signs of appreciable wear.

Influence of Mating Materials. In addition to the foregoing, the problem of wear is complicated by the interaction of different materials on each other. We learn from experience that certain combinations of materials work well together, while other combinations do not. Furthermore, with a given construction, a combination of materials that works well in certain respective positions may not be satisfactory when the materials used for the mating parts are reversed. For example, cast iron and bronze prove by laboratory tests to be an excellent combination. A cast-iron worm mating with a bronze worm gear shows in service a high resistance to wear, much better than a hardened-steel worm and a bronze worm gear. Yet a bronze worm mating with a cast-iron worm gear shows in service a very poor resistance to wear. In this case, our experience with these and other combinations of materials teaches us that the enveloping member or worm gear should always be made of the more plastic material of the combination used, so that it can find itself by plastic flow and cold-working during the initial running-in period.

Other conditions constantly occur where a plausible explanation of cause and effect is not so evident. Thus, for example, why does phosphor bronze operate well as regards resistance to wear with cast iron and hard steel but poorly with soft steel, bronze, or phenolic laminated material? Also, why does soft steel operate well with cast iron, babbitt, soft brasses, and sometimes with hardened steel but not with bronze, soft steel, or phenolic laminated material? Why does hardened steel operate well with the soft bronzes, brass, cast iron, babbitt, phenolic laminated material, and often with hardened steel, but not with the harder heat-treated alloy bronzes? Oil-hardened steel does not always operate well with casehardened steel. Again, two hardened nickel steels do not always operate well together, particularly when any appreciable heat exists. Also difficulties sometimes exist when a hardened nickel steel is mated with a nickel-bronze worm gear. Why does cast iron generally operated well with all other materials? Even so, cast iron in the "as cast" condition is not always as satisfactory as it is when suitably heat-treated.

Importance and Uses of Endurance Limits of Materials. Logically, the working stresses used for design of most machine parts should be based upon the endurance limits of the materials rather than upon their ultimate strength, elastic limit, or yield point, particularly if the parts are to render as long a service as possible. The endurance limit is that unit stress which can be repeatedly imposed indefinitely without causing a fatigue failure. To ensure long life to any mechanism, the stresses imposed in actual service should be kept within the endurance limits of the materials employed in the construction of the mechanism.

In many respects, most materials appear to have much in common with human muscles; when they are stressed within their endurance limits, their strength increases up to some optimum with repeated stressing or exercise; when they are stressed beyond their endurance limit with repeated stressing, they will be permanently weakened, and eventually they will fail by fatigue.

If the endurance limits, both flectional and surface, of materials are known and used as the basis for the working stresses, then the designer can design a given mechanism so that it will stand up against both breakage and excessive wear. Or again, if a limited life is acceptable, the designer can proceed with his task accordingly and with much more assurance than is otherwise possible.

However, at present, in the great majority of cases, the designer is ignorant of one of the most important factors of machine design. This is the knowledge of the actual intensity of the dynamic or maximum momentary loads that exist on various parts of the mechanism under the service conditions. Even the value of the average loads is often unknown. Today, this part of machine design is experimental and empirical almost without exception.

Designs have been developed by trial and experiment until they perform well in certain limited fields. If that mechanism, however successful it may be in its usual application, is used under more severe conditions than before, often it proves to be inadequate for such hew use. For example, designers of motor trucks have developed from years of experience designs that prove adequate for commercial hauling, most of which is done on hard roads. If these same vehicles, however, are used in lumbering operations, or to bring supplies to an army in the field, over poor or wrecked roads or no roads at all, the chances are that the average service conditions here would impose average loads of double, or more than double those of commercial hauling. Under these conditions, with the many hardened-steel parts involved in the construction of the trucks, the useful life of such vehicles would probably be very much less than their life when used commercially.

In the same way, considerable uncertainty exists as to the real load capacity of many other mechanical units. Take, for another example, geared speed reducers. Here may be one that has run successfully for several years transmitting the power from, say, a 10-hp motor, to a given machine. This does not mean, however, that the average transmitted power has been 10 hp. It may have been nearer 2 or 3 hp. Yet the inclination is to rate such a speed reducer at the same power as the rating of the motor to which it has been attached.

The designer may know the extent of the useful or applied load on the several parts of a given machine, but the actual intensity of the dynamic load is generally a profound mystery. In the experimental stages, a part may break. This gives some measure of the stresses existing in that part, but such breakage often causes damage to some other parts of the mechanism, often including the breakage of other parts. Under such conditions it is often difficult or impossible to determine which part broke first.

If the surface fatigue characteristics of the materials used are known and if the surfaces are carefully watched, dynamic loads that cause surface failures can generally be detected before any harm is done to any other part of the mechanism, and a reasonably close measure of the intensity of the dynamic load can be made from the condition and appearance of the surface failure of the particular part. In other words, the surface of the material itself, on which is imposed the dynamic loads, can be used as an indicator or measure of the extent of these loads. In cases where no surface failure is evident, a part of weaker surface strength may be substituted for the original one until the desired information is obtained. To my mind, the most effective research laboratory for the study of such dynamic loads, and many other features of the design, is the performance of the product itself in actual service. Thus, a greater knowledge of the surface-fatigue characteristics of materials will provide a powerful tool for the study of the dynamics of many types of mechanisms and a measure of the actual intensities of the dynamic loads existing there under the actual conditions of use and abuse, a subject of which today our knowledge is very incomplete.

Surface-endurance Limits of Metals. Since 1931, the ASME Special Research Committee on the Strength of Gear Teeth has been conducting tests to determine the surface-fatigue characteristics of various metals that are used in the construction of gears. Such tests take time, and to date but a very small part of the field has been covered. The following is a summary of the work covered to date:

Tests on Cast-iron Alloys. The most extensive series of tests to date on the surface-endurance limits of metals has been made on several cast-iron alloys.   The results of these tests, referred to a common basis of two rolls in contact, each with a 1-in. radius (2-in. diameter) and a 1-in. face are shown in the following charts.   These are plotted against logarithmic coordinates.   The applied load is plotted against the number of cycles of stress.   It appears from these tests that the endurance limit is reached at about 30,000,000 cycles of stress.   All tests were run in combination with a hardened and ground steel roll unless otherwise noted.

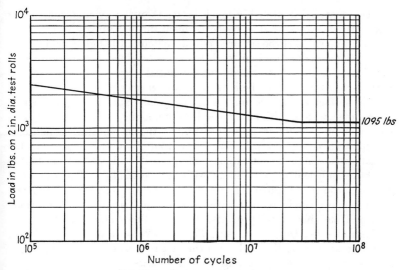

Fig. 23-1.   Cast iron with steel scrap.

Cast Iron with Steel Scrap.   Figure 23-1 shows the results of tests on gray iron with 30 per cent steel scrap.   The chemical analysis of a sample of this material is as follows:

Element	Per cent
Silicon	1.84
Sulfur	0.136
Manganese	0.65
Phosphorus	0.387
Total carbon	3.25
Graphite	2.80
Combined carbon	0.45

The results of tests of the physical properties of this material, both as cast and heat-treated, are as follows:

	As cast	Heat-treated*
Ultimate strength...............	35,200 lb/in.^{2}	45,950
Elastic limit....................	15,000 lb/in.^{2}	35,750
Brinell hardness.................	223	255
Flectional-endurance limit.........	21,000 lb/in.^{2}	25,000

* The heat-treatment was as follows: Heat to 1500°F and quench in oil.   Draw at 950 to 1000°F.

This material is probably much better than the average run of cast iron with steel scrap.   The heat-treatment did not appear to have any effect on the value of the surface-endurance limit, although it did increase the flectional-endurance limit slightly.   It is possible that the cold-working of the "as cast" material during the initial stages of the tests accomplished the same purpose here as the heat-treatment.

The surface-endurance limit of this material, under pure rolling conditions, appears to be with an applied load of about 1,095 lb.   The equivalent maximum specific compressive stress, on the basis of static conditions, is equal to about 87,000 psi.   We know from photoelastic tests that the actual stresses here are somewhat higher, possible 10 per cent, but we do not have sufficient data to speak with any certainty.   Hence we shall use for purposes of comparison the equivalent static load stresses.   These samples showed an increase in hardness of surface because of cold-working varying from about 5 to 20 points, Brinell hardness number.

NICKEL CAST IRON, AS CAST.   Figure 23-2 shows the test results on gray iron alloyed with nickel, with the test rolls as cast.   The chemical analysis of a sample of this material is as follows:

Element	*Per cent*
Silicon...	1.42
Sulfur...	0.117
Manganese...	0.37
Phosphorus..	0.448
Total carbon...	3.36
Graphite...	2.50
Combined carbon.....................................	0.86
Nickel...	1.52

The results of tests of the physical properties of this material are as follows:

Ultimate strength................................	35,400 lb/in.^{2}
Elastic limit......................................	12,600 lb/in.^{2}
Brinell hardness...................................	217
Flectional endurance limit..........................	16,000 lb/in.^{2}

The surface-endurance limit of this material under pure rolling conditions appears to be with an applied load of about 685 lb. The equivalent specific compressive stress, based on static loading, is about 69,200 psi. These samples showed an increase in surface hardness because of cold-working of from 5 to 15 points, Brinell hardness number.

Samples of this material were tested with about 9 per cent sliding action and gave substantially the same results as for pure rolling.

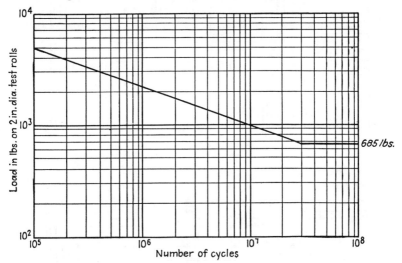

Fig. 23-2. Nickel cast iron as cast.

NICKEL CAST IRON, HEAT-TREATED. Another series of tests was made with this same nickel cast iron when heat-treated as follows: Heat to 1500°F and quench in oil; draw to 980°F.

The results of tests of the physical properties of this material are as follows:

Ultimate strength.................................. 41,700 lb/in.²
Elastic limit...................................... 24,000 lb/in.²
Brinell hardness..................................   246

The surface-endurance limit of this material under pure rolling conditions appears to be with an applied load of about 822 lb. The equivalent maximum specific compressive stress, based on static loading, is about 75,800 psi. The results of these tests are shown in Fig. 23-3.

Another series of tests on this material, heat-treated as before, but drawn to about 350 and 400 Brinell hardness, was made. The results were about the same for both degrees of hardness. The surface-endurance limit appears to be with a test load of about 960 lb. The equivalent

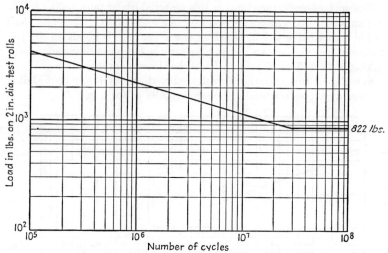

FIG. 23-3.   Heat-treated nickel cast iron, 300 Brinell.

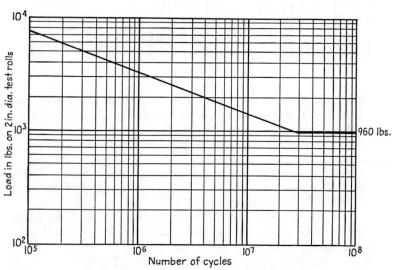

FIG. 23-4.   Heat-treated nickel cast iron, 350 to 400 Brinell.

maximum specific compressive stress, based on static loading, is about 81,900 psi. These samples showed no definite increase in surface hardness because of cold-working. The results of these tests are shown in Fig. 23-4.

NICKEL CAST IRON, HOT-QUENCH TREATMENT. While the foregoing tests on heat-treated nickel cast iron were being made, experiments were being conducted by the company supplying the material on different

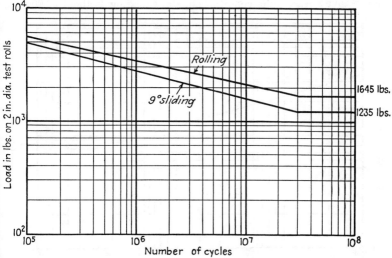

FIG. 23-5. Hot-quenched nickel cast iron.

heat-treatments. The following heat-treatment[1] was developed, which changed the structure of the material, particularly the form and dispersion of the graphite, and gave much higher physical properties to the material: Heat to 1500°F and hold until thoroughly heated; quench to 650°F and hold until heat is uniform; cool in boiling soda water. For the test rolls, the parts are held at temperature for about 30 min.

This interrupted-quench treatment gave the following physical properties to the test samples:

Ultimate strength.................................... 46,500 psi
Elastic limit.. 25,000 psi
Brinell hardness.....................................   287
Flectional endurance limit........................... 19,000 psi

The results of the surface-endurance tests for this material are sʰ ᴧ in Fig. 23-5. The surface-endurance limit under pure rolling cᴏˑ   ⠆s

---

[1] Patents have been applied for on this heat-treatment.

appears to be with a load of about 1,645 lb. The equivalent maximum specific compressive stress, under conditions of static loading, is about 107,300 psi. These samples showed an increase in surface hardness because of cold-working of from 5 to 40 points, Brinell hardness number.

Some of these samples were tested with about 9 per cent sliding action. The surface-endurance limit under these conditions appears to be with an applied load of about 1,235 lb. The equivalent maximum specific compressive stress, under conditions of static loading, is about 92,900 psi.

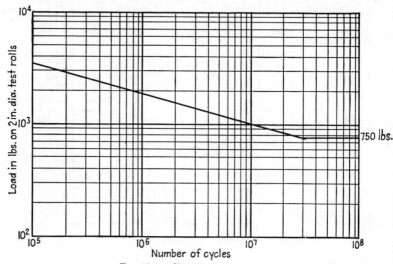

Fig. 23-6. Chrome-nickel cast iron.

CHROME-NICKEL CAST IRON. Figure 23-6 shows the results of tests on chrome-nickel cast iron, both as cast and heat-treated. The results in both cases were substantially the same. The chemical analysis of a sample of this material is as follows:

Element	Per cent
Silicon	1.24
Sulfur	0.130
Manganese	0.44
Phosphorus	0.297
Total carbon	3.39
Graphite	2.50
Combined carbon	0.89
Nickel	1.44
Chrome	0.50

Tests of the physical properties of this material gave the following results:

	As cast	Heat-treated*
Ultimate strength..................	39,000 psi	44,730 psi
Elastic limit......................	15,100 psi	31,600 psi
Brinell hardness..................	234	243

* The heat-treatment was as follows: Heat to 1550°F; quench in oil; draw to desired hardness.

The surface-endurance limit under pure rolling action appears to be with an applied load of about 750 lb. The equivalent maximum specific compressive stress, based on static loading, is about 72,400 psi. These

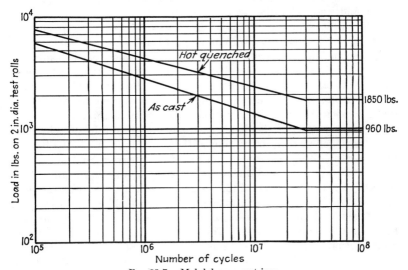

Fig. 23-7. Molybdenum cast iron.

samples showed an increase in surface hardness because of cold-working of from 0 to 15 points, Brinell hardness number.

MOLYBDENUM CAST IRON, HOT-QUENCH TREATMENT. Figure 23-7 shows the results of tests on molybdenum cast iron, heat-treated. The heat-treatment used was the same interrupted quench as was used on the nickel cast iron. The chemical analysis of a sample of this material is as follows:

Element	Per cent
Silicon...	1.77
Manganese..	0.59
Total carbon..	3.12
Molybdenum..	0.66

The physical properties of this material, heat-treated as noted, are as follows:

Ultimate strength...................................... 49,000 psi
Elastic limit... 22,400 psi
Brinell hardness...................................... 290
Flectional endurance limit............................ 22,000 psi

The results of the surface-endurance tests on this material, both as cast and heat-treated, are shown in Fig. 23-7. As cast, the surface-endurance limit under pure rolling action appears to be with an applied load of about 960 lb. The equivalent maximum specific compressive stress, under static loading, is about 82,200 psi.

The surface-endurance limit of the heat-treated material under pure rolling action appears to be with an applied load of about 1,850 lb. The equivalent maximum specific compressive stress, under static loading, is about 113,700 psi. These samples showed an increase in surface hardness because of cold-working of from 15 to 40 points, Brinell hardness number.

This material is the best of all the cast irons we have tested. It has given excellent results in service on a wide variety of applications.

Tests on Bronze. Tests have been made on several lots of phosphor gear bronze, all of them being substantially the same as SAE-65 bronze. One lot was cast against a chill; all the other lots were sand-cast. One lot was cast in an iron foundry, and the metal was overheated when poured. Some tests were run against hot-quenched nickel-cast-iron rolls; all the other tests were run against hardened and ground steel rolls.

PHOSPHOR BRONZE. The test results of all the tests of phosphor bronze are shown in Fig. 23-8. The surface-endurance limit, under pure rolling action, of the sand-cast bronze appears to be with an applied load of about 590 lb. The equivalent maximum specific compressive stress, under static loading, is about 67,600 psi. These samples showed an increase in surface hardness because of the cold-working of from 50 to 80 points, Brinell hardness number. The initial Brinell hardness number was about 80.

The surface-endurance limit of the bronze cast against a chill under pure rolling action appears to be with an applied load of about 1,025 lb. The equivalent maximum specific compressive stress, under static loading, is about 82,800 psi. These samples showed an increase in surface hardness because of cold-working of from 80 to 100 points, Brinell hardness number. The initial Brinell hardness number was about 80.

The surface-endurance limit of the overheated bronze, sand-cast, under pure rolling action was with an applied load of about 100 lb. The equivalent maximum specific compressive stress, under static loading, is about 25,000 psi.   These samples were etched and broken after the tests. Very large crystals were apparent.   On the outside surface where destructive pitting was present, no crystal structure could be seen, indicating amorphous material where the surface had started to disintegrate.   This is an indication of the care and technique required in the casting of bronze.   These samples showed an increase in surface hardness because of cold-working of from 10 to 70 points, Brinell hardness number.

The surface-endurance limit of the sand-cast bronze running against a hot-quenched nickel-cast-iron roll appears to be with an applied load

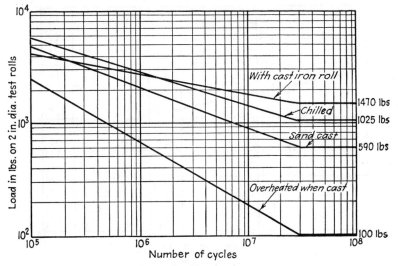

Fig. 23-8.   Phosphur bronze.

of about 1,470 lb.   The equivalent maximum specific compressive stress, under static loading, is about 83,300 psi.   This load is much higher than can be accounted for because of the lower modulus of elasticity of the mating cast-iron roll as compared with that of the hardened-steel roll. It is possible that the more effective lubrication because of the free graphite released from the cast-iron roll was largely responsible for this increase in load-carrying ability.   Current tests on soft steel indicate that when more effective lubrication exists, the influence of sliding and creep, because of the elastic and plastic wave at the contact area, is reduced, and the surface-endurance limit is increased materially.   These bronze samples showed an increase in surface hardness because of cold-working of from 50 to 100 points, Brinell hardness number.

NICKEL BRONZE. Tests were also made on a nickel bronze, sand-cast, running with a hardened and ground steel test roll. The surface-endurance limit of this material under pure rolling action appears to be with an applied load of about 820 lb. The equivalent maximum specific compressive stress, under static loading, is about 75,700 psi. These samples showed an increase in surface hardness because of cold-working of from 5 to 20 points, Brinell hardness number. The initial Brinell hardness number was about 80. The results of these tests are shown in Fig. 23-9.

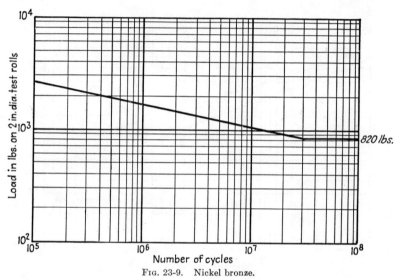

FIG. 23-9.   Nickel bronze.

Soft Steel. No comprehensive tests, as yet, have been completed on soft steel. From the tests on such materials to date, we obtain some indication of the difficulties that will be met here. On the softer steels, under 200 Brinell hardness number, for example, corrugations or waves are developed on the surface, particularly if any sliding action is present, under relatively light loads. When the loads are light enough to avoid this plastic flow of the surface material, they are generally well within the surface-endurance limits of the materials. After a preliminary run under a light load, the surface material becomes hardened by cold-working. Then heavier loads can be imposed without developing corrugations on the surface. This process can be repeated until the surface hardness of the material has been increased 100 points or more, Brinell hardness number. Then if these preworked samples are tested for their endurance limits, this value will depend more upon the cold-worked conditions than they do upon the original condition of the material. It may thus be necessary to establish surface-endurance limits for such soft steels on the basis of various degrees of running-in or cold-working. If so, when such values are applied to wear loads for gear teeth, these gear teeth must receive an equivalent running-in before they will be able to carry the full rated load.

Tests are now under way on these materials. Such tests take time, so that progress is slow. It will probably be several years before many definite values can be reported. In the meantime, tentative values have been established on the basis of actual service performance and have been used successfully in gear design for some years past. Many or most of them may be more conservative than necessary. These values are listed in a table in Chap. 24. Some degree of running-in is probably present in these values, as the full rated load on gears is seldom applied continuously in service. Without such running-in, the use of steels below 200 Brinell hardness number for gears is always hazardous.

Hardened Steels. Two series of tests on hardened steels have been made: one on casehardened steel and the other on induction-hardened steel. This is but a small start on this project. As noted before, hardened steel does not appear to have any definite endurance limit. Tests up to 400,000,000 cycles of stress show the limit load reducing, following the same line of the tests at the heavier loads and fewer cycles, with the increasing number of cycles.

Both of these two series of tests on hardened steel were run with similar material in the mating rolls and with about 10 per cent sliding action between them.

CASEHARDENED STEEL. The results of the tests on casehardened steel are shown in Fig. 23-10. This steel was substantially SAE 2515. Sets of samples were prepared with three depths of case: about 0.030 in., about 0.040 in., and about 0.055 in. The heat-treatment was as follows: Carburize at 1650°F; quench in oil at 1425°F; draw one hour at 290°F. The physical properties of the core material tested as follows:

	0.030-in. case	0.040-in. case	0.055-in. case
Ultimate strength, psi	179,000	182,000	181,000
Yield point, psi	161,000	164,500	166,000
Elongation, per cent	15.2	16.8	16.8
Reduction of area, per cent	63	62	62
Core hardness, Rockwell C	38	37	37
Case hardness, Rockwell C	60	56	56

All the samples of the 0.030-in. case failed at the line of case and core. All the other samples failed in the case. The results of these tests, based

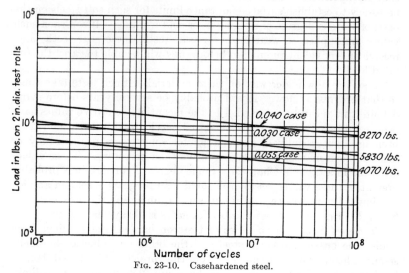

Fig. 23-10. Casehardened steel.

on the applied loads on a pair of 2-in.-diameter test rolls with a 1-in. face are as follows:

Number of cycles	Applied loads		
	0.030-in. case	0.040-in. case	0.050-in. case
1,000,000	8,360	12,890	5,920
10,000,000	6,960	9,840	4,880
100,000,000	5,830	8,270	4,090

The equivalent maximum specific compressive stresses, based on static loading, of these samples is as follows:

Number of cycles	Maximum specific compressive stress		
	0.030-in. case	0.040-in. case	0.055-in. case
1,000,000	296,000	362,000	249,300
10,000,000	270,300	321,400	226,300
100,000,000	247,400	294,600	207,200

The surfaces of these test rolls were ground and lapped. There is reason to believe that the slope of these surface-fatigue graphs for hardened steel is influenced by the character of the surface finish, the smoother hardened-steel surfaces having the lesser slope.

In these tests, the 0.040-in. case depth gave the best results. The shallower case failed at the line of case and core, and did not develop the full strength of the case. Here the region of high shear stress extended below the case depth. The samples of the 0.055-in. case depth showed a shattered appearance around the pits, a possible indication that the surface had become brittle because of the conditions of the heat-treatment.

INDUCTION-HARDENED STEEL. The tests on induction-hardened steel were made on an experimental lot of substantially SAE-1040 steel. These samples were finish-turned before hardening. They were not finished after hardening. No physical tests were made of these samples.

The results of these tests, based on the applied loads on a pair of 2-in.-diameter test rolls with a 1-in. face, are shown below. The tabulated maximum specific compressive stresses are based on conditions of static loading.

Number of cycles	Load, lb	Maximum specific compressive stress, psi
1,000,000	8,950	306,500
10,000,000	6,950	270,100
100,000,000	5,390	237,900

These results are plotted in Fig. 23-11. These test results show a steeper slope on the graph than those of the casehardened steel, possibly because of the difference in the smoothness of the finish of the surfaces.

Summary of Surface-endurance Tests. Although the complex behavior of these materials under the repeated surface stresses set up by the rolling and sliding action is still far from being understood, and even though the actual intensities of the several stresses, compressive, tensile, and shear, are thus far indeterminate, yet the actual test load results may be applied safely to design. These tests were started in 1931, and the test results have been successfully applied in the design of cams and their roller followers, and in the design of gears ever since they have been available. The comparative results between different materials as found in these tests are confirmed by the behavior of these materials in service as elements of automatic machines operating in production. The limiting loads as determined from these tests have thus far resulted in designs with no appreciable wear.

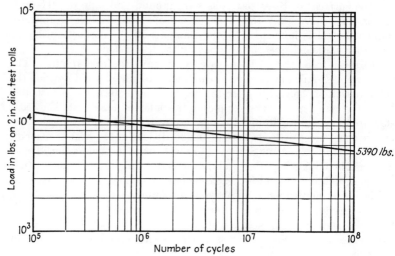

Fig. 23-11.   Induction-hardened steel.

Load-diameter-stress Factor.   For the purpose of setting up a load-diameter-stress factor, we shall start with the Hertz equation for the stresses set up between two loaded cylinders in contact.   Thus when

s = maximum specific compressive stress, psi

w = applied load on cylinders, lb/(in. length)

r_1, r_2 = radii of cylinders, in.

E_1, E_2 = modulus of elasticity of materials

$$s^2 = \frac{0.35w[(1/r_1) + (1/r_2)]}{(1/E_1) + (1/E_2)} \tag{23-1}$$

We shall now introduce an experimental factor of load stress based upon the test values.

K_1 = experimental load-stress factor for two cylinders

$$K_1 = w\left(\frac{1}{r_1} + \frac{1}{r_2}\right) \qquad \text{by definition} \tag{23-2}$$

Then

$$w = \frac{K_1}{(1/r_1) + (1/r_2)} \tag{23-3}$$

Referring now to the tests on the cast iron with 30 per cent steel scrap, we have as the surface-endurance limit load on two 2-in.-diameter test rolls, one of steel, an applied load of 1,095 lb.   This gives the following factors for the solution of Eq. (23-2):

$$w = 1,095 \qquad r_1 = 1.000 \qquad r_2 = 1.000$$
$$K_1 = 1,095(\tfrac{1}{1} + \tfrac{1}{1}) = 2,190$$

The values of these factors K_1 may be used to determine the limiting surface loads between two curved surfaces of the same combination of materials. These limiting loads are the ones that can be carried indefinitely without appreciable wear. If abrasive particles are present between the surfaces, these values do not apply. For example, if the minimum radius of curvature of a cam is 4 in. and the cam roll is 2 in. in diameter, and the parts are made of these same two materials, then

$$r_1 = 1.000 \qquad r_2 = 4.000 \qquad K_1 = 2,190$$
$$w = \frac{2,190}{\tfrac{1}{1} + \tfrac{1}{4}} = \frac{2,190}{1.25} = 1,752 \text{ lb/(in. face)}$$

In other words, this cam can be loaded up to some 1,752 lb per inch of face width.

Load-stress Factor for Gear Teeth. For involute spur-gear teeth when

K = load-stress factor for gear teeth

ϕ = pressure angle of gears

D_1 = pitch diameter of pinion, in.

D_2 = pitch diameter of gear, in.

$$r_1 = \frac{D_1 \sin \phi}{2}$$
$$r_2 = \frac{D_2 \sin \phi}{2}$$

We shall let

$$K = \frac{s^2 \sin \phi \, [(1/E_1) + (1/E_2)]}{4 \times 0.35} \tag{23-4}$$

The value of K_1 for cylinders is as follows:

$$K_1 = w\left(\frac{1}{r_1} + \frac{1}{r_2}\right) = \frac{s^2[(1/E_1) + (1/E_2)]}{0.35}$$

Whence

$$\frac{K}{K_1} = \frac{\sin \phi}{4}$$

and

$$K = \frac{K_1 \sin \phi}{4} \tag{23-5}$$

Values for the load-stress factors for cylinders, K_1, and the load-stress factors for gear teeth, K, established from the foregoing test data, are tabulated in Table 23-1.

TABLE 23-1. EXPERIMENTAL LOAD-STRESS FACTORS
(Mated with hardened steel unless otherwise noted)

Material	K_1, cylinders	K, gears	
		14½-deg	20-deg
Cast iron with 30% steel scrap	2,190*	137*	187*
Nickel cast iron	1,369	85	117
Nickel cast iron, heat-treated to 300 Br.	1,643	102	140
Nickel cast iron, heat-treated to 350–400 Br.	1,917	120	163
Nickel cast iron, hot-quench	3,286	205	280
Nickel cast iron, hot-quench, 9% sliding	2,465	154	210
Chrome-nickel cast iron	1,506	94	128
Molybdenum cast iron	1,917	120	163
Molybdenum cast iron, hot-quench	3,697	231	316
Phosphor bronze, sand-cast	1,177	73	100
Phosphor bronze, chilled	2,054	128	175
Phosphor bronze, overheated when cast	205	12	17
Phosphor bronze, sand-cast, with nickel cast iron, hot-quench	2,730	171	234
For 1,000,000 cycles			
SAE-2515 steel, 0.030-in. case	16,720	1,045	1,430
SAE-2515 steel, 0.040-in. case	25,780	1,610	2,204
SAE-2515 steel, 0.055-in. case	11,840	740	1,012
SAE-1040 steel, induction-hardened	17,900	1,118	1,530
For 10,000,000 cycles			
SAE-2515 steel, 0.030-in. case	13,920	870	1,190
SAE-2515 steel, 0.040-in. case	19,680	1,230	1,682
SAE-2515 steel, 0.055-in. case	9,760	610	834
SAE-1040 steel, induction-hardened	13,900	868	1,188
For 100,000,000 cycles			
SAE-2515 steel, 0.030-in. case	11,660	728	996
SAE-2515 steel, 0.040-in. case	16,540	1,033	1,414
SAE-2151 steel, 0.055-in. case	8,180	511	699
SAE-1040 steel, 0.055-in. case	10,780	673	921

* Probably much higher than for average of this material.

CHAPTER 24

LIMITING LOADS FOR WEAR ON GEARS

Charles H. Logue, in the "American Machinists' Gear Book," published in 1910, suggested the use of the radius of curvature of the gear-tooth profile as a measure of the stresses on gear teeth that affect the wear. Joseph Jandesek followed this same thought in articles published in 1920 to 1927, giving numerous formulas, diagrams, and calculations based on the Hertz equation and using the maximum surface pressure or compressive stress as a measure of the wearing qualities.

About 1920, the writer first used the Hertz equation as a measure of gear-tooth wear and in May, 1926, presented before the American Gear Manufacturers' Association a paper in which were given constants that had proved generally satisfactory during about 7 years' use.

LIMITING LOADS FOR WEAR ON SPUR GEARS

The contact conditions between spur-gear-tooth profiles are similar to those between two cylinders, except that on gear-tooth profiles the radius of curvature is constantly changing. If we use the contact and pressure conditions between two cylinders as a measure of the stresses on the surfaces of gear teeth, we must first select some definite part of the gear-tooth profile for use as a basis of comparison.

In many cases, wear on gear teeth first becomes apparent at or near the pitch line. Possibly one contributing cause for this effect is the fact that one pair of teeth is usually carrying the entire load when contact exists on this part of the tooth profile; when contact takes place near the top or the bottom of the active profile, two pairs of mating teeth are usually sharing the load. Again, the impact or dynamic load is usually imposed on the gear teeth near the pitch-line area. And it is the intensity of this dynamic load that is largely responsible for the surface fatigue of the gear material. Hence we have reasonable cause to select the radius of curvature of the gear-tooth profile at the pitch line as the one to use as a basis of comparison with the Hertz equation, and to apply there the results of tests for surface endurance on cylindrical test rolls. Thus when

s = maximum specific compressive stress, psi

W_w = limiting load for wear, lb

F = face width of gears, in.

r_1, r_2 = radii of contacting cylinders, in.

E_1, E_2 = modulus of elasticity of materials

we have the Hertz equation as follows:

$$s^2 = \frac{0.35 W_w[(1/r_1) + (1/r_2)]}{F[(1/E_1) + (1/E_2)]} \qquad (24\text{-}1)$$

When D_1 = pitch diameter of pinion, in.

D_2 = pitch diameter of gear, in.

N_1 = number of teeth in pinion

N_2 = number of teeth in gear

K = load-stress factor for materials

Q = ratio factor

ϕ = pressure angle of gears

$$r_1 = \frac{D_1 \sin \phi}{2}$$

$$r_2 = \frac{D_2 \sin \phi}{2}$$

$$\frac{1}{r_1} + \frac{1}{r_2} = \frac{2}{\sin \phi}\left(\frac{1}{D_1} + \frac{1}{D_2}\right)$$

Substituting this value into Eq. (24-1), we obtain

$$s^2 = \frac{0.70 W_w[(1/D_1) + (1/D^2)]}{F \sin \phi[(1/E_1) + (1/E_2)]} \qquad (24\text{-}2)$$

Solving Eq. (24-2) for W_w, we obtain

$$W_w = \frac{s^2 F \sin \phi[(1/E_1) + (1/E_2)]}{0.70[(1/D_1) + (1/D_2)]} \qquad (24\text{-}3)$$

$$\frac{1}{(1/D_1) + (1/D_2)} = \frac{D_1 D_2}{D_1 + D_2}$$

We shall let

$$Q = \frac{2N_2}{N_1 + N_2} = \frac{2D_2}{D_1 + D_2} \qquad (24\text{-}4)$$

$$\frac{D_1 Q}{2} = \frac{D_1 D_2}{D_1 + D_2}$$

Substituting this into Eq. (24-3), we obtain

$$W_w = D_1 F Q \frac{s^2 \sin \phi[(1/E_1) + (1/E_2)]}{1.40} \qquad (24\text{-}5)$$

But we already have

$$K = \frac{s^2 \sin \phi[(1/E_1) + (1/E_2)]}{1.40} \qquad (23\text{-}4)$$

Substituting this value of K into Eq. (24-5), we obtain the following equation for the limiting wear load:

$$W_w = D_1 F K Q \qquad (24\text{-}6)$$

This limiting load for wear should be equal to or slightly greater than the dynamic load. No appreciable margin of safety is needed here, because an occasional overload will have but little effect. This wear is a matter of fatigue, and repeated loads are required to develop the fatigue of the materials.

Tentative Load-stress Values for K. Tentative values for the load-stress factor K for gear teeth, based primarily upon service data, are given in Table 24-1. These are for use until more definite experimental values can be established. These values have been used for several years successfully for general machine design. Many successful drives appear to use values in excess of the tabulated ones. On the other hand, many drives that have shown excessive wear in service have used values only slightly in excess of the tabulated values. With the softer steels in particular, it is possible that the specific conditions of the initial operation and running-in has had a pronounced influence on the results.

Depth to Point of Maximum Shear. When casehardened steel gears are used, it is necessary to have some measure of the depth to the point of maximum shear stress so that the depth of case will extend below the region of high shear stress, else the full surface strength of the case material will not be effective. Experience and tests indicate that the depth of case should not be less than about double the depth to the point of maximum shear. Thus when

Z = depth to point of maximum shear stress, in.

w = load per inch of face, lb

E = modulus of elasticity of material

r_1, r_2 = radii of curvature of mating profiles, in.

ϕ = pressure angle of gears

R_1, R_2 = pitch radii of gears, in.

$$r_1 = R_1 \sin \phi \qquad r_2 = R_2 \sin \phi$$
$$Z = 1.19 \sqrt{w(r_1 r_2)/E(r_1 + r_2)} \qquad (24\text{-}7)$$

The maximum shear stress is equal to about

$$0.304 \times \text{maximum specific compressive stress}$$

Limiting Wear Load on Spur Gears. *First Example.* As a definite example we shall determine the limiting wear load for a pair of 7-DP soft-steel gears, 20-deg full-depth form, of 28 and 56 teeth, with a 3-in. face width. The pinion will be of 250

TABLE 24-1. VALUES OF LOAD-STRESS FACTOR K

Brinell number		s, psi	K	
Pinion	Gear		14½-deg, lb	20-deg, lb
Steel pinion and steel gear				
150	150	50,000	30	41
200	150	60,000	43	58
250.	150	70,000	58	79
200	200	70,000	58	79
250	200	80,000	76	103
300	200	90,000	96	131
250	250	90,000	96	131
300	250	100,000	119	162
350	250	110,000	144	196
300	300	110,000	144	196
350	300	120,000	171	233
400	300	125,000	186	254
350	350	130,000	201	275
400	350	140,000	233	318
400	400	150,000	268	366
Steel pinion and cast iron gear				
150	. . .	50,000	44	60
200 and over	. . .	70,000	87	119
Steel pinion and nickel cast iron, hot-quenched				
150	. . .	50,000	44	60
200	. . .	70,000	87	119
250	. . .	90,000	144	196
300 and over	. . .	93,000	154	210
Steel pinion and phosphor-bronze gear				
150	. . .	50,000	46	62
200 and over	. . .	65,000	73	100
250 and over	*	83,000	128	175
Cast-iron pinion and cast-iron gear				
. . .	. . .	80,000	152	208
Hot-quenched nickel-cast-iron pinion and gear				
. . .	. . .	93,000	206	281
Hot-quenched nickel-cast-iron pinion and phosphor-bronze gear				
. . .	. . .	83,000	171	234

* Chilled bronze.

Brinell hardness number, and the gear will be of 200 Brinell hardness number. This gives the following values:

$$N_1 = 28 \qquad N_2 = 56 \qquad D_1 = 4.000 \qquad F = 3.000$$

From Table 24-1 we obtain

$$K = 103$$
$$Q = \frac{2 \times 56}{28 + 56} = 1.333$$

Whence

$$W_w = 4.00 \times 3.00 \times 103 \times 1.333 = 1,648 \text{ lb}$$

Second Example. As a second example we shall determine the limiting wear load for a pair of 8-DP casehardened steel gears, 20-deg full-depth form, of 24 and 56 teeth with a 2.50-in. face width. This gives the following values:

$$N_1 = 24 \qquad N_2 = 56 \qquad D_1 = 3.000 \qquad F = 2.500$$

From Table 23-1 for SAE-2515 steel, 0.040-in. case depth, and 100,000,000 cycles of stress, we have

$$K = 1,414$$
$$Q = \frac{2 \times 56}{24 + 56} = 1.40$$

Whence

$$W_w = 3.00 \times 2.50 \times 1,414 \times 1.40 = 14,857 \text{ lb}$$

We shall determine the depth to the point of maximum shear to see if the depth of case of 0.040 in. is adequate. For this we have

$$r_1 = 1.5 \times 0.34202 = 0.513 \qquad r_2 = 3.5 \times 0.34202 = 1.043$$
$$w = \frac{14,847}{2.500} = 5,939 \text{ lb}$$
$$Z = 1.19 \sqrt{\frac{5,939 \times 0.535}{30,000,000 \times 1.556}} = 0.0097 \text{ in.}$$

As the depth of case is more than double the depth to the point of maximum shear, this depth of case is adequate.

LIMIT LOADS FOR WEAR ON INTERNAL GEARS

The conditions on internal gears with straight teeth are the same as those on spur gears with straight teeth except that the form of the internal-gear-tooth profile is concave instead of convex. Hence the sign of the radius of curvature for that profile is minus instead of plus. Thus when

W_w = limiting load for wear, lb

F = face width, in.

D_1 = pitch diameter of spur pinion, in.

N_1 = number of teeth in spur pinion

N_2 = number of teeth in internal gear

K = load-stress factor for materials
Q = ratio factor

$$Q = 2N_2/(N_2 - N_1) \tag{24-8}$$
$$W_w = D_1FKQ \tag{24-6}$$

Example of Limiting Load for Wear on Internal-gear Drive.   As a definite example we shall determine the limiting wear load for an internal-gear drive, of 7 DP, gear teeth of 20-deg full-depth form, 21 and 56 teeth, of soft steel, with a face width of 2 in.   The pinion will be of 250 Brinell hardness number, and the internal gear will be of 200 Brinell hardness number.   This gives the following values:

$$N_1 = 21 \qquad N_2 = 56 \qquad D_1 = 3.000 \qquad F = 2.000$$

From Table 24-1 we have

$$K = 103$$
$$Q = \frac{2 \times 56}{56 - 21} = 3.200$$
$$W_w = 3.000 \times 2.000 \times 103 \times 3.20 = 1,977 \text{ lb}$$

LIMIT LOADS FOR WEAR ON HELICAL AND HERRINGBONE GEARS

The contact line of mating helical-gear teeth is at an angle to the trace of the pitch surface, and hence the equivalent radius of curvature of the mating cylinders is larger than that for the same diameter and pressure angle of spur gears.   We shall therefore modify the wear-load equations for spur gears accordingly and use the same load-stress factors here as for spur gears.   The form of the teeth will be referred to the normal basic-rack form.   Thus when

W_w = limiting load for wear, lb
F_a = active face width of gears, in.
D_1 = pitch diameter of helical or herringbone pinion, in.
N_1 = number of teeth in pinion
N_2 = number of teeth in gear
ψ = helix angle at pitch line
K = load-stress factor for materials
Q = ratio factor

For external helical gears

$$Q = 2N_2/(N_1 + N_2) \tag{24-4}$$

For internal helical gears

$$Q = 2N_2/(N_2 - N_1) \tag{24-8}$$

For all helical and herringbone gears

$$W_w = D_1F_aKQ/\cos^2 \psi \tag{24-9}$$

Example of Limiting Wear Load on Helical Gears.   As a definite example we shall use the following values: 48-tooth pinion, 240-tooth gear, 8-DP normal, $14\frac{1}{2}$-deg normal tooth form, 30-deg helix angle, 10-in. active face width, steel pinion of 250 Brinell hardness number, steel gear of 200 Brinell hardness number. Whence

$$N_1 = 48 \qquad N_2 = 240 \qquad D_1 = 6.928 \qquad F_a = 10.000 \qquad \psi = 30°$$

From Table 24-1 we have

$$K = 76$$
$$Q = \frac{2 \times 240}{48 + 240} = 1.667$$
$$W_w = \frac{6.928 \times 10 \times 76 \times 1.667}{0.7500} = 8,775 \text{ lb}$$

LIMIT LOADS FOR WEAR ON SPIRAL GEARS

The contact conditions on the teeth of spiral gears are similar to the contact conditions between two cylinders with crossed axes. In this case, relatively small applied loads set up high compressive stresses at the point of contact, and the load-carrying capacities of these gears are very limited. In addition, we have relatively high sliding velocities, so that the possibilities of galling are always present. The beam strength of these gear teeth is seldom a limiting factor.

The limiting load for wear for spiral gears is a very uncertain value. If the gears are allowed to operate for a greater or lesser period of time in their actual working position under a light load until the contacting surfaces have been cold-worked and polished along the lines where contact occurs, they can then carry appreciably greater loads than they could have carried if they were assembled and loaded without the preliminary polishing run. Again, if the gears are carefully run in under increasing loads until a polished band of appreciable width is developed on the tooth surfaces of the gears where contact takes place, they can then carry very much greater loads without excessive wear than they can after a short polishing run. In fact, the longer a pair of spiral gears are operated without abrasive wear or galling on the tooth surfaces, the greater the loads will be that they can carry without excessive wear.

It should be apparent, therefore, that any load factors for spiral gears are dependent upon the care with which they have been run in after their assembly in their operating position. With proper care at the start, a load that would cause excessive wear on the gears when they are first assembled may often be increased to several times the original limiting load because of the influence of careful running-in, and not result in appreciable wear. On the other hand, when abrasive wear or galling has once started, it is almost impossible to stop it without shifting the relative positions of the gears and so bringing new portions of the tooth surfaces into contact.

It should also be apparent that smooth tooth surfaces are essential for spiral gears. A roughly finished surface, particularly if the material of one gear is harder than that of the other, or if the material of either does not cold-work readily, will act much the same as a rotary file.

Any analysis of limit wear loads on spiral gears will always be indeterminate to some degree. We may assume point contact, but cold-working and wear may develop a definite width of contact, which will reduce the calculated value of the surface stresses. The results must therefore always be checked against service data and experience. We shall, however, start off with the analysis of the compressive stresses set up between two crossed cylinders under load. For this we have the following when the axes of the cylinders are at right angles to each other:

Let W = applied load on crossed cylinders, lb

s = maximum specific compressive stress, psi

R_{c1} = smaller radius of curvature, in.

R_{c2} = larger radius of curvature, in.

E_1, E_2 = modulus of elasticity of materials

m_1, m_2 = Poisson's ratio for materials

A = value depending upon value of R_{c2}/R_{c1} (see Table 24-2)

B = value depending upon value of R_{c2}/R_{c1} (see Table 24-2)

ψ = helix angle of gear

R = pitch radius of gear

ϕ = pressure angle of normal basic rack

TABLE 24-2

R_{c2}/R_{c1}	A	B
1.000	0.908	1.000
1.500	1.045	0.765
2.000	1.158	0.632
3.000	1.350	0.482
4.000	1.505	0.400
6.000	1.767	0.308
10.000	2.175	0.221

With this notation we have

$$R_c = \frac{R \sin \phi}{\cos^2 \psi} \tag{22-13}$$

The values of R_c for both gears must be determined first in order to establish the smaller and larger values.

$$s = \frac{1.5W}{\pi cd} \tag{24-10}$$

where

$$c = A \sqrt[3]{\frac{2W(R_{c1}R_{c2})}{R_{c1} + R_{c2}} \left(\frac{1 - m_1^2}{E_1} + \frac{1 - m_2^2}{E_2}\right)} \qquad (24\text{-}11)$$

We shall use the approximation $1 - m^2 = 0.900$.

$$d = Bc \qquad (24\text{-}12)$$

Compressive Stresses on Spiral Gears. *First Example.* As a definite example, to obtain some measure of the intensity of the compressive stresses on spiral gears, we shall use a pair of hardened-steel spiral gears of the following values:

$$N_1 = 12 \qquad N_2 = 48 \qquad P_n = 10 \qquad \psi_1 = 60° \qquad \psi_2 = 30°$$
$$\phi = 14.500° \qquad W = 20 \text{ lb}$$
$$R_1 = \tfrac{6}{10} \cos 60° = 1.200$$
$$R_2 = \tfrac{24}{10} \cos 30° = 2.7713$$
$$R_{c2} = \frac{1.20 \times 0.25038}{0.25} = 1.202$$
$$R_{c1} = \frac{2.7713 \times 0.25038}{0.75} = 0.925$$
$$\frac{R_{c2}}{R_{c1}} = 1.3 \qquad A = 0.990 \qquad B = 0.859$$
$$c = 0.990 \sqrt[3]{\frac{40(0.925 \times 1.202)}{0.925 + 1.202} \frac{1.800}{30,000,000}} = 0.01069$$
$$d = 0.859 \times 0.01069 = 0.00918$$
$$s = \frac{1.5 \times 20}{3.1416 \times 0.01069 \times 0.00918} = 97,400 \text{ psi}$$

In this example, a load of 20 lb develops surface compressive stresses of nearly 100,000 psi.

Second Example. As a second example we shall use a pair of cast-iron spiral gears of the same sizes as before. This gives the following values:

$$N_1 = 12 \qquad N_2 = 48 \qquad P_n = 10 \qquad \psi_1 = 60° \qquad \psi_2 = 30° \qquad W = 20$$
$$\phi = 14.500° \qquad R_1 = 1.200 \qquad R_2 = 2.7713 \qquad R_{c1} = 0.925 \qquad R_{c2} = 1.202$$
$$\frac{R_{c2}}{R_{c1}} = 1.3 \qquad A = 0.990 \qquad B = 0.859$$
$$c = 0.99 \sqrt[3]{\frac{40(0.925 \times 1.202)}{0.925 + 1.202} \frac{1.800}{15,000,000}} = 0.01346$$
$$d = 0.859 \times 0.01346 = 0.01156$$
$$s = \frac{1.5 \times 20}{3.1416 \times 0.01346 \times 0.01156} = 61,500 \text{ psi}$$

With the lower modulus of elasticity of cast iron, and all other factors the same, the compressive stresses are reduced to about two-thirds of those in the steel gears.

Limit Load for Wear on Spiral Gears. In order to determine the limit load for wear on spiral gears, we must rearrange these equations to solve for the load that will develop the endurance limit stress of the materials.   Combining Eqs. (24-10) and (24-12), we obtain

$$s = \frac{1.5W}{\pi Bc^2}$$

Solving for c, we have

$$c = \sqrt{\frac{1.5W}{\pi Bs}} \qquad (24\text{-}13)$$

Introducing the approximation $1 - m^2 = 0.900$ into Eq. (24-11), we have

$$c = A \sqrt[3]{\frac{1.8W(R_{c1}R_{c2})}{R_{c1} + R_{c2}} \left(\frac{1}{E_1} + \frac{1}{E_2} \right)} \qquad (24\text{-}14)$$

Equating Eqs. (24-13) and (24-14), raising them to the sixth power, and solving for W, we obtain

$$W = \frac{A^6 \left[\dfrac{1.8(R_{c1}R_{c2})}{R_{c1} + R_{c2}} \left(\dfrac{1}{E_1} + \dfrac{1}{E_2} \right) \right]^2}{(1.5/\pi Bs)^3} \qquad (24\text{-}15)$$

We shall introduce a ratio factor and a load-stress factor to simplify this equation as follows:

When W_w = limiting load for wear, lb

   Q = ratio factor

   K = load-stress factor

we shall let

$$Q = \left(\frac{R_{c1}R_{c2}}{R_{c1} + R_{c2}} \right)^2 \qquad (24\text{-}16)$$

$$R_c = \frac{R \sin \phi}{\cos^2 \psi} \qquad (22\text{-}13)$$

$$K = \frac{(1.8)^2 \, \pi^3 s^3}{(1.5)^3} \left(\frac{1}{E_1} + \frac{1}{E_2} \right)^2$$

Whence

$$K = 29.7662s^3 \left(\frac{1}{E_1} + \frac{1}{E_2} \right)^2 \qquad (24\text{-}17)$$

Then

$$W_w = A^6 B^3 KQ \qquad (24\text{-}18)$$

Tentative values of K, based on service data, are given in Table 24-4. The values of K for the hardened-steel combination show little or no increase with running-in.   The values for the softer materials, however, show considerable increases with running-in.   The values for a short running-in period may be used in all cases except when a definite running-in operation under increasing loads is made as a definite part of the

TABLE 24-3

R_{c2}/R_{c1}	A^6	B^3	A^6B^3
1.000	0.560	1.000	0.560
1.500	1.302	0.449	0.583
2.000	2.411	0.252	0.609
3.000	6.053	0.112	0.678
4.000	11.620	0.064	0.744
6.000	30.437	0.0292	0.889
10.000	106.069	0.0108	1.141

assembling and testing process. Also, when the contact ratio is two or more, then two pairs of teeth are sharing the load after a short running-in period, and the limit wear load will be double that for a single pair of mating teeth.

For spiral gears, as with all other types of gears, the limiting wear load should be equal to the dynamic load.

Example of Limiting Wear Load on Spiral Gears. As a definite example we shall use the same values as before, for both the hardened-steel and the cast-iron combinations. This gives the following values:

$$N_1 = 12 \quad N_2 = 48 \quad P_n = 10 \quad \psi_1 = 60° \quad \psi_2 = 30° \quad \phi = 14.500°$$
$$R_1 = 1.200 \quad R_2 = 2.7713 \quad R_{c1} = 0.925 \quad R_{c2} = 1.202$$
$$\frac{R_{c2}}{R_{c1}} = 1.3 \quad A^6 = 1.005 \quad B^3 = 0.690$$

For hardened steel and hardened steel

$$K = 446$$

For cast iron and cast iron

$$K = 770$$
$$Q = \left(\frac{0.925 \times 1.202}{0.925 + 1.202}\right)^2 = 0.272$$

For the pair of hardened-steel spiral gears

$$W_w = 1.005 \times 0.690 \times 446 \times 0.272 = 84 \text{ lb}$$

When the contact ratio is two or more, $W_w = 168$ lb.

For the pair of cast-iron spiral gears

$$W_w = 1.005 \times 0.690 \times 770 \times 0.272 = 145 \text{ lb}$$

When the contact ratio is two or more, $W_w = 290$ lb.

These values should be equal to or slightly larger than the value of the dynamic load.

TABLE 24-4. LOAD-STRESS FACTORS FOR SPIRAL GEARS

Pinion (driver)	Gear (follower)	s, psi	K, lb
With initial point contact			
Hardened steel	Hardened steel	150,000	446
Hardened steel	Bronze	83,000	170
Cast iron	Bronze	83,000	302
Cast iron	Cast iron	90,000	385
With short running-in period			
Hardened steel	Hardened steel	. . . .	446
Hardened steel	Bronze	. . . .	230
Cast iron	Bronze	. . . .	600
Cast iron	Cast iron	. . . .	770
With extensive running-in period			
Hardened steel	Hardened steel	. . . .	446
Hardened steel	Bronze	. . . .	300
Cast iron	Bronze	. . . .	1,200
Cast iron	Cast iron	. . . .	1,500

LIMIT LOADS FOR WEAR ON WORM GEARS

The contact on worm-gear drives is line contact. In effect, the action is that of a rack and a gear. The changing form of the worm across the face of the gear makes it impossible to derive a simple mathematical expression for the contact curvatures. In addition, the combination of the lead angle of the worm and the position of the pitch plane of the worm in reference to the worm-thread profile has a pronounced influence on the position and form of the actual contact lines. We are therefore forced to use empirical values for the load-stress values.

The values given are based on worms of low lead angles. When the pitch plane is near the root of the worm thread in the axial section, as the lead angles increase, these values increase also because of the more favorable position of the contact lines. Thus the tabulated values given in Table 24-5 are for use with lead angles below 10 deg. For lead angles from 10 to about 25 deg, these values may be increased to 125 per cent of the tabulated values. For lead angles above 25 deg, these values may be increased to 150 per cent of the tabulated values.

When complete contact analyses of the worm-gear-tooth contacts are made, the average length of the actual contact line would be used. Here, more definite load-stress values may be established from experience. For general purposes, the width of the effective face at the pitch line of the worm gear will be used as the average length of the contact line.

The final generation of the worm-gear-tooth profiles is obtained by cold-working the surface material of the worm gear in actual service. Experiments indicate that these surfaces, on the softer gear bronzes, will cold-work a maximum of about 0.002 in. normal to the tooth surfaces without developing abrasive wear. For example, if a worm-gear drive in service shows about one-half of its tooth surface cold-worked, and the drive is operating without excessive wear, this is conclusive evidence that the load-stress value employed is only about one-half of the limiting value for the specific conditions of operation. The careful observation of these drives in service offers many opportunities of establishing more accurate load-stress factors for the particular operating conditions.

The worm, or helicoid member, should be made of the harder material, and the worm gear, or enveloping member, should be made of the more plastic material. The most common combination of materials for worm-gear drives is hardened steel for the worm and bronze for the worm gear. If soft steel is used for the worm, the minimum hardness of the steel used should be about 250 Brinell hardness number, and as much harder as possible. In any event, the thread surfaces on the worm should be as smooth as it is practical to make them. Rough worm threads will act as a rotary file on the worm gears and will develop excessive cutting or scoring.

Undercut should be avoided on worm-gear drives. This means, in general, that the sum of the numbers of teeth in the worm and worm gear should not be less than about 40. When undercut is present on a worm-gear drive and the loads are appreciable, the outer edge or corner of the worm thread may cut off flakes of appreciable size from the worm-gear-tooth profiles, particularly if any measurable deflection of the worm or rim of the worm gear exists.

The wear-load capacity of a worm-gear drive depends largely upon the diameter of the worm gear. The major influence of the diameter of the worm is to limit the effective face width of the worm gear. The thread angle of the worm has some influence, but the lead angle has more. The thread angle must be increased with an increased lead angle to avoid conditions of undercut. Hence we shall ignore the minor influence of the thread angle. Thus when

W_w = limiting load for wear, lb

D_2 = pitch diameter of worm gear, in.

F = effective face width of worm gear, in. (assumed length of contact line)

K = load-stress factor for materials

TABLE 24-5. LOAD-STRESS FACTORS FOR WORM GEARS

Worm	Gear	K, lb
Steel of 250 Br. hardness number ..	Phosphor bronze	60
Hardened steel..................	Phosphor bronze	80
Hardened steel..................	Chilled phosphor bronze	120
Hardened steel..................	Antimony bronze	120
Cast iron......................	Phosphor bronze	150

$$W_w = D_2 F K \tag{24-19}$$

Values for K are given in Table 24-5.

Example of Limiting Wear Load on Worm-gear Drive. As a definite example we shall use a hardened and ground steel worm and a phosphor-bronze worm gear with the following values: 6-start worm and 48-tooth worm gear, 1-in. axial pitch. This gives the following values:

$$D_1 = 3.820 \qquad D_2 = 15.278 \qquad \lambda = 26.565° \qquad \phi_n = 30° \qquad L = 6.000$$
$$p_x = 1.000 \qquad F = 2.250 \qquad K = 1.50 \times 80 = 120$$
$$W_w = 15.278 \times 2.250 \times 120 = 4,125 \text{ lb}$$

This value should be equal to or slightly larger than the dynamic load.

LIMIT LOADS FOR WEAR ON BEVEL GEARS

The contact on bevel gears is line contact and is very similar to that on spur gears. We shall therefore use the equivalent spur gears from Tregold's approximation to determine the limiting wear loads for bevel gears.

Because of the overhung pinion and the deflection under load of both the bevel pinion and the bevel gear, only about three-quarters of the full face of the bevel gears is generally effectively in contact. We shall therefore assume that only this part of the gear face is available to resist the surface fatigue of the materials. Thus when

W_w = limiting load for wear, lb

N_p = number of teeth in bevel pinion

N_g = number of teeth in bevel gear

N_{vp} = virtual number of teeth in bevel pinion

N_{vg} = virtual number of teeth in bevel gear

D_p = pitch diameter of bevel pinion at large end, in.

D_{vp} = virtual pitch diameter of bevel pinion at middle of face, in.

P = diametral pitch of bevel gears at large ends

F = face width of bevel gears, in.

γ_p = pitch-cone angle of bevel pinion

γ_g = pitch-cone angle of bevel gear

ϕ = pressure angle of crown rack

Q = ratio factor

K = load-stress factor for materials (same as for spur gears)

$$D_p = N_p/P \tag{24-20}$$

$$D_{vp} = (D_p - F \sin \gamma_p)/\cos \gamma_p \tag{24-21}$$

$$N_{vp} = N_p/\cos \gamma_p$$

$$N_{vg} = N_g/\cos \gamma_g \tag{15-9}$$

$$Q = 2N_{vg}/(N_{vp} + N_{vg}) \tag{24-22}$$

$$W_w = 0.75D_{vp}FKQ \tag{24-23}$$

Example of Limiting Wear Load for Bevel Gears. As a definite example we shall use a pair of 6-DP bevel gears of 24 and 48 teeth, 20-deg full-depth form, with a face width of 1 in. Both gears are of steel, 250 Brinell hardness number. This gives the following values:

$N_p = 24$	$\gamma_p = 26.565°$	$\cos \gamma_p = 0.89442$	$D_p = 4.000$
$N_g = 48$	$\gamma_g = 63.435°$	$\cos \gamma_g = 0.44721$	$D_g = 8.000$
	$F = 1.000$	$\phi = 20°$	

From Table 24-1 we obtain $K = 131$.

$$D_{vp} = \frac{4.00 - (1 \times 0.44721)}{0.89442} = 3.9755$$

$$N_{vp} = \frac{24}{0.89442} = 26.83$$

$$N_{vg} = \frac{48}{0.44721} = 107.33$$

$$Q = \frac{2 \times 107.33}{26.83 + 107.33} = 1.60$$

$$W_w = 0.75 \times 3.9755 \times 1.00 \times 131 \times 1.60 = 625 \text{ lb}$$

This value should be equal to or slightly larger than the dynamic load.

LIMIT LOADS FOR WEAR ON SPIRAL BEVEL GEARS

The relationship between bevel gears with straight teeth and spiral bevel gears is practically the same as the relationship between spur gears with straight teeth and helical gears. We shall therefore set up equations for the limiting wear loads on spiral bevel gears based on those for bevel gears with straight teeth, but adjusted to the spiral angle. These same equations will also be used for the limiting wear loads on hypoid gears when the gear member is substantially the same as that for a spiral-bevel-gear drive. With the increased sliding action on hypoid gears, the problem of lubrication is more critical than that for spiral bevel gears. When both members are made of hardened steel, as in rear-axle drives for automobiles, an extreme-pressure lubricant is generally necessary. Thus when

W_w = limiting load for wear, lb
N_p = number of teeth in spiral bevel pinion
N_g = number of teeth in spiral bevel gear
N_{vp} = virtual number of teeth in spiral bevel pinion
N_{vg} = virtual number of teeth in spiral bevel gear
D_p = pitch diameter of spiral bevel pinion at large end, in.
D_{vp} = virtual pitch diameter of spiral bevel pinion at middle of gear face, in.
P = diametral pitch at large end of gears
F = face width of gears, in.
γ_p = pitch-cone angle of spiral bevel pinion
γ_g = pitch-cone angle of spiral bevel gear
ϕ_n = normal pressure angle at middle of gear face
ψ = spiral angle at middle of gear face
Q = ratio factor
K = load-stress factor for materials (same as for spur gears)

$$D_p = N_p/P \tag{24-20}$$
$$D_{vp} = (D_p - F \sin \gamma_p)/\cos \gamma_p \tag{24-21}$$
$$N_{vp} = N_p/\cos \gamma_p \tag{15-8}$$
$$N_{vg} = N_g/\cos \gamma_g \tag{15-9}$$
$$Q = 2N_{vg}/(N_{vp} + N_{vg}) \tag{24-22}$$
$$W_w = (0.75D_{vp}FKQ)/\cos^2 \psi \tag{24-24}$$

Example of Limiting Wear Load for Spiral Bevel Gears. As a definite example we shall use a pair of 6-DP spiral bevel gears of 24 and 48 teeth, 20-deg normal pressure angle, full-depth form, with a face width of 1 in. and a spiral angle of 30 deg. Both gears are of steel, 250 Brinell hardness number. This gives the following values:

$N_p = 24$ $\gamma_p = 26.565°$ $\cos \gamma_p = 0.89442$ $D_p = 4.000$
$N_g = 48$ $\gamma_g = 63.435°$ $\cos \gamma_g = 0.44721$ $D_g = 8.000$
$F = 1.000$ $\phi_n = 20°$ $\psi = 30°$ $\cos \psi = 0.86603$

From Table 24-1 we have $K = 131$.

$$D_{vp} = \frac{4.00 - (1.00 \times 0.44721)}{0.89442} = 3.9755$$

$$N_{vp} = \frac{24}{0.89442} = 26.83$$

$$N_{vg} = \frac{48}{0.44721} = 107.33$$

$$Q = \frac{2 \times 107.33}{26.83 + 107.33} = 1.60$$

$$W_w = \frac{0.75 \times 3.9755 \times 1.00 \times 131 \times 1.60}{0.7500} = 833 \text{ lb}$$

This value for the limiting wear load should be equal to or slightly greater than the dynamic tooth load.

INDEX

A CATALOG OF SELECTED
DOVER BOOKS
IN ALL FIELDS OF INTEREST

A CATALOG OF SELECTED DOVER
BOOKS IN ALL FIELDS OF INTEREST

DRAWINGS OF REMBRANDT, edited by Seymour Slive. Updated Lippmann, Hofstede de Groot edition, with definitive scholarly apparatus. All portraits, biblical sketches, landscapes, nudes. Oriental figures, classical studies, together with selection of work by followers. 550 illustrations. Total of 630pp. 9⅛ × 12¼.
21485-0, 21486-9 Pa., Two-vol. set $25.00

GHOST AND HORROR STORIES OF AMBROSE BIERCE, Ambrose Bierce. 24 tales vividly imagined, strangely prophetic, and decades ahead of their time in technical skill: "The Damned Thing," "An Inhabitant of Carcosa," "The Eyes of the Panther," "Moxon's Master," and 20 more. 199pp. 5⅜ × 8½.   20767-6 Pa. $3.95

ETHICAL WRITINGS OF MAIMONIDES, Maimonides. Most significant ethical works of great medieval sage, newly translated for utmost precision, readability. Laws Concerning Character Traits, Eight Chapters, more. 192pp. 5⅜ × 8½.
24522-5 Pa. $4.50

THE EXPLORATION OF THE COLORADO RIVER AND ITS CANYONS, J. W. Powell. Full text of Powell's 1,000-mile expedition down the fabled Colorado in 1869. Superb account of terrain, geology, vegetation, Indians, famine, mutiny, treacherous rapids, mighty canyons, during exploration of last unknown part of continental U.S. 400pp. 5⅜ × 8½.   20094-9 Pa. $6.95

HISTORY OF PHILOSOPHY, Julián Marías. Clearest one-volume history on the market. Every major philosopher and dozens of others, to Existentialism and later. 505pp. 5⅜ × 8½.   21739-6 Pa. $8.50

ALL ABOUT LIGHTNING, Martin A. Uman. Highly readable non-technical survey of nature and causes of lightning, thunderstorms, ball lightning, St. Elmo's Fire, much more. Illustrated. 192pp. 5⅜ × 8½.   25237-X Pa. $5.95

SAILING ALONE AROUND THE WORLD, Captain Joshua Slocum. First man to sail around the world, alone, in small boat. One of great feats of seamanship told in delightful manner. 67 illustrations. 294pp. 5⅜ × 8½.   20326-3 Pa. $4.50

LETTERS AND NOTES ON THE MANNERS, CUSTOMS AND CONDITIONS OF THE NORTH AMERICAN INDIANS, George Catlin. Classic account of life among Plains Indians: ceremonies, hunt, warfare, etc. 312 plates. 572pp. of text. 6⅛ × 9¼.   22118-0, 22119-9 Pa. Two-vol. set $15.90

ALASKA: The Harriman Expedition, 1899, John Burroughs, John Muir, et al. Informative, engrossing accounts of two-month, 9,000-mile expedition. Native peoples, wildlife, forests, geography, salmon industry, glaciers, more. Profusely illustrated. 240 black-and-white line drawings. 124 black-and-white photographs. 3 maps. Index. 576pp. 5⅜ × 8½.   25109-8 Pa. $11.95

THE BOOK OF BEASTS: Being a Translation from a Latin Bestiary of the Twelfth Century, T. H. White. Wonderful catalog real and fanciful beasts: manticore, griffin, phoenix, amphivius, jaculus, many more. White's witty erudite commentary on scientific, historical aspects. Fascinating glimpse of medieval mind. Illustrated. 296pp. 5⅜ × 8¼. (Available in U.S. only) 24609-4 Pa. $5.95

FRANK LLOYD WRIGHT: ARCHITECTURE AND NATURE With 160 Illustrations, Donald Hoffmann. Profusely illustrated study of influence of nature—especially prairie—on Wright's designs for Fallingwater, Robie House, Guggenheim Museum, other masterpieces. 96pp. 9¼ × 10¾. 25098-9 Pa. $7.95

FRANK LLOYD WRIGHT'S FALLINGWATER, Donald Hoffmann. Wright's famous waterfall house: planning and construction of organic idea. History of site, owners, Wright's personal involvement. Photographs of various stages of building. Preface by Edgar Kaufmann, Jr. 100 illustrations. 112pp. 9¼ × 10.
23671-4 Pa. $7.95

YEARS WITH FRANK LLOYD WRIGHT: Apprentice to Genius, Edgar Tafel. Insightful memoir by a former apprentice presents a revealing portrait of Wright the man, the inspired teacher, the greatest American architect. 372 black-and-white illustrations. Preface. Index. vi + 228pp. 8¼ × 11. 24801-1 Pa. $9.95

THE STORY OF KING ARTHUR AND HIS KNIGHTS, Howard Pyle. Enchanting version of King Arthur fable has delighted generations with imaginative narratives of exciting adventures and unforgettable illustrations by the author. 41 illustrations. xviii + 313pp. 6⅛ × 9¼. 21445-1 Pa. $5.95

THE GODS OF THE EGYPTIANS, E. A. Wallis Budge. Thorough coverage of numerous gods of ancient Egypt by foremost Egyptologist. Information on evolution of cults, rites and gods; the cult of Osiris; the Book of the Dead and its rites; the sacred animals and birds; Heaven and Hell; and more. 956pp. 6⅛ × 9¼.
22055-9, 22056-7 Pa., Two-vol. set $20.00

A THEOLOGICO-POLITICAL TREATISE, Benedict Spinoza. Also contains unfinished *Political Treatise*. Great classic on religious liberty, theory of government on common consent. R. Elwes translation. Total of 421pp. 5⅜ × 8½.
20249-6 Pa. $6.95

INCIDENTS OF TRAVEL IN CENTRAL AMERICA, CHIAPAS, AND YUCATAN, John L. Stephens. Almost single-handed discovery of Maya culture; exploration of ruined cities, monuments, temples; customs of Indians. 115 drawings. 892pp. 5⅜ × 8½. 22404-X, 22405-8 Pa., Two-vol. set $15.90

LOS CAPRICHOS, Francisco Goya. 80 plates of wild, grotesque monsters and caricatures. Prado manuscript included. 183pp. 6⅝ × 9⅜. 22384-1 Pa. $4.95

AUTOBIOGRAPHY: The Story of My Experiments with Truth, Mohandas K. Gandhi. Not hagiography, but Gandhi in his own words. Boyhood, legal studies, purification, the growth of the Satyagraha (nonviolent protest) movement. Critical, inspiring work of the man who freed India. 480pp. 5⅜ × 8½. (Available in U.S. only)
24593-4 Pa. $6.95

ILLUSTRATED DICTIONARY OF HISTORIC ARCHITECTURE, edited by Cyril M. Harris. Extraordinary compendium of clear, concise definitions for over 5,000 important architectural terms complemented by over 2,000 line drawings. Covers full spectrum of architecture from ancient ruins to 20th-century Modernism. Preface. 592pp. 7½ × 9⅜. 24444-X Pa. $14.95

THE NIGHT BEFORE CHRISTMAS, Clement Moore. Full text, and woodcuts from original 1848 book. Also critical, historical material. 19 illustrations. 40pp. 4⅝ × 6. 22797-9 Pa. $2.25

THE LESSON OF JAPANESE ARCHITECTURE: 165 Photographs, Jiro Harada. Memorable gallery of 165 photographs taken in the 1930's of exquisite Japanese homes of the well-to-do and historic buildings. 13 line diagrams. 192pp. 8⅜ × 11¼. 24778-3 Pa. $8.95

THE AUTOBIOGRAPHY OF CHARLES DARWIN AND SELECTED LETTERS, edited by Francis Darwin. The fascinating life of eccentric genius composed of an intimate memoir by Darwin (intended for his children); commentary by his son, Francis; hundreds of fragments from notebooks, journals, papers; and letters to and from Lyell, Hooker, Huxley, Wallace and Henslow. xi + 365pp. 5⅜ × 8. 20479-0 Pa. $5.95

WONDERS OF THE SKY: Observing Rainbows, Comets, Eclipses, the Stars and Other Phenomena, Fred Schaaf. Charming, easy-to-read poetic guide to all manner of celestial events visible to the naked eye. Mock suns, glories, Belt of Venus, more. Illustrated. 299pp. 5¼ × 8¼. 24402-4 Pa. $7.95

BURNHAM'S CELESTIAL HANDBOOK, Robert Burnham, Jr. Thorough guide to the stars beyond our solar system. Exhaustive treatment. Alphabetical by constellation: Andromeda to Cetus in Vol. 1; Chamaeleon to Orion in Vol. 2; and Pavo to Vulpecula in Vol. 3. Hundreds of illustrations. Index in Vol. 3. 2,000pp. 6⅛ × 9¼. 23567-X, 23568-8, 23673-0 Pa., Three-vol. set $36.85

STAR NAMES: Their Lore and Meaning, Richard Hinckley Allen. Fascinating history of names various cultures have given to constellations and literary and folkloristic uses that have been made of stars. Indexes to subjects. Arabic and Greek names. Biblical references. Bibliography. 563pp. 5⅜ × 8½. 21079-0 Pa. $7.95

THIRTY YEARS THAT SHOOK PHYSICS: The Story of Quantum Theory, George Gamow. Lucid, accessible introduction to influential theory of energy and matter. Careful explanations of Dirac's anti-particles, Bohr's model of the atom, much more. 12 plates. Numerous drawings. 240pp. 5⅜ × 8½. 24895-X Pa. $4.95

CHINESE DOMESTIC FURNITURE IN PHOTOGRAPHS AND MEASURED DRAWINGS, Gustav Ecke. A rare volume, now affordably priced for antique collectors, furniture buffs and art historians. Detailed review of styles ranging from early Shang to late Ming. Unabridged republication. 161 black-and-white drawings, photos. Total of 224pp. 8⅜ × 11¼. (Available in U.S. only) 25171-3 Pa. $12.95

VINCENT VAN GOGH: A Biography, Julius Meier-Graefe. Dynamic, penetrating study of artist's life, relationship with brother, Theo, painting techniques, travels, more. Readable, engrossing. 160pp. 5⅜ × 8½. (Available in U.S. only) 25253-1 Pa. $3.95

HOW TO WRITE, Gertrude Stein. Gertrude Stein claimed anyone could understand her unconventional writing—here are clues to help. Fascinating improvisations, language experiments, explanations illuminate Stein's craft and the art of writing. Total of 414pp. 4⅝ × 6⅜. 23144-5 Pa. $5.95

ADVENTURES AT SEA IN THE GREAT AGE OF SAIL: Five Firsthand Narratives, edited by Elliot Snow. Rare true accounts of exploration, whaling, shipwreck, fierce natives, trade, shipboard life, more. 33 illustrations. Introduction. 353pp. 5⅜ × 8½. 25177-2 Pa. $7.95

THE HERBAL OR GENERAL HISTORY OF PLANTS, John Gerard. Classic descriptions of about 2,850 plants—with over 2,700 illustrations—includes Latin and English names, physical descriptions, varieties, time and place of growth, more. 2,706 illustrations. xlv + 1,678pp. 8½ × 12¼. 23147-X Cloth. $75.00

DOROTHY AND THE WIZARD IN OZ, L. Frank Baum. Dorothy and the Wizard visit the center of the Earth, where people are vegetables, glass houses grow and Oz characters reappear. Classic sequel to *Wizard of Oz*. 256pp. 5⅜ × 8.
24714-7 Pa. $4.95

SONGS OF EXPERIENCE: Facsimile Reproduction with 26 Plates in Full Color, William Blake. This facsimile of Blake's original "Illuminated Book" reproduces 26 full-color plates from a rare 1826 edition. Includes "The Tyger," "London," "Holy Thursday," and other immortal poems. 26 color plates. Printed text of poems. 48pp. 5¼ × 7. 24636-1 Pa. $3.50

SONGS OF INNOCENCE, William Blake. The first and most popular of Blake's famous "Illuminated Books," in a facsimile edition reproducing all 31 brightly colored plates. Additional printed text of each poem. 64pp. 5¼ × 7.
22764-2 Pa. $3.50

PRECIOUS STONES, Max Bauer. Classic, thorough study of diamonds, rubies, emeralds, garnets, etc.: physical character, occurrence, properties, use, similar topics. 20 plates, 8 in color. 94 figures. 659pp. 6⅛ × 9¼.
21910-0, 21911-9 Pa., Two-vol. set $14.90

ENCYCLOPEDIA OF VICTORIAN NEEDLEWORK, S. F. A. Caulfeild and Blanche Saward. Full, precise descriptions of stitches, techniques for dozens of needlecrafts—most exhaustive reference of its kind. Over 800 figures. Total of 679pp. 8¾ × 11. Two volumes. Vol. 1 22800-2 Pa. $10.95
Vol. 2 22801-0 Pa. $10.95

THE MARVELOUS LAND OF OZ, L. Frank Baum. Second Oz book, the Scarecrow and Tin Woodman are back with hero named Tip, Oz magic. 136 illustrations. 287pp. 5⅜ × 8½. 20692-0 Pa. $5.95

WILD FOWL DECOYS, Joel Barber. Basic book on the subject, by foremost authority and collector. Reveals history of decoy making and rigging, place in American culture, different kinds of decoys, how to make them, and how to use them. 140 plates. 156pp. 7⅞ × 10¾. 20011-6 Pa. $7.95

HISTORY OF LACE, Mrs. Bury Palliser. Definitive, profusely illustrated chronicle of lace from earliest times to late 19th century. Laces of Italy, Greece, England, France, Belgium, etc. Landmark of needlework scholarship. 266 illustrations. 672pp. 6⅛ × 9¼. 24742-2 Pa. $14.95

ILLUSTRATED GUIDE TO SHAKER FURNITURE, Robert Meader. All furniture and appurtenances, with much on unknown local styles. 235 photos. 146pp. 9 × 12. 22819-3 Pa. $7.95

WHALE SHIPS AND WHALING: A Pictorial Survey, George Francis Dow. Over 200 vintage engravings, drawings, photographs of barks, brigs, cutters, other vessels. Also harpoons, lances, whaling guns, many other artifacts. Comprehensive text by foremost authority. 207 black-and-white illustrations. 288pp. 6 × 9.
24808-9 Pa. $8.95

THE BERTRAMS, Anthony Trollope. Powerful portrayal of blind self-will and thwarted ambition includes one of Trollope's most heartrending love stories. 497pp. 5⅜ × 8½. 25119-5 Pa. $8.95

ADVENTURES WITH A HAND LENS, Richard Headstrom. Clearly written guide to observing and studying flowers and grasses, fish scales, moth and insect wings, egg cases, buds, feathers, seeds, leaf scars, moss, molds, ferns, common crystals, etc.—all with an ordinary, inexpensive magnifying glass. 209 exact line drawings aid in your discoveries. 220pp. 5⅜ × 8½. 23330-8 Pa. $3.95

RODIN ON ART AND ARTISTS, Auguste Rodin. Great sculptor's candid, wide-ranging comments on meaning of art; great artists; relation of sculpture to poetry, painting, music; philosophy of life, more. 76 superb black-and-white illustrations of Rodin's sculpture, drawings and prints. 119pp. 8⅜ × 11¼. 24487-3 Pa. $6.95

FIFTY CLASSIC FRENCH FILMS, 1912–1982: A Pictorial Record, Anthony Slide. Memorable stills from Grand Illusion, Beauty and the Beast, Hiroshima, Mon Amour, many more. Credits, plot synopses, reviews, etc. 160pp. 8¼ × 11.
25256-6 Pa. $11.95

THE PRINCIPLES OF PSYCHOLOGY, William James. Famous long course complete, unabridged. Stream of thought, time perception, memory, experimental methods; great work decades ahead of its time. 94 figures. 1,391pp. 5⅜ × 8½.
20381-6, 20382-4 Pa., Two-vol. set $19.90

BODIES IN A BOOKSHOP, R. T. Campbell. Challenging mystery of blackmail and murder with ingenious plot and superbly drawn characters. In the best tradition of British suspense fiction. 192pp. 5⅜ × 8½. 24720-1 Pa. $3.95

CALLAS: PORTRAIT OF A PRIMA DONNA, George Jellinek. Renowned commentator on the musical scene chronicles incredible career and life of the most controversial, fascinating, influential operatic personality of our time. 64 black-and-white photographs. 416pp. 5⅜ × 8¼. 25047-4 Pa. $7.95

GEOMETRY, RELATIVITY AND THE FOURTH DIMENSION, Rudolph Rucker. Exposition of fourth dimension, concepts of relativity as Flatland characters continue adventures. Popular, easily followed yet accurate, profound. 141 illustrations. 133pp. 5⅜ × 8½. 23400-2 Pa. $3.50

HOUSEHOLD STORIES BY THE BROTHERS GRIMM, with pictures by Walter Crane. 53 classic stories—Rumpelstiltskin, Rapunzel, Hansel and Gretel, the Fisherman and his Wife, Snow White, Tom Thumb, Sleeping Beauty, Cinderella, and so much more—lavishly illustrated with original 19th century drawings. 114 illustrations. x + 269pp. 5⅜ × 8½. 21080-4 Pa. $4.50

SUNDIALS, Albert Waugh. Far and away the best, most thorough coverage of ideas, mathematics concerned, types, construction, adjusting anywhere. Over 100 illustrations. 230pp. 5⅜ × 8½. 22947-5 Pa. $4.00

PICTURE HISTORY OF THE NORMANDIE: With 190 Illustrations, Frank O. Braynard. Full story of legendary French ocean liner: Art Deco interiors, design innovations, furnishings, celebrities, maiden voyage, tragic fire, much more. Extensive text. 144pp. 8⅜ × 11¾. 25257-4 Pa. $9.95

THE FIRST AMERICAN COOKBOOK: A Facsimile of "American Cookery," 1796, Amelia Simmons. Facsimile of the first American-written cookbook published in the United States contains authentic recipes for colonial favorites—pumpkin pudding, winter squash pudding, spruce beer, Indian slapjacks, and more. Introductory Essay and Glossary of colonial cooking terms. 80pp. 5⅜ × 8½. 24710-4 Pa. $3.50

101 PUZZLES IN THOUGHT AND LOGIC, C. R. Wylie, Jr. Solve murders and robberies, find out which fishermen are liars, how a blind man could possibly identify a color—purely by your own reasoning! 107pp. 5⅜ × 8½. 20367-0 Pa. $2.00

THE BOOK OF WORLD-FAMOUS MUSIC—CLASSICAL, POPULAR AND FOLK, James J. Fuld. Revised and enlarged republication of landmark work in musico-bibliography. Full information about nearly 1,000 songs and compositions including first lines of music and lyrics. New supplement. Index. 800pp. 5⅜ × 8¼. 24857-7 Pa. $14.95

ANTHROPOLOGY AND MODERN LIFE, Franz Boas. Great anthropologist's classic treatise on race and culture. Introduction by Ruth Bunzel. Only inexpensive paperback edition. 255pp. 5⅜ × 8½. 25245-0 Pa. $5.95

THE TALE OF PETER RABBIT, Beatrix Potter. The inimitable Peter's terrifying adventure in Mr. McGregor's garden, with all 27 wonderful, full-color Potter illustrations. 55pp. 4¼ × 5½. (Available in U.S. only) 22827-4 Pa. $1.75

THREE PROPHETIC SCIENCE FICTION NOVELS, H. G. Wells. *When the Sleeper Wakes, A Story of the Days to Come* and *The Time Machine* (full version). 335pp. 5⅜ × 8½. (Available in U.S. only) 20605-X Pa. $5.95

APICIUS COOKERY AND DINING IN IMPERIAL ROME, edited and translated by Joseph Dommers Vehling. Oldest known cookbook in existence offers readers a clear picture of what foods Romans ate, how they prepared them, etc. 49 illustrations. 301pp. 6⅛ × 9¼. 23563-7 Pa. $6.00

SHAKESPEARE LEXICON AND QUOTATION DICTIONARY, Alexander Schmidt. Full definitions, locations, shades of meaning of every word in plays and poems. More than 50,000 exact quotations. 1,485pp. 6½ × 9¼. 22726-X, 22727-8 Pa., Two-vol. set $27.90

THE WORLD'S GREAT SPEECHES, edited by Lewis Copeland and Lawrence W. Lamm. Vast collection of 278 speeches from Greeks to 1970. Powerful and effective models; unique look at history. 842pp. 5⅜ × 8½. 20468-5 Pa. $10.95

THE BLUE FAIRY BOOK, Andrew Lang. The first, most famous collection, with many familiar tales: Little Red Riding Hood, Aladdin and the Wonderful Lamp, Puss in Boots, Sleeping Beauty, Hansel and Gretel, Rumpelstiltskin; 37 in all. 138 illustrations. 390pp. 5⅜ × 8½. 21437-0 Pa. $5.95

THE STORY OF THE CHAMPIONS OF THE ROUND TABLE, Howard Pyle. Sir Launcelot, Sir Tristram and Sir Percival in spirited adventures of love and triumph retold in Pyle's inimitable style. 50 drawings, 31 full-page. xviii + 329pp. 6½ × 9¼. 21883-X Pa. $6.95

AUDUBON AND HIS JOURNALS, Maria Audubon. Unmatched two-volume portrait of the great artist, naturalist and author contains his journals, an excellent biography by his granddaughter, expert annotations by the noted ornithologist, Dr. Elliott Coues, and 37 superb illustrations. Total of 1,200pp. 5⅜ × 8.
Vol. I 25143-8 Pa. $8.95
Vol. II 25144-6 Pa. $8.95

GREAT DINOSAUR HUNTERS AND THEIR DISCOVERIES, Edwin H. Colbert. Fascinating, lavishly illustrated chronicle of dinosaur research, 1820's to 1960. Achievements of Cope, Marsh, Brown, Buckland, Mantell, Huxley, many others. 384pp. 5¼ × 8¼. 24701-5 Pa. $6.95

THE TASTEMAKERS, Russell Lynes. Informal, illustrated social history of American taste 1850's–1950's. First popularized categories Highbrow, Lowbrow, Middlebrow. 129 illustrations. New (1979) afterword. 384pp. 6 × 9.
23993-4 Pa. $6.95

DOUBLE CROSS PURPOSES, Ronald A. Knox. A treasure hunt in the Scottish Highlands, an old map, unidentified corpse, surprise discoveries keep reader guessing in this cleverly intricate tale of financial skullduggery. 2 black-and-white maps. 320pp. 5⅜ × 8½. (Available in U.S. only) 25032-6 Pa. $5.95

AUTHENTIC VICTORIAN DECORATION AND ORNAMENTATION IN FULL COLOR: 46 Plates from "Studies in Design," Christopher Dresser. Superb full-color lithographs reproduced from rare original portfolio of a major Victorian designer. 48pp. 9¼ × 12¼. 25083-0 Pa. $7.95

PRIMITIVE ART, Franz Boas. Remains the best text ever prepared on subject, thoroughly discussing Indian, African, Asian, Australian, and, especially, Northern American primitive art. Over 950 illustrations show ceramics, masks, totem poles, weapons, textiles, paintings, much more. 376pp. 5⅜ × 8. 20025-6 Pa. $6.95

SIDELIGHTS ON RELATIVITY, Albert Einstein. Unabridged republication of two lectures delivered by the great physicist in 1920–21. *Ether and Relativity* and *Geometry and Experience*. Elegant ideas in non-mathematical form, accessible to intelligent layman. vi + 56pp. 5⅜ × 8½. 24511-X Pa. $2.95

THE WIT AND HUMOR OF OSCAR WILDE, edited by Alvin Redman. More than 1,000 ripostes, paradoxes, wisecracks: Work is the curse of the drinking classes, I can resist everything except temptation, etc. 258pp. 5⅜ × 8½. 20602-5 Pa. $3.95

ADVENTURES WITH A MICROSCOPE, Richard Headstrom. 59 adventures with clothing fibers, protozoa, ferns and lichens, roots and leaves, much more. 142 illustrations. 232pp. 5⅜ × 8½. 23471-1 Pa. $3.95

PLANTS OF THE BIBLE, Harold N. Moldenke and Alma L. Moldenke. Standard reference to all 230 plants mentioned in Scriptures. Latin name, biblical reference, uses, modern identity, much more. Unsurpassed encyclopedic resource for scholars, botanists, nature lovers, students of Bible. Bibliography. Indexes. 123 black-and-white illustrations. 384pp. 6 × 9. 25069-5 Pa. $8.95

FAMOUS AMERICAN WOMEN: A Biographical Dictionary from Colonial Times to the Present, Robert McHenry, ed. From Pocahontas to Rosa Parks, 1,035 distinguished American women documented in separate biographical entries. Accurate, up-to-date data, numerous categories, spans 400 years. Indices. 493pp. 6½ × 9¼. 24523-3 Pa. $9.95

THE FABULOUS INTERIORS OF THE GREAT OCEAN LINERS IN HISTORIC PHOTOGRAPHS, William H. Miller, Jr. Some 200 superb photographs capture exquisite interiors of world's great "floating palaces"—1890's to 1980's: *Titanic, Ile de France, Queen Elizabeth, United States, Europa,* more. Approx. 200 black-and-white photographs. Captions. Text. Introduction. 160pp. 8⅞ × 11¾. 24756-2 Pa. $9.95

THE GREAT LUXURY LINERS, 1927-1954: A Photographic Record, William H. Miller, Jr. Nostalgic tribute to heyday of ocean liners. 186 photos of Ile de France, Normandie, Leviathan, Queen Elizabeth, United States, many others. Interior and exterior views. Introduction. Captions. 160pp. 9 × 12. 24056-8 Pa. $9.95

A NATURAL HISTORY OF THE DUCKS, John Charles Phillips. Great landmark of ornithology offers complete detailed coverage of nearly 200 species and subspecies of ducks: gadwall, sheldrake, merganser, pintail, many more. 74 full-color plates, 102 black-and-white. Bibliography. Total of 1,920pp. 8⅝ × 11¼. 25141-1, 25142-X Cloth. Two-vol. set $100.00

THE SEAWEED HANDBOOK: An Illustrated Guide to Seaweeds from North Carolina to Canada, Thomas F. Lee. Concise reference covers 78 species. Scientific and common names, habitat, distribution, more. Finding keys for easy identification. 224pp. 5⅜ × 8½. 25215-9 Pa. $5.95

THE TEN BOOKS OF ARCHITECTURE: The 1755 Leoni Edition, Leon Battista Alberti. Rare classic helped introduce the glories of ancient architecture to the Renaissance. 68 black-and-white plates. 336pp. 8⅜ × 11¼. 25239-6 Pa. $14.95

MISS MACKENZIE, Anthony Trollope. Minor masterpieces by Victorian master unmasks many truths about life in 19th-century England. First inexpensive edition in years. 392pp. 5⅜ × 8½. 25201-9 Pa. $7.95

THE RIME OF THE ANCIENT MARINER, Gustave Doré, Samuel Taylor Coleridge. Dramatic engravings considered by many to be his greatest work. The terrifying space of the open sea, the storms and whirlpools of an unknown ocean, the ice of Antarctica, more—all rendered in a powerful, chilling manner. Full text. 38 plates. 77pp. 9¼ × 12. 22305-1 Pa. $4.95

THE EXPEDITIONS OF ZEBULON MONTGOMERY PIKE, Zebulon Montgomery Pike. Fascinating first-hand accounts (1805-6) of exploration of Mississippi River, Indian wars, capture by Spanish dragoons, much more. 1,088pp. 5⅜ × 8½. 25254-X, 25255-8 Pa. Two-vol. set $23.90

A CONCISE HISTORY OF PHOTOGRAPHY: Third Revised Edition, Helmut Gernsheim. Best one-volume history—camera obscura, photochemistry, daguerreotypes, evolution of cameras, film, more. Also artistic aspects—landscape, portraits, fine art, etc. 281 black-and-white photographs. 26 in color. 176pp. 8⅜ × 11¼. 25128-4 Pa. $12.95

THE DORÉ BIBLE ILLUSTRATIONS, Gustave Doré. 241 detailed plates from the Bible: the Creation scenes, Adam and Eve, Flood, Babylon, battle sequences, life of Jesus, etc. Each plate is accompanied by the verses from the King James version of the Bible. 241pp. 9 × 12. 23004-X Pa. $8.95

HUGGER-MUGGER IN THE LOUVRE, Elliot Paul. Second Homer Evans mystery-comedy. Theft at the Louvre involves sleuth in hilarious, madcap caper. "A knockout."—Books. 336pp. 5⅜ × 8½. 25185-3 Pa. $5.95

FLATLAND, E. A. Abbott. Intriguing and enormously popular science-fiction classic explores the complexities of trying to survive as a two-dimensional being in a three-dimensional world. Amusingly illustrated by the author. 16 illustrations. 103pp. 5⅜ × 8½. 20001-9 Pa. $2.00

THE HISTORY OF THE LEWIS AND CLARK EXPEDITION, Meriwether Lewis and William Clark, edited by Elliott Coues. Classic edition of Lewis and Clark's day-by-day journals that later became the basis for U.S. claims to Oregon and the West. Accurate and invaluable geographical, botanical, biological, meteorological and anthropological material. Total of 1,508pp. 5⅜ × 8½.
21268-8, 21269-6, 21270-X Pa. Three-vol. set $25.50

LANGUAGE, TRUTH AND LOGIC, Alfred J. Ayer. Famous, clear introduction to Vienna, Cambridge schools of Logical Positivism. Role of philosophy, elimination of metaphysics, nature of analysis, etc. 160pp. 5⅜ × 8½. (Available in U.S. and Canada only) 20010-8 Pa. $2.95

MATHEMATICS FOR THE NONMATHEMATICIAN, Morris Kline. Detailed, college-level treatment of mathematics in cultural and historical context, with numerous exercises. For liberal arts students. Preface. Recommended Reading Lists. Tables. Index. Numerous black-and-white figures. xvi + 641pp. 5⅜ × 8½.
24823-2 Pa. $11.95

28 SCIENCE FICTION STORIES, H. G. Wells. Novels, *Star Begotten* and *Men Like Gods,* plus 26 short stories: "Empire of the Ants," "A Story of the Stone Age," "The Stolen Bacillus," "In the Abyss," etc. 915pp. 5⅜ × 8½. (Available in U.S. only)
20265-8 Cloth. $10.95

HANDBOOK OF PICTORIAL SYMBOLS, Rudolph Modley. 3,250 signs and symbols, many systems in full; official or heavy commercial use. Arranged by subject. Most in Pictorial Archive series. 143pp. 8⅛ × 11. 23357-X Pa. $5.95

INCIDENTS OF TRAVEL IN YUCATAN, John L. Stephens. Classic (1843) exploration of jungles of Yucatan, looking for evidences of Maya civilization. Travel adventures, Mexican and Indian culture, etc. Total of 669pp. 5⅜ × 8½.
20926-1, 20927-X Pa., Two-vol. set $9.90

DEGAS: An Intimate Portrait, Ambroise Vollard. Charming, anecdotal memoir by famous art dealer of one of the greatest 19th-century French painters. 14 black-and-white illustrations. Introduction by Harold L. Van Doren. 96pp. 5⅜ × 8½.
25131-4 Pa. $3.95

PERSONAL NARRATIVE OF A PILGRIMAGE TO ALMANDINAH AND MECCAH, Richard Burton. Great travel classic by remarkably colorful personality. Burton, disguised as a Moroccan, visited sacred shrines of Islam, narrowly escaping death. 47 illustrations. 959pp. 5⅜ × 8½.   21217-3, 21218-1 Pa., Two-vol. set $17.90

PHRASE AND WORD ORIGINS, A. H. Holt. Entertaining, reliable, modern study of more than 1,200 colorful words, phrases, origins and histories. Much unexpected information. 254pp. 5⅜ × 8½.   20758-7 Pa. $4.95

THE RED THUMB MARK, R. Austin Freeman. In this first Dr. Thorndyke case, the great scientific detective draws fascinating conclusions from the nature of a single fingerprint. Exciting story, authentic science. 320pp. 5⅜ × 8½. (Available in U.S. only)   25210-8 Pa. $5.95

AN EGYPTIAN HIEROGLYPHIC DICTIONARY, E. A. Wallis Budge. Monumental work containing about 25,000 words or terms that occur in texts ranging from 3000 B.C. to 600 A.D. Each entry consists of a transliteration of the word, the word in hieroglyphs, and the meaning in English. 1,314pp. 6⅜ × 10.
23615-3, 23616-1 Pa., Two-vol. set $27.90

THE COMPLEAT STRATEGYST: Being a Primer on the Theory of Games of Strategy, J. D. Williams. Highly entertaining classic describes, with many illustrated examples, how to select best strategies in conflict situations. Prefaces. Appendices. xvi + 268pp. 5⅜ × 8½.   25101-2 Pa. $5.95

THE ROAD TO OZ, L. Frank Baum. Dorothy meets the Shaggy Man, little Button-Bright and the Rainbow's beautiful daughter in this delightful trip to the magical Land of Oz. 272pp. 5⅜ × 8.   25208-6 Pa. $4.95

POINT AND LINE TO PLANE, Wassily Kandinsky. Seminal exposition of role of point, line, other elements in non-objective painting. Essential to understanding 20th-century art. 127 illustrations. 192pp. 6½ × 9¼.   23808-3 Pa. $4.50

LADY ANNA, Anthony Trollope. Moving chronicle of Countess Lovel's bitter struggle to win for herself and daughter Anna their rightful rank and fortune— perhaps at cost of sanity itself. 384pp. 5⅜ × 8½.   24669-8 Pa. $6.95

EGYPTIAN MAGIC, E. A. Wallis Budge. Sums up all that is known about magic in Ancient Egypt: the role of magic in controlling the gods, powerful amulets that warded off evil spirits, scarabs of immortality, use of wax images, formulas and spells, the secret name, much more. 253pp. 5⅜ × 8½.   22681-6 Pa. $4.00

THE DANCE OF SIVA, Ananda Coomaraswamy. Preeminent authority unfolds the vast metaphysic of India: the revelation of her art, conception of the universe, social organization, etc. 27 reproductions of art masterpieces. 192pp. 5⅜ × 8½.
24817-8 Pa. $5.95

CHRISTMAS CUSTOMS AND TRADITIONS, Clement A. Miles. Origin, evolution, significance of religious, secular practices. Caroling, gifts, yule logs, much more. Full, scholarly yet fascinating; non-sectarian. 400pp. 5⅜ × 8½.
23354-5 Pa. $6.50

THE HUMAN FIGURE IN MOTION, Eadweard Muybridge. More than 4,500 stopped-action photos, in action series, showing undraped men, women, children jumping, lying down, throwing, sitting, wrestling, carrying, etc. 390pp. 7⅞ × 10⅝.
20204-6 Cloth. $19.95

THE MAN WHO WAS THURSDAY, Gilbert Keith Chesterton. Witty, fast-paced novel about a club of anarchists in turn-of-the-century London. Brilliant social, religious, philosophical speculations. 128pp. 5⅜ × 8½.
25121-7 Pa. $3.95

A CEZANNE SKETCHBOOK: Figures, Portraits, Landscapes and Still Lifes, Paul Cezanne. Great artist experiments with tonal effects, light, mass, other qualities in over 100 drawings. A revealing view of developing master painter, precursor of Cubism. 102 black-and-white illustrations. 144pp. 8¾ × 6⅝.
24790-2 Pa. $5.95

AN ENCYCLOPEDIA OF BATTLES: Accounts of Over 1,560 Battles from 1479 B.C. to the Present, David Eggenberger. Presents essential details of every major battle in recorded history, from the first battle of Megiddo in 1479 B.C. to Grenada in 1984. List of Battle Maps. New Appendix covering the years 1967–1984. Index. 99 illustrations. 544pp. 6½ × 9¼.
24913-1 Pa. $14.95

AN ETYMOLOGICAL DICTIONARY OF MODERN ENGLISH, Ernest Weekley. Richest, fullest work, by foremost British lexicographer. Detailed word histories. Inexhaustible. Total of 856pp. 6½ × 9¼.
21873-2, 21874-0 Pa., Two-vol. set $17.00

WEBSTER'S AMERICAN MILITARY BIOGRAPHIES, edited by Robert McHenry. Over 1,000 figures who shaped 3 centuries of American military history. Detailed biographies of Nathan Hale, Douglas MacArthur, Mary Hallaren, others. Chronologies of engagements, more. Introduction. Addenda. 1,033 entries in alphabetical order. xi + 548pp. 6½ × 9¼. (Available in U.S. only)
24758-9 Pa. $11.95

LIFE IN ANCIENT EGYPT, Adolf Erman. Detailed older account, with much not in more recent books: domestic life, religion, magic, medicine, commerce, and whatever else needed for complete picture. Many illustrations. 597pp. 5⅜ × 8½.
22632-8 Pa. $8.50

HISTORIC COSTUME IN PICTURES, Braun & Schneider. Over 1,450 costumed figures shown, covering a wide variety of peoples: kings, emperors, nobles, priests, servants, soldiers, scholars, townsfolk, peasants, merchants, courtiers, cavaliers, and more. 256pp. 8⅜ × 11¼.
23150-X Pa. $7.95

THE NOTEBOOKS OF LEONARDO DA VINCI, edited by J. P. Richter. Extracts from manuscripts reveal great genius; on painting, sculpture, anatomy, sciences, geography, etc. Both Italian and English. 186 ms. pages reproduced, plus 500 additional drawings, including studies for *Last Supper, Sforza* monument, etc. 860pp. 7⅞ × 10¾. (Available in U.S. only) 22572-0, 22573-9 Pa., Two-vol. set $25.90

THE ART NOUVEAU STYLE BOOK OF ALPHONSE MUCHA: All 72 Plates from "Documents Decoratifs" in Original Color, Alphonse Mucha. Rare copyright-free design portfolio by high priest of Art Nouveau. Jewelry, wallpaper, stained glass, furniture, figure studies, plant and animal motifs, etc. Only complete one-volume edition. 80pp. 9⅜ × 12¼. 24044-4 Pa. $8.95

ANIMALS: 1,419 COPYRIGHT-FREE ILLUSTRATIONS OF MAMMALS, BIRDS, FISH, INSECTS, ETC., edited by Jim Harter. Clear wood engravings present, in extremely lifelike poses, over 1,000 species of animals. One of the most extensive pictorial sourcebooks of its kind. Captions. Index. 284pp. 9 × 12.
23766-4 Pa. $9.95

OBELISTS FLY HIGH, C. Daly King. Masterpiece of American detective fiction, long out of print, involves murder on a 1935 transcontinental flight—"a very thrilling story"—NY Times. Unabridged and unaltered republication of the edition published by William Collins Sons & Co. Ltd., London, 1935. 288pp. 5⅜ × 8½. (Available in U.S. only) 25036-9 Pa. $4.95

VICTORIAN AND EDWARDIAN FASHION: A Photographic Survey, Alison Gernsheim. First fashion history completely illustrated by contemporary photographs. Full text plus 235 photos, 1840–1914, in which many celebrities appear. 240pp. 6½ × 9¼. 24205-6 Pa. $6.00

THE ART OF THE FRENCH ILLUSTRATED BOOK, 1700–1914, Gordon N. Ray. Over 630 superb book illustrations by Fragonard, Delacroix, Daumier, Doré, Grandville, Manet, Mucha, Steinlen, Toulouse-Lautrec and many others. Preface. Introduction. 633 halftones. Indices of artists, authors & titles, binders and provenances. Appendices. Bibliography. 608pp. 8⅜ × 11¼. 25086-5 Pa. $24.95

THE WONDERFUL WIZARD OF OZ, L. Frank Baum. Facsimile in full color of America's finest children's classic. 143 illustrations by W. W. Denslow. 267pp. 5⅜ × 8½. 20691-2 Pa. $5.95

FRONTIERS OF MODERN PHYSICS: New Perspectives on Cosmology, Relativity, Black Holes and Extraterrestrial Intelligence, Tony Rothman, et al. For the intelligent layman. Subjects include: cosmological models of the universe; black holes; the neutrino; the search for extraterrestrial intelligence. Introduction. 46 black-and-white illustrations. 192pp. 5⅜ × 8½. 24587-X Pa. $6.95

THE FRIENDLY STARS, Martha Evans Martin & Donald Howard Menzel. Classic text marshalls the stars together in an engaging, non-technical survey, presenting them as sources of beauty in night sky. 23 illustrations. Foreword. 2 star charts. Index. 147pp. 5⅜ × 8½. 21099-5 Pa. $3.50

FADS AND FALLACIES IN THE NAME OF SCIENCE, Martin Gardner. Fair, witty appraisal of cranks, quacks, and quackeries of science and pseudoscience: hollow earth, Velikovsky, orgone energy, Dianetics, flying saucers, Bridey Murphy, food and medical fads, etc. Revised, expanded In the Name of Science. "A very able and even-tempered presentation."—The New Yorker. 363pp. 5⅜ × 8.
20394-8 Pa. $5.95

ANCIENT EGYPT: ITS CULTURE AND HISTORY, J. E Manchip White. From pre-dynastics through Ptolemies: society, history, political structure, religion, daily life, literature, cultural heritage. 48 plates. 217pp. 5⅜ × 8½. 22548-8 Pa. $4.95

SIR HARRY HOTSPUR OF HUMBLETHWAITE, Anthony Trollope. Incisive, unconventional psychological study of a conflict between a wealthy baronet, his idealistic daughter, and their scapegrace cousin. The 1870 novel in its first inexpensive edition in years. 250pp. 5⅜ × 8½. 24953-0 Pa. $4.95

LASERS AND HOLOGRAPHY, Winston E. Kock. Sound introduction to burgeoning field, expanded (1981) for second edition. Wave patterns, coherence, lasers, diffraction, zone plates, properties of holograms, recent advances. 84 illustrations. 160pp. 5⅜ × 8¼. (Except in United Kingdom) 24041-X Pa. $3.50

INTRODUCTION TO ARTIFICIAL INTELLIGENCE: SECOND, EN-LARGED EDITION, Philip C. Jackson, Jr. Comprehensive survey of artificial intelligence—the study of how machines (computers) can be made to act intelligently. Includes introductory and advanced material. Extensive notes updating the main text. 132 black-and-white illustrations. 512pp. 5⅜ × 8½. 24864-X Pa. $8.95

HISTORY OF INDIAN AND INDONESIAN ART, Ananda K. Coomaraswamy. Over 400 illustrations illuminate classic study of Indian art from earliest Harappa finds to early 20th century. Provides philosophical, religious and social insights. 304pp. 6⅜ × 9⅜. 25005-9 Pa. $8.95

THE GOLEM, Gustav Meyrink. Most famous supernatural novel in modern European literature, set in Ghetto of Old Prague around 1890. Compelling story of mystical experiences, strange transformations, profound terror. 13 black-and-white illustrations. 224pp. 5⅜ × 8½. (Available in U.S. only) 25025-3 Pa. $5.95

ARMADALE, Wilkie Collins. Third great mystery novel by the author of *The Woman in White* and *The Moonstone*. Original magazine version with 40 illustrations. 597pp. 5⅜ × 8½. 23429-0 Pa. $7.95

PICTORIAL ENCYCLOPEDIA OF HISTORIC ARCHITECTURAL PLANS, DETAILS AND ELEMENTS: With 1,880 Line Drawings of Arches, Domes, Doorways, Facades, Gables, Windows, etc., John Theodore Haneman. Sourcebook of inspiration for architects, designers, others. Bibliography. Captions. 141pp. 9 × 12. 24605-1 Pa. $6.95

BENCHLEY LOST AND FOUND, Robert Benchley. Finest humor from early 30's, about pet peeves, child psychologists, post office and others. Mostly unavailable elsewhere. 73 illustrations by Peter Arno and others. 183pp. 5⅜ × 8½. 22410-4 Pa. $3.95

ERTÉ GRAPHICS, Erté. Collection of striking color graphics: *Seasons, Alphabet, Numerals, Aces* and *Precious Stones*. 50 plates, including 4 on covers. 48pp. 9⅜ × 12¼. 23580-7 Pa. $6.95

THE JOURNAL OF HENRY D. THOREAU, edited by Bradford Torrey, F. H. Allen. Complete reprinting of 14 volumes, 1837–61, over two million words; the sourcebooks for *Walden*, etc. Definitive. All original sketches, plus 75 photographs. 1,804pp. 8½ × 12¼. 20312-3, 20313-1 Cloth., Two-vol. set $80.00

CASTLES: THEIR CONSTRUCTION AND HISTORY, Sidney Toy. Traces castle development from ancient roots. Nearly 200 photographs and drawings illustrate moats, keeps, baileys, many other features. Caernarvon, Dover Castles, Hadrian's Wall, Tower of London, dozens more. 256pp. 5⅜ × 8¼. 24898-4 Pa. $5.95

AMERICAN CLIPPER SHIPS: 1833–1858, Octavius T. Howe & Frederick C. Matthews. Fully-illustrated, encyclopedic review of 352 clipper ships from the period of America's greatest maritime supremacy. Introduction. 109 halftones. 5 black-and-white line illustrations. Index. Total of 928pp. 5⅜ × 8½.
25115-2, 25116-0 Pa., Two-vol. set $17.90

TOWARDS A NEW ARCHITECTURE, Le Corbusier. Pioneering manifesto by great architect, near legendary founder of "International School." Technical and aesthetic theories, views on industry, economics, relation of form to function, "mass-production spirit," much more. Profusely illustrated. Unabridged translation of 13th French edition. Introduction by Frederick Etchells. 320pp. 6⅛ × 9¼. (Available in U.S. only) 25023-7 Pa. $8.95

THE BOOK OF KELLS, edited by Blanche Cirker. Inexpensive collection of 32 full-color, full-page plates from the greatest illuminated manuscript of the Middle Ages, painstakingly reproduced from rare facsimile edition. Publisher's Note. Captions. 32pp. 9⅜ × 12¼. 24345-1 Pa. $4.50

BEST SCIENCE FICTION STORIES OF H. G. WELLS, H. G. Wells. Full novel *The Invisible Man,* plus 17 short stories: "The Crystal Egg," "Aepyornis Island," "The Strange Orchid," etc. 303pp. 5⅜ × 8½. (Available in U.S. only)
21531-8 Pa. $4.95

AMERICAN SAILING SHIPS: Their Plans and History, Charles G. Davis. Photos, construction details of schooners, frigates, clippers, other sailcraft of 18th to early 20th centuries—plus entertaining discourse on design, rigging, nautical lore, much more. 137 black-and-white illustrations. 240pp. 6⅛ × 9¼.
24658-2 Pa. $5.95

ENTERTAINING MATHEMATICAL PUZZLES, Martin Gardner. Selection of author's favorite conundrums involving arithmetic, money, speed, etc., with lively commentary. Complete solutions. 112pp. 5⅜ × 8½. 25211-6 Pa. $2.95

THE WILL TO BELIEVE, HUMAN IMMORTALITY, William James. Two books bound together. Effect of irrational on logical, and arguments for human immortality. 402pp. 5⅜ × 8½. 20291-7 Pa. $7.50

THE HAUNTED MONASTERY and THE CHINESE MAZE MURDERS, Robert Van Gulik. 2 full novels by Van Gulik continue adventures of Judge Dee and his companions. An evil Taoist monastery, seemingly supernatural events; overgrown topiary maze that hides strange crimes. Set in 7th-century China. 27 illustrations. 328pp. 5⅜ × 8½. 23502-5 Pa. $5.00

CELEBRATED CASES OF JUDGE DEE (DEE GOONG AN), translated by Robert Van Gulik. Authentic 18th-century Chinese detective novel; Dee and associates solve three interlocked cases. Led to Van Gulik's own stories with same characters. Extensive introduction. 9 illustrations. 237pp. 5⅜ × 8½.
23337-5 Pa. $4.95

Prices subject to change without notice.

Available at your book dealer or write for free catalog to Dept. GI, Dover Publications, Inc., 31 East 2nd St., Mineola, N.Y. 11501. Dover publishes more than 175 books each year on science, elementary and advanced mathematics, biology, music, art, literary history, social sciences and other areas.